Lecture Notes in Artificial Intelligence   11177

Subseries of Lecture Notes in Computer Science

More information about this series at http://www.springer.com/series/1244

Michelangelo Ceci · Nathalie Japkowicz
Jiming Liu · George A. Papadopoulos
Zbigniew W. Raś (Eds.)

Foundations of Intelligent Systems

24th International Symposium, ISMIS 2018
Limassol, Cyprus, October 29–31, 2018
Proceedings

 Springer

Editors
Michelangelo Ceci 🆔
Università degli Studi di Bari Aldo Moro
Bari, Italy

George A. Papadopoulos
University of Cyprus
Nicosia, Cyprus

Nathalie Japkowicz
American University
Washington, DC, USA

Zbigniew W. Raś
University of North Carolina
Charlotte, NC, USA

Jiming Liu 🆔
Hong Kong Baptist University
Kowloon, Hong Kong

ISSN 0302-9743   ISSN 1611-3349   (electronic)
Lecture Notes in Artificial Intelligence
ISBN 978-3-030-01850-4   ISBN 978-3-030-01851-1   (eBook)
https://doi.org/10.1007/978-3-030-01851-1

Library of Congress Control Number: 2018956816

LNCS Sublibrary: SL7 – Artificial Intelligence

This Springer imprint is published by the registered company Springer Nature Switzerland AG
The registered company address is: Gewerbestrasse 11, 6330 Cham, Switzerland

Preface

This volume contains the papers selected for presentation at the 24th International Symposium on Methodologies for Intelligent Systems (ISMIS 2018), which was held in Limassol, Cyprus, October 29–31, 2018. The symposium was organized by the Department of Computer Science at the University of Cyprus. ISMIS is a conference series that started in 1986. Held twice every three years, it provides an international forum for exchanging scientific, research, and technological achievements in building intelligent systems. In particular, major areas selected for ISMIS 2018 include bioinformatics and health informatics, graph mining, image analysis, intelligent systems, mining complex patterns, novelty detection and class imbalance, social data analysis, spatio-temporal analysis, topic modeling and opinion mining. This year, three special sessions were organized: namely, Special Session on Intelligent Methodologies for Traffic Data Analysis and Mining, Special Session on Advanced Methods in Machine Learning for Modeling Complex Data, and Special Session on Granular and Soft Clustering for Data Science. Moreover, this year, the ISMIS conference was co-located with the 21st International Conference on Discovery Science (DS 2018).

We received 59 submissions that were carefully reviewed by three or more Program Committee members or external reviewers. Papers submitted to special sessions were subject to the same reviewing procedure as those submitted to regular sessions. After a rigorous reviewing process, 32 regular papers, eight short papers, and four application papers were accepted for presentation at the conference and publication in the ISMIS 2018 proceedings volume.

It is truly a pleasure to thank all the people who helped this volume come into being and made ISMIS 2018 a successful and exciting event. In particular, we would like to express our appreciation for the work of the ISMIS 2018 Program Committee members and external reviewers who helped assure the high standard of accepted papers. We would like to thank all authors of ISMIS 2018, without whose high-quality contributions it would not have been possible to organize the conference. We are grateful to the organizers of special sessions of ISMIS 2018: Pawan Lingras, Georg Peters, Richard Weber, and Hong Yu (Special Session on Granular and Soft Clustering for Data Science), Fabio Liuzzi and Fulvio Rotella (Special Session on Intelligent Methodologies for Traffic Data Analysis and Mining), Yang Liu, Jiming Liu, and Keith C. C. Chan (Special Session on Advanced Methods in Machine Learning for Modeling Complex Data). We also thank the DS program chairs, Larisa Soldatova and Joaquin Vanschoren, for ensuring the smooth coordination with DS and myriad other organizational aspects.

Invited talks were shared between the two meetings (ISMIS and DS). The invited talks for ISMIS 2018 were "Mining Big and Complex Data" by Sašo Džeroski from Jozef Stefan Institute, Slovenia; "Artificial Intelligence and the Industrial Knowledge Graph" by Michael May from Siemens, Germany, and "Bridging the Gap Between Data Diversity and Data Dependencies" by Jean-Marc Petit from INSA Lyon and

Université de Lyon, France. The DS invited talks were "Automating Predictive Modeling and Knowledge Discovery" by Ioannis Tsamardinos from the University of Crete, Greece, and "Emojis, Sentiment, and Stance in Social Media" by Petra Kralj Novak from Jozef Stefan Institute, Slovenia. Abstracts of all five invited talks are included in these proceedings. We wish to express our thanks to all the invited speakers for accepting our invitation to give plenary talks.

We are thankful to Alfred Hofmann of Springer for supporting the ISMIS 2018 Best Paper and Best Student Paper awards and for his continuous support. We also thank Anna Kramer from Springer for her work on the proceedings.

We believe that the proceedings of ISMIS 2018 will become a valuable source of reference for your ongoing and future research activities.

July 2018

Michelangelo Ceci
Nathalie Japkowicz
Jiming Liu
George A. Papadopoulos
Zbigniew W. Raś

Organization

Symposium Chair

George Papadopoulos — University of Cyprus, Cyprus

Program Committee Co-chairs

Michelangelo Ceci — University of Bari, Italy
Nathalie Japkowicz — American University, USA
Jiming Liu — Hong Kong Baptist University, SAR China

Special Sessions Chair

Stefano Ferilli — University of Bari, Italy

Steering Committee Chair

Zbigniew Ras — UNC-Charlotte, USA and Polish-Japanese Academy of IT, Poland

Steering Committee

Troels Andreasen — Roskilde University, Denmark
Annalisa Appice — Università degli Studi di Bari, Italy
Jaime Carbonell — CMU, USA
Li Chen — Hong Kong Baptist University, SAR China
Henning Christiansen — Roskilde University, Denmark
Juan Carlos Cubero — University of Granada, Spain
Floriana Esposito — Università degli Studi di Bari, Italy
Alexander Felfernig — Graz University of Technology, Austria
Mohand-Saïd Hacid — Université Claude Bernard Lyon 1, France
Marzena Kryszkiewicz — Warsaw University of Technology, Poland
Jiming Liu — Hong Kong Baptist University, SAR China
Olivier Pivert — Université Rennes 1, France
Zbigniew Ras — UNC-Charlotte, USA and Polish-Japanese Academy of IT, Poland
Henryk Rybinski — Warsaw University of Technology, Poland
Lorenza Saitta — University Piemonte Orientale, Italy
Andrzej Skowron — University of Warsaw, Poland
Dominik Slezak — University of Warsaw, Poland
Maria Zemankova — NSF, USA

Program Committee

Annalisa Appice	University of Bari, Italy
Martin Atzmueller	Tilburg University, The Netherlands
Colin Bellinger	University of Alberta, Canada
Robert Bembenik	Warsaw University of Technology, Poland
Salima Benbernou	Université Paris Descartes, France
Petr Berka	University of Economics, Czech Republic
Marenglen Biba	University of New York in Tirana, Albania
Maria Bielikova	Slovak University of Technology in Bratislava, Slovakia
Jerzy Błaszczyński	Poznań University of Technology, Poland
Gloria Bordogna	National Research Council, Italy
Jose Borges	University of Porto, Portugal
Ivan Bratko	University of Ljubljana, Slovenia
Jianhua Chen	Louisiana State University, USA
Roberto Corizzo	University of Bari, Italy
Germàn Creamer	Stevens Institute of Technology, USA
Bruno Cremilleux	Université de Caen, France
Claudia d'Amato	University of Bari, Italy
Marcilio De Souto	LIFO/University of Orleans, France
Luigi Di Caro	University of Turin, Italy
Nicola Di Mauro	University of Bari, Italy
Stephan Doerfel	University of Kassel, Germany
Brett Drury	Scicrop, Brazil
Saso Dzeroski	Jozef Stefan Institute, Slovenia
Christoph F. Eick	University of Houston, USA
Tapio Elomaa	Tampere University of Technology, Finland
Floriana Esposito	University of Bari, Italy
Hadi Fanaee-T	University of Oslo, Norway
Nicola Fanizzi	University of Bari, Italy
Stefano Ferilli	University of Bari, Italy
Sebastien Ferre	Université de Rennes 1, France
Carlos Ferreira	LIAAD INESC Porto LA, Portugal
Jørgen Fischer Nilsson	Technical University of Denmark, Denmark
Naoki Fukuta	Shizuoka University, Japan
Fabio Fumarola	University of Bari, Italy
Mohamed Gaber	Birmingham City University, UK
Chao Gao	Southwest University, USA
Paolo Garza	Politecnico di Torino, Italy
Laura Giordano	DISIT, Università del Piemonte Orientale, Italy
Michael Granitzer	University of Passau, Germany
Jacek Grekow	Bialystok Technical University, Poland
Hakim Hacid	Zayed University, UAE
Mohand-Said Hacid	Université Claude Bernard Lyon 1 – UCBL, France
Allel Hadjali	LIAS/ENSMA, France

Maria Halkidi   University of Piraeus, Greece
Jaakko Hollmén   Aalto University, Finland
Lothar Hotz   University of Hamburg, Germany
Dino Ienco   IRSTEA, France
Mieczysław Kłopotek   Institute of Computer Science, Polish Academy
                                 of Sciences, Poland
Dragi Kocev   Jozef Stefan Institute, Slovenia
Bozena Kostek   Gdansk University of Technology, Poland
Lars Kotthoff   University of Wyoming, USA
Marzena Kryszkiewicz   Warsaw University of Technology, Poland
Dominique Laurent   Université Cergy-Pontoise, France
Anne Laurent   LIRMM – UM, France
Marie-Jeanne Lesot   LIP6 – UPMC, France
Rory Lewis   University of Colorado, USA
Antoni Ligeza   AGH University of Science and Technology, Poland
Pawan Lingras   Saint Mary's University, Nova Scotia, Canada
Yang Liu   Hong Kong Baptist University, SAR China
Zhen Liu   American University, USA
Corrado Loglisci   University of Bari, Italy
Henrique Lopes Cardoso   University of Porto, Portugal
Donato Malerba   University of Bari, Italy
Yannis Manolopoulos   Aristotle University of Thessaloniki, Greece
Elio Masciari   ICAR-CNR, Italy
Florent Masseglia   Inria, France
Paola Mello   University of Bologna, Italy
Corrado Mencar   University of Bari, Italy
João Mendes-Moreira   University of Porto, Portugal
Anna Monreale   University of Pisa, Italy
Luis Moreira-Matias   NEC Laboratories Europe, Portugal
Neil Murray   ILS Institute, University at Albany – SUNY, USA
Mirco Nanni   KDD-Lab ISTI-CNR Pisa, Italy
Amedeo Napoli   LORIA Nancy (CNRS – Inria – Université de
                                 Lorraine), France
Pance Panov   Jozef Stefan Institute, Slovenia
Ioannis Partalas   Expedia LPS, USA
Ruggero G. Pensa   University of Turin, Italy
Jean-Marc Petit   Université de Lyon, INSA Lyon, France
Gianvito Pio   University of Bari, Italy
Olivier Pivert   IRISA-ENSSAT, France
Marc Plantevit   LIRIS – Université Claude Bernard Lyon 1, France
Lubos Popelinsky   Masaryk University, Czech Republic
Henri Prade   IRIT – CNRS, France
Ronaldo Prati   Universidade Federal do ABC, Brazil

Additional Reviewers

Baghoussi, Yassine
Buono, Paolo
Chen, Hechang
Del Signore, Emiliano
Dridi, Amna
Eick, Christoph F.
Elomaa, Tapio
Fumarola, Fabio
Gao, Chao
He, Tiantian

Ko, Tobey H.
Koo, Allen
Koperwas, Jakub
Kozłowski, Marek
Lanotte, Fabiana
Leuzzi, Fabio
Li, Jinhai
Lodi, Giorgia
Novielli, Nicole
Pazienza, Andrea

Peters, Georg
Rossiello, Gaetano
Shi, Benyun
Susmaga, Robert
Weber, Richard
Xu, Weihua
Yao, Yiyu
Yu, Hong

Invited Talks

Mining Big and Complex Data

Sašo Džeroski

Jozef Stefan Institute and Jozef Stefan International Postgraduate School,
Slovenia

Abstract. Increasingly often, data mining has to learn predictive models from big data, which may have many examples or many input/output dimensions and may be streaming at very high rates. Contemporary predictive modeling problems may also be complex in a number of other ways: they may involve (a) structured data, both as input and output of the prediction process, (b) incompletely labelled data, and (c) data placed in a spatio-temporal or network context.

The talk will first give an introduction to the different tasks encountered when learning from big and complex data. It will then present some methods for solving such tasks, focusing on structured-output prediction, semi-supervised learning (from incompletely annotated data), and learning from data streams. Finally, some illustrative applications of these methods will be described, ranging from genomics and medicine to image annotation and space exploration.

Artificial Intelligence and the Industrial Knowledge Graph

Michael May

Siemens, Munich, Germany

Abstract. In the context of digitalization Siemens is leveraging various technologies from artificial intelligence and data analytics connecting the virtual and physical world to improve the entire customer value chain. The internet of things has made it possible to collect vast amount of data about the operation of physical assets in real time, as well as storing them in cloud-based data lakes. This rich set of data from heterogeneous sources allows addressing use cases that have been impossible only a few years ago. Using data analytics e.g. for monitoring and predictive maintenance is nowadays in wide-spread use.

We also find an increasing number of use cases based on Deep Learning, especially for imaging applications. In my talk I will argue that these techniques should be complemented by AI-based approaches that have originated in the knowledge representation & reasoning communities.

Especially industrial knowledge graphs play an important role in structuring and connecting all the data necessary to make our digital twins smarter and more effective. The talk gives an overview of existing and planned application scenarios incorporating AI technologies, data analytics and knowledge graphs within Siemens, e.g. building digital companions for product design and configuration or capturing the domain knowledge of engineering experts from service reports using Natural Language Processing.

Bridging the Gap Between Data Diversity and Data Dependencies

Jean-Marc Petit

Université de Lyon, CNRS
INSA de Lyon, LIRIS, UMR5205
F-69621, Villeurbanne, France
`jean-marc.petit@liris.cnrs.fr`

1 Motivation

Data are one of the four pillars of computer science, along with algorithms, languages and machines, and play a major role in the development of our digital societies. Data are diverse, they can be very simple or arbitrarily complex. Moreover, "real-life" data are known to be inconsistent, uncertain, heterogeneous and therefore often qualified as dirty. In addition, their diversity comes also from their implicit intended meaning related to the applications which produce them. Data diversity is one of the main problem in practise to deal with large amount of data.

Around the notion of data, data dependencies are declarative statements allowing to express constraints that data have to satisfy in every possible instance. By essence, data dependencies can be qualified as clean.

They turn out to be useful in many applications: for years, we know their importance in database design where functional dependencies and many others like inclusion dependencies or multi-valued dependencies have played a major role [1] to explain why database schemas have to be normalized to avoid update, insertion and suppression anomalies. Another, more recent, application of data dependencies is related to data quality where many dependencies such as conditional functional dependencies, matching dependencies [7] or denial dependencies [6] have been proposed to partially bring answers to dirty data problems.

For the former application, dependencies are specified before data production occurs whereas for the latter, dependencies are defined whereas the data are already around.

Roughly speaking, many data dependencies have been proposed to take into account some aspects of data diversity. Moreover "clean" data dependencies for dirty data have been extensively studied by different communities, for instance in databases [4, 7], fuzzy systems [8] or formal concept analysis [2, 3] to mention a few.

2   Keynote Overview

In this talk, my ambition is to bridge the gap between data diversity and data dependencies. I will come back to the very simple notion of logical implication, a concept introduced for years in logics and used by humankind since the beginning of humanity. Then I will introduce the main ingredients to represent most of data dependencies proposed in the literature. Focusing on dependencies similar to functional dependencies, a declarative query language, called RQL, will be presented. RQL turns out to be a user-friendly SQL-like query language devoted to data dependencies [5]. Finally, I will consider how lattice theory based structural properties of attribute domains give an original angle to revisit existing data dependencies [9].

Acknowledgments. Most of the material presented here follows from intense (and diverse !) discussions with Prof. Lhouari Nourine (Université Clermont Auvergne), without forgetting those with Marie Pailloux, Brice Chardin and Emmanuel Coquery. This work is partially supported by the Qualisky project (http://projets.isima.fr/qualisky/), funded by the CNRS under the Mastodon program.

References

1. Abiteboul, S., Hull, R., Vianu, V.: Fondements des bases de données. Addison Wesley (2000)
2. Baixeries, J., Codocedo, V., Kaytoue, M., Napoli, A.: Characterizing approximate-matching dependencies in formal concept analysis with pattern structures. Discrete Applied Mathematics (2018)
3. Baklouti, F., Levy, G., Emilion, R.: A fast algorithm for general galois lattices building. Elec. J. Symb. Data Anal. **2**(1), 19–31 (2005)
4. Caruccio, L., Deufemia, V., Polese, G.: Relaxed functional dependencies a survey of approaches. IEEE Trans. Knowl. Data Eng. **28**(1), 147–165 (2016)
5. Chardin, B., Coquery, E., Pailloux, M., Petit, J.: RQL: a query language for rule discovery in databases. Theor. Comput. Sci. **658**, 357–374 (2017)
6. Chu, X., Ilyas, I.F., Papotti, P.: Discovering denial constraints. PVLDB **6**(13), 1498–1509 (2013)
7. Fan, W., Jia, X., Li, J., Ma, S.: Reasoning about record matching rules. PVLDB **2**(1), 407–418 (2009)
8. Jezková, L., Cordero, P., Enciso, M.: Fuzzy functional dependencies: a comparative survey. Fuzzy Sets Syst. **317**, 88–120 (2017)
9. Nourine, L., Petit, J.M.: A structural look at data dependencies. In: 12th International Workshop on Information, Search, Integration and Personalization (workshop on invitation, abstract)

Automating Predictive Modeling and Knowledge Discovery

Ioannis Tsamardinos

University of Crete, Greece

Abstract. There is an enormous, constantly increasing need for data analytics (collectively meaning machine learning, statistical modeling, pattern recognition, and data mining applications) in a vast plethora of applications and including biological, biomedical, and business applications. The primary bottleneck in the application of machine learning is the lack of human analyst expert time and thus, a pressing need to automate machine learning, and specifically, predictive and diagnostic modeling. In this talk, we present the scientific and algorithmics problems arising from trying to automate this process, such as appropriate choice of the combination of algorithms for preprocessing, transformations, imputation of missing values, and predictive modeling, tuning of the hyper-parameter values of the algorithms, and estimating the predictive performance and producing confidence intervals. In addition, we present the problem of feature selection and how it fits within an automated analysis pipeline, arguing that feature selection is the main tool for knowledge discovery in this context.

Emojis, Sentiment and Stance in Social Media

Petra Kralj Novak

Jozef Stefan Institute, Slovenia

Abstract. Social media are computer-based technologies that provide means of information and idea sharing, as well as entertainment and engagement handly available as mobile applications and websites to both private users and businesses. As social media communication is mostly informal, it is an ideal environment for the use of emoji. We have collected Twitter data and engaged 83 human annotators to label over 1.6 million tweets in 13 European languages with sentiment polarity (negative, neutral, or positive). About 4% of the annotated tweets contain emojis. We have computed the sentiment of the emojis from the sentiment of the tweets in which they occur. We observe no significant differences in the emoji rankings between the 13 languages. Consequently, we propose our Emoji Sentiment Ranking as a European language-independent resource for automated sentiment analysis. In this talk, several emoji, sentiment and stance analysis applications will be presented, varying in data source, topics, language, and approaches used.

Contents

Bioinformatics and Health Informatics

Fully Automatic Classification of Flow Cytometry Data. 3
  Bartosz Paweł Piotrowski and Miron Bartosz Kursa

Positive Unlabeled Link Prediction via Transfer Learning
for Gene Network Reconstruction 13
  Paolo Mignone and Gianvito Pio

Early Detection of Heart Symptoms with Convolutional Neural Network
and Scattering Wavelet Transformation 24
  Mariusz Kleć

Rough Sets: Visually Discerning Neurological Functionality During
Thought Processes. 32
  Rory Lewis, Chad A. Mello, Yanyan Zhuang, Martin K.-C. Yeh,
  Yu Yan, and Dan Gopstein

Graph Mining

Solving the Maximal Clique Problem on Compressed Graphs. 45
  Jocelyn Bernard and Hamida Seba

Clones in Graphs 56
  Stephan Doerfel, Tom Hanika, and Gerd Stumme

Knowledge-Based Mining of Exceptional Patterns in Logistics Data:
Approaches and Experiences in an Industry 4.0 Context 67
  Eric Sternberg and Martin Atzmueller

An Intra-algorithm Comparison Study of Complete Search FSM
Implementations in Centralized Graph Transaction Databases 78
  Rihab Ayed, Mohand-Saïd Hacid, Rafiqul Haque,
  and Abderrazek Jemai

Critical Link Identification Based on Bridge Detection for Network
with Uncertain Connectivity 89
  Kazumi Saito, Kouzou Ohara, Masahiro Kimura, and Hiroshi Motoda

Image Analysis

Deep Neural Networks for Face Recognition: Pairwise Optimisation   103
  Elitsa Popova, Athanasios Athanasopoulos, Efraim Ie,
  Nikolaos Christou, and Ndifreke Nyah

Mobile Application with Image Recognition for Persons with Aphasia.   111
  Jan Gonera, Krzysztof Szklanny, Marcin Wichrowski,
  and Alicja Wieczorkowska

A Comparative Study on Soft Biometric Approaches to Be Used in
Retail Stores.   120
  Berardina De Carolis, Nicola Macchiarulo, and Giuseppe Palestra

Low Cost Intelligent System for the 2D Biomechanical Analysis
of Road Cyclists   130
  Camilo Salguero, Sandra P. Mosquera, Andrés F. Barco,
  and Élise Vareilles

Intelligent Systems

An Approach for the Police Districting Problem Using
Artificial Intelligence.   141
  José Manuel Rodríguez-Jiménez

Unsupervised Vehicle Recognition Using Incremental Reseeding
of Acoustic Signatures.   151
  Justin Sunu, Allon G. Percus, and Blake Hunter

A Big Data Framework for Analysis of Traffic Data in Italian Highways. . . .   161
  Claudia Diamantini, Domenico Potena, and Emanuele Storti

Traffic Data Classification for Police Activity.   169
  Stefano Guarino, Fabio Leuzzi, Flavio Lombardi,
  and Enrico Mastrostefano

Multipurpose Web-Platform for Labeling Audio Segments Efficiently
and Effectively   179
  Ayman Hajja, Griffin P. Hiers, Pierre Arbajian, Zbigniew W. Raś,
  and Alicja A. Wieczorkowska

A Description Logic of Typicality for Conceptual Combination   189
  Antonio Lieto and Gian Luca Pozzato

Mining Complex Patterns

Sparse Multi-label Bilinear Embedding on Stiefel Manifolds.   203
    Yang Liu, Guohua Dong, and Zhonglei Gu

Learning Latent Factors in Linked Multi-modality Data   214
    Tiantian He and Keith C. C. Chan

Researcher Name Disambiguation: Feature Learning and Affinity
Propagation Clustering.   225
    Zhizhi Yu and Bo Yang

Hierarchical Clustering of High-Dimensional Data Without Global
Dimensionality Reduction   236
    Ilari Kampman and Tapio Elomaa

Exploiting Order Information Embedded in Ordered Categories for Ordinal
Data Clustering.   247
    Yiqun Zhang and Yiu-ming Cheung

User-Emotion Detection Through Sentence-Based Classification Using
Deep Learning: A Case-Study with Microblogs in Albanian   258
    Marjana Prifti Skenduli, Marenglen Biba, Corrado Loglisci,
    Michelangelo Ceci, and Donato Malerba

A Novel Personalized Citation Recommendation Approach
Based on GAN.   268
    Ye Zhang, Libin Yang, Xiaoyan Cai, and Hang Dai

Novelty Detection and Class Imbalance

Unsupervised LSTMs-based Learning for Anomaly Detection in Highway
Traffic Data   281
    Nicola Di Mauro and Stefano Ferilli

SCUT-DS: Learning from Multi-class Imbalanced Canadian Weather Data. . .   291
    Olubukola M. Olaitan and Herna L. Viktor

An Efficient Algorithm for Network Vulnerability Analysis Under
Malicious Attacks   302
    Toni Mancini, Federico Mari, Igor Melatti, Ivano Salvo,
    and Enrico Tronci

Social Data Analysis

An Instrumented Methodology to Analyze and Categorize Information
Flows on Twitter Using NLP and Deep Learning: A Use Case
on Air Quality   315
   B. Juanals and J. L. Minel

Market-Aware Proactive Skill Posting   323
   Ashiqur R. KhudaBukhsh, Jong Woo Hong, and Jaime G. Carbonell

Evidential Multi-relational Link Prediction Based on Social Content   333
   Sabrine Mallek, Imen Boukhris, Zied Elouedi, and Eric Lefevre

Spatio-temporal Analysis

Predicting Temporal Activation Patterns via Recurrent Neural Networks   347
   Giuseppe Manco, Giuseppe Pirrò, and Ettore Ritacco

Handling Multi-scale Data via Multi-target Learning for Wind
Speed Forecasting   357
   Annalisa Appice, Antonietta Lanza, and Donato Malerba

Temporal Reasoning with Layered Preferences   367
   Luca Anselma, Alessandro Mazzei, Luca Piovesan,
   and Paolo Terenziani

Granular and Soft Clustering

An Adaptive Three-Way Clustering Algorithm for Mixed-Type Data   379
   Jing Xiong and Hong Yu

Three-Way Spectral Clustering   389
   Hong Shi, Qiang Liu, and Pingxin Wang

The Granular Structures in Formal Concept Analysis   399
   Ruisi Ren and Ling Wei

Fuzzy RST and RST Rules Can Predict Effects of Different Therapies
in Parkinson's Disease Patients   409
   Andrzej W. Przybyszewski

From Knowledge Discovery to Customer Attrition   417
   Katarzyna Tarnowska and Zbigniew Ras

Initial Analysis of Multivariate Factors for Prediction of Shark Presence
and Attacks on the Coast of North Carolina   426
   Sonal Kaulkar, Lavanya Vinodh, and Pamela Thompson

Topic Modelling and Opinion Mining

An Experimental Evaluation of Algorithms for Opinion Mining in
Multi-domain Corpus in Albanian   439
   Nelda Kote, Marenglen Biba, and Evis Trandafili

Predicting Author's Native Language Using Abstracts of Scholarly Papers . . .   448
   Takahiro Baba, Kensuke Baba, and Daisuke Ikeda

Identifying Exceptional Descriptions of People Using Topic Modeling
and Subgroup Discovery   454
   Andrew T. Hendrickson, Jason Wang, and Martin Atzmueller

Author Index   463

Bioinformatics and Health Informatics

Fully Automatic Classification of Flow Cytometry Data

Bartosz Paweł Piotrowski and Miron Bartosz Kursa[✉][iD]

Interdisciplinary Centre for Mathematical and Computational Modelling,
University of Warsaw, Warsaw, Poland
M.Kursa@icm.edu.pl

Abstract. Flow cytometry is a powerful analytical method, allowing to measure several properties individually for even hundreds of thousands of particles contained in some sample. Their joint distribution is a highly informative descriptor, yet directly unusable for standard machine learning methods.

Hence, such data is traditionally pre-processed into numerical features, which is often a manual or semi-automatic process. This paper introduces flowForest, an ensemble classifier capable of directly processing flow cytomtery data, modelled after the popular Random Forest method. We demonstrate that it can achieve high classification performance in a fully automatic way.

Keywords: Flow cytometry · Random forest · Bioinformatics

1 Introduction

Flow cytometry is a modern, high-throughput laboratory technique for a thorough analysis of various particle suspensions. In a flow cytometer, the said suspension is forced though a thin capillary, in a way that only a single particle at a time passes through a system of detectors. This way, the machine can in principle analyse each particle separately, and generate a number of independent measurements corresponding to its various properties; some systems may even use this data to sort particles of certain classes into separate containers for further investigation.

Most often, the method is used to analyse suspensions of biological cells and cell fragments (usually blood), while detectors measure light scattering properties and fluorescence emissions of a number of fluorophore dies that can be selectively attached to particles based on their characteristics, usually by using modified antibodies.

Flow cytometry is widely used in biomedical research and applications; among others in agriculture [6,15], ecology [7,16], immunology [4,9] and medicine [1,13]. Still, the volume and complexity of the data it provides makes it an important field for applications of computational data analysis techniques [11,14].

© Springer Nature Switzerland AG 2018
M. Ceci et al. (Eds.): ISMIS 2018, LNAI 11177, pp. 3–12, 2018.
https://doi.org/10.1007/978-3-030-01851-1_1

In this paper we investigate a problem of supervised classification of suspensions, each quantified by a single or a series of flow cytometry experiments. Traditionally, this is done in two stages [3, 8, 12]: first, the raw flow cytometry data is somehow, often manually, clustered in order to be reduced into a set of particle populations. Then, population cardinalities and other properties are used as features, yielding a fairly standard information system which can be used with numerous supervised classification methods.

In contrast, we aim to develop a fully automatic, generic method which would not rely on an explicit notion of particle populations, but rather some intrinsic mapping established from the training data. To this end, we propose a novel algorithm called flowForest, which is a modification of Random Forest, a popular, powerful and versatile model for classification and regression proposed by Breiman in 2001 [5].

1.1   Background

For a sake of flowForest definition, we have assumed a certain model of the information system representing a generic flow cytometry experiment; it does not cover all possibilities, however we believe it is generic enough for flowForest to be widely applicable in practice. The said information system is organised as follows; at the top level we have N objects, each corresponding to one sample and annotated with a binary decision. Either certain class can be featured or they may be equivalent, which allows for controlling on which aspects of a problem the algorithm shall focus; this aspect will be further discussed in Sect. 2.

While particles in a sample may often be stained to expose more properties than a cytometer may measure, it is common that one sample is split into T parts (further call *tubes*), which are then stained according to different protocols to expose various sets of properties, and finally, separately fed to a flow cytometer. To facilitate this, we assume that each object contains T flow cytometry measurements numbered from 1 to T, and that t-th measurement in each object corresponds to a sample stained according to a same protocol t, consequently that the cytometer measured a set of p_t parameters in that case.

Finally, each measurement is a set of some number of events, each of which is a vector of numbers, measured values of parameters p_t. Ideally, any event should correspond to a single particle, but in practice it may as well correspond to a small group of particles, some contaminant or be a measurement artefact; hence, a modelling method must be robust to such inputs. Anyhow, p_t defines a multidimensional phase space in which events are located according to some probability density function representing a composition of various populations of particles in the sample. The flow cytometry analysis heavily relies on a concept of a *gate*, a compact subset of this space of an arbitrary shape which naturally describes a certain subset of events which it contains. In particular, gates are often understood strictly as corresponding to a certain particle population; yet, in this paper we will adhere to a general definition, as, quite obviously, inference on such meaningful gates is logically equivalent to an inference on a combination of seemingly nonsense gates (which is actually what flowForest does). While in general standard gate is relevant only in context of a single tube and we consider multi-tube data, it is handy to extend its notion in this regard; i.e., we will further assume that a gate also denotes on which tube it operates.

2   Materials and Methods

FlowForest, similar to Random Forest, is an ensemble classifier composed of decision trees; weak, weakly correlated and non-biased classifiers, predictions of which are agreed via simple voting. Still, flowForest uses a specialised form of decision trees—which branches operate directly on flow cytometry data rather than through a previously defined or calculated descriptors. In this regard, it can be considered to be a special case of a First Order Random Forest [17]; still, we believe that the applied specialisations are crucial for its robustness and computational feasibility. Alternatively, according to the Generalised Random Forest framework proposed in [10], flowForest is identical to Random Forest except uses pivot classifiers composed of a gate G, event statistic E, a function mapping event subset into a single real value, finally a threshold value mapping such statistic's value into a binary outcome, L or R. Such a 3-tier structure is motivated by the flow cytometry data structure, and adheres to the variability of construction process of each tier. The threshold can be easily and explicitly optimised via simple sweep. The set of reasonable event statistics is both small and basically forced by the data structure motivating an easy best-of-few selection. On the other hand, the gate is well fitted to be selected randomly (provided certain quality requirements). Such a substantial source of randomness lowers correlations between trees, which is essential for voting to bring performance improvements.

Precisely, flowForest is built using the following procedure, presented in a pseudocode form as Algorithm 1. Every tree is grown using different N-element bootstrap sample of the training N-element set, called *bag*. This way roughly one-third of training objects are not used in the bootstrap sample for a given tree; they form a so-called *out-of-bag* subset (OOB).

The tree is built according to a standard, recursive partitioning scheme. For an input set of objects we first check if they belong to a single class; if so, a leaf with this class is generated. Otherwise, we build a split. To this end, m gates are randomly generated and for each of them and each object a table of statistics is calculated, containing the count and the average value of each of k parameters for events enclosed by the gate. For these $m \cdot (k + 1)$ constructed statistics, we consider all possible N threshold values and assess their performance of splitting the decision into possibly homogeneous parts; Gini impurity is used as a criterion. Finally, we select a single best triple of gate, statistic and threshold as the pivot, and recursively apply this procedure for each part of the split input. As in the case of the Random Forest method, we aim at growing trees to a maximum extent possible, that is the maximal depth is set large enough to be only a technical safeguard, and we do not trim the final tree.

After repeating this randomized procedure n times, a collection of n trees is obtained, constituting a flowForest. Classification of an object is done by simple

---

Algorithm 1. flowForest model training

Given: training set $\mathcal{T}$, tree count n, tried gates count m, maximal depth d
function PARTITION(training subset $\mathcal{X}$, recursion depth left d)
    if decision($\mathcal{X}$) is purely one class **or** $d = 0$ **then**
        return most frequent class in decision($\mathcal{X}$)   ▷ Generates a leaf
    else
        $\mathcal{S} \leftarrow \emptyset$
        for $i = 1, m$ **do**
            $G \leftarrow$ random gate (on a random tube)
            $\Xi \leftarrow$ for each element of $\mathcal{X}$, a set of events within G
            $\mathcal{S} \leftarrow \mathcal{S} \cup \{(G, \text{event count of each element of } \Xi)\}$
            for $j = 1$, number of event parameters in Ξ **do**
                $\mathcal{S} \leftarrow \mathcal{S} \cup \{(G, \text{average value of parameter } j \text{ in each element of } \Xi)\}$
            end for
        end for
        $S \leftarrow$ best pivot among features in $\mathcal{S}$, completed with an optimal threshold
        $(\mathcal{X}_l, \mathcal{X}_r) \leftarrow \mathcal{X}$ split according to S
        return $(S, \text{PARTITION}(\mathcal{X}_l, d-1), \text{PARTITION}(\mathcal{X}_r, d-1))$ ▷ Generates a branch
    end if
end function
for $i \leftarrow 1, n$ **do**
    $\mathcal{B} \leftarrow$ bootstrap sample of $\mathcal{T}$
    forest$[i] \leftarrow$ PARTITION($\mathcal{B}, d$)
end for
return forest

---

voting of all trees in the ensemble; the fraction of votes on certain class is returned as well, which can be used as a continuous estimate of the model confidence.

The crucial part of this approach is the gate generation step; here we have used a simple idea of making a cuboid centred on a random event and sized randomly, though proportional to the dimensions of the cloud of all events. Naturally, we do not expect such gates to be meaningful as understood by a flow cytometry expert; they are rather probes which results become a description of event populations only after being combined by the overall model structure. Precisely, we use a following algorithm. First, we select a random tube. One event C is randomly picked from randomly chosen object from that tube. Let's now assume that event C is characterized by k parameters $(c_1, c_2, \ldots c_k)$. Subsequently, k numbers $e_1, e_2, \ldots, e_k$ are sampled from uniform distribution $\mathcal{U}(0.05, 1)$. Let a_i and b_i be, respectively, 5-th and 95-th percentile of events distribution on i-th parameter. We define

$$r_i = \frac{e_i \cdot (b_i - a_i)}{2}$$

for all k parameters. Finally, we construct a cuboid rectangular gate defined as follows $[c_1 - r_1, c_1 + r_1] \times [c_2 - r_2, c_2 + r_2] \times \ldots \times [c_k - r_k, c_k + r_k]$. As mentioned earlier, not all possible random gates are fit for further consideration; some may be trivially nonsense, like being to small or unfortunately located to contain any event with reasonable odds. On the other hand, gates containing very small numbers of events (several per object on average) will have a substantial entropy and may lead to overfitting, similar to how categorical variables with a number of levels close to the number of objects would behave in a regular Random Forest model.

This is why we employ a simple *quality control* (QC) procedure of dropping any gate which contains less than two events for some object. One should note, however, that gates covering almost all events are not problematic, as they catch global properties of the sample. Additionally, in case there is a featured positive class, we also reject gates for which the average count of contained events is higher for the objects of the negative class than for of the positive class. This way, splits are substantially more likely to be based on the representative populations of cells for the positive class.

Obviously, it is possible that some other method of generating gates may be more appropriate, especially in certain cases; such modifications are the most natural path of extending flowForest in the future. One should also note that such a framework allows for incorporating existing expert knowledge, either by also considering gates given *a priori*, or by evaluating simple splits on some meta attributes. These extensions are not considered in this work, however, as we focus on formulating a possibly generic and autonomous method.

Similar to Random Forest, flowForest can establish an OOB estimate of an error. This procedure works as follows. Each object is predicted by a subset of trees for which this object was OOB, and only those predictions are combined by a regular voting. Such predictions are then compared with true classes and summarized into an OOB estimation of a prediction error, or any other possible performance metric (precision, recall, F-score, etc.).

FlowForest is implemented as an open source R package, available from https://github.com/flowforest/flowforest.

3   Results and Discussion

We have assessed the algorithm using three real-world flow cytometry data sets, earlier used in the FlowCAP II challenge [2]:

- AML, discrimination between patients affected by acute myeloid leukaemia and healthy donors, based on blood and bone marrow aspirate samples. 359 objects (316 normal, negative, and 43 AML, positive), 8 tubes, 7 parameters per each tube, effectively 42 parameters total.
- HVTN, discrimination between post-HIV vaccine blood samples stimulated by two HIV antigens. 96 objects (48 GAG-, 28 ENV-stimulated, equivalent), single tube, 11 parameters (we discarded the technical, non-informational *time* parameter, corresponding to the time of an event detection).
- HEU, discrimination between infants exposed to HIV but uninfected and not exposed to HIV, based on blood samples. 44 objects (24 unexposed, negative, and 20 HIV-exposed, positive), 7 tubes, 10 parameters (as in case of HVTN,

we discarded the time parameter) per each tube, effectively 70 parameters total.

During the challenge, all data sets were split into a training and testing subsets; we have performed our analysis both using this split, to obtain accuracy numbers comparable with those achieved by the participants. Next, we have performed a 10-fold cross-validation to investigate the robustness of the flowForest method; while classes are not balanced, we employed a slight variation of a canonical methodology, forcing all folds to have the same class distribution as the whole set. Finally, we have trained a model on a full data set to gather the OOB error estimate.

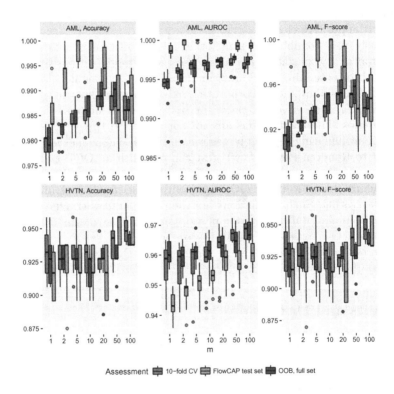

Fig. 1. FlowForest classification performance on AML and HVTN problems, measured by accuracy, area under ROC curve and F-score, for $n = 200$ and different values of m. Three types of assessments were applied—cross-validation, test set as used in FlowCAP, finally OOB error estimate.

While flowForest is a stochastic algorithm, all these assessments have been replicated 10 times with different random seeds. We have investigated a number of gate tries parameter m values, namely 1, 2, 5, 10, 20, 50 and 100. The tree count parameter n was fixed to 200, which we found enough for ensemble predictions to stabilise. Performance was measured as raw accuracy, F-score and

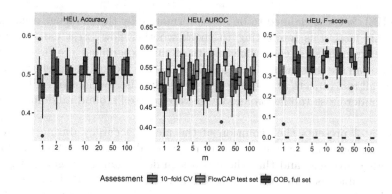

Fig. 2. FlowForest classification performance on a non-informative HVTN problem, for $n = 200$ and different values of m. Three types of assessments were applied—cross-validation, test set as used in FlowCAP, finally OOB error estimate.

the area under ROC curve (AUROC). As the model confidence score required for AUROC, we have employed the fraction of trees in the ensemble which voted on the positive class.

The collected results for AML and HVTN problems have been presented on Fig. 1. One can see that the model was quite accurate, yielding over 90% accuracy in all investigated cases; there is also a good agreement between cross-validation and OOB estimates of error, which is important provided that the latter can be calculated in roughly one tenth of the CPU time. Both had relatively low variance across realisations, proving that the algorithm has converged and is robust to small disturbances.

Neither FlowCAP test subset seem to be representative for its corresponding full data set, however they allow us to compare flowForest with other methods, described in the after-challenge paper [2]. Using this metric, in case of the AML set, the method is able to reliably provide perfect accuracy for $m = 5$ or 10, namely achieve perfect classification in 65% of realisations, and a perfect ordering (AUROC=1) in 85%. This is consistent with the results of FlowCAP, in which 9 out of 23 submissions were perfect and the median F-score was .97; also the FlowCAP organisers concluded that the problem was relatively easy, and the only issue was a single outlier object.

On the other hand, HVTN was more challenging; flowForest required $m = 50$ to achieve best test result, and the average F-score was .95; this is worse than 6 (out of 9) methods which obtained perfect classification, and equal to one other method. However, only one of them, flowType+feaLect, was not a part of an ad hoc pipeline tuned to the problem and did not used biological knowledge.

It is also interesting to focus on $m = 1$, which corresponds to a forest built on random gates, selected only to fulfil QC criteria but not optimal with respect to the split. While, as expected, it never was an optimal value, it lead to a surprisingly good performance, showing the capabilities of a tree ensemble structure.

The HEU set was claimed to be non-informative by the challenge organisers; despite this fact, it is interesting to check whether flowForest can provide a negative answer in such case. The results of this experiment are presented on Fig. 2. One can see that in all investigated set-ups the model has degenerated and predicted all test objects as not-exposed; similarly, distributions of AUROC and accuracy under cross-validation and OOB were centred around random-guessing values.

The Fig. 3 presents the evolution of the model accuracy during the forest construction; one can see that in case of meaningful AML and HVTN problems the process converges, and that there is a significant gain from gathering trees into an ensemble, as expected from a Random Forest-based method. In case of HEU, AUROC converges to a nonsense value, and its course is substantially more variable.

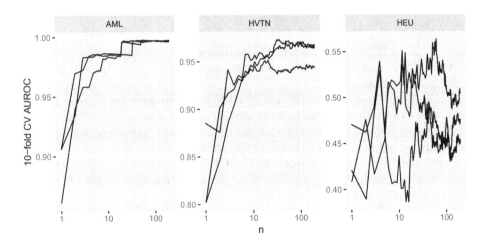

Fig. 3. Evolution of the flowForest accuracy, expressed as a cross-validation estimate of AUROC, with the growing ensemble size ($n \in [1, 200]$). Each of three random replications is shown as a separate line; parameter m is fixed to 10.

4   Conclusions

The undoubtful, rising role of flow cytometry in biomedical research and medicine, as well as the volume and complexity of data it provides, calls for a novel, more automated analytical tools.

In this paper, we propose flowForest, a novel classifier based on the popular Random Forest method. We show that it can provide an accurate, robust classification on a pure flow cytometry outputs, without specialised pre-processing, injected biological knowledge, ad hoc assumptions nor any other prior information, and with minimal hyper-parameter tuning. To our knowledge, this is the only available method of a said capability, with virtually all other approaches following a two-step scenario in which data is first gated, and the inference is done only on such obtained, simplified representation. This way flowForest may discover more subtle aspects of the problem, by effectively investigating gates that are only relevant in specific, local contexts.

We also demonstrate, based on the HEU dataset, that it can reliably detect an inconclusive problem, avoiding over-fitting despite highly dimensional representation of data it uses. Quite obviously, this is a crucial property for a reliability of any potential methodology involving it, which is especially important given numerous uses of machine learning in biomedical research had suffered from such issues.

Thus, we believe it may be useful to assist manual studies, as it provides a perfectly unbiased view of data, or become a part of an automatic analysis pipeline.

Acknowledgements. This work has been financed by the National Science Centre, grant 2011/01/N/ST6/07035. Computations were performed in ICM, grant G48-6.

References

1. Aebisher, D., Bartusik, D., Tabarkiewicz, J.: Laser flow cytometry as a tool for the advancement of clinical medicine. Biomed. Pharmacother. **85**, 434–443 (2017)
2. Aghaeepour, N., et al.: Critical assessment of automated flow cytometry data analysis techniques. Nat. Methods **10**(3), 228–238 (2013)
3. Aghaeepour, N., Nikolic, R., Hoos, H.H., Brinkman, R.R.: Rapid cell population identification in flow cytometry data. Cytom. Part A **79A**(1), 6–13 (2011)
4. Bacher, P., Scheffold, A.: New technologies for monitoring human antigen-specific T cells and regulatory T cells by flow-cytometry. Curr. Opin. Pharmacol. **23**, 17–24 (2015)
5. Breiman, L.: Random forests. Mach. Learn. **45**, 5–32 (2001)
6. Czeh, A., et al.: A flow cytometry based competitive fluorescent microsphere immunoassay (CFIA) system for detecting up to six mycotoxins. J. Immunol. Methods **384**(1–2), 71–80 (2012)
7. Dashkova, V., Malashenkov, D., Poulton, N., Vorobjev, I., Barteneva, N.S.: Imaging flow cytometry for phytoplankton analysis. Methods **112**, 188–200 (2017)
8. Ge, Y., Sealfon, S.C.: flowpeaks: a fast unsupervised clustering for flow cytometry data via k-means and density peak finding. Bioinformatics **28**(15), 2052–2058 (2012)
9. Kanegane, H., et al.: Flow cytometry-based diagnosis of primary immunodeficiency diseases. Allergol. Int. **67**(1), 43–54 (2018)
10. Kursa, M.B.: Generalised random forest space overview (2015). https://arxiv.org/abs/1501.04244
11. Lizard, G.: Flow cytometry analyses and bioinformatics: interest in new softwares to optimize novel technologies and to favor the emergence of innovative concepts in cell research. Cytom. A **71A**, 646–647 (2007)
12. Lo, K., Hahne, F., Brinkman, R.R., Gottardo, R.: flowclust: a bioconductor package for automated gating of flow cytometry data. BMC Bioinform. **10**(1), 1–8 (2009)
13. Maguire, O., Tario, J.D., Shanahan, T.C., Wallace, P.K., Minderman, H.: Flow cytometry and solid organ transplantation: a perfect match. Immunol. Investig. **43**, 756–774 (2014)

14. O'Neill, K., Aghaeepour, N., Špidlen, J., Brinkman, R.: Flow cytometry bioinformatics. PLoS Comput. Biol. **9**, 1–10 (2013)
15. Sliwinska, E.: Flow cytometry - a modern method for exploring genome size and nuclear DNA synthesis in horticultural and medicinal plant species. Folia Hortic. **30**(1), 103–128 (2018)
16. Takahashi, T.: Life cycle analysis of endosymbiotic algae in an endosymbiotic situation with paramecium bursaria using capillary flow cytometry. Energies **10**(9), 1413 (2017)
17. Van Assche, A., Vens, C., Blockeel, H., Džeroski, S.: First order random forests: Learning relational classifiers with complex aggregates. Mach. Learn. **64**(1), 149–182 (2006)

Positive Unlabeled Link Prediction via Transfer Learning for Gene Network Reconstruction

Paolo Mignone and Gianvito Pio[✉]

Department of Computer Science, University of Bari Aldo Moro,
Via Orabona, 4, 70125 Bari, Italy
{paolo.mignone,gianvito.pio}@uniba.it

Abstract. Transfer learning can be employed to leverage knowledge from a source domain in order to better solve tasks in a target domain, where the available data is exiguous. While most of the previous papers work in the supervised setting, we study the more challenging case of positive-unlabeled transfer learning, where few positive labeled instances are available for both the source and the target domains. Specifically, we focus on the link prediction task on network data, where we consider known existing links as positive labeled data and all the possible remaining links as unlabeled data. In many real applications (e.g., in bioinformatics), this usually leads to few positive labeled data and a huge amount of unlabeled data. The transfer learning method proposed in this paper exploits the unlabeled data and the knowledge of a source network in order to improve the reconstruction of a target network. Experiments, conducted in the biological field, showed the effectiveness of the proposed approach with respect to the considered baselines, when exploiting the *Mus Musculus* gene network (source) to improve the reconstruction of the *Homo Sapiens Sapiens* gene network (target).

1 Introduction

The link prediction task aims at estimating the probability of the existence of an interaction between two entities on the basis of the available set of known interactions, which belong to the same data distribution, the same context and described according to the same features. However, in many real cases, the identical data distribution assumption does not hold. For example, in the study of biological networks, collecting training data is very expensive and it is necessary to build link prediction models on the basis of data regarding different (even if related) contexts. At this regard, *transfer learning* strategies can be adopted to leverage knowledge from a source domain to improve the performance of a task solved on a target domain, for which we have few labeled data (see Fig. 1).

In the literature, we can find several applications where transfer learning approaches have proved to be beneficial. For example, in the classification of Web documents, where the goal is to assign a category to a certain Web document,

© Springer Nature Switzerland AG 2018
M. Ceci et al. (Eds.): ISMIS 2018, LNAI 11177, pp. 13–23, 2018.
https://doi.org/10.1007/978-3-030-01851-1_2

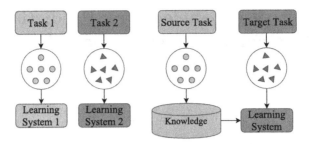

Fig. 1. Exploitation of the knowledge acquired on a source task to solve the target task (right-side), compared to solving two machine learning tasks independently (left-side).

transfer learning approaches can be exploited to classify newly created Web sites which follow a different data distribution [3] (e.g., the content is related to new subtopics). Another example is the work proposed in [11], where the authors exploit transfer learning approaches in a situation where data become easily outdated. In particular, the authors aim to adapt a WiFi localization model trained in one time period (source domain) to a new time period (target domain), where the available data possibly follow a different data distribution.

Focusing on the link prediction task, in the literature we can find several works in the biological field, since biological entities and their relationships can be naturally represented as a network. In the specific field of genomics, recent studies have significantly relied on high throughput technologies and on computational methods, which led to an improved understanding of the working mechanisms in several organisms. Such mechanisms are usually modeled through gene interaction networks, where nodes represent genes and edges represent regulation activities. The direct observation of the real structure of these interaction networks would require expensive in-lab experiments. Since gene expression data are easy to obtain, several methods have been proposed in the literature that exploit this kind of data [9]. These approaches analyze the expression level of the genes under different conditions (e.g., with a specific disease or after a treatment with a specific drug) or, alternatively, under a single condition in different time instants. Therefore, most machine learning approaches aiming to solve the link prediction task generally analyze gene expression data. In this context, the goal is to reconstruct the whole network structure (Gene Network Reconstruction - GNR), providing the biologists with a general overview of the interactions among the genes. However, while existing methods generally work effectively on sufficiently large training data, a transfer learning approach could favor the GNR of specific organisms which are not well studied, by exploiting the knowledge acquired about different, related organisms.

The main contribution of this paper is to evaluate the possible benefits that transfer learning techniques can provide to the task of link prediction. In particular, we exploit the available information about a source network for the reconstruction of a target network with poor available data. Moreover, we study the more challenging case of Positive-Unlabeled (PU) transfer learning, where few positive labeled examples are available for both the source and the target domains, and no negative example is available. PU learning setting holds in many real context (e.g., text categorization [8], bioinformatics [5]) where it is very expensive or unfeasible to obtain negative examples for the concept that we intend to model. As described in [16], PU learning methods can be divided into three classes: *(a)* the first, called two-step strategy, tries to identify some reliable negative examples in the unlabeled data, and then applies supervised learning algorithms; *(b)* the second assigns different weights to positive and unlabeled examples, by estimating the conditional probability of an example of being positive; *(c)* the third just treats the unlabeled data as highly noisy negative data.

In this paper, we consider the link prediction task as a PU learning task of class *(b)*, that, according to previous studies [2,13], allows us to avoid the strong assumptions about the negative examples made by methods relying on classes *(a)* and *(c)*. In particular, we propose a link prediction method which aims at building a binary classifier for all the possible links, where each link $\langle v', v'' \rangle$ between two nodes v' and v'' is represented as the concatenation of the feature vectors of v' and v''. The training set is built by considering the set of vectors associated to known (i.e., validated) links as positive examples and the vectors associated to all the possible remaining links, excluding self-links, as unlabeled examples. Methodologically, in a first stage we build a clustering model for the source domain and a clustering model for the target domain. In both cases, this is performed only on the positive labeled examples in order to catch several different viewpoints of the underlying concept of positive interactions. In a second stage, the unlabeled data of both source and target domains are weighted according to the similarities with respect to the clusters' centroids. In a third stage, we exploit the positive examples and all the (weighted) unlabeled examples, coming from both the source and the target domains, by training a classifier which is able to handle weights on instances. According to [10,14], our method belongs to the category of *homogeneous* transfer learning approaches, where the source and target domains are described in the same feature space, with possibly different data distributions. This setting is in contrast with the *heterogeneous* transfer setting, which assumes different feature spaces.

In order to evaluate the performance of the proposed method, we performed experiments in the biological domain. In particular, our experiments focused on the reconstruction of the human (Homo Sapiens Sapiens) gene network guided by the gene network of another, related organism, i.e., the mouse (Mus Musculus).

In Sect. 2, we describe in details our method, while in Sect. 3 we show the experimental evaluation and report some comments about the results. Finally, in Sect. 4, we draw some conclusions and outline possible future works.

2   The Proposed Method

In this section, we describe our transfer learning approach to solve link prediction tasks in network data. Before describing it in details, we introduce some useful notions and formally define the link prediction task for a single domain. Let:

- V be the set of nodes of the network;
- $x = \langle v', v'' \rangle \in (V \times V)$ be a (possible) link between two nodes v' and v'', where $v' \neq v''$;
- $e(v) = [e_1(v), e_2(v), \ldots, e_n(v)]$ be the vector of features related to the node v, where $e_i(v) \in \mathbb{R}, \; \forall i \in \{1, 2, \ldots, n\}$;
- $e(x) = [e_1(v'), e_2(v'), \ldots, e_n(v'), e_1(v''), e_2(v''), \ldots, e_n(v'')]$ be the vector of features related to the link $x = \langle v', v'' \rangle$;
- $sim(a, b) \in [0, 1]$ be a similarity function between the vectors a and b;
- $l(x)$ a function that returns 1 if the link x is a known existing, and 0 if its existence is unknown;
- $L = \{x \mid x \in (V \times V) \wedge l(x) = 1\}$ be the set of labeled links;
- $U = (V \times V) \setminus L$ be the set of unlabeled links;
- $D = \{\widetilde{X}, P(X)\}$ be the domain described by the feature space $\widetilde{X} = \mathbb{R}^{2n}$, with a specific marginal data distribution $P(X)$, where $X = L \cup U$;
- $w(x) \; (0 \leq w(x) \leq 1)$ be a computed weight for the link $x \in U$;
- $f(x)$ be an ideal (target) function which returns 1 if x is an existing link, and 0 otherwise.

The task we intend to solve is then defined as follows:

Given: a set of training examples $\{\langle e(x), w(x) \rangle\}_x$, each of which described by a feature vector and a weight;

Find: a function $f' : \mathbb{R}^{2n} \to [0, 1]$ which takes as input a vector of features $e(x)$ and returns the probability that the link x exists. Therefore, $f'(e(x)) \approx \mathcal{P}(f(x) = 1)$ or, in other terms, f' approximates the probability distribution over the values of the ideal function f.

Our method works with two different domains: the source domain $D_s = \{\widetilde{X_s}, P(X_s)\}$, and the target domain $D_t = \{\widetilde{X_t}, P(X_t)\}$. We remind that our method works with homogeneous feature spaces, that is $\widetilde{X_s} = \widetilde{X_t}$, while the marginal data distributions is generally different, that is $P(X_s) \neq P(X_t)$.

Given the two sets of labeled examples L_s and L_t, regarding the source and the target domain respectively, the method consists of three stages, that are summarized in Fig. 2 and detailed in the following subsections.

Stage I - Clustering. The first stage of our method consists in the identification of a clustering model for the positive examples of each domain (i.e., on L_s and L_t). The application of a clustering method is motivated by the necessity to distinguish among possible multiple viewpoints of the underlying concept of positive interactions. Moreover, a summarization in terms of clusters' centroids becomes useful also from a computational viewpoint, since in the subsequent stages we can compare centroids instead of single instances. In this paper, we adopt the classical *k-means* algorithm, since it is well established in the literature. However, any other prototype-based clustering algorithm, possibly able to catch specific peculiarities of the data at hand, could be plugged into our method.

Stage II - Instance Weighting. Although an unlabeled link could be either a positive or a negative example, we consider all the unlabeled examples as positive examples and compute a weight representing the degree of certainty in $[0, 1]$ of

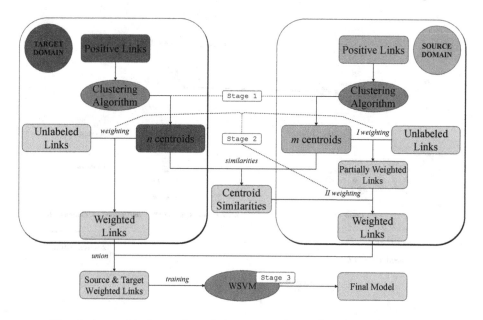

Fig. 2. A graphical overview of the proposed transfer learning approach.

being a positive example: a value close to 0 means that the example is likely to be a negative example, while a value close to 1 means that the example is likely to be a positive example. The weight associated to the unlabeled instances of both the source and the target domains are computed according to their similarities with respect to the centroids obtained in the first stage. In particular, we identify a different weighting function for the source and target domains, in order to smooth the contribution provided by instances coming from the source domain.

Specifically, an unlabeled link x belonging to the target network (i.e., $x \in V_t \times V_t$) is weighted according to its similarity with respect to the centroid of its closest cluster, among the clusters identified from the target network. Formally:

$$w(x) = max_{c_t \in C_t}(sim(e(x), c_t)), \tag{1}$$

where C_t are the clusters identified from positive examples of the target network.

On the other hand, an unlabeled link x' belonging to the source network (i.e., $x' \in V_s \times V_s$) is weighted by considering two similarity values: *(i)* the similarity with respect to the centroid of its closest cluster, computed among the clusters identified from the source network, and *(ii)* the similarity between such a centroid and the closest centroid identified on the target network. Formally, let $c' = argmax_{c_s \in C_s}(sim(e(x'), c_s))$ be the closest centroid with respect to x' among the possible centroids C_s identified in the source network. Then:

$$w(x') = sim(e(x'), c') \cdot max_{c'' \in C_t}(sim(c'', c')). \tag{2}$$

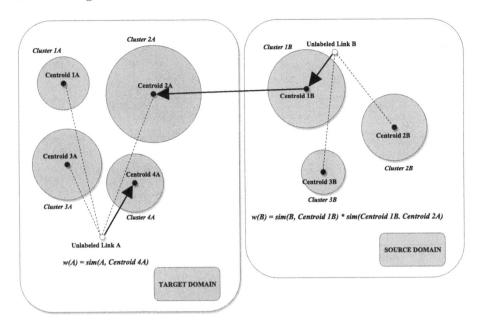

Fig. 3. Example of unlabeled link weighting process.

As a similarity function, we exploit the Euclidean distance, after applying a min-max normalization (in the range $[0, 1]$) to all the features of the feature vectors. Formally, $sim(e(x'), e(x'')) = 1 - \sqrt{\sum_{k=1}^{n} (e_k(x') - e_k(x''))^2}$.

An overview of the weighting strategy can be graphically observed in Fig. 3.

Stage III - Training the Classifier. In the third stage, we train a probabilistic classifier, based on linear Weighted Support Vector Machines (WSVM) [15] with Platt scaling [1], from the weighted unlabeled instances coming from both the source and the target networks. We selected an SVM-based classifier mainly because *(i)* it has a (relatively) good computational efficiency, especially in the prediction phase, and *(ii)* it already proved to be effective (with Platt scaling) in the semi-supervised setting [4]. At the end of the training phase, the WSVM classifier produces a model in the form of an hyperplane function h. In the specific PU setting, while h is not class discriminatory, we can consider as positive examples those appearing close to the identified hyperplane[1]. At this respect, by exploiting the Platt scaling, for each unlabeled link x, we compute the probability of being a positive example as $f'(e(x)) = \frac{1}{1+e^{-h(e(x))}}$, where $h(e(x))$ is the score obtained by the learned WSVM. Finally, we rank all the predicted links in descending ordering with respect to the their probability of being positive. A pseudo-code representation of the proposed method is shown in Algorithm 1.

---

[1] The less the distance between an unlabeled example and the hyperplane, the higher the probability of the existence of the link.

Algorithm 1: PU Link Prediction via Transfer Learning

Data:
- $L_s = \{x_s \mid x_s \in (V_s \times V_s) \wedge l(x_s) = 1\}$: positive links of the source network
- $U_s = (V_s \times V_s) \setminus L_s$: unlabeled links of the source network
- $L_t = \{x_t \mid x_t \in (V_t \times V_t) \wedge l(x_t) = 1\}$: positive links of the target network
- $U_t = (V_t \times V_t) \setminus L_t$: unlabeled links of the target network
- $e(x)$: feature vector of the link x
- $sim(e(x'), e(x'')) = 1 - \sqrt{\sum_{k=1}^{n} (e_k(x') - e_k(x''))^2}$
- k_1, k_2: number of positive clusters for the source and the target network, respectively

Result:
- $ranked\_links$: predicted links ordered according to their likelihood

1 **begin**
2      $C_s \leftarrow kmeans(L_s, k_1)$; $C_t \leftarrow kmeans(L_t, k_2)$;
3      $source\_training\_set \leftarrow \emptyset$; $target\_training\_set \leftarrow \emptyset$; $ranked\_links \leftarrow \emptyset$;
4   **foreach** $x_t \in U_t$ **do**
5          $w(x_t) \leftarrow max_{(c_t \in C_t)}(sim(e(x_t), c_t))$;
6          $target\_training\_set \leftarrow target\_training\_set \cup \{\langle e(x_t), w(x_t)\rangle\}$;
7   **foreach** $x_s \in U_s$ **do**
8          $source\_centroid \leftarrow argmax_{(c_s \in C_s)}(sim(e(x_s), c_s))$;
9          $partial\_weight \leftarrow sim(e(x_s), source\_centroid)$;
10          $centroid\_sim \leftarrow max_{(c_t \in C_t)}(sim(source\_centroid, c_t))$;
11          $w(x_s) \leftarrow partial\_weight \cdot centroid\_sim$;
12          $source\_training\_set \leftarrow source\_training\_set \cup \{\langle e(x_s), w(x_s)\rangle\}$;
13      $training\_set \leftarrow source\_training\_set \cup target\_training\_set$;
14      $h(\cdot) \leftarrow WSVM(training\_set)$;
15      $f'(\cdot) = \frac{1}{1 + e^{-h(\cdot)}}$;
16   **foreach** $x \in U_t$ **do**
17          $ranked\_links \leftarrow ranked\_links \cup \{\langle x, f'(e(x))\rangle\}$;
18      $ranked\_links \leftarrow sort\_by\_score(ranked\_links)$;
19   **return** $ranked\_links$;

3 Experiments

Our experiments have been performed in the biological field. In particular, the specific task we intend to solve is the reconstruction of the human (*Homo Sapiens Sapiens - HSS*) gene network. As a source task, we will exploit the gene network of another, related organism, i.e., the mouse (*Mus Musculus - MM*).

3.1 Dataset

The considered dataset consists of gene expression data related to specific organs. In particular, we analyzed the gene expression levels of 6 organs (lung, liver, skin, brain, bone marrow, heart), obtained by the samples available at Gene Expression Omnibus (GEO), a public functional genomics repository. On overall, 161 and 174 samples were considered respectively for MM and HSS.

All the samples of each organism were processed according to the data acquisition workflow adopted for the DREAM5 challenge [9]. In particular, we processed the samples through the *Affymetrix Expression Console Software*, which led to produce a dataset consisting of a total of 45, 101 genes over the 161 samples for MM and a dataset of 54, 675 genes over the 174 samples for HSS.

Although we originally had a different number of features for the considered organisms, we built two homogeneous datasets by aggregating the features according to the organs. In particular, for each organisms, we represented their genes by means of 6 features (one for each organ), by averaging the expression levels measured within the same organ. Accordingly, the datasets representing the interactions among genes were built by considering all the possible pairs of genes, i.e., by concatenating the feature vectors associated to the genes, leading to 12-dimensional feature vectors. The set of validated gene interactions was extracted from BioGRID[2], which is an interaction repository containing data compiled through comprehensive curation efforts. This set represents our ground truth in terms of positive links. As for the unlabeled instances, we performed a random sampling without replacement from all the other possible links involving at least one gene that appears in the BioGRID ground truth. This procedure led us to build a balanced dataset between positive and unlabeled examples.

3.2   Experimental Setting

Since our method exploits the k-means clustering algorithm, we performed the experiments with different values for k_1 (i.e., the number of clusters for the MM organism) and k_2 (i.e., the number of clusters for the HSS organism), in order to evaluate the possible effect of such parameters on the results. In particular, we considered the following parameter values: $k_1 \in \{2, 3\}, k_2 \in \{2, 3\}$.

We remind that we work in the PU learning setting (i.e., the dataset does not contain any negative example). Therefore, inspired by the experiments performed in [12], we evaluated the results in terms of recall@k. The adoption of this measure allows us to avoid the estimation of possible negative examples in the ground truth, which could lead to a wrong evaluation of the results. In particular, in order to quantitatively compare the obtained results, we draw the recall@k curve, by varying the value of k, and compute the area under the curve.

The experiments have been performed according to the 10 fold cross validation (10 fold CV) on the positive labeled data. In particular, for each iteration of the 10 fold CV, we considered: *(a)* a portion of the positive labeled data (9 out of 10 folds) to build the clustering models through k-means; *(b)* all the unlabeled data and a portion of positive labeled data (1 out of 10 folds) as test set.

We compared our method, indicated as **transfer**, with approaches:

- **no_transfer**, which corresponds the WSVM with Platt scaling learned only from the target network (i.e., from the HSS network). This baseline allows us to evaluate the contribution of the source domain.
- **union**, which is the WSVM with Platt scaling learned from a single dataset consisting of the union of the instances coming from both MM and HSS. This baseline allows us to evaluate the effect of our weighing strategy.

Since we are interested in observing the contribution provided by our approach with respect to the non-transfer approach, results will be evaluated in terms of improvement with respect to the **no_transfer** baseline.

---

[2] https://thebiogrid.org.

3.3   Results

In Fig. 4, we show the results obtained with different values of k_1 and k_2. In particular, we considered different percentages of the recall@k curve and measured the area under each sub-curve. This evaluation is motivated by the fact that biologists usually focus their in-lab studies on the analysis of the top-ranked predicted interactions. Therefore, a better result in the first part of the recall@k curve (i.e., at 1%, 2%) appears to be more relevant for the real biological application. The graphs show that both the **union** baseline and the proposed method (**transfer**) are able to obtain a better result with respect to the variant without any transfer of knowledge (**no_transfer**). This confirms that, in this case, the external source of knowledge (the MM gene network) can be exploited to improve the reconstruction of the target network (the HSS gene network).

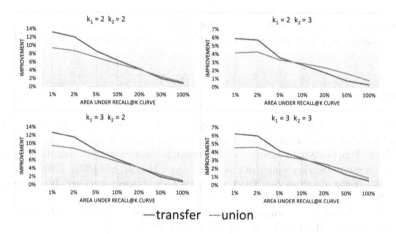

Fig. 4. Improvement over no_transfer, with different values of k_1 and k_2.

By comparing the **union** baseline with our method, we can observe that the proposed weighting strategy was effective in assigning the right contribution to each unlabeled instance (coming either from the source or from the target network) in the learning phase. This is even more evident in the first part of the recall@k curve, where our method was able to retrieve about 120 additional true interactions at the top 1% of the ranking with respect to the baseline approaches.

Finally, by analyzing the results with respect to the values of k_1 and k_2, we can conclude that the highest improvement over the baseline approaches has been obtained with $k_1 = 2$ and $k_2 = 2$. This means that clustering can affect the results, and that even higher improvements could be obtained by adopting smarter clustering strategies that can, for example, catch and exploit the distribution, in terms of density, of the examples in the feature space.

4   Conclusion and Future Work

In this paper, we proposed a transfer learning method to solve the link prediction task in the PU learning setting. By resorting to a clustering-based strategy, our method is able to exploit unlabeled data as well as labeled and unlabeled data of a different, related domain, identifying a different weight for each training instance. Focusing on biological networks, we evaluated the performance of the proposed method in the reconstruction of the Human gene network, supported by the knowledge about the mouse gene network. Results show that the proposed method was able to improve the accuracy of the reconstruction, if compared to two baseline approaches. As future work, we plan to implement a distributed version of the proposed method, and to adopt some ensemble-based approaches [6,7] to exploit multiple clusters in the prediction. We also plan to perform an extensive comparison with state-of-the-art methods in the biological field.

Acknowledgments. We would like to acknowledge the European project MAESTRA - Learning from Massive, Incompletely annotated, and Structured Data (ICT-2013-612944).

References

1. Platt, J.C.: Probabilistic outputs for support vector machine and comparisons to regularized likelihood methods. In: Advances in Large Margin Classifiers (1999)
2. Ceci, M., Pio, G., Kuzmanovski, V., Džeroski, S.: Semi-supervised multi-view learning for gene network reconstruction. Plos One, **10**(12), e0144031 (2015)
3. Dai, W., Yang, Q., Xue, G., Yu, Y.: Boosting for transfer learning. In: Proceedings of ICML, pp. 193–200 (2007)
4. Elkan, C., Noto, K.: Learning classifiers from only positive and unlabeled data. In: Proceedings of ACM SIGKDD, pp. 213–220 (2008)
5. Jowkar, G., Mansoori, E.: Perceptron ensemble of graph-based positive unlabeled learning for disease gene identification. Comput. Biol. Chem. **64**, 263–270 (2016)
6. Levatic, J., Ceci, M., Kocev, D., Dzeroski, S.: Self-training for multi-target regression with tree ensembles. Knowl. Based Syst. **123**, 41–60 (2017)
7. Levatic, J., Kocev, D., Ceci, M., Dzeroski, S.: Semi-supervised trees for multi-target regression. Inf. Sci. **450**, 109–127 (2018)
8. Liu, B., Lee, W.S., Yu, P.S., Li, X.: Partially supervised classification of text documents. In: Proceedings of ICML, pp. 387–394 (2002)
9. Marbach, D., et al.: Wisdom of crowds for robust gene network inference. Nat. Meth. **9**(8), 796–804 (2016)
10. Pan, S.J., Yang, Q.: A survey on transfer learning. IEEE Trans. Knowl. Data Eng. **22**(10), 1345–1359 (2010)
11. Pan, S.J., Zheng, V.W., Yang, Q., Hu, D.H.: Transfer learning for wifi-based indoor localization. In: Workshop on Transfer Learning for Complex Task AAAI (2008)
12. Pio, G., Ceci, M., Malerba, D., D'Elia, D.: ComiRNet:a web-based system for the analysis of miRNA-gene regulatory networks. BMC Bioinform. **16**(S-9), S7 (2015)
13. Pio, G., Malerba, D., D'Elia, D., Ceci, M.: Integrating microRNA target predictions for the discovery of gene regulatory networks: a semi-supervised ensemble learning approach. BMC Bioinform. **15**(S-1), S4 (2014)

14. Weiss, K.R., Khoshgoftaar, T.M., Wang, D.: A survey of transfer learning. J. Big Data **3**, 9 (2016)
15. Yang, X., Song, Q., Wand, Y.: A weighted support vector machine for data classification. Int. J. Pattern Recogn. **21**, 961–976 (2007)
16. Zhang, B., Zuo, W.: Learning from positive and unlabeled examples: a survey. In: ISIP/WMWA, pp. 650–654 (2008)

Early Detection of Heart Symptoms with Convolutional Neural Network and Scattering Wavelet Transformation

Mariusz Kleć[(✉)]

Multimedia Department, Polish-Japanese Academy of Information Technology,
Warsaw, Poland
mklec@pjwstk.edu.pl

Abstract. The paper utilizes Convolutional Neural Network (CNN) for preliminary screening of cardiac pathologies by classifying the signal of heartbeat, recorded by digital stethoscope and mobile devices. The Scattering Wavelet Transformation (SWT) was used for the heartbeat representation. The experiments revealed the optimum concatenation size of SWT windows to obtain the state-of-the-art in the majority of metrics, coming from the PASCAL Classifying Heart Sounds Challenge.

Keywords: Heartbeat classification · Convolutional Neural Network
Scattering Wavelet Transformation

1 Introduction

The World Health Organization (WHO) states that mortality from heart disease is a plague of the 21st century. Annually, 17.5 million people die due to cardiovascular diseases.[1] In Poland, the Central Statistical Office (GUS)[2] reports that cardiovascular diseases are also the most common cause of mortality which states to be 46% of all deaths in Poland. Therefore, early diagnosis becomes a huge challenge for the medical community in the field of implementation of preventive care, also in Third World countries where the access to medical care and medical devices is limited.

Every effort to delay or prevent the morbid events is worth considering. Correct assessment of the heart function, recorded by electrocardiogram (ECG), attracts more attention from the community of computer scientists [1,14,15,17]. Convolutional Neural Networks (CNN) have found their application for ECG data classification. The Arrhythmia Detection (AD) is the most popular field of research among the cardiology fields. The comprehensive survey of ECG-based heartbeat classification for AD is carried out in [12]. They report that a lot of research rely on publicly available MIT-BIH[3] database (which is

---

[1] http://new.who.int/news-room/fact-sheets/detail/the-top-10-causes-of-death.

[2] https://stat.gov.pl/obszary-tematyczne/ludnosc/ludnosc/statystyka-zgonow-i-umieralnosci-z-powodu-chorob-ukladu-krazenia,22,1.html.

[3] http://ecg.mit.edu/.

© Springer Nature Switzerland AG 2018
M. Ceci et al. (Eds.): ISMIS 2018, LNAI 11177, pp. 24–31, 2018.
https://doi.org/10.1007/978-3-030-01851-1_3

recommended by ANSI/AAMI [7] for validation of medical equipment) and AHA.[4] However, the main disadvantage of using popular benchmarks is that they do not represent very big volume of data, what in the domain of deep learning could be utilized efficiently and could presumably boost performance results. This is exactly what Andrew Ng's scientific group did. They obtained state-of-the-art in both recall and precision in cardiology performance using CNN with 34-layers and the dataset with more than 500 times of data than the previously studied corpora [16]. In another paper [14], the authors used CNN and a large volume of raw ECG time-series data to obtain the feature representation for identifying patients with paroxysmal atrial fibrillation (PAF) (life threatening cardiac arrhythmia). They experimentally verified that the learned representation can effectively replace the user's hand-crafted features, as CNN learned the key ones, unique to the PAF. They have also conducted the comparison with several conventional machine learning classifiers and indicated that combining the learned features with other classifiers significantly improves the performance of the patient screening systems. Their findings were verified by many researchers who claim that the problem of ECG classification heavily depends on the appropriate features that represent the data [15–17].

The Wavelet Transform (WT) is very often used by researchers for representing ECG signals [8,10,11]. The WT allows information extraction from both frequency and time domains, different from what is usually achieved by the traditional Fourier transform, which permits the analysis of only the frequency domain [6]. Within the types of WT, the Discrete Wavelet Transform (DWT) is the most popular [10]. Apart from the DWT, Continuous Wavelet Transform (CWT) has also been used to extract features from the ECG signals [2], since it overcomes some of the DWT drawbacks, such as the coarseness of the representation and instability.

This paper extends these methods by using Scattering Wavelet Transform (SWT) [3] as it was not used so far for the problem of heartbeat classification, especially with the data coming from the PASCAL Classifying Heart Sounds Challenge [4]. In [20], the authors classified this data directly from its Fourier transformation with CNN, omitting the segmentation phase to detect the fundamental physical characteristics of the heartbeat (WSCNN). Similar approach is presented in [5] where the authors described the framework based on the auto-correlation feature and diffusion maps, that are further provided to the SVM classifier (SVM-DM). Another paper describes a scaled spectrogram and partial least squares regression (SS-PLSR) for classifying the heartbeat signal [21]. In [22], they used tensor decomposed features (SS-TD) for the same purpose. All the results from these papers are aggregated in Table 2. The method described in this paper is abbreviated to CSWT for convenience.

The paper is organized as follows. Section 1 presents the recent works and introduction to the field. In Sect. 2 the data and pre-processing are described. The Sect. 3 contains the description of performed experiments. Finally, the results and conclusions are presented in Sects. 4 and 5 respectively.

---

[4] http://www.ahadata.com/.

2   Data Description and Pre-processing

Two datasets were provided for the PASCAL Classifying Heart Sounds Challenge [4]. Dataset A comprises data recorded by iPhone app called iStethoscope. Dataset B comprises data collected from a clinical trial in hospitals using the digital stethoscope DigiScope. The data was gathered in real-world situations and frequently contained the background noise such as speech, traffic, brushing the microphone against cloth or skin etc. The audio files were of different lengths, between 1 and 30 s. Datasets were divided into training and testing sets with 4 categories for Dataset A (Normal, Murmur, Extra Heart Sound and Artifact) and 3 categories for Dataset B (Normal, Murmur and Extrasystole). Table 1 contains statistics of both datasets used in the experiments. The meaning of the categories is described in [19] in more detail.

The audio files were re-sampled to the equal value of 22050 Hz in both Datasets. The high-pass filter was applied below 200 Hz as the heartbeat information exists in the low frequencies.

Table 1. The number of files in Dataset A and B with respect to categories

Dataset A, 44100 Hz, 16bit			Dataset B, 4000 Hz, 16bit		
Category	Train	Test	Category	Train	Test
Normal	31	14	Normal	200	136
Murmur	34	14	Murmur	66	39
Extrasound	19	8	Extrasystole	46	20
Artifact	40	16	–	–	–

In this paper, the SWT was computed to a depth of 2 with 8 wavelets per octave of the Gabor kind in the first layer and 2 wavelets per octave of the Morlet type in the second layer. The author decided to use the same settings as used for Automatic Genre Recognition [3], assuming that music and heartbeat share the similar subset of structures (like sudden changes and short-term instabilities) which can be well captured by SWT [3]. The longest SWT window length T was 0.74 s. The two additional window lengths were half size of the previous ones: $T = 0.37$ s and $T = 0.185$ s. Eventually, the data was standardized to have zero mean and unit variance.

3   Experiments

The experiments were on training CNNs to classify the real heartbeat recordings described in Sect. 2. The use of the three SWT window lengths was evaluated: $T = 0.74$ s, $T = 0.37$ s and $T = 0.185$ s. Training and testing examples were constructed by the concatenation of several SWT windows into frames. The overall frame length was not longer than the shortest testing file (1.75 s for

Dataset A and 1 s for Dataset B). The frames half overlapped one another within the overall length of audio file.

The CNN contained 4 convolutional layers with max-pooling layers in between. The *stride* value was set to $1x1$, apart from the last max-pooling layer, where the *stride* was set to $2x1$ in order to reduce the output dimensions from the previous layer. The detailed architecture of CNN is presented in Fig. 1.

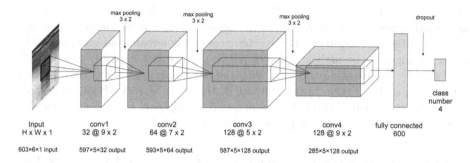

Fig. 1. The picture presents the architecture of CNN used for all experiments. The input size differed depending on the size of the input frames. The numbers below the name of layers indicate: the number of filters @ height x width of the filters. Furthermore, the instance of input frame is presented ($603 \times 6 \times 1$) together with the size of output from each subsequent convolutional layer. The instance of input frame represents the concatenation of six SWT windows with 0.185 s length each.

The validation accuracy has been monitored in subsequent epoch during training. The training was terminated when the validation accuracy has not been larger or equal to the previously highest accuracy for 10 epochs. The evaluation metrics were derived from the state of the network when the accuracy was the highest. However, the author noticed a slight deviation in results when running the same experiment several times. To provide a fair comparison, the final values were averaged after running the same experiment ten times. Additionally, the author observed that the highest validation accuracy in Dataset B always occurred after the first epoch. The model becomes over-fitted every time training is being continued beyond the first epoch.

For the loss function optimization, Adaptive Moment Estimation (ADAM) was chosen [9] with the following options: $\beta_1 = 0.9$, $\beta_2 = 0.999$ and $\alpha = 0.0005$. The initial learning rate was reduced by the factor of 0.9 every 5 epoch. The decision of choosing ADAM was dictated by its documented superior performance over Stochastic Gradient Descent [9]. The majority voting was used to derive the final category of audio file from the sequence of frame labels, outputted from the model.

The evaluation follows the same work-flow as used for the PASCAL Classifying Heart Sounds Challenge [4]. The results are compared to those presented in Table 2. The evaluation takes into account Precision per category, the Youden Index, the F-score (only for the Dataset A) the Discriminant Power (only for the Dataset B) and Normalized Precision. The more detailed definitions of the evaluation criteria are described in [22].

4   Results

The Table 2 highlights the best values for each metric, gathered from the literature. There is no strong conclusion which method overcomes the other ones in all possible aspects. However, the Normalized Precision (NP) can be treated as a good comparative determinant as it aggregates the precision of recognizing all heart categories together, taking into account the imbalanced number of files in each category. The highest value of NP in the Dataset A (0.80) was possible to obtain by representing the data with the concatenation of six SWT frames, 0.185 s long each. In turn, in the Dataset B, the best value of NP (0.75) belongs to the concatenation of two SWT frames, 0.37 s long each. In both cases the NP exceeds all the other results reported so far (see Table 2).

Table 2. The table presents the results on the Datasets A and B, gathered from the literature: SS-PLSR [21], SVM-DM [5], SS-TD [22], WSCNN [20]. The method proposed in this paper is abbreviated to CSWT. The values in column CSWT also contain the standard deviation from the results being averaged after running the same experiment ten times.

Results on Dataset A					
Evaluation criteria	SS-PLSR	SVM-DM	SS-TD	WSCNN	**CSWT** T = 0.185 s, 6F
Precision of Normal (PN)	0.60	0.62	**0.67**	0.61	0.57 ± 0.03
Precision of Murmur (PM)	0.91	0.91	**1.00**	0.91	**1.00 ± 0.00**
Precision of Extra Heart Sound (PE)	0.44	**1.00**	0.43	0.50	0.57 ± 0.19
Precision of Artifact (PA)	**0.94**	0.64	0.80	**0.94**	**0.94 ± 0.05**
Artifact Sensitivity (ASe)	**1.00**	**1.00**	**1.00**	**1.00**	0.80 ± 0.20
Artifact Specificity (ASp)	0.64	0.58	0.64	**0.67**	0.66 ± 0.03
Youden Index of Artifact (YIx)	0.64	0.58	0.64	**0.67**	0.66 ± 0.03
F-score (FS)	0.30	0.31	0.30	**0.67**	0.30 ± 0.02
Total Precision (TP)	2.89	**3.17**	2.90	2.96	3.07 ± 0.19
Normalized Precision (NP)	0.76	0.76	0.76	0.77	**0.80 ± 0.03**
Results on Dataset B					
Evaluation criteria	SS-PLSR	SVM-DM	SS-TD	WSCNN	**CSWT** T = 0.37 s, 2F
Precision of Normal (PN)	0.76	0.77	**0.83**	0.81	0.78 ± 0.01
Precision of Murmur (PM)	0.65	0.76	0.70	0.67	**0.96 ± 0.05**
Precision of Extrasystole (PE)	0.33	**0.50**	0.15	0.14	0.15 ± 0.33
Heart Problem Sensitivity (HPSe)	0.34	0.34	**0.49**	0.51	0.34 ± 0.04
Heart Problem Specificity (HPSp)	0.90	0.95	0.84	0.80	**0.99 ± 0.01**
Youden Index of heart problem (YIxp)	0.24	0.29	**0.33**	0.31	**0.33 ± 0.03**
Discriminant Power (DP)	0.36	0.54	0.39	0.34	**1.39 ± 0.91**
Total Precision	1.75	**2.03**	1.68	1.62	1.89 ± 0.35
Normalized Precision (NP)	0.69	0.74	0.74	0.71	**0.75 ± 0.04**

Analyzing each metric separately, we can conclude that the proposed method (CSWT) performs very well in recognizing Murmur. The method returns the high values of PM in both Datasets (1.00 on the Dataset A and 0.96 on the Dataset B), resulting the state-of-the-art in this metric, especially on the Dataset B (see Table 2). The DP evaluates how well the algorithm distinguishes between normal and problematic heartbeats (Murmur and Extrasystole categories combined). The high value of PM boosted the DP to the value of 1.39 and the HPSp to the value of 0.99, resulting the state-of-the-art also in these metrics on Database B (see Table 2).

Table 3. The first table presents the evaluation results on the Dataset A: Precision of Normal (PN), Precision of Murmur (PM), Precision of Extrasound (PE), Precision of Artifact (PA), Artifact Sensitivity (ASe), Artifact Specificity (ASp), The Youden Index of Artifact (YIx), F-Score of Heartproblem (FS), Total Precision (TP) and Normalized Precision (NP). The second table presents the results on the Dataset B: The Sensitivity of heart problems (HPSe), The Specificity of heart problems (HPSp), The Youden Index of Heartproblem (YIxp), Discriminant Power (DP). The columns refer to the SWT window size T and the number of concatenated frames F.

Results on the Dataset A

	$T = 0.74s$		$T = 0.37s$			$T = 0.185s$				
	1F	2F	1F	2F	4F	1F	2F	4F	6F	8F
PN	0.49	0.52	0.57	0.58	0.55	0.53	0.57	0.56	**0.57**	0.54
PM	0.82	0.86	0.98	0.96	0.90	0.82	0.95	0.97	**1.00**	0.90
PE	0.43	0.35	0.47	0.46	0.50	0.43	0.49	**0.64**	0.57	0.54
PA	**0.96**	0.95	0.93	0.90	0.94	0.92	0.92	0.89	0.94	0.94
ASe	0.8	0.8	0.80	**0.96**	0.8	0.75	0.84	0.8	0.8	0.8
ASp	0.57	0.57	0.65	0.64	0.62	0.57	0.64	0.64	**0.66**	0.62
YIx	0.57	0.57	0.65	0.64	0.62	0.56	0.64	0.64	**0.66**	0.62
FS	0.28	0.27	0.30	0.30	0.30	0.27	0.30	**0.31**	0.30	0.29
TP	2.70	2.69	2.95	2.91	2.89	2.7	2.92	3.05	**3.07**	2.92
NP	0.71	0.72	0.77	0.76	0.76	0.71	0.77	0.78	**0.8**	0.76

Results on the Dataset B

	T=0.74s	T=0.37s		T=0.185s		
	1f	1f	2f	1f	2f	4f
PN	**0.79**	0.77	0.78	0.75	0.76	0.76
PM	0.85	0.88	**0.96**	0.90	0.92	0.86
PEs	**0.21**	0.14	0.15	0.07	0	0.12
HPSe	**0.39**	0.33	0.34	0.24	0.27	0.29
HPSp	0.96	0.96	**0.99**	0.98	**0.99**	0.98
YIxp	**0.34**	0.28	0.33	0.22	0.25	0.26
DP	0.66	0.63	**1.39**	0.77	0.84	0.8
TP	1.84	1.80	**1.89**	1.72	1.79	1.68
NP	0.74	0.73	**0.75**	0.71	0.72	0.71

The model's ability to detect Artifacts from the heartbeat signal is important to inform the user to repeat the recording and avoid failure. Interestingly, the state-of-the-art in the PA (0.96) is possible to obtain with the use of only one and wide singular SWT window, 0.74 s long (see Table 3). It might be caused by the high ability of SWT to capture more short-term instabilities (that exist in the Artifacts) with a wider window [13]. The high ability of avoiding failures is highlighted by the high value of YIx on the Dataset A (see Table 3). This finding can guide the future solutions to treat the Artifact recognition differently than other categories.

However, there is still room for improvement of Precision of Normal, Precision of Extra Heart Sound and the Precision of Extrasystole. The highest values still belong to the other methods (see Table 2).

5   Conclusions

In this paper Convolutional Neural Network is utilized for preliminary screening of cardiac pathologies by classifying the heartbeat signals recorded by the digital stethoscope and the mobile phone. The Scattering Wavelet Transformation (SWT) is used to represent the data coming from the PASCAL Classifying Heart Sounds Challenge. The experiments reveal the optimum concatenation size of SWT windows to obtain the state-of-the-art in the Normalized Precision, Precision of Murmur, Precision of Artifact, Heart Problem Specificity and Discriminant Power. However, the Datasets used for these experiments are quite limited in terms of the number of training and testing examples. The PhysioNet/Computing in Cardiology Challenge 2016 (CinC)[5] addresses this problem by assembling the largest public heart sound database, aggregated from eight sources obtained by seven independent research groups around the world. The author plans to run CSWT method on that data and aims to fully validate the findings by running additional experiments and incorporate comparative statistical inference.

As it is pointed out in [18], data captured with off-the-person based devices (like mobile and wearable devices or electronic stethoscopes) can be highly correlated to those captured with traditional on-the-person based equipment (ECG systems). The author believes that the off-the-person approach is worth to further research as it can extend preventive medicine practices by allowing the heartbeat monitoring without interference on daily routine. It could help people to avoid serious problems and hopefully significantly improve the health statistics.

References

1. Acharya, U.R., et al.: A deep convolutional neural network model to classify heartbeats. Comput. Biol. Med. **89**, 389–396 (2017)
2. Addison, P.S.: Wavelet transforms and the ECG: a review. Physiol. Measur. **26**(5), R155 (2005)
3. Andén, J., Mallat, S.: Deep scattering spectrum. IEEE Trans. Signal Proces. **62**(16), 4114–4128 (2014)
4. Bentley, P., Nordehn, G., Coimbra, M., Mannor, S.: The PASCAL Classifying Heart Sounds Challenge 2011 (CHSC2011) Results (2011). http://www.peterjbentley.com/heartchallenge/index.html

---

[5] https://www.physionet.org/challenge/.

5. Deng, S.W., Han, J.Q.: Towards heart sound classification without segmentation via autocorrelation feature and diffusion maps. Future Gener. Comput. Syst. **60**, 13–21 (2016)
6. Dokur, Z., Ölmez, T.: ECG beat classification by a novel hybrid neural network. Comput. Methods Programs Biomed. **66**(2–3), 167–181 (2001)
7. AAMI/ANSI EC57: Testing and reporting performance results of cardiac rhythm and ST segment measurement algorithms. Association for the Advancement of Medical Instrumentation, Arlington (1998)
8. Güler, İ., Übeylı, E.D.: ECG beat classifier designed by combined neural network model. Pattern Recognit. **38**(2), 199–208 (2005)
9. Kingma, D.P., Ba, J.: Adam: a method for stochastic optimization. arXiv preprint arXiv:1412.6980 (2014)
10. Kutlu, Y., Kuntalp, D.: Feature extraction for ECG heartbeats using higher order statistics of WPD coefficients. Comput. Methods Programs Biomed. **105**(3), 257–267 (2012)
11. Lin, C.H., Du, Y.C., Chen, T.: Adaptive wavelet network for multiple cardiac arrhythmias recognition. Expert Syst. Appl. **34**(4), 2601–2611 (2008)
12. Luz, E.J.d.S., Schwartz, W.R., Cámara-Chávez, G., Menotti, D.: ECG-based heartbeat classification for arrhythmia detection: a survey. Comput. Methods Programs Biomed. **127**, 144–164 (2016)
13. Mallat, S.: Group invariant scattering. Commun. Pure Appl. Math. **65**(10), 1331–1398 (2012)
14. Pourbabaee, B., Roshtkhari, M.J., Khorasani, K.: Deep convolutional neural networks and learning ECG features for screening paroxysmal atrial fibrillation patients. IEEE Trans. Syst. Man Cybern. Syst. **99**, 1–10 (2017)
15. Pyakillya, B., Kazachenko, N., Mikhailovsky, N.: Deep learning for ECG classification. J. Phys. Conf. Ser. **913**, 012004 (2017)
16. Rajpurkar, P., Hannun, A.Y., Haghpanahi, M., Bourn, C., Ng, A.Y.: Cardiologist-level arrhythmia detection with convolutional neural networks. arXiv preprint arXiv:1707.01836 (2017)
17. Rubin, J., Abreu, R., Ganguli, A., Nelaturi, S., Matei, I., Sricharan, K.: Recognizing abnormal heart sounds using deep learning. arXiv preprint arXiv:1707.04642 (2017)
18. da Silva, H.P., Carreiras, C., Lourenço, A., Fred, A., das Neves, R.C., Ferreira, R.: Off-the-person electrocardiography: performance assessment and clinical correlation. Health Technol. **4**(4), 309–318 (2015)
19. Yuenyong, S., Nishihara, A., Kongprawechnon, W., Tungpimolrut, K.: A framework for automatic heart sound analysis without segmentation. Biomed. Eng. Online **10**(1), 13 (2011)
20. Zhang, W., Han, J.: Towards heart sound classification without segmentation using convolutional neural network. Computing **44**, 1 (2017)
21. Zhang, W., Han, J., Deng, S.: Heart sound classification based on scaled spectrogram and partial least squares regression. Biomed. Signal Process. Control **32**, 20–28 (2017)
22. Zhang, W., Han, J., Deng, S.: Heart sound classification based on scaled spectrogram and tensor decomposition. Expert Syst. Appl. **84**, 220–231 (2017)

Rough Sets: Visually Discerning Neurological Functionality During Thought Processes

Rory Lewis[1(✉)], Chad A. Mello[1], Yanyan Zhuang[1], Martin K.-C. Yeh[2], Yu Yan[3], and Dan Gopstein[4]

[1] Department of Computer Science, University of Colorado at Colorado Springs, Colorado Springs, CO 80918, USA
rlewis5@uccs.edu
[2] School of Information Sciences and Technology, Penn State Brandywine, Media, PA 19063, USA
[3] Department of Learning and Performance Systems, Penn State University, University Park, PA 168023, USA
[4] Tandon School of Engineering, New York University, New York, NY 10012, USA

Abstract. The central aim of this paper is to test and illustrate the viability of utilizing Rough Set Theory to visualize neurological events that occur when a human is thinking very intensely to solve a problem or, conversely, solving a trivial problem with little to no effort. Since humans solve complex problems by leveraging synapses from a distributed neural network in the frontal and parietal lobe, which is a difficult portion of the brain to research, it has been a challenge for the neuroscience community to functionally measure how intensely a subject is thinking while trying to solve a problem. Herein, we present our research of optimizing machine intelligence to visually illustrate when members of our cohort experienced misunderstandings and challenges during periods where they read and comprehended short code snippets. This research is a continuation of the authors' research efforts to use Rough Sets and artificial intelligence to deliver a system that will eventually visually illustrate deception.

1 Introduction

Although it is generally understood that the prefrontal cortex and parietal lobes are responsible for solving problems, there remains a need to validate specific functional specializations within the prefrontal cortex and parietal lobes [11,28]. This paper presents how we researched whether or not machine intelligence, in the form of the authors' neuroClustering[TM] system [22–24], based on rough set theory and fuzzy logic, could validate the role the prefrontal cortex and parietal lobes play in solving complex problems by visually displaying the synaptic activity therein. In Sect. 2.1 we first review previous computational neuroscience and how we utilize neuroClustering[TM] and Rough Set Theory in the neuroscience domain. In Sect. 2.2 we review human cognition and confusion. In Sect. 3 we discuss the experiments and analysis of our results in. Lastly, we discuss our future

© Springer Nature Switzerland AG 2018
M. Ceci et al. (Eds.): ISMIS 2018, LNAI 11177, pp. 32–41, 2018.
https://doi.org/10.1007/978-3-030-01851-1_4

work in Sect. 6 to predict when humans experience more complex neurological events such as deception versus truth and emotion.

2   Related Work

2.1   Review I: Computational Neuroscience

Computational neuroscience is the field of study in which mathematical tools and theories are used to investigate brain functionality [30] and, in some cases, synthetically recreate the brain as is being done in the European Human Brain Project [1,17,32]. The field of visually modeling brain intelligence, however, is a relatively new one where the early models, for example, synchronized distributed neurological assemblies based upon the visual cortex of a cat [12], or used neural networks to model visual pattern recognition [14] or modeled electroencephalographic (EEG) dynamics [26]. The current state-of-the-art of computational neuroscience focuses on visually representing and modeling the most functionally relevant brain regions for binding information over time [10]. For example, Forstmann, *et al.* describes a fusion of neuroanatomy, physics, mathematical, cognitive and clinical neuroscience techniques to study the human subcortex [13]. Poldrack researches fMRI, which measures blood oxygenation to image neural activity, to track when particular regions of the brain are activated, in other words, what portions of the brain are used [29]. Similarly, as in this case, to visualize and validate the role of the prefrontal cortex and parietal lobes in solving complex problems, the authors have infused machine intelligence into computational neuroscience by using their neuroClustering^{TM} system comprised of elements of fuzzy logic and Rough Set Theory, to see whether this could validate the role the prefrontal cortex and parietal lobes play in solving complex problems.

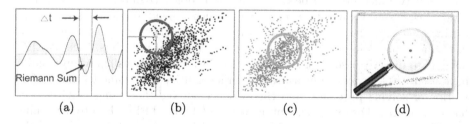

(a)   (b)   (c)   (d)

Fig. 1. NeuroClustering^{TM} [Fuzzy & Rough Sets]: (a) EEG is split with a spline to extract temporal duration and Riemann values. (b) Each point is one instance of temporal duration. (c) Centroid of *b*'s cluster is instantiated to one dot in (d) which is a moving dialog of EEGs for artificial intelligence. See video 1 at https://goo.gl/Mojejy.

The goal of neuroClustering^{TM} is to transform millions of EEGs being emitted each second from the prefrontal cortex and parietal lobes, into a malleable 2D array of x, y members suitable for an ecosystem of Rough Set tools. It is known in the field of neurodiagnostic studies, that proper interpretation of EEGs relies on identifying specific oscillatory states [19]. In Fig. 1 we discretized large sets of EEG into centroids in a moving window. To do this let $\mathbf{X} = x_1, x_2, ..., x_N$ represent the finite set and $2 \leq c \leq N$ be an integer. The objective is to partition data set $\mathbf{X}$ into c clusters where one assumes that c is known [31]. To ensure that μ_{ik} retains real values in [0,1] we let $\mathbf{X} = [x_1, x_1, ..., x_N]$ to represent the finite set where $2 \leq c \leq N$ as an integer, with the 'fuzzy' partitioning space for $\mathbf{X}$ represented by the set $M_{fc} = \mathbf{u}_{ij} \in \mathbf{R}^{N \times c} | \mu_{ik} \in [0,1], \forall i, k$; where the i-th column of $\mathbf{U}$ contains values of the membership function of the i-th fuzzy subset of $\mathbf{X}$ and constrains the sum of each column to 1. This means that the total membership of each x_k in $\mathbf{X}$ equals 1, thus making the distribution of the memberships (thinking very hard or trivially) amongst the c fuzzy subsets flexible.

For discretization, the authors hypothesized that rough sets could capture the average degree of completeness of knowledge of all the clustered windows in Fig. 1(b) [25]. We were able to apply the clustered membership to a series of incoming EEGs as indicated in Fig. 1(a), where one can see a split series of EEGs with a spline that enabled us to extract the temporal duration by calculating the $\triangle$t between each intersection of the EEG signal and the spline. Next, we calculated the area under the curve using a Left Riemann sum. Now, looking at the x, y point in the circle in Fig. 1(b), we see that if the $\triangle$t and Riemann Sum were at the respective x, y points in Fig. 1(a) that one point would be represented by the dot in the circle. Accordingly, each point represents one instance of temporal duration versus Riemann integral during 0.333 s. Figure 1(c) is the centroid of the cluster illustrated in Fig. 1(b) and instantiated onto Fig. 1(d). See video 1 of 3 at https://goo.gl/Mojejy.

2.2 Review II: Human Cognition and Confusion

Typically, two methods can be used to assess people's cognitive effort. One traditional way is asking questions in surveys, which depends on people's subjective justification [21]. The *NASA Task Load Index* [18] is an example instrument used in this method. Another method is using physiological measures, such as EEG devices, to directly assess cognitive load and awareness status [2]. Many studies have used EEG devices to measure learner's cognitive load while learning information or solving problems, and the evidence showed that using EEG devices has some merits. For example, Antonenko *et al.* [3] used EEG data to determine the effect of hypertext leads on subjects' cognitive load and learning. Recently, EEG data have been used to improve subjects learning performance. For example, Beal and Galan [15] used EEG to measure students' attention and cognitive workload while solving math problems. Additionally, Crk, Kluthe and Stefik [8] used EEG devices to investigate different levels of expertise in programming by recording programmers' brain activities when solving Java code snippets.

The authors' goal was to set up source code, that when read by coders, would trigger confusion. We did this being well aware that not all software is equally comprehensible, in that some sections of code are readily understood, while different sections of code take considerable effort to comprehend. We set out to understand whether programmers reacted differently to short C/C++ code snippets containing different levels of obfuscation through analyzing their brain activity and whether the brain activity measure is consistent with the type of code snippet. Our first hypothesis was that the EEG collected when a programmer is trying to understand confusing code snippets would be, in one way, significantly different from times when the programmer was interacting with non-confusing code snippets. In preparation of testing the hypotheses, we identified sets of confusing snippets that were kept as small as possible while still producing confusion to ensure *generalizability* [5,6] to other code bases. We call these types of patterns *atoms of confusion*. These minimal obfuscated code snippets were paired with control samples to highlight the specific cause of confusion. For example, if a = b++; is confusing, it is important to test whether a=b; b++; is also confusing. If both assignment or post-increment are confusing if they appear alone, then the confusion is not due to the combination of the two. Each obfuscated snippet, often only as long as a single line of code, was then compared against its corresponding control snippet in a human subjects experiment with 73 programmers [16].

3   Experiment Set up

Subjects consisted of undergraduate or graduate students who had taken at least one semester of C/C++ coursework. After the experiment was explained to the subjects and consent forms were signed, the first step was to fit the EEG (EPoc +) device on the subject's head. Adjustments were made when necessary to make sure all fourteen electrodes were green, indicating the strength of the signal was strong. When the fitting was completed, the subject used a web-based application that we created using jsPsych [9] to record their answers and the time-stamp of each page. After the subject read the instructions, performed a practice round and had no further questions, they were shown one code snippet, followed by one self-report on the difficulty of the question and the confidence of their answer.

During the experiment, the experimenter used another laptop to run Test-Bench, a program created by Emotiv, the manufacturer of EPoc+, to record the subject's EEG signals wirelessly. TestBench's output was in both EDF (European Data Format) and CSV (Common Separate Value) format and displayed in real time. EPoc+ has 14 channels: AF3, F7, F3, FC5, T7, P7, O1, O2, P8, T8, FC6, F4, F8, AF4 as shown in Fig. 2(b) with 128 Hz or 256 Hz sampling rate. We imported the EDF files into R for analysis. Signals were processed by first using a band pass filter between 0.16 and 13 Hz. The lower frequency is recommended by Emotiv to remove DC offset. The higher frequency of the band pass filter is because 13 Hz is the highest frequency we use in the study. We then marked all amplitudes that are either greater than 200 µv or less than −200 µv as NA because signals outside of this range represent high noise [8]. Due to the large volume of questions, we decided not to show every question to every subject.

To reduce mental fatigue, we aimed to constrain the length of each session to approximately 60 min for the average participant [4]. Since each question, on average, took just under a minute to answer in our pilot, we chose to show each subject only 2 of each group of 3 question pairings. Each subject always saw both the confusing code (which we call obfuscated code in the following) and its clarified version, and assignment of the two pairs for each participant was accomplished by cycling through each permutation. We controlled for the possibility of a learning effect [27] in two distinct ways. Firstly, we randomized the order of every question, so that any bias inherent in the question ordering was distributed evenly among all participants. Secondly, between each obfuscated/clarified pair of questions we enforced a minimum distance of 11 intermediate questions. This number was chosen by extrapolating from our pilot experiment results. In the pilot, we identified the optimum distance after which learning effects diminished, and then scaled this value by the number of new questions added for the main experiment.

For reasons outside the scope of this paper the authors decided to create a utility program to run a portion of the C# code on a Mac and use Python's Matplotlib [20] to better illustrate the motion of the neuroClustering vectors. First, we had the original C# analyze the channels that we have selected on the frontal lobe FC5, Temporal Lobe T7 and Parietal lobe P7, as positioned on the subject's head is correlated to our C# code in Fig. 2(b). Once the code is compiled it renders a CSV Log File. We then had our Python code read it and send it to Matplotlib for animation where we have coded a call-back function that recursively calls the Log File over and over creating the animation. Essentially, each time we call the next row of data in the Log File we refresh that frame with the new data from the next row. The output of the Matplotlib is shown in Fig. 2(a), along with the Log File and the Python code. Looking at Fig. 2(b) we see the code rendering the animation as illustrated on the video. See video 2 of 3 at https://goo.gl/sUpLwV.

4   Results of Atom of Confusion Experiments

Atoms of confusion caused considerable confusion among our sampled programmers. The difference in subject's performance in predicting outcome for code with atoms, as compared to code with the atoms removed, is displayed in Fig. 3(a) [16]. The authors' null hypothesis is that atoms do not impact hand evaluation accuracy. When the results of all questions from all proposed atoms are collected together, the null hypothesis can be rejected with a p-value of $p < .01$ and an effect size of $\phi = 0.36$. Both these values include the atoms that individually we cannot accept as confusing. The difference in subject's performance in predicting outcome for code with atoms, as compared to code with the atoms removed, is displayed in Fig. 3(a). For the 15 accepted atoms, the authors calculated the effect size using the Phi coefficient; Accepted values for small, medium, and large sizes are $\phi = \{0.1, 0.3, 0.5\}$ respectively [7]. Figure 3(c) addresses the issue of whether obfuscated programs are evaluated more accurately than clarified programs to which we found that our obfuscated question were on average,

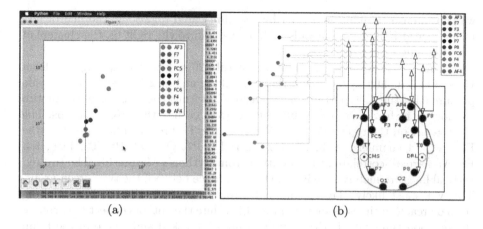

(a)   (b)

Fig. 2. All ten EEG locations are excited when a subject begins to focus and similar patterns of excitement are found on all subjects **(a) Python Matplotlib** log file procured by the C# portion of the neuroCustering code and the Matplotlib window with the dots representing clusters from each of the EEG locations. **(b) EPoc+'s 14 channels** are all correlated to the output from Python's Matplotlib. See video 3 of 3 at https://goo.gl/yvfMJD (Color figure online)

harder to interpret than every clarified question. *Conclusion;* Subjects' mean correctness rates increased by over 50% between clarified (0.47) and obfuscated (0.73) programs.

5   Results of NeuroClustering Experiments

The overall results show that neuroClustering was able to discern between clarified and obfuscated programs using both the original C# neuroClustering and the newer version using Python. As shown in video 2 of 3 at https://goo.gl/sUpLwV, we ran the neuroClustering in a double blind test not knowing where the authors who carried out the atom of confusion tests identified obfuscated and clarified programs and the match was 100% for all the subjects. The basis of the neuroClustering showed that when subjects were thinking intensely the $\triangle t$ between each intersection of the EEG signal and the spline as illustrated in Fig. 1(a), and the Riemann Sum of the area under the curve of each EEG wave form was larger than when a subject was not thinking intensely. As expected, when the $\triangle t$ was greater, the centroid shifted to the right hand side of the x-axis. Additionally, when the area under the curve was indicated by a larger Riemann sum, the centroid moved up on the y-axis. This meant that when both the $\triangle t$ and the Riemann sum increased when the subject was thinking intensely, then centroid should have a larger x-axis and y-axis, and this explains why it continuously shot up towards the upper right corner of the neuroClustering Window as seen in See video 3 of 3 at https://goo.gl/yvfMJD. In essence, every time there was an intense thinking indicated by the obfuscated questions, the neuroClustering placed K-mean's centroids high above, and to the right of, the clusters of easy thinking centroids.

Figures 2(a & b) are all Python Matplotlib [20] taken from screengrabs from video 3 of 3 at https://goo.gl/yvfMJD. Looking at Fig. 2 (a) we see the log file procured by the C# portion of the neuroCustering code and the Matplotlib window with the dots representing clusters from each of the EEG locations. The black panel on the right side is the Python code. Figure 2(b) illustrates how the EPoc+'s 14 channels: AF3, F7, F3, FC5, T7, P7, O1, O2, P8, T8, FC6, F4, F8, AF4 are all correlated to the output from Python's Matplotlib. The two red vertical lines are merely showing the time distance between the two screen-grabs. This will most likely play an important role as we have the machine learning train on reaction times as shown in Fig. 3(b). Here the authors expect to correlate the reaction time with both the type of question asked and where in the brain different reactions occurred. The dots on the left side of Fig. 2(b) represent a point where the subject had begun to think intensely, one will notice one blue dot, the lowest dot on the panel does not have an arrow on it. This is because that blue dot representing AF3 is the lower bound, and the upper bound of AF3 is located one dot to the right and above the lower bound blue dot. Following the light grey lines we see the that each grey lines extends from an excited state of the EEG when the subject was thinking intensely trying to so solve the code he was reading. Next we see the inset square with the EEG placement on the patient's head.

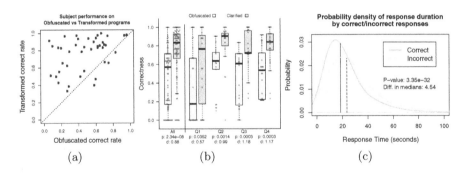

(a)   (b)   (c)

Fig. 3. (a) Subject's performance on obfuscated code vs. its clarified version snippets. Subjects above the diagonal did better on clarified code, while those below the diagonal did better on obfuscated code. **(b) Average score by question type** Rate of correct answers for each program type, **(c) Kernel density estimate** of subject response times [16]

6   Conclusions

First Finding. It is a known that the prefrontal cortex and parietal lobes excite when subjects think intensely on a problem, This research proves that rough sets is a powerful tool to separate the excitation in the prefrontal cortex and parietal lobes depending on the subjects thinking activity See EEGs P8 and P7 jump when subjects start concentrating on video 3 of 3 at https://goo.gl/yvfMJD

Second Finding. Rough sets used in the neuroClustering prove that Intense thinking is captured on every subject at every instance they were thinking intensely and never missed one period when the subject was thinking intensely. All ten EEG locations are excited when a subject begins to focus.

Third Finding. Rough sets used in the neuroClustering prove that similar patterns of excitement are found on all subjects as they encountered certain questions.

6.1   Unexpected Results Show Novel Finding

As the authors studied the results an unexpected novel finding was discovered. It immediately became visible, to the naked eye, that neuroClustering centroids excited at the parietal lobes, P7 and P8, were on average, much greater than excitations at other locations such as the AF3 location. Looking at Fig. 2(b) one sees these distinct patterns of the shooting further for the P7 and P8, while not shooting up as far for the AF3 connections. While this is out of the scope of this study, this will certainly be a problem we will research in the near future. This is exciting that psychologists and psychiatrists will soon have the ability to see when a patient is struggling to overcome a particular problem in their life.

The fact that for the first time, using rough sets and fuzzy logic, we have procured an intelligent device that can essentially look inside the brain and know when a person is struggling with a thought is exciting. This is the first paper that has shown such an event and the authors are excited to think about the future work that lies ahead. Even at the infantile stage that the neuroClustering is at now, it can, as of now, have a psychologist, psychiatrist or neurologist simply view an imaging device and see, on the screen, whether a patient is struggling with a new problem, a familiar one, or one which they are indifferent to. Another issue that has already been mentioned is that there are many papers and research efforts that have argued or questioned whether or not the prefrontal cortex and parietal lobes are involved when subjects struggle to solve problems. Figure 2 clearly illustrates, the fact, that the prefrontal cortex and parietal lobes were indeed becoming more and more excited as subjects would continue to struggle comprehending code.

Acknowledgment. This work was supported in part by NSF grant 1444827.

References

1. Amunts, K., Ebell, C., Muller, J., Telefont, M., Knoll, A., Lippert, T.: The human brain project: creating a european research infrastructure to decode the human brain. Neuron **92**(3), 574–581 (2016)
2. Antonenko, P., Paas, F., Grabner, R., Van Gog, T.: Using electroencephalography to measure cognitive load. Educ. Psychol. Rev. **22**(4), 425–438 (2010)
3. Antonenko, P.D., Niederhauser, D.S.: The influence of leads on cognitive load and learning in a hypertext environment. Comput. Hum. Behav. **26**(2), 140–150 (2010)
4. Boksem, M.A., Meijman, T.F., Lorist, M.M.: Effects of mental fatigue on attention: an ERP study. Cogn. Brain Res. **25**(1), 107–116 (2005)
5. Bordin, E.S.: The generalizability of the psychoanalytic concept of the working alliance. Psychother. Theory Res. Pract. **16**(3), 252 (1979)
6. Brennan, R.L.: Generalizability theory. Educ. Meas. Issues Pract. **11**(4), 27–34 (1992)
7. Cohen, J.: Statistical Power Analysis for the Behavioral Sciences. 2nd edn. L. Erlbaum Associates, Hillsdale (1988)
8. Crk, I., Kluthe, T., Stefik, A.: Understanding programming expertise: an empirical study of phasic brain wave changes. ACM Trans. Comput. Hum. Interact. (TOCHI) **23**(1), 2 (2016)
9. De Leeuw, J.R.: jsPsych: a javascript library for creating behavioral experiments in a web browser. Behav. Res. Methods **47**(1), 1–12 (2015)
10. Deco, G., Van Hartevelt, T.J., Fernandes, H.M., Stevner, A., Kringelbach, M.L.: The most relevant human brain regions for functional connectivity: evidence for a dynamical workspace of binding nodes from whole-brain computational modelling. NeuroImage **146**, 197–210 (2017)
11. Duncan, J., Owen, A.M.: Common regions of the human frontal lobe recruited by diverse cognitive demands. Trends Neurosci. **23**(10), 475–483 (2000)
12. Dung, P.M.: On the acceptability of arguments and its fundamental role in non-monotonic reasoning, logic programming and n-person games. Artif. Intell. **77**(2), 321–357 (1995)
13. Forstmann, B.U., de Hollander, G., van Maanen, L., Alkemade, A., Keuken, M.C.: Towards a mechanistic understanding of the human subcortex. Nat. Rev. Neurosci. **18**(1), 57–65 (2017)
14. Fukushima, K., Miyake, S., Ito, T.: Neocognitron: a neural network model for a mechanism of visual pattern recognition. IEEE Trans. Syst. Man Cybern. **5**, 826–834 (1983)
15. Cirett Galán, F., Beal, C.R.: EEG estimates of engagement and cognitive workload predict math problem solving outcomes. In: Masthoff, J., Mobasher, B., Desmarais, M.C., Nkambou, R. (eds.) UMAP 2012. LNCS, vol. 7379, pp. 51–62. Springer, Heidelberg (2012). https://doi.org/10.1007/978-3-642-31454-4_5
16. Gopstein, D.: Understanding misunderstandings in source code. In: Proceedings of the 2017 11th Joint Meeting on Foundations of Software Engineering. ACM (2017)
17. Grillner, S., et al.: Worldwide initiatives to advance brain research. Nat. Neurosci. **19**(9), 1118–1122 (2016)
18. Hart, S.G.: Nasa-task load index (NASA-TLX); 20 years later. In: Proceedings of the Human Factors and Ergonomics Society Annual Meeting, vol. 50, pp. 904–908. Sage Publications Sage CA, Los Angeles (2006)
19. Hogan, R.: Automated eeg detection algorithms and clinical semiology in epilepsy: importance of correlations. Epilepsy Behav. **22**, S4–S6 (2011)

20. Hunter, J.D.: Matplotlib: a 2D graphics environment. Comput. Sci. Eng. **9**(3), 90–95 (2007)
21. Kalyuga, S., Ayres, P., Chandler, P., Sweller, J.: The expertise reversal effect. Educ. Psychol. **38**(1), 23–31 (2003)
22. Lewis, R., Mello, C.A., Ellenberger, J., White, A.M.: Domain adaptation for pathologic oscillations. In: Ciucci, D., Inuiguchi, M., Yao, Y., Ślęzak, D., Wang, G. (eds.) RSFDGrC 2013. LNCS (LNAI), vol. 8170, pp. 374–379. Springer, Heidelberg (2013). https://doi.org/10.1007/978-3-642-41218-9_40
23. Lewis, R., Mello, C.A., Carlsen, J., Grabenstatter, H., Brooks-Kayal, A., White, A.M.: Autonomous neuroclustering of pathologic oscillations using discretized centroids. In: 8th International Conference on Mass Data Analysis of Images and Signals with Applications in Medicine, 13–16 July 2013, New York, USA (2013)
24. Lewis, R.A., Waziri, A.: minedICE: a knowledge discovery platform for neurophysiological artificial intelligence. In: Kryszkiewicz, M., Rybinski, H., Skowron, A., Raś, Z.W. (eds.) ISMIS 2011. LNCS (LNAI), vol. 6804, pp. 575–580. Springer, Heidelberg (2011). https://doi.org/10.1007/978-3-642-21916-0_61
25. Lingras, P., West, C.: Interval set clustering of web users with rough k-means. J. Intell. Inf. Syst. **23**(1), 5–16 (2004)
26. Makeig, S., Debener, S., Onton, J., Delorme, A.: Mining event-related brain dynamics. Trends in cognitive sciences **8**(5), 204–210 (2004)
27. Neely, J.H.: Semantic priming effects in visual word recognition: A selective review of current findings and theories. Basic processes in reading: Visual word recognition **11**, 264–336 (1991)
28. Newman, S.D., Carpenter, P.A., Varma, S., Just, M.A.: Frontal and parietal participation in problem solving in the tower of london: fmri and computational modeling of planning and high-level perception. Neuropsychologia **41**(12), 1668–1682 (2003)
29. Poldrack, R.: Neuroscience: The risks of reading the brain. Nature **541**(7636), 156–156 (2017)
30. E. L. Schwartz. Computational neuroscience. Mit Press, 1993
31. Trinidad, J.F., Shulcloper, J.R., Cortes, M.S.: Structuralization of universes. Fuzzy Sets and Systems **112**(3), 485–500 (2000)
32. Yuste, R., Bargmann, C.: Toward a global brain initiative. Cell **168**(6), 956–959 (2017)

Graph Mining

Solving the Maximal Clique Problem
on Compressed Graphs

Jocelyn Bernard and Hamida Seba[(⊠)]

Université de Lyon, Universié Lyon 1, CNRS, LIRIS, UMR5205, Villeurbanne, France
{jocelyn.bernard,hamida.seba}@univ-lyon1.fr

Abstract. The Maximal Clique Enumeration problem (MCE) is a graph problem encountered in many applications such as social network analysis and computational biology. However, this problem is difficult and requires exponential time. Consequently, appropriate solutions must be proposed in the case of massive graph databases. In this paper, we investigate and evaluate an approach that deals with this problem on a compressed version of the graphs. This approach is interesting because compression is a staple of massive data processing. We mainly show, through extensive experimentations, that besides reducing the size of the graphs, this approach enhances the efficiency of existing algorithms.

Keywords: Maximal clique enumeration · Graph compression
Modular decomposition

1 Introduction

Several real world applications, particularly in the fields of social networks, computer communication or genetics, rely on a graph representation of objects. A graph $G = (V, E)$ consists of two finite sets V and E. The elements of V are called the vertices and the elements of E the edges of G. Each edge is a pair of vertices. A clique C, in a graph $G = (V, E)$, is a subset of V such that the induced subgraph $G[C]$, i.e., the graph whose vertex set is C and whose edge set consists of all of the edges in E that have both endpoints in C, is a complete graph, i.e., a graph where all vertices are pairwise adjacent. A clique C is said to be maximal if there exists no vertex in G that can be added to C so that C still remains a clique. The goal of the maximal clique enumeration (MCE) problem is to enumerate all maximal cliques in a given graph. MCE is a fundamental problem in graph theory that is encountered in several graph based applications such as detecting protein-protein interactions and particular motifs in computational biology [26], network analysis applications like community detection [9] and, social network analysis [25]. MCE is equivalent to finding all maximal independent sets in the complimentary graph (The complement of a

This work received a support from Département Info-Bourg, IUT Lyon 1 and from PHC TASSILI 17MDU984.

M. Ceci et al. (Eds.): ISMIS 2018, LNAI 11177, pp. 45–55, 2018.
https://doi.org/10.1007/978-3-030-01851-1_5

graph $G = (V, E)$ is a graph with vertex set V and edge set E' such that $e \in E'$ if and only if $e \notin E$) and is shown to be NP-hard [15]. Therefore, MCE is a challenging problem for large graphs as well as for massive graph databases. To tackle this problem, several approaches are proposed such as non-exact solutions that use heuristics to approach optimal solutions [8] and distributed and parallel algorithms [16,20].

In this paper, we explore a completely different idea that consists to work on smaller data by compressing the input graph. Compression is an old data processing paradigm which is gaining importance with big data but with new tricks as we aim to process data without decompression. This work comes to strengthen this trend by investigating the MCE problem on a compressed graph. The aim is to save storage space for graphs and enhance their time processing. We implement this approach by revisiting exiting algorithms and show how they can be implemented on compressed graphs.

2   Graph Compression

Graph compression is gaining importance with the increasing volume of data in several applications and services based on graph representation of this data. There are several ways to compress a graph, depending on the type of the graph but also on what we aim to do with the compressed graph. We also have both lossy and lossless graph compression methods. We highlight here the most known compression algorithms and refer to [17] for a more detailed survey. In [5], Boldi et al. compress the web graph by taking advantage of the fact that most hyperlinks on most Web pages point to other pages on the same host as the page itself (i.e., locality property) and that many pages on the same host tend to point to a common set of pages (i.e., similarity property). In [19], the author proposes a lossless graph compression by grouping vertices that have close neighborhoods. The obtained summary reduces the number of edges and vertices and allows to retrieve the original graph. For weighted graphs, Toivonen et al. [22] propose a lossy compression method where vertices and edges are compressed into super-vertices and super-edges. The weight of a super-edge is approximated by the weights of the edges that comprise it. Using a similar approach, Fan et al. [11] summarize graphs in a manner that preserves reachability queries, i.e., a query of this class returns the same result when applied to a graph G and when applied to the compression of G. In [14], the authors use a more general approach to compress graphs using modular decomposition of graphs. Modular decomposition is a concept of graph theory introduced by Galai [12] and further studied by Möhring and Radermacher [18] and Habib et al. [13,21]. Modular decomposition highlights set of vertices, called modules, that have the same neighborhood in the graph.

Definition 1. *A module of a graph $G = (V, E)$ is a set $m \subseteq V$ of vertices where all vertices in m have the same neighbors in $V \setminus m$.*

We consider four types of modules:

- A leaf is a simple vertex of the original graph,
- A series module is a set of vertices that form a clique,
- A parallel module is a set of vertices that have no edges between them, and
- A neighborhood module is a set S of vertices such that both $G[S]$ and its complement ($\overline{G}[S]$) are connected.

With modular decomposition, we can compress a graph by recursively compacting the subgraphs induced by modules into super-vertices. Figure 1 illustrates the compression of a graph using modular decomposition. In each compression step, illustrated in the figure, a set of vertices are compacted into a super-vertex. This process is iterated recursively until no further compaction is possible.

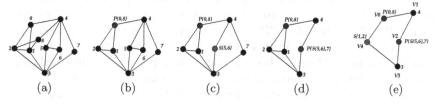

(a) Initial Graph. (b)Compacting the Parallel module $P\{0, 8\}$. (c) Compacting the Series module $S\{5, 6\}$. (d) Compacting the Parallel module $P\{S\{5, 6\}, 7\}$. (e) Compacting the Series module $S\{1, 2\}$.

Fig. 1. Graph summarizing with modular decomposition: S: series module. P: parallel module.

In the final compressed graph each super-vertex can render the original structure of the vertices within it. The label given to each super-vertex reveals this structure as well as the successive compacting steps that have been used to obtain it. Compacting steps can be represented by a tree, called the modular decomposition tree [13], where simple vertices are the leaves. For compression, series and parallel modules are the most important modules as they encompass the adjacency information related to their components, i.e., we do not need to store edges for them. However, a neighborhood module is stored with the edges that connect its vertices. Thus, the final compressed graph in the example given in Fig. 1 is the graph corresponding to the final *Neighborhood* module $N(3, 4, S(1, 2), P(0, 8), P(S(5, 6), 7))$. This graph has the set of vertices $V = \{V0, V1, V2, V3, V4\}$ and the set of edges $E = \{V0V1, V0V4, V1V2, V2V3, V3V4\}$. To distinguish super-vertices from simple ones, we denote them with majuscules and for simplicity, we will use the terms super-vertex and module interchangeably to denote the set of vertices and the graph induced by them. The number of neighborhood modules and the number of edges within these modules are the main properties that determine the degree of compression of a graph. In the remaining of the paper, we will use the ratio between the number of edges in the compressed graph on the number of edges in the original graph to measure how well a graph is compressible.

3   Related Work on MCE Algorithms

We describe in this section the main exact in-memory algorithms proposed in the literature to solve the MCE problem. Most algorithms for MCE involve the construction of some kind of search tree and are often extension of the basic algorithm due to Bron and Kerbosch [7]. The algorithm of Bron and Kerbosch works with 3 sets: a set R which is the partial constructed clique, a set P which is the set of vertices to grow R, i.e., vertices adjacent to the last added vertex to R, and a set X of forbidden vertices, i.e., vertices that were already considered to grow the clique. Each vertex of P is considered in a loop, where the arbitrarily chosen vertex v is eliminated from P. Vertex v is added to R and a recursion call with $P \cap N(v)$ and $X \cap N(v)$ takes place. On return v is added to X, because all maximal cliques in this subtree of the search including v have been generated. Whenever P and X become empty we report C as a maximal clique. The algorithm involves a pivot vertex u chosen randomly from P. This pivot is intended to prune the tree search by reducing the number of recursive calls. Any maximal clique must include either the pivot u or one of its non-neighbors, otherwise the clique could be augmented by adding u to it. Therefore, only u and its non-neighbors need to be tested as the choices for the vertex v that is added to R in each recursive call of the algorithm. Among the extensions of Bron and Kerbosch's algorithm, Tomita's algorithm [23] is considered as a reference algorithm. It has a time complexity of $\mathcal{O}(3^{(n/3)})$. It differs from Bron and Kerbosch's algorithm by the choice of the pivot that allows a more efficient pruning. In Tomita's algorithm, the pivot u is selected in $P \cup X$ such as to maximize $|P \cap N(u)|$. By maximizing $|P \cap N(u)|$ during the choice of the pivot we are minimizing $P \setminus N(u)$, i.e., the number of recursive calls made from the current call. In [10], Eppstein et al. provide another improvement to the basic algorithm using an ordering of the vertices rather than pivoting. This ordering is based on graph degeneracy.

Definition 2. *A graph G is said to be k-degenerate if for every induced subgraph S of G there exists a vertex with degree at most k in S.*

A *degeneracy ordering* is an ordering of vertices, such that every vertex v has k or fewer neighbors that come later in the ordering. k-degeneracy assures such an ordering. A *degeneracy ordering* can be calculated in linear time with respect to $|G|$ [4]. The algorithm of Eppstein et al. differs from the basic algorithm by this vertex ordering. For every vertex v_i in the degeneracy ordering the basic algorithm is called with $P = N(v_i) \cap \{v_{i+1} \ldots v_n\}$, $R = \{v_i\}$ and $X = N(v_i) \cap \{v_1 \ldots v_{i-1}\}$. So, P is restricted to the neighbors of v_i that come after v_i in the degenerate ordering and, X to the vertices that come before v_i in the degenerate ordering. Eppstein et al. show that this algorithm can be implemented with $\mathcal{O}(kn3^{k/3})$ running time and this is nearly optimal to a constant factor as there are graphs with $\Theta((n-k)3^{k/3})$ maximal cliques. We note here that degeneracy is a measure of graph sparseness. Sparse graphs have low degeneracy, the method works particularly well for sparse graph and consequently for large real world graphs [10].

4   Solving MCE on a Compressed Graph

This section presents our contribution which settles the MCE problem on summarized graphs. We mainly present SumMCE our maximal clique enumeration algorithm on summarized graphs. SumMCE is a framework that enables the use of an existing MCE algorithm on compressed graphs. We rely on the work of Tedder et al. [21] on algorithmic aspect of modular decomposition to compress our graphs. Their algorithm computes the modular decomposition of a graph in linear time. So, we compress an input graph in $\mathcal{O}(n+m)$ time, where n is the number of vertices and m the number of edges of the graph.

SumMCE is a framework that allows to use any existing MCE algorithm on a compressed graph. It takes as input a summarized graph as defined in Sect. 3, which can be computed from the original input graph $G = (V, E)$ in $\mathcal{O}(n+m)$. SumMCE (see Algorithm 1) is a recursive algorithm that explores the compressed graph and finds cliques according to the type of modules as follows:

- Leaf: the algorithm returns the corresponding vertex as a clique (see lines 3–4).
- Series module: a series module is a clique. So, the algorithm merges the maximum cliques from the components of the module. This merging is a simple set union (see lines 6–7) achieved by Function *CliqueSetUnion* which performs a union operation between the sets of cliques that it takes as input. Figure 2a illustrates this case with the series module $S\{1,2\}$ which returns the clique $\{1,2\}$ resulting from the union of the sets $\{1\}$ and $\{2\}$ given by its two components which are in this case simple vertices, i.e., leaves.
- Parallel module: as the elements of a parallel module are not connected, the algorithm returns the maximum cliques of each of the components of the parallel module without modifying them (see lines 8–14). Figure 2a illustrates this case with the parallel module $P\{S\{5,6\},7\}$. For this module, the algorithm returns the two cliques of its components, i.e., the clique $\{5,6\}$ obtained from the series module $S\{5,6\}$ and the clique $\{7\}$ from the simple vertex 7.
- Neighborhood module: a Neighborhood module needs a clique searching algorithm. Any existing algorithm can be used (see lines 15–16). The cliques returned here are cliques of modules. Each clique of modules can be transformed to a clique of vertices by replacing each module by the set of vertices returned by the algorithm when applied to this module. Function *Leaves* achieves this task. Figure 2b illustrates this case with the Neighborhood module defined on the set of super-vertices $\{V0, V1, V2, V3, V4\}$. A clique searching algorithm applied on this graph of super-vertices will return as set of

cliques $\{\{V0, V1\}, \{V1, V2\}, \{V2, V3\}, \{V3, V4\}\}$. Each clique of super-vertices is then turned into a set of cliques of simple vertices by function *Leaves*. For example the clique $\{V0, V1\}$ will give the cliques $\{\{0, 4\}, \{8, 4\}\}$. Function *Leaves* uses function *CliqueSetUnion* to achieve a union of two cliques.

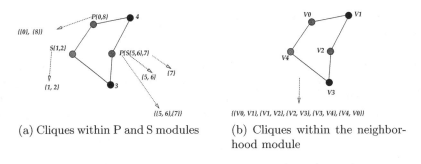

(a) Cliques within P and S modules   (b) Cliques within the neighbor-hood module

Fig. 2. Clique enumeration on the compressed graph.

Algorithm 1. SumMCE(m)

---

1 **begin**
2 | **switch** *type of m* **do**
3 | | **case** *a leaf*
4 | | | **return** $\{m\}$
5 | | **case** *a series module*
6 | | | Let $m = \{A_1, A_2, \cdots, A_t\}$;
7 | | | **return** $CliqueSetUnion(SumMCE(A_1), SumMCE(A_2), \cdots, SumMCE(A_t))$
8 | | **case** *a parallel module*
9 | | | $C \leftarrow \emptyset$;
10 | | | Let $m = \{A_1, A_2, \cdots, A_t\}$;
11 | | | **foreach** $A_i \in m$ **do**
12 | | | | $C \leftarrow C \cup \{SumMCE(A_i)\}$;
13 | | | **end**
14 | | | **return** C
15 | | **case** *a Neighborhood module*
16 | | | **return** $Leaves(MCE\text{-}Algorithm(m))$
17 | **endsw**
18 **end**

---

5   Complexity Analysis

Let G be a graph of n vertices and m edges. The time complexity of SumMCE depends on the complexity of its clique enumeration function on the four main types of modules and the complexity of the compression algorithm. According to [21], we obtain the compressed graph of G in $\mathcal{O}(n + m)$ time steps. Note that compression is achieved offline and the graphs are stored in their compressed format. Consequently, we will focus on the complexity of clique enumeration.

Let $\mathcal{N}$, $\mathcal{P}$ and $\mathcal{S}$ be respectively the number of neighborhood, parallel and series modules in the compressed version of G and let t, p, and s be respectively the average number of vertices in a neighborhood, parallel and series module. SumMCE needs $\mathcal{O}(1)$ time steps to process a simple vertex and it processes the n simple vertices of the original graph. The complexity of processing a series or a parallel module is prevailed by the complexity of set union. If we consider that set union is linear on the total number of elements, processing series (resp. parallel) modules is $\mathcal{O}(\mathcal{S}s)$ (resp. $\mathcal{O}(\mathcal{P}p)$). Processing a neighborhood module depends on the maximal clique enumeration algorithm used to handle it. Let consider that the time complexity of this algorithm is $c(k)$ where k is the number of vertices of the considered graph. Handling neighborhood modules needs $\mathcal{O}(\mathcal{N}c(t))$ time steps. This yields to a global complexity of $\mathcal{O}(n + \mathcal{P}p + \mathcal{S}s + \mathcal{N}c(t))$ which is mainly prevailed by the term $\mathcal{N}c(t)$, i.e., by the number of neighborhood modules and their size. For example, if we use Tomita's algorithm for the neighborhood module we obtain $\mathcal{O}(n + \mathcal{P}p + \mathcal{S}s + \mathcal{N}3^{t/3})$.

To improve the speed of our algorithms, each module of the compressed graph stores the maximal cliques upon the first recursive call on the module. This allows to reduce the number of recursive calls by processing each module once. In fact, a module may be found in several cliques and therefore the recursive function is called several times on it.

6 Evaluation

In this section, we describe the experiments that we conducted to evaluate our approach. For this, we have implemented our algorithms and all the state of the art algorithms presented in Sect. 3 within the same framework in C ++ using the SNAP (Stanford Network Analysis Platform) library [2]. This allows us to use the same structure for all algorithms. We worked on a 64-bit machine running Ubuntu v14.04 LTS, with an i5-4690 3.5 GHz processor having 8G RAM. The algorithms were compiled with g++ compiler version 4.8.2 with options -std=c++ 11 and optimization -O3. We used several graphs to compare the algorithms and evaluate our approach. These graphs come from three main datasets: DIMACS [1], the RI project [6] that contains mainly graphs representing DNA, RNA, and proteins, and Stanford large graph networks [3] for testing scalability issues. Table 1 summarises the characteristics of these graphs.

We launched our algorithms together with the state of the art algorithms on the set of graphs. The first experiments concern MCE on the biological graphs of the RI database and their aim is to determine which algorithm to use with SumMCE to deal with neighborhood modules. For this, we implemented two versions SumMCE-Tomita and SumMCE-BK with respectively Tomita's algorithm and Bron and Kerbosch's algorithm for neighborhood modules. The results are depicted in Fig. 3(a). This Figure plots the execution time of each algorithm on each graph. It shows clearly that SumMCE-Tomita is the most efficient algorithm for almost all the graphs with a compression ratio greater than 15%, except for TRANSFERASE that shows better performance with SumMCE-BK.

Table 1. Datasets

Source	Name	$|V|$	$|E|$	cr	ct	#Cliques
RI	HUMAN	4,675	86,282	69.43	0.06	6,980
	CELL_ADHESION	33,067	30,773	17.29	3.89	30,773
	TRANSFERASE	20,030	17,263	17.60	1.02	17,263
	ISOMERASE	11,147	8,459	14.77	0.87	8,459
	OXIDOREDUCTASE	15,593	12,010	15.58	1.63	15,242
	RATTUS	8,763	39,932	6.64	0.61	41,118
DIMACS	16pk	4,919	4,972	15.56	0.11	4,972
	3djd	11,862	12,010	16.94	0.12	12,010
	c-fat500-5	500	23,191	99.93	0.01	16
	6pfk	17,123	17,263	17.59	0.41	17,263
	3dmk	30,386	30,773	17.24	1.62	30,773
	c-fat500-10	500	46,627	99.98	4.69	8
SNAP	web-NotreDame	325,729	1,497,134	29.85	1,037	495,947
	com-Amazon	334,863	925,872	11.9	3,309	258,830
	web-Google	875,713	5,105,039	42.64	22,275	939,059
	com-dblp	317,080	1,049,866	29.65	1,547	257,551

cr: Compression ratio (%). ct: Compression time

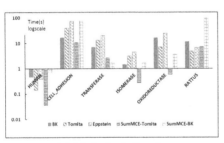

(a) Biological Graphs

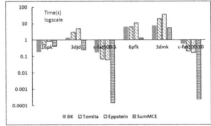

(b) Dimacs Graphs

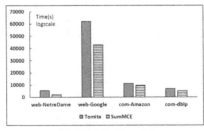

(c) SNAP Graphs

Fig. 3. Maximal clique enumeration time performance (results are in log scale).

So, according to this, we used Tomita's Algorithm within SumMCE for the rest of the experiments. We note also that Bron and Kerbosch's algorithm has not a constant time execution on a graph because of the random choice of the pivot. So, we run the algorithm 10 times on each graph and retained the smallest time execution for each graph.

Figure 3(b) shows the time performance of the different algorithms on Dimacs graphs. This figure confirms the results obtained with the biological graphs as SumMCE outperforms all the other algorithms for almost all the graphs. We can see here that the exception is the smallest time obtained by Bron and Kerbosch's algorithm on the graph 16pk.

For the SNAP graphs that are larger, we compared Tomita's algorithm with SumMCE. The results are depicted in Fig. 3(c) that shows clearly that SumMCE outperforms Tomita's algorithm for the selected graphs. We note also that the SNAP Graphs have a compression ratio greater than 11%.

7   Conclusion

Listing maximal cliques is a complex and difficult problem especially when the graphs we work on are large or numerous. There are many algorithms and heuristics proposed to solve the MCE problem, more or less suitable for massive graphs. In this paper, we proposed a new approach that consists to solve the MCE problem on compressed graphs to deal with massive data. Our experimentations on the Dimacs benchmark and on real world graphs from biology and social networks show that existing algorithms for maximal clique enumeration perform better on compressed graphs than on non compressed graphs especially when the graphs are well compressible (compression ratio beyond 11%). Our study shows mainly that we can save space with compression while having good time performance. The obtained results can be enhanced by storing information about cliques during compression mainly for parallel and series modules. Defining pruning rules adapted to compressed graphs may also be an interesting issue as we have not used any. Another possible extension is to deal with directed graphs and also with weighted graphs. Finally, it will be interesting to investigate how to avoid listing overlapping cliques in a compressed graph. This issue has been investigated by Wang et al. in [24]. Their algorithm is an enhancement of Bron and Kerbosch's algorithm that relies on the concept of *Visibility* and τ-*Visibility* of a clique [24] to verify that the clique being built is not similar to a previous one. A similar concept may be defined on modules.

References

1. Dimacs challenge. http://dimacs.rutgers.edu/Challenges/
2. Snap. https://snap.stanford.edu/snap-2.3/
3. Stanford large networks. http://snap.stanford.edu/data/
4. Batagelj, V., Zaversnik, M.: An o (m) algorithm for cores decomposition of networks. arXiv preprint cs/0310049 (2003)

5. Boldi, P., Vigna, S.: The webgraph framework i: compression techniques. In: Proceedings of the 13th International Conference on World Wide Web, pp. 595–602. ACM (2004)

6. Bonnici, V., Giugno, R., Pulvirenti, A., Shasha, D., Ferro, A.: A subgraph isomorphism algorithm and its application to biochemical data. BMC Bioinf. **14(Suppl 7)**, (S13) (2013)

7. Bron, C., Kerbosch, J.: Algorithm 457: finding all cliques of an undirected graph. Commun. ACM **16**(9), 575–577 (1973)

8. Conte, A., Virgilio, R.D., Maccioni, A.: Finding all maximal cliques in very large social networks. In: 19th International Conference on Extending Database Technology (EDB), 15–18 March, Bordeaux, France (2016)

9. Du, N., Wu, B., Pei, X., Wang, B., Xu, L.: Community detection in large-scale social networks. In: Proceedings of the 9th WebKDD and 1st SNA-KDD 2007 Workshop on Web Mining and Social Network Analysis, WebKDD/SNA-KDD 2007, pp. 16–25. ACM, New York (2007)

10. Eppstein, D., Strash, D.: Listing all maximal cliques in large sparse real-world graphs. In: Pardalos, P.M., Rebennack, S. (eds.) SEA 2011. LNCS, vol. 6630, pp. 364–375. Springer, Heidelberg (2011). https://doi.org/10.1007/978-3-642-20662-7_31

11. Fan, W., Li, J., Wang, X., Wu, Y.: Query preserving graph compression. In: Proceedings of the 2012 ACM SIGMOD International Conference on Management of Data, pp. 157–168 (2012)

12. Gallai, T.: Transitiv orientierbare graphen. Acta Mathematica Hungarica **18**(1), 25–66 (1967)

13. Habib, M., de Montgolfier, F., Paul, C.: A simple linear-time modular decomposition algorithm for graphs, using order extension. In: Hagerup, T., Katajainen, J. (eds.) SWAT 2004. LNCS, vol. 3111, pp. 187–198. Springer, Heidelberg (2004). https://doi.org/10.1007/978-3-540-27810-8_17

14. Lagraa, S., Seba, H., Khennoufa, R., M'Baya, A., Kheddouci, H.: A distance measure for large graphs based on prime graphs. Pattern Recognit. **47**(9), 2993–3005 (2014)

15. Lawler, E., Lenstra, J., Rinnooy Kan, A.: Generating all maximal independent sets: NP-hardness and polynomial-time algorithms. SIAM J. Comput. **9**(3), 558–565 (1980)

16. Lessley, B., Perciano, T., Mathai, M., Childs, H., Bethel, E.W.: Maximal clique enumeration with data-parallel primitives. In: 2017 IEEE 7th Symposium on Large Data Analysis and Visualization (LDAV), pp. 16–25, October 2017

17. Liu, Y., Dighe, A., Safavi, T., Koutra, D.: A graph summarization: A survey. CoRR abs/1612.04883 (2016). http://arxiv.org/abs/1612.04883

18. Möhring, R., Radermacher, F.: Substitution decomposition and connection with combinatorial optimization. Ann. Discrete Math. **19**, 257–356 (1984)

19. Navlakha, S., Rastogi, R., Shrivastava, N.: Graph summarization with bounded error. In: Proceedings of the 2008 ACM SIGMOD International Conference on Management of Data, SIGMOD 2008, pp. 419–432 (2008)

20. Segundo, P.S., Artieda, J., Strash, D.: Efficiently enumerating all maximal cliques with bit-parallelism. Comput. Oper. Res. **92**, 37–46 (2018)

21. Tedder, M., Corneil, D.G., Habib, M., Paul, C.: Simpler linear-time modular decomposition via recursive factorizing permutations. In: 35th International Colloquium on Automata, Languages and Programming, Iceland, 7–11 July 2008, pp. 634–645 (2008)

22. Toivonen, H., Zhou, F., Hartikainen, A., Hinkka, A.: Compression of weighted graphs. In: Proceedings of the 17th International Conference on Knowledge Discovery and Data Mining, KDD 2011, pp. 965–973 (2011)
23. Tomita, E., Tanaka, A., Takahashi, H.: The worst-case time complexity for generating all maximal cliques and computational experiments. Theor. Comput. Sci. **363**(1), 28–42 (2006)
24. Wang, J., Cheng, J., Fu, A.W.C.: Redundancy-aware maximal cliques. In: Proceedings of the 19th ACM SIGKDD International Conference on Knowledge Discovery and Data Mining, KDD 2013, pp. 122–130 (2013)
25. Wasserman, S., Faust, K.: Social Network Analysis Methods and Applications, vol. 8, January 1993
26. Zhang, B., Park, B.H., Karpinets, T.V., Samatova, N.F.: From pull-down data to protein interaction networks and complexes with biological relevance. Bioinformatics **24**(7), 979–86 (2008)

Clones in Graphs

Stephan Doerfel[1], Tom Hanika[2,3(✉)], and Gerd Stumme[2,3]

[1] Micromata GmbH Kassel, Kassel, Germany
stephan.doerfel@doerfel.info
[2] Knowledge and Data Engineering Group, University of Kassel, Kassel, Germany
{tom.hanika,stumme}@cs.uni-kassel.de
[3] Interdisciplinary Research Center for Information System Design,
University of Kassel, Kassel, Germany

Abstract. Finding structural similarities in graph data, like social networks, is a far-ranging task in data mining and knowledge discovery. A (conceptually) simple reduction would be to compute the automorphism group of a graph. However, this approach is ineffective in data mining since real world data does not exhibit enough structural regularity. Here we step in with a novel approach based on mappings that preserve the maximal cliques. For this we exploit the well known correspondence between bipartite graphs and the data structure formal context (G, M, I) from Formal Concept Analysis. From there we utilize the notion of clone items. The investigation of these is still an open problem to which we add new insights with this work. Furthermore, we produce a substantial experimental investigation of real world data. We conclude with demonstrating the generalization of clone items to permutations.

Keywords: Social network analysis · Formal concept analysis · Clones

1 Introduction

The identification of (structural) similar entities in graph data sets is a particularly relevant task in data analysis: it provides insights into entities in the data (e.g., in members of social networks); it allows grouping entities and even reducing data sets by removing redundant (structurally equivalent) elements (factorization). For *bipartite* graph data, a notion of structural similarity that suggests itself is that of *clone items*, known from the realm of Formal Concept Analysis (FCA). The latter is a mathematical toolset for qualitative data analysis, relying on algebraic notions s.a. lattices and closure systems. Here, clone items are entities from the same partition that are completely interchangeable within the family of that partition's closed subsets.

In this paper, we follow up on a long-standing open problem of FCA, collected at ICFCA 2006,[1] regarding the meaning of clone items in real world graph data.

---

[1] http://www.upriss.org.uk/fca/problems06.pdf.

© Springer Nature Switzerland AG 2018
M. Ceci et al. (Eds.): ISMIS 2018, LNAI 11177, pp. 56–66, 2018.
https://doi.org/10.1007/978-3-030-01851-1_6

The notion of clones was initially proposed[2] in "Clone items: a pre-processing information for knowledge discovery" by Medina and Nourine. Subsequently, a plethora of desirable properties of clone items has been shown, such as, "hidden combinatorics" [7] that allow factorizations of data structures containing clones, computational properties investigations, like [14], or the use of clones in association rule mining [13]. Finally, the question of semantics was addressed by [11], who investigated clones in three well-known data sets (*Mushroom, Adults,* and *Anonymous* from the UCI Machine Learning Repository [9]). Following the observation that two data sets were free of clones whereas the mushroom data set had only few, [11] introduced *nearly clones* relying rather on statistical than on structural properties. However, despite these previous efforts, the question – are clone items frequent in natural graph data sets – in particular in social network data – has not yet been answered in general.

The contributions of this paper are threefold: First, we provide a prove for the characterization of clone items on the level of formal contexts that allows us to easily compute clone items in large data sets. Second, we investigate a diverse variety of public realworld data sets coming from different domains and exhibiting different properties. We show that clones are not common in these data sets and conclude that in their present form, clones are not as useful as one would have hoped, regarding the efforts made in previous literature. Third, to resolve this dilemma, we point out a more general notion of clones. For this we fall back to permutations on the set of attributes in a formal context, providing a natural extension of the clone property. These *higher order clones* are able to identify more complicated "clone structures" and should be the next step in the investigation of relational data structures.

This work is structured as follows. In Sect. 2, we recall basic notations of FCA and show the correspondence to graphs. Then, in Sect. 3, we provide a characterization of clone items on the level of formal contexts. Following this, in Sect. 3.1 we demonstrate how the notion of clones can be applied in the realm of graphs. Subsequent to experiments on various data sets, in Sect. 4, we extend the notion of clone items to higher order clones. Eventually, we conclude our work with Sect. 5.

2   Preliminaries

We give a short recollection of the ideas from formal concept analysis as introduced in [5,18] that are relevant in this work. We use the common presentation of formal contexts by $\mathbb{K} = (G, M, I)$, where G and M are sets and $I \subseteq G \times M$. The elements of G are called objects, those of M are called attributes, and $(g, m) \in I$ signifies that object g has the attribute m. The correspondence to a bipartite graph (network) is at hand. Let $H = (U \cup W, E)$ be such an undirected bipartite graph with $U \cap W = \emptyset$ where U is a set of entities (often users), W some set of common properties, and $E \subseteq \{\{u, w\} \mid u \in U, w \in W\}$ the set of edges between U and W. There are two natural ways of identifying H as a formal context. In

---

[2] This work is noted to be submitted (e.g., in [7]), but has never been published.

the following, we choose $\mathbb{K}(H) = (U, W, I)$ as the to H associated formal context,[3] where for $u \in U$ and $w \in W$, we have $(u, w) \in I :\Leftrightarrow \exists e \in E : u \in e \wedge w \in e$. For the case of a non-bipartite Graph $G = (V, E)$ we simply construct the formal context by $\mathbb{K} = (V, V, I)$ with $(u, v) \in I \Leftrightarrow \{u, v\} \in E$ for all $u, v \in V$. In the following we use the terms network, (bipartite) graph, and formal context interchangeably in the sense above.

We will utilize the common *derivation* operators $\cdot' : \mathcal{P}(G) \to \mathcal{P}(M), A \mapsto B := \{m \in M \mid \forall g \in A : (g, m) \in I\}$ and $\cdot' : \mathcal{P}(M) \to \mathcal{P}(G), B \mapsto A := \{g \in G \mid \forall m \in B : (g, m) \in I\}$. Having those operations we call a formal context $\mathbb{K} = (G, M, I)$ *object clarified* iff $\forall g, h \in G, g \neq h : g' \neq h'$, *attribute clarified* iff $\forall m, n \in M, m \neq n : m' \neq n'$ and *clarified* iff it is both. In this definition we used g' as shorthand for $\{g\}'$. Clarification will later on correspond to a particular trivial kind of clones. Similarly we call a clarified context $\mathbb{K}$ *object reduced* if for all $g \in G$ there is no $S \subseteq G \setminus \{g\}$ such that $g' = S'$. We call $\mathbb{K}$ *attribute reduced* iff for all $m \in M$ there is no $S \subseteq M \setminus \{m\}$ such that $m' = S'$. And, we call this $\mathbb{K}$ *reduced* iff $\mathbb{K}$ is attribute and object reduced.

A pair (A, B) where $A \subseteq G$, $B \subseteq M$ with $A' = B$ and $B' = A$ is called a *formal concept*. Here, A is called the *concept extent* and B is called the *concept intent*. The set of all these formal concepts, i.e., $\mathfrak{B}(\mathbb{K}) := \{(A, B) \mid A \subseteq G, B \subseteq M, A' = B, B' = A\}$ gives rise to an order structure $(\mathfrak{B}, \leq)$ using $(A, B) \leq (C, D) :\Leftrightarrow A \subseteq C$, called *concept lattice*. For clone items we are particularly interested in the two entailed closure systems, i.e., in the *object closure system* $\mathfrak{G}(\mathbb{K}) := \{A \in G \mid (A, B) \in \mathfrak{B}(\mathbb{K})\}$ and the *attribute closure system* $\mathfrak{M}(\mathbb{K}) := \{B \in M \mid (A, B) \in \mathfrak{B}(\mathbb{K})\}$. We may denote those by $\mathfrak{G}$ and $\mathfrak{M}$ whenever the according context is implicitly given.

Clones. Besides the original definition of what clone items are there will be some graduations useful to graphs. We start with the common definition. Given a formal context $\mathbb{K} = (G, M, I)$ and two items $a, b \in M$, we say a *is clone to* b in $\mathfrak{M}$ if $\forall X \in \mathfrak{M} : \varphi_{a,b}(X) \in \mathfrak{M}$, with:

$$\varphi_{a,b}(X) := \begin{cases} X \setminus \{a\} \cup \{b\} & \text{if } a \in X \wedge b \notin X \\ X \setminus \{b\} \cup \{a\} & \text{if } a \notin X \wedge b \in X \\ X & \text{else} \end{cases}$$

We may denote this property by $a \sim_{\mathbb{K}} b$ and whenever the context is distinctive $a \sim b$. It is obvious that $\sim$ is a reflexive and symmetric relation on $M \times M$. Actually, it is also transitive, which can be shown easily, hence $\sim$ is an equivalence relation. Since every $a \in M$ is a clone to itself we say an a is a *proper clone* iff there is a $b \in M \setminus \{a\}$ such that $a \sim b$. In a not-clarified formal context there might be some $m, n \in M, m \neq n$ such that $m' = n'$. Those elements are proper clones. However, this is obvious and not revealing any hidden structure besides the fact that two identical copies are present. Therefore we call a proper clone $a \in M$ *trivial* iff there is a $b \in M \setminus \{a\}$ with $a' = b'$.

---

[3] The second way yields the dual context $\mathbb{K}(H) = (W, U, I)$.

A this point one may ask if it is hard to construct a formal context having a significant number of non-trivial clones. This is very easy as the following example discloses.

Example 2.1. The nominal scales, i.e., $(\{1, \ldots, n\}, \{1, \ldots, n\}, =)$ and the contra-nominal-scale $(\{1, \ldots, n\}, \{1, \ldots, n\}, \neq)$ provide formal contexts where every attribute element is a non-trivial clone. Furthermore, the union of two formal contexts, i.e., $\mathbb{K}_1 := (G_1, M_1, I_1)$ and $\mathbb{K} := (G_2, M_2, I_2)$ becomes $\mathbb{K}_1 \cup \mathbb{K}_2 := (G_1 \cup G_2, M_1 \cup M_2, I_1 \cup I_2)$, preserves the clones from $\mathbb{K}_1$ and $\mathbb{K}_2$.

All the above can be defined similarly for elements of G using the dual-context, i.e., the context where objects and attributes are interchanged. We therefore omit the explicit definitions and continue assuming the necessary definitions are made. However, we may provide some wording to differentiate between clones in $\mathfrak{M}$ and clones in $\mathfrak{G}$ for some formal context (G, M, I). When necessary we call the former *attribute clone* and the latter *object clone*.

3 Theoretical Observations

In this section, we derive some crucial properties of clones as well as a characterization of the clone property on the level of the context table. These theoretical results allow a fast computation of clones and help understanding the nature of clones in data. The first shows that for attributes with $a \sim b$ the object sets a' and b' are incomparable.

Lemma 3.1 (Clones are incomparable). *Let $\mathbb{K} = (G, M, I)$ be a formal context and $a, b \in M$. If $a \sim b$, then from $a' \subseteq b'$ follows $a' = b'$.*

Proof. Using $a' \subseteq b'$ we show $b' \subseteq a'$. We examine the mapping

$$\varphi_{ab}(b'') = \begin{cases} b'' & \text{if } a \in b'' \\ b'' \setminus \{b\} \cup \{a\} & \text{if } a \notin b''. \end{cases}$$

We show that the second case is invalid. From $a \sim b$ and $\varphi_{ab}(b'')$ being a closure we deduce $a'' \subseteq b'' \setminus \{b\} \cup \{a\}$. Since $a' \subseteq b'$, we have $b'' \subseteq a''$ and together we yield $b \in b'' \subseteq a'' \subseteq b'' \setminus \{b\} \cup \{a\}$ contradicting the case. Hence, only the first case can exist, meaning $a \in b''$, thus obviously $b' \subseteq a'$. □

The next results indicates, that reducible elements of a formal context can be ignored in the search for clones.

Lemma 3.2 (Clone irreducability). *Let $\mathbb{K} = (G, M, I)$ be a clarified formal context and attributes $a, b \in M : a \neq b$ with $a \sim b$. Then a is irreducible in $\mathbb{K}$.*

Proof. Assume a is reducible, i.e., there exists a set of attributes $N \subseteq M$ with $a \notin N$ and $\bigcap_{n \in N} n' = a'$. As $\mathbb{K}$ is clarified, we have $a' \neq b'$, thus from Lemma 3.1

follows $b \notin a''$. Therefore $\varphi_{a,b}(a'') = a'' \setminus \{a\} \cup \{b\}$. From the reducibility assumption follows

$$\forall n \in N : n' \supseteq a' \implies n \in a'' \overset{n \neq a}{\implies} n \in a'' \setminus \{a\} \cup \{b\} = \varphi_{a,b}(a'').$$

Thus, $a' = \bigcap_{n \in N} n' \supseteq \varphi_{a,b}(a'')'$, which means $a'' \subseteq \varphi_{a,b}(a'') = a'' \setminus \{a\} \cup \{b\}$. Clearly, this means $a = b$ contradicting the lemma's assumption.   □

While clarifying a context removes the non-trivial clones, additionally reducing that context does not change the clone relationship any further. Therefore, for finding non-trivial clones it suffices considering reduced contexts. Next, we describe for such contexts how clones can be identified directly from the context's table. [7] already found it is sufficient to check join-irreducible intents to check the clone property. The respective result there (Proposition 1) is formulated for the dual version of formal contexts, i.e., where G and M are interchanged. Also, for the proof the authors of [7] refer to a manuscript that had been submitted (at the time) but appears to have never been published. For the sake of completeness, we present a variation of their result in the common notion of a formal context and present a proof. Here, we already use the fact that in a reduced context, the join irreducible concepts are exactly the object concepts.

Theorem 3.1. *Let* $\mathbb{K} = (G, M, I)$ *be a reduced formal context and* $a, b \in M$ *with* $a \neq b$. *The following are equivalent:*

1. $a \sim b$
2. *For each object* $g \in G$, *there is an object* $h \in G$ *such that* $\varphi_{a,b}(g') = h'$.

Proof. First we show, 1. $\implies$ 2. For $a, b \in g'$ or $a, b \notin g'$, the claim is obvious (using $h := g$). Without loss of generality, we can assume $a \in g'$ and $b \notin g'$, thus $\varphi_{a,b}(g') = g' \setminus \{a\} \cup \{b\}$.

As $\varphi_{a,b}(g')$ is an intent, there exists a set of objects $H \subseteq G$ with $H' = \varphi_{a,b}(g') = g' \setminus \{a\} \cup \{b\}$. We can partition H into $H_a := \{h \in H \mid a \in h'\}$ and $H_{\bar{a}} := \{h \in H \mid a \notin h'\}$. As clearly $a \notin \varphi_{a,b}(g')$, $H_{\bar{a}}$ cannot be empty. We yield:

$$g' \setminus \{a\} \cup \{b\} = \bigcap_{h \in H_a} h' \cap \bigcap_{h \in H_{\bar{a}}} h'$$

$$H_{\bar{a}} \neq \emptyset, b \notin g' \implies \quad g' \setminus \{a\} = \bigcap_{h \in H_a} h' \cap \bigcap_{h \in H_{\bar{a}}} (h' \setminus \{b\})$$

$$a \in g' \implies \quad g' = \bigcap_{h \in H_a} h' \cap \bigcap_{h \in H_{\bar{a}}} (h' \setminus \{b\} \cup \{a\})$$

$$b \in \varphi_{a,b}(g') = H' \implies \quad g' = \bigcap_{h \in H_a} h' \cap \bigcap_{h \in H_{\bar{a}}} \varphi_{a,b}(h')$$

As g is irreducible, we either have an object $h \in H_a$ with $g' = h'$ or an object $h \in H_{\bar{a}}$ with $g' = \varphi_{a,b}(h')$. Clearly, the former cannot be true, as $b \in h'$ for $h \in H$ and $b \notin g'$. From the latter follows $\varphi_{a,b}(g') = h'$.

Next, we show 2. $\implies$ 1: Let $N \subseteq M$ be an intent of $\mathbb{K}$, i.e., there is a set of objects $H \subseteq G$ such that $N = H'$. We show that $\varphi_{a,b}(N)$ is an intent. This is trivial for the cases $a, b \in N$ and $a, b \notin N$. Without loss of generality, we assume $a \in N$ and $b \notin N$. Then

$$\varphi_{a,b}(N) = \varphi_{a,b}(H') = H' \setminus \{a\} \cup \{b\} = \bigcap_{h \in H_b} (h' \setminus \{a\} \cup \{b\}) \cap \bigcap_{h \in H_{\bar{b}}} (h' \setminus \{a\} \cup \{b\})$$

with H_b and $H_{\bar{b}}$ defined as H_a and $H_{\bar{a}}$. For $h \in H_b$ it holds $h' \setminus \{a\} \cup \{b\} = h' \setminus \{a\}$. As $H_{\bar{b}} \neq \emptyset$ (a.p., $b \notin N = H'$) we yield $\varphi_{a,b}(N) = \bigcap_{h \in H_b} h' \cap \bigcap_{h \in H_{\bar{b}}} (h' \setminus \{a\} \cup \{b\})$. Since $a \in H'$, for $h \in H_{\bar{b}}$: $h' \setminus \{a\} \cup \{b\} = \varphi_{a,b}(h')$, which by 2. is g' for some $g \in G$. Thus $\varphi_{a,b}(N)$ is the intersection of intents and therefore itself an intent. $\square$

The theorem characterizes clones on the context level: Two attributes a and b are clones if for each object $g \in G$ whose row contains only one of the two attributes, there is another object $h \in G$ such that its row contains only the other of the two attributes, while the remaining parts of the rows are identical, i.e., $g' \setminus \{a\} = h' \setminus \{b\}$.

3.1   Clones in Graph Data

Clones in Social Networks. In the following, we identify any given graph with the formal context counterpart $\mathbb{K} = (U, W, I)$, as in Sect. 2. Transferring the definitions from Sect. 2, we obtain what clones in graphs, in particular in social networks, are. For the special case of social networks we call object clones *user clones* and attribute clones are either some *property clone*, in the bipartite case, or also user clones, in the single-mode case.

Example 3.1 (Social Network). In Fig. 1 we show a small artificial example of a possible social network. Represented as context as described in Sect. 2 we get with $M = \{$Swimming, Hiking, Biking, Rafting, Jogging$\}$ the closure system $\mathfrak{M}(\mathbb{K}) = \{\{S\}, \{H\}, \{B\}, \{R\}, \{J\}, \{B,R\}\}, \{B,J\}, \{R,J\}\}$. The associated clone classes are denoted in Fig. 1.

Data Set Description. Almost all of the following data sets can be obtained from the UCI Machine Learning Repository [9]. We consider nine social network graphs and two non social network data sets: **zoo** [9]: 101 animals and seventeen attributes (fifteen Boolean and two numerical). All attributes were nominal scaled, resulting in a set with 43 attributes; **cancer** [12]: 699 instances of breast cancer diagnoses with ten numerical attributes, which were nominal scaled; **face-booklike** [15]: 337 forum users with 522 topics they communicated on; **southern** [17]: classical small world social network consisting of fourteen woman attending eighteen different events; **club** [3]: 25 corporate executive officers and fifteen social clubs in which they are involved in; **movies** [4]: 39 composers of film music and their relations to 62 producers; **aplnm** [1]: 79 participants of the

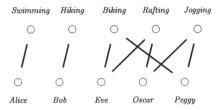

Fig. 1. Example of a social network graph exhibiting various clones items. Edges connect a person with his or her activity. Equivalence classes for attribute clones are {Swimming,Hiking}, {Biking, Rafting, Jogging}.

Lange Nacht der Musik in 2013 and the 188 events they participated in; **jazz** [6]: 198 jazz musicians and their collaborations; **dolphin** [10]: 62 bottlenose dolphins with contacts amongst each other; **hightech** [16]: Some (one-mode) social network with 33 users from within the parameters of a social network but with no further insights provided; **wiki** [8,16]: 764 voters on Wikipedia with 605 users to be voted on.

For comparison, we also investigate randomized versions of all those data sets, generated using a coin draw process. This may imply that the resulting formal contexts are prone to the stegosaurus phenomenon. However, no unbiased method for generating formal context for a given number of objects, attributes, and density is known [2].

Computation. Computing the attribute (object) clones for a given formal context (G, M, I) would imply to know the associated attribute (object) closure system. However, computing those is computational infeasible for contexts of a particular size or greater. To cope with this barrier we utilize Lemma 3.2 and Theorem 3.1. Hence, instead of checking all elements of a closure system we only need to check the irreducibles. Therefore we checked brute force all combinations of attributes (objects) for every given data set by checking the according irreducibles.

In particular we computed for every data set the number of trivial and non-trivial object clones, and attribute clones. The results are shown in Table 1. In addition we also computed the number of trivial and non-trivial clones for the object/attribute-projections for every formal context. However, besides creating more trivial clones no further insights could be grasped from this. Also, the experiment on randomly generated formal contexts had not different outcome. Therefore we omitted presenting the particular results for the latter two.

Discussion. The most obvious result for all data sets alike is that non-trivial clones are very infrequent. Omitting the wiki data set only two data sets have clones at all, in particular a very small number of object clones compared to the size of the network. We investigated the exception by the wiki data set further and discovered a large nominal scale as subcontext responsible for the

Table 1. Properties of the considered (social) networks and data sets and results for clone experiment. With *G*-t we denote trivial clones whereas clones denote non-trivial clones.

Name	$|U|$	$|M|$	density	# G-clones	# M-clones	# G-t-clones	# M-t-clones
zoo	101	43	0.390	0	0	42	2
cancer	699	92	0.110	0	0	236	0
facebooklike	377	522	0.014	7	0	24	83
southern	18	14	0.352	0	0	1	1
aplnm	79	188	0.061	0	0	1	21
club	25	15	0.250	0	0	0	0
movies	62	39	0.079	0	0	1	0
jazz	198	198	0.068	7	7	0	0
dolphins	62	62	0.082	0	0	2	2
hightech	33	33	0.148	0	0	1	1
wiki	764	605	0.006	234	234	73	30

vast amount of clones. Since the wiki data set is the result of a collection of voting processes this would represent single votes. For trivial clones we have diverse observations. Some networks like facebooklike have a significant amount of trivial clones. Others of comparable size, however, do not, like jazz. Since those clones do not reveal any hidden structure but the fact that copies of users or properties are present in the network, we consider these clones uninteresting.

For the object and attribute projections we obtain almost the same results. Almost no non-trivial clones are present. Though, the number of trivial clones has increased in almost all the networks. This could be another indication that simple one-mode projections are insufficient for analyzing bipartite networks.

All in all, the notion of non-trivial clones seems insufficient for the investigation of graphs. The explanation for this is that the structural requirements for two attributes being clone are too strong, cf. theoretical results in Sect. 3. However, it strikes the question if there is a generalization which is softening those requirements while preserving enough structure.

4   Generalized Clones

The results from the previous section motivate finding a more general clone notion for formal contexts. In [7] the authors provided an interesting generalization of clones in a formal context. They proposed *P*-Clones, i.e., clones with respect to the family of pseudo intents, and *A*-Clones, i.e., clones in a particular kind of atomized context. Both approaches are based on using some kind of modified family of sets. Another course of action was taken in [11], in which the author used a measure of "cloneniness" based on the number of incorrect mapped sets. We take a different approach, using the original set of closures – the intents – based on the following observation.

Remark 4.1 (Clone permutation). Every pair (a, b) of elements $a, b \in M$ with $a \sim b$ for a given formal context (G, M, I) gives rise to a permutation $\sigma : M \to M, m \mapsto \sigma(m)$, with $\sigma(a) = b$, $\sigma(b) = a$, and $\sigma(m) = m$ for $m \in M \setminus \{a, b\}$. We denote such permutations as *clone permutations*.

Since for every $a \in M$ we have $a \sim a$, the set of clone permutations S for a given formal context (G, M, I) contains the identity. For any two elements $a, b \in M$ with $a \sim b$ we can represent the associated clone permutation σ by $\sigma := (ab)$ using the reduced cycle notation. From this we note that the set of all pairs of proper clones corresponds to a particular subset of permutations on M where every permutation σ contains exactly one two-cycle. This gives rise to two possible generalizations. Both associated computational problems require sophisticated algorithms to be developed.

Multiple Two-Cycles. We motivate this approach using the lattice for a closure system on $M = \{a, b, c, d\}$ represented in Fig. 2 (left). In this closure system there are no proper clones. However, we can find a permutation σ that preserves the closure system. For example, the permutation $\sigma = (ab)(cd)$, which is a permutation of two disjoint cycles of length two. This permutation is not representable by exactly one cycle of length two. Hence, we propose permutations representable as products of cycles of length two as one generalization of clones. Yet, this immediately gives rise to the idea of higher order permutations.

Higher Order. Again, we want to motivate this generalization by providing an example. In Fig. 2 (middle), we show the lattice for a closure system $M = \{a, b, c, d\}$. This closure system is free of (proper) clones. However, we find a permutation $\sigma = (ab)(cd)$ in the above described manner. In addition we find a permutation of order four, i.e., $\sigma^4 = id$, preserving the closure system, e.g., $\sigma = (acbd)$. In the same figure on the right we observe a permutation of order five, i.e., $\sigma = (acedb)$, answering the natural question for a permutation with odd order.

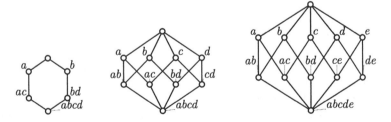

Fig. 2. Example for clone-free closure system on four attributes (left, middle) and on five attributes (right).

5   Conclusion

While starting the investigation the authors of this work were confident to discover clones in graph data sets, at least for graphs of a particular minimal size. In order to cope with the computational complexity of closure systems we utilized results from [7] and expressed them in terms of statements about formal contexts. However, our investigation did reveal the absence of clones in real world graph like data. The only significant observation was the emergence of trivial clones while projecting bipartite social networks to one set of nodes.

This setback, though, led us to discover two more general notions of clones, which can cope with more structural requirements. Investigating those more thoroughly should be the next step in clone related research, building on the theoretical results in Sect. 3. To this end, we finish our work with the following three open questions. **Question 1:** To which graph theoretical notion could the idea of clone permutation correspond to? **Question 2:** Does the set of all valid clone permutations on a closure set always form a group and if no, why not? **Question 3:** If yes, can this group provide new insights into the structure of closure systems or of social networks?

Acknowledgments. This work was funded by the German Federal Ministry of Education and Research (BMBF) in its program "Forschung zu den Karrierebedingungen und Karriereentwicklungen des Wissenschaftlichen Nachwuchses (FoWiN)" under Grant 16FWN016.

References

1. Borchmann, D., Hanika, T.: Individuality in social networks. In: Missaoui, R., Kuznetsov, S.O., Obiedkov, S. (eds.) Formal Concept Analysis of Social Networks, pp. 19–40. Springer International Publishing, Cham (2017). https://doi.org/10.1007/978-3-319-64167-6_2
2. Borchmann, D., Hanika, T.: Some experimental results on randomly generating formal contexts. In: Huchard, M., Kuznetsov, S. (eds.) CLA. CEUR Workshop Proceedings, vol. 1624, pp. 57–69. CEUR-WS.org (2016)
3. Club membership network dataset – KONECT, January 2016
4. Faulkner, R.R.: Music on Demand: Composers and Careers in the Hollywood Film Industry. Composers and Careers in the Hollywood Film Industry. Transaction Publishers, New Brunswick (2003)
5. Ganter, B., Wille, R.: Formal Concept Analysis: Mathematical Foundations. Springer, Heidelberg (1999). https://doi.org/10.1007/978-3-642-59830-2. pp. x+284
6. Gleiser, P., Danon, L.: Community structure in jazz. Adv. Complex Syst. **6**, 565 (2003)
7. Gély, A., Medina, R., Nourine, L., Renaud, Y.: Uncovering and reducing hidden combinatorics in Guigues-Duquenne bases. In: Ganter, B., Godin, R. (eds.) ICFCA 2005. LNCS (LNAI), vol. 3403, pp. 235–248. Springer, Heidelberg (2005). https://doi.org/10.1007/978-3-540-32262-7_16

8. Leskovec, J., Huttenlocher, D., Kleinberg. J.: Signed networks in social media. In: Proceedings of the SIGCHI Conference on Human Factors in Computing Systems, pp. 1361–1370. ACM (2010)

9. Lichman, M.: UCI Machine Learning Repository (2013)

10. Lusseau, D., et al.: The bottlenose dolphin community of Doubtful Sound features a large proportion of long-lasting associations. Behav. Ecol. Sociobiol. **54**(4), 396–405 (2003)

11. Macko, J.: On open problem – semantics of the clone items. In: Domenach, F., Ignatov, D.I., Poelmans, J. (eds.) Contributions to the 10th International Conference on Formal Concept Analysis (ICFCA 2012). CEUR Workshop Proceedings, vol. 876, pp. 130–144. CEUR-WS.org (2012)

12. Mangasarian, O.L., Wolberg, W.H.: Cancer diagnosis via linear programming. SIAM News **23**(5), 1–18 (1990)

13. Medina, R., Nourine, L., Raynaud, O.: Interactive association rules discovery. In: Missaoui, R., Schmidt, J. (eds.) ICFCA 2006. LNCS (LNAI), vol. 3874, pp. 177–190. Springer, Heidelberg (2006). https://doi.org/10.1007/11671404_12

14. Medina, R., et al.: Efficient algorithms for clone items detection. In: CLA 2005, pp. 70–81 (2005)

15. Opsahl, T., Panzarasa, P.: Clustering in weighted networks. Soc. Netw. **31**(2), 155–163 (2009)

16. Rossi, R.A., Ahmed, N.K.: The network data repository with interactive graph analytics and visualization. In: Proceedings of the Twenty-Ninth AAAI Conference on Artificial Intelligence (2015)

17. Wasserman, S., Faust, K.: Social Network Analysis. Methods and Applications. Structural Analysis in the Social Sciences. Cambridge University Press, New York (1994)

18. Wille, R.: Restructuring lattice theory: an approach based on hierarchies of concepts. In: Rival, I. (ed.) Ordered Sets, pp. 445–470. Springer, Dordrecht (1982). https://doi.org/10.1007/978-94-009-7798-3_15

Knowledge-Based Mining of Exceptional Patterns in Logistics Data: Approaches and Experiences in an Industry 4.0 Context

Eric Sternberg[1] and Martin Atzmueller[2(✉)]

[1] University of Kassel (ITeG), Wilhelmshöher Allee 73, 34121 Kassel, Germany
est@cs.uni-kassel.de
[2] Tilburg University (CSAI), Warandelaan 2, 5037 AB Tilburg, The Netherlands
m.atzmuller@uvt.nl

Abstract. In the context of Industry 4.0 and smart production, industrial large-scale enterprise data is applied for enabling data-driven analysis and modeling methods. However, the majority of the currently applied approaches consider the data in isolated fashion such that data from different sources, e.g., from large data warehouses are only considered independently. Furthermore, connections and relations between those data, i.e., relating to semantic dependencies are typically not considered, while these would open up integrated semantic approaches for effective data mining methods. This paper tackles these issues and demonstrates approaches and experiences in the context of a real-world case study in the industrial logistics domain: We propose knowledge-based data analysis applying subgroup discovery for identifying exceptional patterns in a semantic approach using appropriately constructed knowledge graphs.

1 Introduction

In the industrial world of today, the amounts of data are increasing at a rapid pace, enabling large-scale business intelligence and data-driven decision support. Then, exploratory data mining provides a viable tool for obtaining relevant insights. Here, in particular methods for local pattern mining, e.g., subgroup discovery and exceptional model mining enable powerful approaches for detecting interesting, i.e., unexpected, anomalous and exceptional patterns with a broad range of applications for industrial data analytics.

Problem. However, so far the applied approaches consider the data in isolated fashion such that different data types ranging from unstructured to structured data, e.g., tabular and graph-data, are only considered independently. This requires an efficient and effective integrated approach for semantic modeling and data mining, which, however, has not been established at large-scale yet.

Objectives. In this paper, we exemplify an approach tackling these issues: We apply subgroup discovery for identifying exceptional patterns in the context of

© Springer Nature Switzerland AG 2018
M. Ceci et al. (Eds.): ISMIS 2018, LNAI 11177, pp. 67–77, 2018.
https://doi.org/10.1007/978-3-030-01851-1_7

finding inventory differences. For that, a knowledge-based approach is presented, integrating large-scale data into a knowledge graph representation. We discuss experiences in the context of Industry 4.0 and smart production.

In particular, we present results of a real-world case study in a productive logistic environment of a large scale industrial plant. One major goal for the client was to identify specific logistic processes that possibly lead to erroneous financial assessments, so-called inventory differences. The whole process was strongly assisted by domain experts, so another goal was to deliver reasonable explanations and transparency to them. One major issue was to represent the particular domain dependencies in an integrated knowledge graph and the construction of appropriate features to enable the synchronization between the experts domain understanding and the knowledge graph representation.

Contributions. Our contributions are summarized as follows:

1. We describe an integrated approach for large-scale industrial data analytics implemented by knowledge-based data mining utilizing knowledge graphs.
2. We demonstrate the application of local pattern mining in that context, focusing on subgroup discovery implemented in an intelligent system.
3. We present an anonymized real-world case study in the context of Industry 4.0 and smart production. Furthermore, we discuss insights and experiences in the application domain of production logistics.

The rest of the paper is structured as follows: Sect. 2 describes the industrial problem setting in more detail, before Sect. 3 provides related work on knowledge-based approaches and local pattern mining. Next, Sect. 4 summarizes the formal background on subgroup discovery. After that, Sect. 5 presents the applied approaches for knowledge graph construction and pattern mining in the industrial context, and discusses experiences in the context of a real-world industrial case study. Finally, Sect. 6 concludes with a summary and presents interesting directions for future work.

2   Industrial Problem Setting

In the field of logistics in industrial production, Industry 4.0 and smart production are important directions for implementing cost-effective measures. Since data is captured continuously during all the relevant processes, powerful data mining methods are required. The standard automation pyramid cf. [14,15] on industrial processes depicts different systems corresponding to different levels of analysis. Data analytics is mainly performed on the upper levels – corresponding to the operations control level and the enterprise level. Here, one prominent case for data analytics is given by uncovering inventory differences. Basically, inventory differences cause deviations in the financial rating of the plants' current assets. In the past these differences where detected once a year and could reach deviations in the region of about EUR 100 million in our application domain, which corresponds to about one percent of the yearly turnover of a large plant.

As a consequence a team of analysts from different departments permanently investigate these differences which decreases the deviations by a factor of 10. This leads to a trade-off between cost of human resources and the inventory difference because the desired state is to resolve as much as possible deviations using automated analysis, i.e., with a minimum commitment of human resources.

3   Related Work

Related work concerns both knowledge-intensive approaches as well as methods for local pattern mining. Domain knowledge is a natural resource for knowledge-intensive data mining methods, e.g., [8,20,23], for example in the context of prototype-based approaches [12] or knowledge graphs [21,24]. However, in data mining, semantic knowledge is scarcely exploited so far. First approaches for integrating knowledge graphs, i.e., based on ontologies and a set of instance data has been proposed in the area of semantic data mining [20,23]. [9] presents a mixed-initiative approach, for semantic feature engineering using a knowledge graph. In a semi-automatic process, the knowledge graph is engineered and refined. Finally, the engineered features are provided for data mining. A similar approach is applied in [5]. Here, data from heterogeneous data sources is integrated into a knowledge graph, which then provides the basis for data mining.

For data analytics, local pattern mining is a broadly applicable and powerful set of methods for exploratory data mining [3,4,7]. Common methods include those for association rule mining [1], subgroup discovery, e.g., [3,25] and exceptional model mining [3,13,19]. Essentially, subgroup discovery is a flexible method for detecting relations between dependent (characterizing) variables and a dependent target concept, e.g., comparing the share or the mean of a nominal/numeric target variable in the subgroup vs. the share or mean in the total population, respectively [3,18,25]. The interestingness of a pattern can then be flexibly defined, e.g., by a significant deviation from a model that is derived from the total population. In the simplest case, (see the example above) a binary target variable is considered, where the share in a subgroup can be compared to the share in the dataset in order to detect (exceptional) deviations.

In contrast to the approaches discussed above, we focus on an integrated approach, exploiting knowledge-based semantic structures, i.e., sets of knowledge components connected to a local pattern mining method. Then, also the knowledge graph can be incrementally refined during the mining process.

4   Background: Subgroup Discovery

Formally, a *database* $DB = (I, A)$ is given by a set of individuals I and a set of attributes A. For each attribute $a \in A$, a range $dom(a)$ of values is defined. An attribute/value assignment $a = v$, where $a \in A, v \in dom(a)$, is called a *feature*. We define the feature space V to be the (universal) set of all features. Intuitively, a *pattern* describes a *subgroup*, i.e., the subgroup consists of instances that are covered by the respective pattern. A subgroup *pattern* P is then defined as a conjunction $P = s_1 \wedge s_2 \wedge \cdots \wedge s_n$ of (extended) features $s_l \subseteq V$, which are then called selection expressions, where each s_l selects a subset of the range $dom(a)$ of an attribute $a \in A$. A *subgroup (extension)* $I_P := ext(P) := \{i \in I | P(i) = true\}$ is the set of all individuals which are covered by pattern P. The set of all possible subgroup descriptions, and thus the possible search space is then given by 2^Σ, i.e., all combinations of the patterns contained in Σ denoting the set of all possible selection expressions. A *quality function* $q: 2^\Sigma \to \mathbb{R}$ maps every pattern in the search space to a real number that reflects the interestingness of a pattern (or the pattern cover, respectively). The result of a subgroup discovery task is the set of k subgroup descriptions $P_1, \ldots, P_k$ with the highest interestingness according to the selected quality function. For subgroup models there exist various quality functions, cf. e.g., [3] for a detailed discussion. A simple example compares the share p of a binary target in the subgroup pattern P to its share in the database p_0, weighted by the size of the subgroup n, i.e., $q(P) = (p - p_0) \cdot n$. For numeric targets, this can easily be adapted replacing the shares by the respective means.

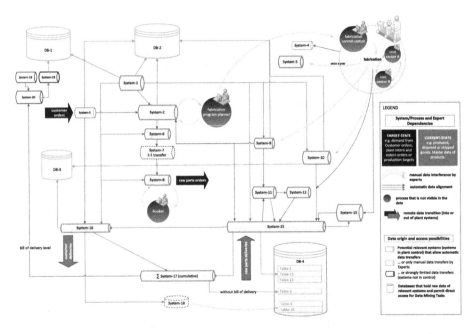

Fig. 1. Overview on the dependencies between systems and processes in our industrial context domain. The blue databases contain raw data accessible for data analytics. (Color figure online)

5   Case Study

This section summarizes results of a real-world case study in industrial logistics. For the task of detecting inventory differences, the domain experts provided background knowledge, including exemplary cases leading to inventory differences in the past, relevant data sources for their analysis and how they detect the causes of those differences. A first step for data analysis was then given by business and data understanding, relating to the CRISP-DM model [11]. As a result, it was possible to identify the most relevant data sources contributing to the inventory difference problem. Figure 1 shows the resulting complex structures. Here, different databases and processes were identified for data preprocessing and integration. This structure then was one of the main outcomes in order to transform the data into a knowledge graph. For data transformation, we implemented a generic framework to load and transform the given data, cf. Fig. 2 for a schematic view, utilizing the VIKAMINE [6] system for subgroup discovery. For the knowledge transformation and modeling we applied Gephi [10] and GraphstreamLib [17]. Here, we distinguish between the *structure data* graph representing information and dependencies taken from the bills of materials, and the *accounting data* graph capturing material flow information within the production processes which are discussed in more detail below.

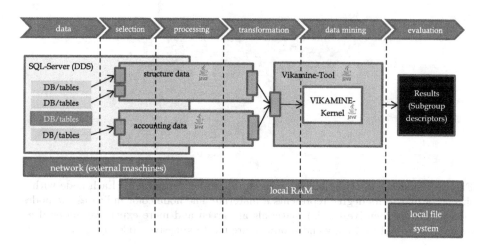

Fig. 2. Schematic view on the data pipeline. The top shows the CRISP steps, the bottom the utilized infrastructure. Gray depicts the present data sources (relates to blue in Fig. 1). Red depicts implementation done for the project where the structure and accounting data boxes essentially depict the knowledge graph components. (Color figure online)

5.1   Knowledge Graph Construction

As outlined above, two logistic data sources emerged in our case study holding essential information for data analytics. The first related to information on logistic bill of material (BOM). This data is stored in several distributed relational databases, and describes the composition of basic parts up into an end-product in a hierarchical way. Utilizing those assembly dependencies we developed a parser

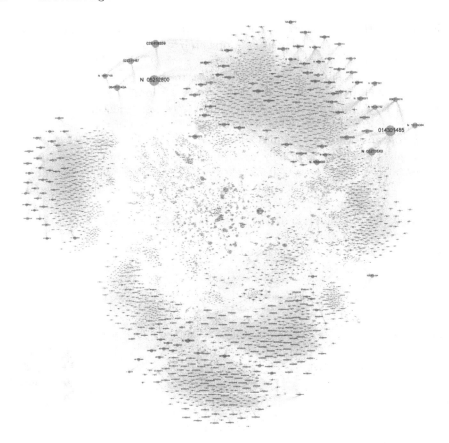

Fig. 3. Visualization of the bill of material (*structure data* graph). Each node with a size according to its degree represents a material. The node color indicates the nodes price where inexpensive (basic) materials are green and more expensive material go over yellow to red (normally end-products are most expensive). (Color figure online)

for constructing a graph structure using the GraphStream framework [17], resulting in the *structure data* graph. An (anonymized) visualization of the (complete) generated graph is shown in Fig. 3. Each dependency is represented by a directed edge from the basic material to the (complex) processed material, the node size corresponds to its degree.

We improved the graph step by step from subgroup discovery in cooperation with the domain knowledge of the experts by additional data sources (e.g., using material prices), in order to represent the relevant dependencies and parts-information according to the experts' judgment, cf. Sect. 5.3. For that, we formulated different dependencies/expectations as target concepts for subgroup discovery and automatically investigated specific patterns that either occurred very rarely or dominated the observed dependencies in the data, applying the quality function described above.

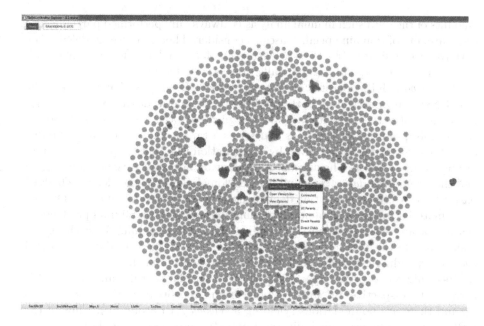

Fig. 4. Screenshot of the interactive BOM explorer. It allows experts to identifi problematic parts and their assembly dependencies in an intuitive and exploratory way. The graph is layouted by a dynamic force layout available from the utilized library [17] and shows only end-products (end/root nodes in the *structure data* graph).

For a better assessment of potentially problematic dependencies indicated by the patterns, we also implemented a visual explorer for the structure data graph, see Fig. 4: This screenshot shows top-level materials (final products), where each material is represented by one node. The user is able to search specific materials (or groups) and to drill down on dependencies via a right-click.

Regarding the inventory differences it also became apparent that additional data sources with information about the logistic material movements were essential to the experts' analysis. We identified a special system that collected booking information from several systems and created daily summaries, e.g., regarding essential attributes of a booking record like sender, material, amount, time, receiver and booking reason. That information basically describes a material flow network (for more information on logistics and logistic material flow networks we refer to, e.g., [22]). Therefore, analogously to the *structure data* graph, we built the *accounting data* graph corresponding to material flow information (see Fig. 2). Both graphs exhibit very significant dependencies which is used in pattern analysis by experts when investigating inventory differences.

5.2 Investigating Expected Relations Using Local Pattern Mining

An essential problem to the experts was to check the relevant bookings of a material and all its dependent materials. Because of the complexity and large amounts of the data such manual analysis is always only performed partially, also there are lots of domain specific cases to consider. Therefore, we modeled a key performance indicator (KPI) as a feature for data analytics for each material-node in the *structure data* graph, called the *relbook* feature.

The *relbook* KPI operates on the two associated graphs that represent the bill of material (structure of parts) and the material flow described by bookings (movement of parts). The feature was designed with the experts expectation of correct interpretation of bookings and set as a target concept for subgroup discovery: *relbook* is essentially given by a calculation rule that outputs a real number between −1 and 1. The calculation is done on both the *structure data* graph and the *accounting data* graph and utilizes domain specific knowledge. The KPI basically traces the construction dependencies in the *structure data* graph from a basic material (a screw, an amount of aluminum etc.) up to all final products (a gearing or an exhaust system) containing that basic material. For each visited material-node the particular information from the *accounting data* graph and the corresponding effective booking amounts are determined.

Basically, the experts expected, that if the calculation rule (and the data) is correct, the respective amounts included from concepts of the graphs will add up and there are no deviations (yielding a *relbook* KPI around 0). Thus, if the KPI diverges from zero this indicates a potential problem in the data.

Overall, we discovered two major reasons for deviations from the experts' expectations. The first is indicated by a positive *relbook* KPI generated by an accumulation in storage of specific materials, i.e., those that find no longer or not yet application in active products. On the other hand a negative deviation from the KPI was mainly observed in small groups of bookings. The reasons here where mainly special cost centers and storages with exotic processes e.g., temporary outsourcing of material for extern handling or a special type of (final) waste-booking of old material without actual use in production.

5.3 Identifying Anomalous Subgroup Patterns

Further interesting findings often showed significant deviations from the experts expectations. For detecting anomalous patterns, for example, we examined subgroup targets for the different working-shifts and found descriptions which identify that there are strong associations between specific types of logistic bookings and the current working shift. As an example, Table 1 shows two significant patterns found using the target variable that identifies bookings done in the normal shift (*SchichtKz=01*). The dataset properties (*defined individuals*) show that there are about 8.18 million bookings in the population, also nearly 45% of all bookings are done in the normal shift. The descriptor #2 *BstArt=NIK* states that the normal shift produces 82% of all waste-bookings which was interesting to the experts because they expected a homogeneous distribution over all four shifts. As a result it was uncovered that only particular personal is allowed to perform return or waste bookings – only available in specific shifts. Furthermore it was uncovered that the shifts handling in the booking system is different to the real shifts execution. Therefore the missing 18% of waste-bookings where found in the following system shift which still falls in the normal execution shift.

Table 1. Population properties and two exemplary subgroup patterns for bookings done in the normal shift. Pattern #2 uncovers the strong dependence between this shift and waste bookings.

Target (nominal)	SchichtKz=01				
Population properties					
Defined individuals		Undefined individuals		Target share in population	
8175938		0		44.9%	
Subgroup descriptors (mid-size groups)					
#	Quality	Group size	Target share	TP rate	Coverage
1	**EmpfKz=V AND ZwAnkTag** $]-\infty; 44, 5[$ **AND AnzBwgTag** $]-\infty; 26, 5[$				
	238.15	1506843	64,3%	26,4%	18,4%
2	**BstArt=NIK**				
	214.75	334828	82%	7,5%	4,1%

All results were discovered evaluating patterns from tables like Table 1, where often interesting patterns occured that lead to further investigation such as further expert interviews, individual data inspection, or further drill-down on the patterns. Using different targets, for example, concerning the kind of material application and the fluctuation of material-prices we discovered technical problems in an import process for data. In Fig. 1, this process would be represented by a blue arrow between a relevant system (green) and a database (blue). The problem was noticed because there were patterns that described bookings with empty storage groups, which was against the experts' expectation. Subgroup descriptions discovered for the target "component price fluctuation" also revealed fragmentary price data for a larger group of parts. In later iterations another systematic problem was uncovered where cost center IDs of physically the same cost center where not equal in different but dependent logistic systems.

6   Conclusions

In this paper, we presented approaches for knowledge-based mining of exceptional patterns in logistics data and discussed experiences in the context of a real-world case study. In particular, we focused on modeling background knowledge in the form of knowledge graphs, and we applied subgroup discovery for identifying exceptional patterns. Overall, the process and the results were very well accepted by the domain experts, which especially favored explainability and transparency of the mined patterns. During iterative sessions, interesting and useful patterns were identified for enabling the automatic monitoring of inventory differences. For future work, we aim to explore network patterns for refinement of knowledge graphs, extending feature engineering methods, e.g., [9]. Here, multiplex approaches, e.g., [2, 16] and pattern-based anomaly detection are further interesting directions.

References

1. Agrawal, R., Srikant, R.: Fast algorithms for mining association rules. In: Proceedings of VLDB, pp. 487–499. Morgan Kaufmann (1994)
2. Atzmueller, M.: Data mining on social interaction networks. JDMDH **29**, 1–21 (2014)
3. Atzmueller, M.: Subgroup discovery. WIREs DMKD **5**(1), 35–49 (2015)
4. Atzmueller, M., Baumeister, J., Puppe, F.: Introspective subgroup analysis for interactive knowledge refinement. In: Proceedings of FLAIRS, pp. 402–407. AAAI (2006)
5. Atzmueller, M., et al.: Big data analytics for proactive industrial decision support: approaches & first experiences in the context of the FEE project. ATP Ed. **58**(9), 62–74 (2016)
6. Atzmueller, M., Lemmerich, F.: VIKAMINE – open-source subgroup discovery, pattern mining, and analytics. In: Flach, P.A., De Bie, T., Cristianini, N. (eds.) ECML PKDD 2012. LNCS (LNAI), vol. 7524, pp. 842–845. Springer, Heidelberg (2012). https://doi.org/10.1007/978-3-642-33486-3_60
7. Atzmueller, M., Puppe, F.: Semi-automatic visual subgroup mining using VIKAMINE. J. Univers. Comput. Sci. **11**(11), 1752–1765 (2005)
8. Atzmueller, M., Puppe, F., Buscher, H.P.: Exploiting background knowledge for knowledge-intensive subgroup discovery. In: Proceedings of IJCAI, pp. 647–652 (2005)
9. Atzmueller, M., Sternberg, E.: Mixed-initiative feature engineering using knowledge graphs. In: Proceedings of K-CAP. ACM (2017)
10. Bastian, M., Heymann, S., Jacomy, M.: Gephi: an open source software for exploring and manipulating networks (2009)
11. Chapman, P., et al.: CRISP-DM 1.0. CRISP-DM consortium (2000)
12. Duch, W., Grudzinski, K.: Prototype based rules - a new way to understand the data. In: Proceedings of IJCNN, vol. 3, pp. 1858–1863. IEEE (2001)
13. Duivesteijn, W., Feelders, A.J., Knobbe, A.: Exceptional model mining. Data Min. Knowl. Disc. **30**(1), 47–98 (2016)
14. Givehchi, O., Trsek, H., Jasperneite, J.: Cloud computing for industrial automation systems - a comprehensive overview. In: Proceedings of EFTA, pp. 1–4. IEEE (2013)
15. Hollender, M.: Collaborative Process Automation Systems. ISA (2010)
16. Kanawati, R.: Multiplex network mining: a brief survey. IEEE Intell. Inform. Bull. **16**(1), 24–27 (2015)
17. Laboratoire d'Informatique, du Traitement de l'Information et des Systmes (LITIS): Graphstream project. http://graphstream-project.org
18. Lemmerich, F., Atzmueller, M., Puppe, F.: Fast exhaustive subgroup discovery with numerical target concepts. DMKD **30**, 711–762 (2016)
19. Lemmerich, F., Becker, M., Atzmueller, M.: Generic pattern trees for exhaustive exceptional model mining. In: Flach, P.A., De Bie, T., Cristianini, N. (eds.) ECML PKDD 2012. LNCS (LNAI), vol. 7524, pp. 277–292. Springer, Heidelberg (2012). https://doi.org/10.1007/978-3-642-33486-3_18

20. Rauch, J., Šimůnek, M.: Learning association rules from data through domain knowledge and automation. In: Bikakis, A., Fodor, P., Roman, D. (eds.) RuleML 2014. LNCS, vol. 8620, pp. 266–280. Springer, Cham (2014). https://doi.org/10.1007/978-3-319-09870-8_20

21. Ristoski, P., Paulheim, H.: Semantic web in data mining and knowledge discovery: a comprehensive survey. Web Semant. **36**, 1–22 (2016)

22. Rushton, A., Croucher, P., Baker, P.: The Handbook of Logistics and Distribution Management: Understanding the Supply Chain. Kogan Page Publishers (2014)

23. Vavpetic, A., Podpecan, V., Lavrac, N.: Semantic subgroup explanations. J. Intell. Inf. Syst. **42**(2), 233–254 (2014)

24. Wilcke, X., Bloem, P., de Boer, V.: The knowledge graph as the default data model for learning on heterogeneous knowledge. Data Sci. **1**, 1–19 (2017)

25. Wrobel, S.: An algorithm for multi-relational discovery of subgroups. In: Komorowski, J., Zytkow, J. (eds.) PKDD 1997. LNCS, vol. 1263, pp. 78–87. Springer, Heidelberg (1997). https://doi.org/10.1007/3-540-63223-9_108

An Intra-algorithm Comparison Study of Complete Search FSM Implementations in Centralized Graph Transaction Databases

Rihab Ayed[1]([✉])[ID], Mohand-Saïd Hacid[1], Rafiqul Haque[1], and Abderrazek Jemai[2,3]

[1] Université de Lyon, CNRS, Université Lyon 1, LIRIS UMR5205, 69622 Lyon, France
{rihab.ayed,mohand-said.hacid}@univ-lyon1.fr
[2] Faculté des Sciences de Tunis, Laboratoire LIP2, Université de Tunis El Manar, Tunis, Tunisia
[3] INSAT, Université de Carthage, 1080 Carthage, Tunisia

Abstract. Frequent subgraph mining (FSM) algorithms are widely used in various areas of data analysis. Several experimental studies about FSM algorithms were reported in literature; however, these experiments lack some clarifications about the most efficient implementation of a specific algorithm for a context of use (*e.g.,* medium size datasets). In this paper, we present an experimental study with available implementations of two well known complete search FSM algorithms namely gSpan and Gaston. Our main purpose of this experimental study is to find a suitable Frequent Subgraph Mining implementation for indexing centralized graphs databases for aggregated search(CAIR home page: www.irit.fr/CAIR). In this paper, we provide details of the experimental results according to the input variation cases. We propose (for end users) a summary, about the most efficient FSM implementations for each algorithm (*i.e.,* gSpan and Gaston), based on real datasets from the literature.

Keywords: Frequent subgraph mining · FSM algorithms
Performance

1 Introduction

Frequent Subgraph Mining (FSM) is a branch of Graph Mining [2] which allows to extract subgraphs, in a given graph database, whose occurrence counts (*aka* frequency) are above a specified threshold (named minimum support threshold). The extracted subgraphs are called frequent subgraphs. Complete search FSM (*vs.* Incomplete search) guarantees to find all frequent subgraphs of the

---

This research is performed within the scope of the CAIR project - funded by ANR (Agence Nationale de la Recherche) - www.irit.fr/CAIR.

© Springer Nature Switzerland AG 2018
M. Ceci et al. (Eds.): ISMIS 2018, LNAI 11177, pp. 78–88, 2018.
https://doi.org/10.1007/978-3-030-01851-1_8

input graphs, above a minimum frequency threshold [3]. For most of the algorithms in the literature, the input graphs are assumed to be *labeled* (vertices and edges), *simple, connected and undirected.* The output is assumed to be *connected* subgraphs. In this paper, we consider only algorithms that perform a complete search in a graph-transaction database setting (*vs.* Single graph setting) [2].

Several studies (see, *e.g.,* [1,7,10]) were devoted to benchmarking FSM algorithms. Our investigation of these studies targets the following shortcomings: (i) the conclusions about algorithms do not explicitly consider the effects of input parameters' variability on performance (*e.g.,* characteristics of datasets, low or high support); (ii) two different implementations of a given algorithm reported different performance results [10] which leads to questioning the choice of an implementation among others; (iii) the implementations of some algorithms are refined without any experimental study of their performances (*e.g.,* release v.6 2009 of gSpan 2002 [12]). The mentioned shortcomings lead to a confusion for determining the most efficient FSM implementations of existing algorithms. In fact, the variability of tested FSM implementations, datasets and support threshold values were not characterized adequately in order to have a complete view of the FSM solutions performance. In the framework of the CAIR[1] project, we conducted a preliminary work of experimenting FSM implementations for the most tested FSM algorithms in literature (*i.e.,* gSpan [12] and Gaston [6]).

1.1 Contributions

Our main purpose in this paper is to provide a performance study of existing FSM implementations for a specified algorithm. We selected two algorithms gSpan [12] and Gaston [6] and justify this choice due to their multiple implementations and high usage in literature (see Sect. 2). Our study is dedicated mainly to *end users* who would choose an as-is existing implementation of an FSM algorithm. Following this, we used existing implementations without making any changes. We conducted an experimental study by using eleven real datasets collected from the literature. We included the most tested ones. We analyzed the performance of the FSM solutions according to the following parameters: (i) the number of returned frequent subgraphs, (ii) execution time and (iii) memory size (consumption). We analyzed performance by varying two input parameters: datasets and minimum support threshold.

The remainder of this paper is organized as follows: Sect. 2 describes the selection of FSM algorithms. Section 3 describes the evaluation of the selected FSM algorithms' implementations and discusses the results. We conclude in Sect. 4.

---

[1] Contextual and Aggregated Information Retrieval: www.irit.fr/CAIR.

2 Selection of Candidate FSM Algorithms: gSpan and Gaston

We identified thirty-two algorithms[2] (in the literature) designed to perform complete search FSM. We investigated the usage of these algorithms. We define the usage of an algorithm in accordance with three facets: (i) the number of experiments performed with the algorithm for centralized graph transaction datasets (column E, Table 1), (ii) the number of real datasets used (column D, see Table 1), and (iii) the year of the most recent experiment (column R, see Table 1).

Table 1. The usage of Centralized FSM algorithms (Complete Search)

Algorithm (year)	E	D	R	Algorithm (year)	E	D	R
gSpan (2002)	25	25	2015	ADI-Mine (2004)	2	3	2006
Gaston (2004)	11	14	2014	FSMA (2008)	2	0	2011
FSG (2001)	9	11	2014	MOLFEA (2001)	2	2	2005
FFSM (2003)	5	10	2014	WARMR (1998)	2	1	2004
AcGM (2002)	4	3	2014	LC-Mine (2014)	1	10	-
MoFa (2002)	3	6	2009	The remaining 20 algorithms	1	< 5	-
FSP (2007)	2	3	2014				

We found that eleven out of the thirty two algorithms are tested more than once. Table 1 shows that the most tested algorithms in the literature are: gSpan [12], Gaston [6], FSG [3]. In fact, we noticed that the most recent algorithms tend to be compared only with the oldest algorithms (*e.g.,* gSpan, FSG, Gaston). Besides, gSpan, Gaston and MoFa/MoSS algorithms have multiple available implementations. The rest of algorithms have one or no available implementation. These observations led us to focus on the implementations of each of the two most tested algorithms in literature: gSpan and Gaston.

3 Experimental Study with gSpan and Gaston

In this section, we compare the performance results of different implementations of one algorithm (gSpan or Gaston). It is worth noting that only some of our results (*i.e.,* some support values and datasets) are reported in this paper. However, the drawn conclusions about implementations are based on all tested datasets and support threshold values (generally from 1% to 90%).[3] Also, we do not define the working principals of the algorithms in this paper. Please refer to [5,12] for theoretical background.

---

[2] Please refer to liris.cnrs.fr/rihab.ayed/ESFSM.pdf for references of algorithms and the list of all implementations.

[3] For all results, see liris.cnrs.fr/rihab.ayed/DFSM.pdf.

3.1 Experimental Setup

In this section, we provide the description of our experimental setting.

Datasets. We conducted our experiments with eleven available and real datasets (out of thirty one in literature). Table 2 shows the characteristics of the eleven datasets[4] where $|D|$ is the number of graphs in the dataset, $|Tv|$ and $|Te|$ are the average size of a graph by counting the average number of vertices and edges in the dataset. We categorized datasets into small, medium, large and dense (densely connected) datasets.

Table 2. Selected available Datasets used in the Literature

Size	Small datasets		Medium datasets				
	PTE	HIV-CA	NCI145	NCI330	CAN2DA99	AIDS	AID2DA99
$|D|$	340	422	19 553	23 050	32 553	56 213	42 682
$|Tv|, |Te|$	27 , 27	40 , 42	30, 32	25 , 27	26 , 28	28 , 30	26 , 28
Size	Large datasets				Dense datasets		
	NCI250		DS3M		PS		DD
$|D|$	250 251		273 324		90		1 178
$|Tv|, |Te|$	21 , 23		22 , 24		67 , 256		284 , 716

Minimum Support Threshold (MST). We noticed that different implementations of FSM algorithms convert differently the minimum support threshold (relative value) into the internal minimum frequency (absolute value). In this paper, we compared the implementations with the same strategy of conversion.(see Footnote 4)

Resources Used. All of our experiments were performed using a machine with 4 GB of RAM memory and a Quad core processor (Intel Quad Core i3 2.40 GHz). We used Linux OS for deploying all FSM solutions.

Evaluation Metrics. In order to evaluate the performance of the implementations, we use the three metrics commonly used in the literature: (i) the execution time, (ii) the memory size (consumption), and (iii) the number of returned frequent subgraphs. The solutions are compared with each other by considering one of the three metrics.

---

[4] For dataset links or MST strategies, see liris.cnrs.fr/rihab.ayed/ESFSM.pdf.

3.2   gSpan Implementations Study

There are six available implementations of gSpan (see Table 3). We tested four implementations: Two original implementations provided by authors of gSpan [12] (gSpan Original v.6, gSpan 64-bit Original v.6) and third-party implementations (gSpan ParMol, gSpan^Z). The two others were removed from the comparison due to technical shortcomings. In fact, gSpan ParSeMis displayed some drawbacks related to the quality of frequent subgraphs mined (redundancy) and errors during the execution, and gSpan (Kudo) required an additional software (*i.e.*, Matlab). Note that the two original implementations of gSpan are provided with no specification of the difference between them.

Table 3. Available Implementations of gSpan algorithm

Available Implementations	Last Release	Abbr.
gSpan [12]		
gSpan Original v.6 [11]	2009	SO
gSpan Original 64-bit v.6 [11]	2009	SO64
gSpan ParMol [10]	2013	SP
gSpan (Zhou)^a: gSpan^Z	2015	SK
gSpan (Kudo) [8]	2004	-
gSpan ParSeMis [9]	2011	-

^{a} No theoretical work. github.com/Jokeren/DataMining-gSpan, [Access 2018-05-30].

Abbreviations of implementations (column Abbr. in Table 3) will be used in the presentation of the experimental results (Tables 4, 5 and 6). It is worth noting that gSpan^Z is able to run with only small datasets (*e.g.*, HIV-CA or PTE). Additionally, it produces a number of frequent subgraphs considerably different from the other gSpan versions. For example, for HIV-CA dataset and MST set to 5%, gSpan^Z generates 723 603 frequent subgraphs compared to 905 298 generated by gSpan Original solutions and 905 299 generated by gSpan ParMol. Besides, gSpan^Z requires considerably more memory than gSpan ParMol with less number of frequent subgraphs. For example, for HIV-CA dataset with MST 5%, gSpan^Z consumes 87 GB to produce 723 603 frequent subgraphs. However, gSpan ParMol consumes 2.01 GB to produce 905 299 subgraphs in the same case. In the following, we compare gSpan Original solutions with the third-party implementation (namely gSpan ParMol).

Number of Frequent Subgraphs. The two original versions of gSpan (v.6 and 64-bit v.6) generate the same number of frequent subgraphs except for two values of support (6% and 8%) for the NCI330. The number of subgraphs produced by gSpan ParMol and gSpan Original v.6 are almost the same, except for some low support thresholds. For these low support thresholds, the difference varies between 1142 and 1 frequent subgraphs. Table 4 shows the cases of difference (denoted by *Diff*) between gSpan ParMol and gSpan Original v.6 with their respective support interval (*e.g.*, 5% - 20%) or support values (*e.g.*, 4% , 6%).

Table 4. Number of Frequent Subgraphs (gSpan Original *vs.* gSpan ParMol)

Support	Comp	Diff	Support	Comp	Diff	Support	Comp	Diff
Small Datasets			Medium Datasets			Large Datasets		
HIV-CA			AIDS, NCI145			NCI250		
5%–20%	SO > SP	1 - 1	1.5% –2%	SO < SP	2 - 1	3%	SO > SP	931
PTE			NCI330			DS3M		
1.5%–3%	SO < SP	53 - 1	4%, 6%	SO > SP	1142, 15	5%, 20%	SO < SP	1, 1

Please note that different implementations of gSpan provided by Original authors and by third-party implementers (ParMol) can compute the frequent subgraphs differently. For example, for NCI330 dataset with 6% MST, the two implementations of gSpan generate 4 subgraphs with close frequency values.

Memory Size. GSpan Original solutions are provided as binary codes and no output about memory calculation is provided by authors. However, some conclusions can be made in case of low support thresholds. GSpan Original v.6 is the only implementation among gSpan versions that is able to reach the lowest[5] minimum support threshold values for all datasets (*e.g.*, 4% for the HIV-CA dataset, see Table 5). For low support thresholds, gSpan-64bit Original v.6 requires significantly more memory than gSpan Original v.6 and gSpan ParMol. GSpan-64bit Original is not able to run with very low support threshold values. For example, for the HIV-CA dataset gSpan Original-64bit reaches an MST equal to 8%, while gSpan Original reaches 4% (see Table 5).

Table 5. Minimal Support threshold value reached by gSpan versions

Dataset	Small		Dense	Medium		Large
	PTE	*HIV-CA*	*DD*	*CAN2DA99*	*NCI330*	*NCI250*
SP	1.5%	5%	4%	2%	4%	2%
SO	1.5%	4%	1.5%	1.5%	3.5%	2%
SO64	3%	8%	20%	3%	5%	4%

Runtime. Our experiments show that gSpan-64bit Original v.6 is faster than gSpan Original v.6 for all the tested datasets (see differences in seconds denoted

---

[5] The lowest MST we tested in Table 5 is 1.5%.

by *Diff*, Table 6). For example, for AIDS dataset, and MST set to 2%, gSpan Original v.6 consumes about 1008 seconds more than gSpan Original v.6 64-bit. However, for some low support threshold values (*e.g.*, 8% - 15% for HIV-CA, see Table 6), gSpan-64 bit Original v.6 is slower than gSpan Original v.6 due to a higher memory consumption. Furthermore, our experiments reveal that gSpan ParMol is faster than gSpan Original v.6 for small and medium datasets (with some exceptions for low support values, *e.g.*, 2% for NCI145). For example, gSpan ParMol consumes about 850 seconds less than gSpan Original for AID2DA99 and MST 1.5% for the same number of frequent subgraphs. For large and dense datasets and with low and medium support values, gSpan ParMol is slower than gSpan Original v.6. For medium and small datasets and low support values, gSpan ParMol is faster than gSpan-64bit Original, except for the NCI330 dataset.

Table 6. Runtime Comparison (gSpan Original versions)

Support	Comp	Diff (sec)	Support	Comp	Diff (sec)
Small & Dense Datasets			**Medium & Large Datasets**		
HIV-CA			**AIDS**		
8%	SO < SO64	4	2% - 90%	SO > SO64	1008 - 6.7
9% - 90%	$SO \approx SO64$	-	**NCI250**		
DD			4% - 90%	SO > SO64	200 - 24
20% - 30%	$SO \approx SO64$	-			
40% - 90%	SO > SO64	2			

In the following, we present our experimental results for Gaston Implementations.

3.3   Gaston Implementations Study

There are four available implementations of Gaston (see Table 7). We removed one of them, Gaston ParSeMis, due to technical shortcomings (errors during the execution). We tested three (as-is) implementations: two (Gaston Original v1.1, RE v1.1) are from original authors [6] and a third-party implementation (Gaston ParMol [10]). Gaston Original RE was proposed in order to reduce the memory of Gaston Original. Abbreviations of implementations (column Abbr. in Table 7) will be used in the presentation of the experimental results (Tables 8 and 9).

Table 7. Available Implementations of Gaston algorithm

Available Implementations	Last Release	Abbr.
Gaston Original v1.1 [4]	2005	GO
Gaston Original RE v1.1 [4]	2005	GR
Gaston ParMol [10]	2013	GP
Gaston ParSeMis [9]	2011	-

Number of Frequent Subgraphs. Typically, Gaston Original versions (v1.1, RE v1.1) generate the same number of frequent subgraphs, except for the two dense datasets DD dataset with MST under 20% and PS dataset with MST under 80%. For example, for the 2% MST of DD dataset, Gaston Original v1.1 produces 1359 more frequent subgraphs than Gaston v1.1 RE.

Table 8. Number of Frequent Subgraphs (Gaston ParMol *vs.* Gaston Original)

Support	Comp	Diff	Support	Comp	Diff
Small Datasets			Medium Datasets		
PTE			NCI330		
1.47% - 2.94%	GP < GO	4324 - 10	4% - 90%	GP > GO	90 - 1
3%	GP = GO	-	Dense Datasets		
4% - 90%	GP > GO	16 - 1	PS		
			80% - 90%	GP > GO	7399 - 8

Gaston Original (v1.1, RE v1.1) and Gaston ParMol produce different numbers of frequent subgraphs for all datasets (see Table 8). Also, it is interesting to notice that Gaston ParMol produces a different number of frequent subgraphs than the other implementations of the same framework (ParMol: gSpan, MoFa and FFSM) for the cases of low support thresholds. This variating difference (positive and negative) requires further explanations by their authors.[6]

Memory Size. We found that Gaston Original RE has a linear memory consumption, for the cases where it was able to run. Also, it consumes less memory than Gaston Original for almost all the tested datasets (except for dense DD and PS datasets, see Table 9). However, for low support threshold values (*e.g.*, 3% for NCI330, see Table 9) or large datasets (*e.g.*, DS3M), Gaston Original RE fails, in contrast to Gaston Original which completes successfully with lower MST values. Typically, Gaston ParMol consumes more memory than Gaston Original versions for all datasets and with a different number of frequent subgraphs. For example, for AID2DA99 and MST set to 2%, Gaston ParMol consumes about three times the memory of Gaston Original v1.1 with 9 more frequent subgraphs.

Table 9. Limits of Memory Consumption (KB) of Gaston for low support threshold

Version	GO		GR	GP
Min Sup	1%	3%	3.5%	4%
NCI330	238 512	-	46 072	676 437
Min Sup	2%		90%	50%
DS3M	3067 400		-	-
Min Sup	1%	1.5%	2%	3.5%
DD	66 944	-	2744 180	2114 479

---

[6] We already contacted the authors about this issue.

Runtime. The results show that for all tested datasets, Gaston Original v1.1 is the fastest among all Gaston versions. It is worth noting that Gaston Original RE v1.1 consumes less memory than Gaston Original v1.1, as a trade-off, it is slower (*e.g.,* AID2DA99, see Fig. 1). For the dense datasets (DD and PS), Gaston Original RE requires more time and memory than Gaston Original with a different number of frequent subgraphs.

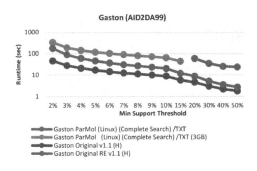

Fig. 1. Gaston Runtime - AID2DA99

3.4   Discussion

The experimental study we conducted allowed the clarification of the performance cases of FSM implementations according to the input variables. In the following, we briefly summarize these cases, based on the datasets of the literature. For low support threshold values, *gSpan Original v.6* consumes the least of memory among all gSpan solutions. GSpan-64bit Original v.6 can be used in a context where execution time is critical and the support threshold values are medium or high. The open source implementation gSpan ParMol can be used for better runtime performance for the cases of small or medium and not dense datasets with not too low support values. *Gaston Original v1.1* is the fastest among all Gaston solutions. Gaston Original RE v1.1 is suitable for memory bound systems, for the following cases: (i) with medium and high support threshold values and (ii) small or medium and not dense datasets. Otherwise, Gaston Original v1.1 should be used. Gaston ParMol consumes the highest amount of memory amongst all Gaston versions (except for small datasets and high support values) and it produces a number of frequent subgraphs different from what Gaston Original versions produce. This last implementation should be further investigated before any use. In this work, we do not discuss the inter-algorithm results.[7]

---

[7] Please refer to liris.cnrs.fr/rihab.ayed/ESFSM.pdf for more information.

4   Conclusion

In this paper, we presented an intra-algorithm study for the implementations of two well known complete search FSM algorithms: gSpan [12] and Gaston [6]. We experimented these algorithms in the case of centralized graph transaction databases. Our selection of these two algorithms comes from the fact that they are widely used and have several implementations. we studied the cases of performance according to input variation. Such a comparison is a preliminary work for assisting end-users (*e.g.,* biochemists) regarding the selection of an FSM implementation for their specific context of use. Future works should target the study and understanding of the difference between results especially for the number of frequent subgraphs. Some authors (of implementations) we contacted justify the differences by referring to their filtering strategies. Authors should inform users about the strategies they integrate with the original algorithm. This study was carried out only on datasets available in the literature in order to relate our results with those of the state of the art. We plan to conduct another study using the most efficient implementations over generic, larger and more diverse real datasets.

Acknowledgments. This work has been elaborated as a part of the CAIR project. Special thanks are addressed to FSM authors especially to Xifeng Yan, Thorsten Meinl, Andrés Gago-Alonso, Christian Borgelt, Mohammad Al Hasan and Sabeur Aridhi for sending us software, datasets, also for providing clarifications and for their availability.

References

1. Gago Alonso, A., Medina Pagola, J.E., Carrasco-Ochoa, J.A., Martínez-Trinidad, J.F.: Mining frequent connected subgraphs reducing the number of candidates. In: Daelemans, W., Goethals, B., Morik, K. (eds.) ECML PKDD 2008. LNCS (LNAI), vol. 5211, pp. 365–376. Springer, Heidelberg (2008). https://doi.org/10.1007/978-3-540-87479-9_42
2. Inokuchi, A., Washio, T., Motoda, H.: An Apriori-based Algorithm for Mining Frequent Substructures from Graph Data, pp. 13–23 (2000)
3. Kuramochi, M., Karypis, G.: Frequent subgraph discovery. In: Proceedings of the 2001 IEEE International Conference on Data Mining, ICDM 2001, pp. 313–320. IEEE Computer Society, Washington, DC (2001)
4. Nijssen, S.: Gaston. http://liacs.leidenuniv.nl/~nijssensgr/gaston/download.html. Accessed 30 May 2016
5. Nijssen, S.: Performance comparison of graph mining algorithms on PTE (2003). http://liacs.leidenuniv.nl/~nijssensgr/farmer/results.html. Accessed 30 May 2016
6. Nijssen, S., Kok, J.N.: A Quickstart in frequent structure mining can make a difference. In: Proceedings of the Tenth ACM SIGKDD International Conference on Knowledge Discovery and Data Mining, KDD 2004, pp. 647–652. ACM, New York (2004). https://doi.org/10.1145/1014052.1014134
7. Nijssen, S., Kok, J.N.: Frequent subgraph miners: Runtime dont say everything. In: Proceedings of the International Workshop on Mining and Learning with Graphs, MLG 2006, pp. 173–180 (2006)

8. Nowozin, S.: gboost graph boosting toolbox for matlab. http://www.nowozin.net/sebastian/gboost/#intro. Accessed 30 May 2016

9. Philippsen, M.: Parsemis. https://www2.cs.fau.de/EN/research/zold/ParSeMiS/index.html. Accessed 30 May 2016

10. Wörlein, M., Meinl, T., Fischer, I., Philippsen, M.: A Quantitative comparison of the subgraph miners MoFa, gSpan, FFSM, and Gaston. In: Jorge, A.M., Torgo, L., Brazdil, P., Camacho, R., Gama, J. (eds.) PKDD 2005. LNCS (LNAI), vol. 3721, pp. 392–403. Springer, Heidelberg (2005). https://doi.org/10.1007/11564126_39

11. Yan, X.: Software - gspan: frequent graph mining package. http://www.cs.ucsb.edu/~xyan/software/gSpan.htm. Accessed 30 May 2016

12. Yan, X., Han, J.: gSpan: graph-based substructure pattern mining. In: 2002 IEEE International Conference on Data Mining, ICDM 2003, Proceedings, pp. 721–724 (2002). https://doi.org/10.1109/ICDM.2002.1184038

Critical Link Identification Based on Bridge Detection for Network with Uncertain Connectivity

Kazumi Saito[1,2], Kouzou Ohara[3(✉)], Masahiro Kimura[4],
and Hiroshi Motoda[5,6]

[1] Faculty of Science, Kanagawa University, Hiratsuka, Japan
k-saito@kanagawa-u.ac.jp
[2] Center for Advanced Intelligence Project, RIKEN, Tokyo, Japan
kazumi.saito@riken.jp
[3] College of Science and Engineering, Aoyama Gakuin University, Sagamihara, Japan
ohara@it.aoyama.ac.jp
[4] Department of Electronics and Informatics, Ryukoku University, Kyoto, Japan
kimura@rins.ryukoku.ac.jp
[5] Institute of Scientific and Industrial Research, Osaka University, Suita, Japan
motoda@ar.sanken.osaka-u.ac.jp
[6] School of Computing and Information Systems, University of Tasmania, Hobart,
Australia

Abstract. Efficiently identifying critical links that substantially degrade network performance if they fail to function is challenging for a large complex network. In this paper, we tackle this problem under a more realistic situation where each link is probabilistically disconnected as if a road is blocked in a natural disaster than assuming that any road is never blocked in a disaster. To solve this problem, we utilize the bridge detection technique in graph theory and efficiently identify critical links in case the node reachability is taken as the performance measure, which corresponds to the number of people who can reach at least one evacuation facility in a disaster. Using two real-world road networks, we empirically show that the proposed method is much more efficient than the other methods that are based on traditional centrality measures and the links our method detected are substantially more critical than those by the others.

1 Introduction

Identifying critical nodes and links that play an important role in a large complex network is one of the essential issues in various fields that include communication network analysis, urban design, evacuation planning, etc. Traditional centrality metrics such as degree centrality and betweenness centrality are often used to quantify the importance of individual nodes and links. However, since these metrics are based only on network topology, other factors that are more realistic, such as geodesic distance and traveling time have also been used to

© Springer Nature Switzerland AG 2018
M. Ceci et al. (Eds.): ISMIS 2018, LNAI 11177, pp. 89–99, 2018.
https://doi.org/10.1007/978-3-030-01851-1_9

assess a certain network performance metric [6–8,10,11,13]. The objective of these studies is identifying the most critical links/nodes in maintaining a desired network performance, that is, identifying those links/nodes that degrade the performance substantially if they are not functioning. For example, suppose that the desired performance is maximizing evacuation or minimizing isolation in evacuation planning. Then, the critical links to be identified are those which maximally reduce the total number of people who can reach one of the evacuation facilities when these links fail to function. This problem can be mathematically formulated as an optimization problem once a quantitative performance measure is defined when a network structure is given.

In reality, every link in a given network is not always available. In a natural disaster, some roads may suffer from serious damage and be blocked. Thus, in this paper, as another kind of realistic factor, we take into account the probability that each link is disconnected and propose a probabilistic link disconnection model. Under this problem setting, we employ, as the performance measure, the contribution value which is defined as the difference, *i.e.*, reduction, of expected number of nodes that are reachable from a given set of target nodes such as evacuation facilities in a road network when a particular link is blocked. To solve the problem of ranking the links according to the contribution values and identifying the most critical links, we focus on the bridge in graph theory and propose an algorithm to efficiently compute the contribution value of each link in a given network by incorporating the standard bridge detection algorithm. A bridge is a link in a connected component such that its deletion divides the component into two disjoint ones. In our problem setting, if one of the disjoint components does not contain any node such as an evacuation facility, the bridge has a positive contribution value and can be a candidate of critical links. Thus, we identify all bridges in a network and compute its contribution value based on the standard bridge detection algorithm whose computational complexity is $O(|\mathcal{E}|)$, where $\mathcal{E}$ is a set of links in the network. To the best of our knowledge, our approach is the first to identify critical links based on the bridge detection algorithm under a probabilistic link disconnection model.

Below we summarize our contributions in this paper. (1) We present a realistic problem of detecting critical links under a probabilistic link disconnection model with disaster evacuation and others in mind. (2) We propose an efficient method to compute the contribution values of each link based on bridge detection, whose computational complexity is $O(H \times |\mathcal{E}|)$, where H is the number of networks generated from an original one based on the probabilistic link disconnection model. (3) We experimentally demonstrate the effectiveness of our proposed method for two real-world road networks in case of disaster evacuation. The results show that our method is much faster than computing the traditional centrality measures, and the links detected by our method are substantially more critical than those by the other methods in terms of the contribution values for the same setting we used for the probabilistic link disconnection model.

The paper is organized as follows. Section 2 briefly explains the related work of this paper. Section 3 formulates the critical link detection problem based on a probabilistic link disconnection model. Section 4 presents the proposed method based on bridge detection. Section 5 reports two datasets used and the experimental results: computational efficiency and criticalness of detected links. Section 6 summarizes the main achievements and future plans.

2 Related Work

Several node-centrality measures have been used to quantify the criticalness of each node/link. These include from traditional ones such as the degree centrality and the betweenness centrality [4] to more recent ones such as PageRank centrality [2]. As noted, these centrality measures are based only on network topology. [3,6,7,9] have used these centralities to analyze spatial networks embedded in the real world from structural viewpoints. Some others have used a more problem specific performance measure to quantify the criticalness of links. For a road network, typical performance measures are based on traveling time for a link or a path in the network [8,13]. We proposed a performance measure based on the node reachability in our past work [10,11], which is close to the one we use in this paper. Our work is different from these existing studies. What we focus on in this paper is the critical link detection problem in a situation where links are probabilistically blocked.

Our proposed method is based on the standard bridge detection algorithm [14] in graph theory. The bridge itself is critical in a network since its removal breaks the connectivity of the network. Thus, bridge detection is embedded into problems encountered in various fields such as wireless sensor networks [1], while it is also utilized to improve computational efficiency of conventional centrality measures [12]. However, these studies assume that the network structure is stable. To the best of our knowledge, there is no work that uses the bridge detection technique to identify critical links under the probabilistic link disconnection model.

In [5], we addressed the contamination minimization problem in information diffusion that is closely related to the critical link detection in this paper. The aim of the problem is minimizing the spread of contamination via a network by blocking a small number of links. In this paper, we adopt the same idea in [5] and sample graphs from a given original network by deciding connectivity of each link according to the disconnection probability to estimate the expected number of nodes reachable from a set of prespecified nodes. However, the difference is that we succeeded in substantially reducing the number of sampled graphs by computing the difference of reachability sizes for both connected and disconnected cases for each link, given a sampled graph, while the method in [5] requires much larger number of samples for each link as it computes the expected value of both when it is connected and disconnected.

3 Problem Formulation

Let $G = (\mathcal{V}, \mathcal{E})$ be a given simple undirected (or bidirectional) network without self-loops, where $\mathcal{V} = \{u, v, w, \cdots\}$ and $\mathcal{E} = \{e, \cdots\}$ are sets of nodes and undirected links, respectively. We also express each link e as a pair of nodes, *i.e.*, $e = (u, v)$. In our problem setting, we assume a fixed group of nodes $\mathcal{U} \subset \mathcal{V}$ such as evacuation facilities on a spatial network. Let $\mathcal{R}(u; G)$ be the set of reachable nodes by following links from a node u over G, where note that $u \in \mathcal{R}(u; G)$. Then, we can define a set $\mathcal{R}(\mathcal{U}; G)$ of reachable nodes by following links from every node $u \in \mathcal{U}$ over G, *i.e.*, $\mathcal{R}(\mathcal{U}; G) = \bigcup_{u \in \mathcal{U}} \mathcal{R}(u; G)$.

For each link $e \in \mathcal{E}$, let x_e be a random variable expressing the link connectivity, *i.e.*, $x_e = 1$ if the link e is disconnected; otherwise $x_e = 0$, where we denote its disconnection probability by $p(x_e = 1) = p_e$. Note that in terms of disaster evacuation over a spatial network, those probabilities are assigned according to some road blockage model based on geographical properties. By using a set of random variables defined by $\mathcal{X} = \{x_e \mid e \in \mathcal{E}\}$, we can define a graph $G_\mathcal{X} = (\mathcal{V}, \mathcal{E}_\mathcal{X})$, where $\mathcal{E}_\mathcal{X} = \{e \mid e \in \mathcal{E}, x_e = 0\}$. Now, by assuming independent Bernoulli trials for all the links, we can compute the occurrence probability of each graph $G_\mathcal{X}$ by

$$p(G_\mathcal{X}) = \prod_{x_e \in \mathcal{X}} p_e^{x_e}(1 - p_e)^{1-x_e}. \tag{1}$$

For each graph $G_\mathcal{X}$, let $G_\mathcal{X}^+(e)$ and $G_\mathcal{X}^-(e)$ be graphs constructed by adding and removing a link e, respectively, *i.e.*, $G_\mathcal{X}^+(e) = (\mathcal{V}, \mathcal{E}_\mathcal{X} \cup \{e\})$ and $G_\mathcal{X}^-(e) = (\mathcal{V}, \mathcal{E}_\mathcal{X} \setminus \{e\})$. Based on $G_\mathcal{X}^+(e)$ and $G_\mathcal{X}^-(e)$, we define the following reachability contribution value of a link $e \in \mathcal{E}$ over $G_\mathcal{X}$.

$$\phi(e) = \langle |\mathcal{R}(\mathcal{U}; G_\mathcal{X}^+(e))| - |\mathcal{R}(\mathcal{U}; G_\mathcal{X}^-(e))| \rangle_{\mathcal{X} \setminus \{x_e\}} \tag{2}$$

where $\langle \cdot \rangle_{\mathcal{X} \setminus \{x_e\}}$ means an expected value taken for all the possible assignments to random variables except for x_e. Note that the expected value $\phi(e)$ can be interpreted as the expected number of nodes, *i.e.*, people, who become unable to move to one of these evacuation facilities when a link e is blocked. Here we should emphasize that by introducing some weight $\rho(v)$ for each node v such as population around the node (junction), we can straightforwardly extend our problem settings to account for population distribution.

Evidently, it is difficult to exactly compute $\phi(e)$ defined in Eq. (2) due to a large number of possible network configurations, which amounts to $2^{|\mathcal{E}|}$. Thus, we employ an approach based on Monte Carlo simulation. Let $\mathcal{H}$ be a set of integers defined by $\mathcal{H} = \{1, \cdots, H\}$. Now, we repeat simulations H times based on the probabilistic model defined in Eq. (1), and sample a set $\mathcal{G}_H$ of H graphs *i.e.*, $\mathcal{G}_H = \{G_h = (\mathcal{V}, \mathcal{E}_h) \mid h \in \mathcal{H}\}$, where $\mathcal{E}_h \subset \mathcal{E}$ is the set of non-disconnected links at the h-th simulation. Then, we can define the reachability contribution value $F(e; \mathcal{G}_H)$ of a link $e \in \mathcal{E}$ for $\mathcal{G}_H$ as follows:

$$F(e; \mathcal{G}_H) = \frac{1}{H} \sum_{h \in \mathcal{H}} (|\mathcal{R}(\mathcal{U}; G_h^+(e))| - |\mathcal{R}(\mathcal{U}; G_h^-(e))|). \tag{3}$$

Hereafter, we denote $F(e; \mathcal{G}_H)$ simply as $F_H(e)$ because the set of generated graphs $\mathcal{G}_H$ is fixed in our experiments. Evidently, when H is a sufficiently large number, $F_H(e)$ can be a close approximation to the expected reachability contribution value $\phi(e)$. In this paper, we focus on the problem of accurately and efficiently calculating $F_H(e)$ for every $e \in \mathcal{E}$.

4   Proposed Method

To develop an effective algorithm for computing contribution value $F_H(e)$ for every $e \in \mathcal{E}$, we focus on the following two facts. For a generated graph $G_h = (\mathcal{V}, \mathcal{E}_h)$, 1) each link $e \in \mathcal{E}_h$ has a positive contribution value, i.e., $|\mathcal{R}(\mathcal{U}; G_h^+(e))| - |\mathcal{R}(\mathcal{U}; G_h^-(e))| > 0$, if e is a bridge in a connected component in G_h, and only if one of the components which are separated after removing e, contains a node $u \in \mathcal{U}$; 2) each link $e \in \mathcal{E} \setminus \mathcal{E}_h$ has a positive contribution value if e connects two connected components in G_h, and only if one of these components contains a node $u \in \mathcal{U}$. Hereafter, the connected components each containing at least one node $u \in \mathcal{U}$ are referred to as PE (possible evacuation) components, while the other components as IE (impossible evacuation) components.

In order to compute all bridges of G_h, we employ the idea of standard bridge detection algorithm originally proposed by Tarjan [14], which constructs a directed, rooted tree for each connected component from an arbitrary selected node $v \in \mathcal{V}$. Here, we propose to select each evacuation node, i.e., $u \in \mathcal{U}$ as the starting node. If a connected component contains other evacuation nodes, these nodes are skipped.

Then, in order to examine whether a bridge e has a positive contribution value or not, we only need to examine by the depth-first search that the descendant part of the component after removing e is an IE component because its ascendant part is guaranteed to be a PE component.

Let $\mu(v)$ and $\lambda(v)$ be the original order and minimum order numbers assigned to each node $v \in \mathcal{V}$ during the depth-first search, just as used in the standard bridge detection algorithm, which yields $\mu(w) = \lambda(w)$ if a link (v, w) is a bridge. Further, let $\eta(v)$ be the induced connected component from descendant nodes of v in the directed, rooted tree obtained by the depth-first search. Also, let $\zeta(v)$ be the connected component containing node v. Here we denote the numbers of nodes in $\eta(v)$ and $\zeta(v)$ by $|\eta(v)|$ and $|\zeta(v)|$, respectively. Then, we can summarize our proposed algorithm as follows:

1: Initialize contribution value $F_H(e)$ as $F_H(e) \leftarrow 0$ for every $e \in \mathcal{E}$.
2: Perform the following steps for every $h \in \mathcal{H}$;
2.1: Initialize the original number as $\mu(v) \leftarrow 0$ for each node.
2.2: Compute PE components by repeatedly doing the depth-first search from a node $u \in \mathcal{U}$ s.t. $\mu(u) = 0$, and after setting $\mu(u) \leftarrow 1$ and $v \leftarrow u$, perform the following steps for each link $e = (v, w)$ obtained during the depth-first search;
2.2.1: Compute $\mu(w)$, $\lambda(w)$ and $\eta(w)$ by the depth-first search.
2.2.2: Update $F_H(e)$ as $F_H(e) \leftarrow F_H(e) + |\eta(w)|$ if the link e is a bridge, i.e., $\mu(w) = \lambda(w)$, and $\eta(w)$ is an IE component.
2.3: Compute IE components by repeatedly doing the depth-first search starting from a node $v \in \mathcal{V}$ s.t. $\mu(v) = 0$.

2.4: Perform the following steps for each link $e = (v, w) \in \mathcal{E} \setminus \mathcal{E}_h$;

2.4.1: Update $F_H(e)$ as $F_H(e) \leftarrow F_H(e) + |\zeta(w)|$ if $\zeta(v)$ and $\zeta(w)$ are PE and IE components, in this order.

2.4.2: Update $F_H(e)$ as $F_H(e) \leftarrow F_H(e) + |\zeta(v)|$ if $\zeta(v)$ and $\zeta(w)$ are IE and PE components, in this order.

3: Output $F_H(e)$ after setting $F_H(e) \leftarrow F_H(e)/H$ for every $e \in \mathcal{E}$ and terminate.

Evidently, for a given $h \in \mathcal{H}$, the computational complexity of the step 2 is $O(|\mathcal{E}|)$ which is the same as the standard bridge detection algorithm. Thus, the total computational complexity of our algorithm becomes $O(H \times |\mathcal{E}|)$. Hereafter, we also refer to each contribution value $F_H(e)$ as criticalness centrality for a link $e \in \mathcal{E}$ (*CRC* for short), and our proposed algorithm described above as the *CRC* method.

5 Experiments

Using real data of road network $G = (\mathcal{V}, \mathcal{E})$ and evacuation facilities $\mathcal{U}$, we evaluated the effectiveness of the proposed method.

5.1 Experimental Settings

We focused on two spatial-network datasets Hamamatsu and Numazu employed in the work [7], where Hamamatsu and Numazu are Japanese cities having risks of tsunami earthquake and volcanic eruption, and the set of evacuation facilities $\mathcal{U}$ are prepared in these cities. Here, the spatial road network of each city $G = (\mathcal{V}, \mathcal{E})$ was extracted from the OSM[1] data. Note that Hamamatsu is larger than Numazu. The numbers of nodes, links and evacuation facilities in Hamamatsu are 104,813, 127,648 and 432, respectively. Those in Numazu are 15,483, 19,053 and 232, respectively. The average degree of a node, *i.e.*, the average number of links attached to a node, is 3.43 for Hamamatsu and 2.46 for Numazu.

To evaluate the fundamental performance of our proposed method we set $p_e = p$ for every $e \in \mathcal{E}$. Thus, p is the only parameter for controlling the disconnection probability.

As mentioned earlier, the problem of finding critical links in a network can be related to node-centrality measures since a link between nodes with high centrality should play a critically important role in information flow. Thus, we compare the proposed CRC with the straightforward extensions of conventional node-centrality measures such as betweenness, degree and PageRank centralities. First, we extend degree centrality and define link-centrality measure DGC by $DGC(e) = deg(v)\, deg(w)$ for a link $e = (v, w) \in \mathcal{E}$, where $deg(v)$ indicates the degree of a node $v \in \mathcal{V}$. Next, we define link-centrality measure PRC from PageRank centrality by $PRC(e) = pgrk(v)\, pgrk(w)$ for a link $e = (v, w) \in \mathcal{E}$, where $pgrk(v)$ is the PageRank score of a node $v \in \mathcal{V}$ and is given by the PageRank algorithm with random jump factor 0.15. Next, by extending betweenness

---

[1] https://openstreetmap.jp/.

centrality, we define two link-centrality measures BWC and DBC in the following ways. BWC is a simple extension of betweenness centrality and is defined by $BWC(e) = bw(v)\,bw(w)$ for a link $e = (v, w) \in \mathcal{E}$, where $bw(v)$ is the betweenness of a node $v \in V$ and is given by

$$bw(v) = \sum_{x \in V} \sum_{y \in V} \{N^{sp}(x, y; v)/N^{sp}(x, y)\}\,.$$

Here, $N^{sp}(x, y)$ denotes the number of the shortest paths from node x to node y, and $N^{sp}(x, y; v)$ denotes the number of those paths passing through a node v. DBC is an extension of betweenness centrality by taking into account the evacuation facilities $\mathcal{U}$, and is defined by $DBC(e) = db(v)\,db(w)$ for a link $e = (v, w) \in \mathcal{E}$, where $db(v)$ is provided by

$$db(v) = \sum_{x \in V} (1/|\mathcal{U}(x)|) \sum_{u \in \mathcal{U}(x)} \{N^{sp}(x, u; v)/N^{sp}(x, u)\}\,.$$

Here, $\mathcal{U}(x)$ indicates the set of nodes $u \in \mathcal{U}$ such that u is the closest node from node x in $\mathcal{U}$ in terms of distance function d over network G.

5.2 Experimental Results

First, we evaluated the computational efficiency of our proposed CRC method in comparison to those for conventional centralities, *i.e.*, BWC, DBC, PRC, and DGC, by the same task of computing the average centralities for $\mathcal{G}_H$ with a setting $H = 10^3$. In our experiments, we set the disconnection probability to $p_e = p = 2^{-k}$ for every link $e \in \mathcal{E}$, and changed $k = 1$ to 9 by adding 1, *i.e.*, $0.0019 < p \le 0.5$. Figure 1 shows our experimental results [2], where Figs. 1(a) and (b) are those for Hamamatsu and Numazu networks, respectively. From Fig. 1, we can observe quite similar tendency to these two networks of different sizes, that is, the processing times for computing CRC were substantially smaller than those for DBC and PRC although they were two or three times larger than those for DGC, most efficiently computable one. DBC needs the processing steps of link traversals for finding the nearest evacuation node from each node, and PRC needs power-iterations for obtaining PageRank scores for each node. BWC is most expensive because its computational complexity approximately becomes a square order of the network size. As notable characteristics, we can observe that the processing times especially for BWC are likely to decrease when the disconnection probability p becomes large. The reason for this must be that the number of disconnected node pairs increases for large p. In short, we can claim that our proposed centrality CRC has a desirable scaling property with respect to the size of networks.

---

[2] We implemented our programs in C, and conducted our experiments on a computer system with a single thread (Xeon X5690 3.47 GHz CPUs) within a 192GB main memory capacity.

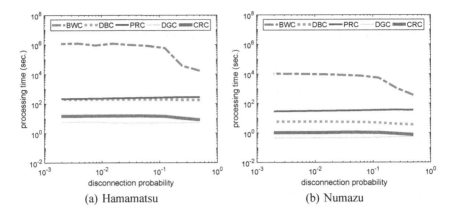

Fig. 1. Evaluation of computational efficiency as a function of the link disconnection probability.

Next, in terms of criticalness centrality, we evaluated the performance of the top-1, top-10 and top-100 links according to $F_H(e)$ for each probability setting of $p \in \{2^{-1}, \cdots, 2^{-9}\}$. Figure 2 shows our experimental results. From Fig. 2, we can also observe quite similar tendency, that is, the criticalness centralities were relatively large in the middle range of disconnection probability p, but become relatively small in the high range. This can be naturally explained by the fact that the original networks are separated to large numbers of connected components, and thus the reachability would not change by adding (or removing) a single link.

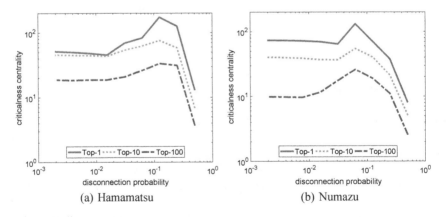

Fig. 2. Evaluation of criticalness centrality of the top-1, top-10 and top-100 links as a function of the link disconnection probability.

We further examined the performance of highly ranked links by the standard centralities, *i.e.*, *BWC*, *DBC*, *PRC*, and *DGC* in terms of criticalness centrality $F_H(e) = CRC(e)$. More specifically, we evaluated the performance defined by $F_H(e_i^{(C)})$, where C stands for one of $\{BWC, DBC, PRC, DGC, CRC\}$ and $e_i^{(C)}$ means the link with the i-th rank according to each centrality measure $C(e)$. Figure 3 shows our experimental results, where the horizontal and vertical axes stand for the rank up to top-100 and the performance evaluated by the critical-ness centrality measure $F_H(e)$, respectively, and Figs. 3(a) and (b) are those for Numazu network with $p = 2^{-4}$ and 2^{-9}, respectively. We selected these two cases because the case of $p = 2^{-4}$ produced the largest criticalness centrality measure, and the case of $p = 2^{-9}$ is the closest to the original network of Numazu. We add that quite similar experimental results were obtained in the other discon-nection probability settings for the two networks. From Fig. 3, we can see that the performance for all of the top-100 links by the standard centralities were almost zeros for all the ranges of p. In short, these experimental results suggest that our criticalness centrality has unique properties.

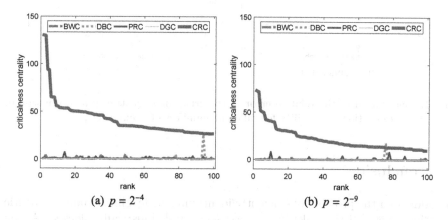

(a) $p = 2^{-4}$ (b) $p = 2^{-9}$

Fig. 3. Evaluation of criticalness centrality of the top-100 links obtained by each cen-trality measure for different link disconnection probabilities ($p = 2^{-4}$ and 2^{-9}).

Finally, we evaluated the approximation accuracy by changing H. More specifically, we computed the criticalness centrality $F_H(e)$ for each link e by setting $H = 1,000,000$, and regarded them as true ones by setting $F^*(e) \leftarrow F_H(e)$. Then, we computed each set of 100 estimation results, $\{F_H^{(i)}(e) \mid 1 \leq i \leq 100\}$, which are obtained by a series of independent Bernoulli trials for $H = 10, 10^2, 10^3$, and 10^4, and evaluated them in terms of the relative error $RE_H(e)$ defined by

$$RE_H(e) = \frac{1}{100} \sum_{i=1}^{100} \left| \frac{F_H^{(i)}(e) - F^*(e)}{F^*(e)} \right|. \tag{4}$$

Figure 4 shows our experimental results of the top-1 link according to $F^*(e)$ for each probability setting of $p \in \{2^{-1}, \cdots, 2^{-9}\}$, where Figs. 4(a) and (b) are those for Hamamatsu and Numazu networks, respectively. Here we should note that we obtained similar experimental results for the other links of lower ranks. From Fig. 4, we can observe for these two networks that the relative errors were less than 10% for any disconnection probability p even by setting $H = 10^3$. In short, these experimental results suggest that we can stably compute our criticalness centrality.

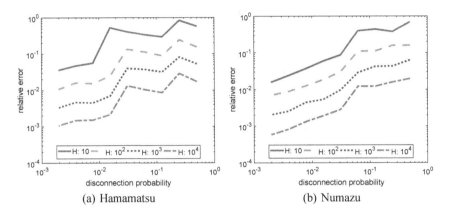

(a) Hamamatsu   (b) Numazu

Fig. 4. Fluctuation of the relative error of the criticalness centrality as a function of the link disconnection probability for different numbers of simulations.

6   Conclusion

We addressed the problem of efficiently identifying critical links under a realistic situation where all the links in a network are probabilistically blocked. A link is critical if it has a large contribution value, that is a measure of performance degradation when the link is disconnected. We formalized this situation as the probabilistic link disconnection model and proposed a novel algorithm that can efficiently identify critical links by incorporating the bridge detection technique to the search algorithm in case the node reachability is taken as the performance measure. We empirically showed through the experiments on two real-world road networks that the proposed method can identify links that are substantially more critical than those detected by the other methods based on the traditional centrality measures with much less computation time. In reality, there exist many factors that have influence on the criticalness of each link, which would depend on the problem at hand. We are planning to extend our method so as to take into account other factors such as the road capacity and investigate the usefulness of the resultant critical links more in depth.

Acknowledgments. This material is based upon work supported by JSPS Grant-in-Aid for Scientific Research (C) (No. 17K00314).

References

1. Akram, V.K., Dagdeviren, O.: Breadth-first search-based single-phase algorithms for bridge detection in wireless sensor networks. Sensors **13**(7), 8786–8813 (2013)
2. Brin, S., Page, L.: The anatomy of a large-scale hypertextual web search engine. Comput. Netw. ISDN Syst. **30**, 107–117 (1998)
3. Crucitti, P., Latora, V., Porta, S.: Centrality measures in spatial networks of urban streets. Phys. Rev. E **73**(3), 036125 (2006)
4. Freeman, L.: Centrality in social networks: conceptual clarification. Soc. Netw. **1**, 215–239 (1979)
5. Kimura, M., Saito, K., Motoda, H.: Blocking links to minimize contamination spread in a social network. ACM Trans. Knowl. Discov. Data **3**, 9:1–9:23 (2009)
6. Ohara, K., Saito, K., Kimura, M., Motoda, H.: Accelerating computation of distance based centrality measures for spatial networks. In: Calders, T., Ceci, M., Malerba, D. (eds.) DS 2016. LNCS (LNAI), vol. 9956, pp. 376–391. Springer, Cham (2016). https://doi.org/10.1007/978-3-319-46307-0_24
7. Ohara, K., Saito, K., Kimura, M., Motoda, H.: Maximizing network performance based on group centrality by creating most effective k-links. In: Proceedings of the 4th IEEE International Conference on Data Science and Advanced Analytics (DSAA 2017), pp. 561–570 (2017)
8. Oliveira, E.L., Portugal, L.S., Junior, W.P.: Determining critical links in a road network: vulnerability and congestion indicators. Procedia Soc. Behav. Sci. **162**, 158–167 (2014)
9. Opsahl, T., Agneessens, F., Skvoretz, J.: Node centrality in weighted networks: Generalizing degree and shortest paths. Social Networks **32**(3), 245–251 (2010)
10. Saito, K., Kimura, M., Ohara, K., Motoda, H.: Detecting critical links in complex network to maintain information flow/reachability. In: Proceedings of the 14th Pacific Rim International Conference on Artificial Intelligence (PRICAI2016), pp. 419–432 (2016)
11. Saito, K., Kimura, M., Ohara, K., Motoda, H.: An accurate and efficient method to detect critical links to maintain information flow in network. In: Proceedings of the 23th International Symposium on Methodologies for Intelligent Systems (ISMIS2017), pp. 116–126 (2017)
12. Sariyüce, A.E., Kaya, K., Saule, E., Çatalyürek, U.V.: Graph manipulations for fast centrality computation. ACM Trans. Knowl. Discov. Data (TKDD) **11**(3), 26 (2017)
13. Shen, Y., Nguyen, N.P., Xuan, Y., Thai, M.T.: On the discovery of critical links and nodes for assessing network vulnerability. IEEE/ACM Trans. Networking **21**(3), 963–973 (2013)
14. Tarjan, R.E.: A note on finding the bridges of a graph. Inf. Process. Lett. **2**(6), 160–161 (1974)

Image Analysis

Deep Neural Networks for Face Recognition: Pairwise Optimisation

Elitsa Popova, Athanasios Athanasopoulos, Efraim Ie, Nikolaos Christou, and Ndifreke Nyah[✉]

School of Computer Science and Technology, University of Bedfordshire, Luton, UK
ndifreke.nyah@study.beds.ac.uk

Abstract. Such factors as lighting conditions, head rotations and view angles affect the reliability of face recognition and make the recognition task difficult. Recognition of multiple subjects requires to learn class boundaries whose complexities quickly grow with the number of subjects. Artificial Neural Networks (ANNs) have provided efficient solutions, although their performances need to be improved. Multiclass and convolutional ANNs require massive computations and finding *ad-hoc* parameters in order to maximise the performance. Pairwise ANN structure has outperformed the multiclass ANNs on some face recognition tasks. We propose the pairwise optimisation for ANN, which requires a significantly smaller number of *ad-hoc* parameters and substantially fewer computations than the multiclass and convolutional networks.

Keywords: Deep neural networks · Random search optimisation
Face recognition

1 Introduction

There are many factors which affect the efficiency of face recognition algorithms. The algorithms need to deal with human faces which naturally reflect emotions and moods. Besides, human faces are often captured under variable lighting conditions, head rotations, view angles, hair styles, that makes the recognition task difficult. For extraction of biometric features from images, methods such as Principal Component Analysis (PCA), Independent Component Analysis, and Linear Discriminant Analysis have been proposed. Feasibility of using PCA for the statistical representation of facial images has been explored in [6,9,24]. The PCA has been found to be efficient for representing the recognition tasks, although the statistical projections become unreliable if images have a poor quality, [25].

Face recognition algorithms have been efficiently developed by using Bayesian methods which employ Markov chain Monte Carlo (MCMC) for integration over posterior parameters [4,17,21]. Bayesian methodology of probabilistic inference has provided highly competitive solutions, as shown in [8,16,18], whilst Markov

© Springer Nature Switzerland AG 2018
M. Ceci et al. (Eds.): ISMIS 2018, LNAI 11177, pp. 103–110, 2018.
https://doi.org/10.1007/978-3-030-01851-1_10

models have been used for designing a highly competitive solution on the Cambridge face database [14].

The above challenges of face recognition have motivated the use of Artificial Neural Networks (ANNs) with fully-connected 3-layer structures including input, hidden, and output layers [26]. Such a structure has to be predefined in order to use back-propagation learning and to minimise an error function on given training data. The minimisation of the error function is achieved by adjusting neural synaptic weights. ANNs with a predefined structure are named "Shallow" [20]. In contrast, "Deep" Neural Networks include multiple layers which generate new features [11,19,20].

Integral transformation, known as convolution, defines feature maps within so-called Convolutional Neural Networks (CNNs) [7,10]. Feature maps are typically described by 3-dimensional tensors. Each map is determined by its width, depth, and height. It has been shown in [2] that a multiclass problem can be represented by a set of 2-class discriminators which have significantly better conditions for recognition of difficult patterns. For a c-class problem, the discriminators are trained to recognise $c(c-1)/2$ pairs of classes. A Pairwise ANN proposed in [25] has significantly outperformed the multiclass ANN on the Yale face recognition benchmark.

Pairwise ANNs reduce a multiclass problem to a set of 2-class discriminators trained to distinguish samples representing pairs of classes. Such 2-class class boundaries are significantly simpler than those of the given multiclass problem [25]. According to [13], the pairwise discriminators have provided outputs within the probabilistic framework for human handwriting recognition. Employing a typical 3-layer ANN for 2-class discrimination, the pairwise architecture has outperformed the multiclass ANN. The use of pairwise ANNs has been also reported to be efficient in [27]. The pairwise architecture however has not guaranteed an improvement in the accuracy if boundaries between pairs are oversimplified.

For some face recognition problems in our experiments the pairwise discrimination has minimised the training error to zero. However validation errors have still happened. We assume that the pairwise discrimination is prone to oversimplify the boundaries between patterns which are difficult for recognition. This motivated us to explore factors which can oversimplify the decision boundaries and find a solution to this problem. Analysis of the validation errors shows that a pairwise discriminator, trained as a 2-class ANN, makes class boundaries dependent on initialisation of ANN as well as on a distribution of the image samples used for training. Both factors arbitrarily change the location of class boundary within an area where the training error is kept to be zero, whilst the validation errors can happen. We will test this hypothesis on the Yale face recognition benchmark [1], and finally we will compare our approach with alternative multiclass ANN and CNN solutions.

2   Related Work: Convolution Neural Networks

The image data received by a CNN are extracted by its convolutional layers. Pooling layers involved in CNN downsample the data in order to reduce the dimensionality of feature map, and thus to decrease processing time. Typically a pooling algorithm extracts subregions of the feature map (eg 2 × 2-pixel tiles) and finds the maximum value which will replace all other values in the map. There are dense (or fully connected) layers which perform classification on the features extracted by the convolutional layers and downsampled by the pooling layers. Nodes in the dense layer are connected to nodes in the previous layer, [10].

The overfitting problem in CNN is reduced by using the Dropout technique which defines random selection of neurons at layers. Data augmentation can also mitigate the problem if some training data will be slightly transformed so as to generate variations in the data. This technique diffuses specific patterns that can exist in the training data but can be absent in the unseen data, [7].

The back-propagation learning is typically used to adjust CNN parameters so as to minimise the error function. An example of CNN described in [12] uses a 5-layer network of convolutional and pooling (or subsampling) layers. The first layer takes a $k \times k$-image convoluted by using a 5 × 5 filter. The second layer receives a set of 4-feature maps, whose data were subsampled in a 2 × 2 window. Another example [23] describes a CNN providing a high recognition accuracy, which uses several layers of inhibitory neurons with different connections between layers, including those which generate feature maps.

3   Method

3.1   Pairwise Neural Networks

Let pairwise discrimination functions be fi/j, where $i = 1, \ldots, c - 1$, and $j = i + 1, \ldots, c$, and c is the number of classes. Figure 1 illustrates a 3-class task represented by samples shown in Green, Red, and Blue. Components x_1 and x_2 represent the input data. The given 3-class problem is transformed to a set of 2-class discriminators $f1/2$, $f1/3$ and $f2/3$. Discriminators fi/j are defined having a tansigmoid function with output y: $y = (1, -1)$. The output $y > 0$ for input x belonging to class i, and $y < 0$ for x belonging to class j. Each discriminator fi/j is trained independently on a set of training samples taken from the classes i and j.

The output layer includes neurons Σ_k where $k = 1, \ldots, c$. Their coefficients are assigned to be +1 for $k = i$ and −1 for $k = j$. The maximum $\max_{k=1}^{c}(y_k)$ determines the class to which the given input x is assigned.

Figure 2 illustrates our approach to the problem, outlined in Sect. 1, associated with oversimplification of pairwise class boundaries. Each discriminant function fi/j can be trained with zero errors as shown in Fig. 2. However the validation errors, denoted by the circles, are observed. The errors happen at random because (i) the 2-class ANNs are initialised with random weights and (ii) the validation samples are randomly distributed. Thus it is reasonable to average the outputs of discriminant functions $f_1, \ldots, f_k$ in order to reduce a side effect of the initialisation. An open question still remains: how many principal components are needed to achieve the best generalisation?

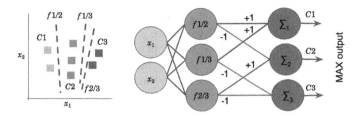

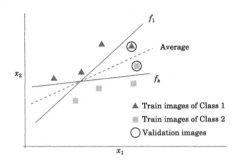

Fig. 1. Pairwise ANN for a 3-class problem.

Fig. 2. Discriminant functions $f_1, \ldots, f_k$ and the expected average class boundary.

3.2 Random Search Optimisation of Hyper-Parameters

To find a solution to the above problem, we propose an algorithm for optimisation of principal components which are used by discriminant functions having parameters Θ. The algorithm makes a random change of Θ and then evaluates the proposed change in terms of entropy calculated on the training data. This search strategy is closely related to Simulated Annealing and Metropolis-Hastings algorithms [5, 15, 22].

The proposed algorithm uses the following settings:

(1) $[X, T]$ is the training data for a pairwise discriminator fi/j, where X are a $n \times m_{max}$ matrix of m_{max} principal components for n data samples, and $\{T_k \in \{1, -1\}\}_{k=1}^{n}$ are the labels of classes i and j, respectively.
(2) m_0 is the prior on the number of principal components, assigned to be $m_0 \sim U(m_{min}, m_{max})$, where m_{min} is the minimal number of components, and U is a uniform distribution function.
(3) v_m is the proposal deviation of the number of components.
(4) E is the entropy of an ANN model with a given number of components, specified by function $f(X, T)$.
(5) K is the number of pairwise functions to be averaged.

The algorithm makes a proposal within a given prior in order to change the number of principal components required by an ANN. The proposals are drawn from a uniform distribution U. The ANN is trained with a new set of principal components to be evaluated in terms of entropy E. The proposed change is accepted according to the following rule:

(1) accept with a probability $p = 1$, if the new entropy E' is smaller than the current entropy E
(2) accept with a probability $p < 1$, if $E' > E$.

During search, proposals are made K times, and finally the Algorithm returns an ensemble of the trained ANNs.

4   Experiments

4.1   Face Recognition Benchmark

In our experiments we use the Yale Extended B database which includes 2280 images of 38 individuals, represented by 60 images, [1]. The pose and the position of the face in the image are kept constant. However, illumination conditions can significantly vary from the bright frontal illumination to the dark images, causing shadows on large part of face.

4.2   CNN Setup

For our experiments CNNs have been build with the following parameters.

(1) The CNNs have been made capable to slightly change the input data by adding Gaussian noise in order to moderate the network overfitting.
(2) The first convolution layer has been designed with 64 filters.
(3) The batch normalization has been used in order to limit the range of hidden unit outputs. Padding has enhanced the process by keeping the outputs within given boundaries. This procedure is applied at the next two layers.
(4) After the first three layers, Maximum pooling is set to be (5, 5) for the feature maps. This procedure continues for the next four layers.
(5) At the final convolutional layers, flattening parameters are added in order to transform the final feature maps into a 1-dimensional vector. This allows the fully connected layers to be applied to the convolutional layers.
(6) For optimisation Adam algorithm has been used for averaging over the gradient second moments. Root Mean Square Propagation has been used to adjust the network weights based on the average first moments.
(7) The loss function was the cross entropy.

4.3   Pairwise ANN Setup

The Random Search Algorithm described in Sect. 3.2 has been applied to the benchmark problem with the following parameters:

(i) The ensemble size has been set in the range $3 \leq K \leq 7$. The training time is linearly depended on K.

(ii) The proposal deviation has been set to range $0 \leq v_m \leq 4$: setting a small v_m limits the search area, and so can reduce ensemble diversity, whilst a greater v_m reduces the acceptance rate. However a small acceptance rate is critical for the efficiency of the search strategy [3].

(iii) The minimal and maximal numbers of principal components have been set $m_{min} = 20$ and $m_{max} = 300$, respectively.

4.4 Performance Comparison

For comparative experiments on the benchmark problem, we run the multiclass (MANN), CNN, and the proposed Pairwise (PANN). MANN has been trained by the scaled conjugate gradient back-propagation method which does not require large memory when training data are large, and the number of principal components $m_{max} = 300$. In our experiments, a MANN with the number of hidden neurons around 200 and 150 inputs has provided the maximal performance. A learning rate, which is another optimisation parameter, has been set to be 0.1.

For recognition of $c = 38$ persons, the proposed PANN has 435 ANN-based discriminant function fi/j. Each of these ANNs is trained to discriminate 2-class problems which are represented by $2n_1(k - 1/k) = 96$ samples, given the number of images per person, $n_1 = 60$, and $k = 5$-fold cross-validation. It has been found that the maximal performance is achieved with one hidden neuron in 2-class ANNs: such ANNs have learnt quickly. Table 1 shows the performances of the MANN, CNN, and proposed PANN within 5-fold cross-validation. Both the CNN and PANN significantly outperform the MANN, whilst the performances of the CNN and PANN are competitive.

Table 1. Performances of the Multi-class ANN, CNN, and proposed Pairwise ANN.

MANN,%	CNN,%	PANN,%
85.2 ± 2.5	97.2 ± 1.9	97.5 ± 2.3

5 Discussion and Conclusion

There are many factors which make face recognition extremely difficult. Multi-class ANNs which employ a conventional architecture including input, hidden, and output layers have a limited ability to solve the recognition problems. The difficulties increase with the number of persons involved in the recognition. To provide the maximal performance, multi-class ANNs require to find a proper structure including the sufficient numbers of inputs and hidden neurons. The back-propagation learning rate must be also optimised. In practice these parameters are estimated *ad-hoc*, by running massive experiments.

In this paper we analysed the potential of recent convolution networks (CNN) and implemented a typical CNN structure for experiments. We have found that settings for CNN are critical for achieving the maximal recognition performance. Massive computations are required for the optimisation. An efficient approach to solving the multi-class recognition problems is a pairwise architecture including 2-class ANNs trained to discriminate pairwise patterns. Although the number of pairwise discriminators increases exponentially with the number of classes, the network settings become significantly simpler than that for both multi-class ANN and CNNs. The decision boundary between 2 classes becomes simple, and so an ANN can quickly learn to recognise a given pair of classes.

In our comparative experiments on the Yale face recognition benchmark data, the proposed pairwise ANN has demonstrated the performance competitive with that provided by the conventional CNN. However, our approach requires a significantly less number of *ad-hoc* parameters in order to maximise the performance. It is also important that the proposed ANN does not require massive computations such as required by CNNs.

Acknowledgements. The authors would like to thank Dr Livija Jakaite, a member of the supervisory team at the School of Computer Science of University of Bedfordshire, for useful and constructive comments.

References

1. Georghiades, A.S., Belhumeur, P.N., Kriegman, D.J.: From few to many: illumination cone models for face recognition under variable lighting and pose. IEEE Trans. Pattern Anal. Mach. Intell. **23**(6), 643–660 (2001)
2. Hastie, T., Tibshirani, R.: Classification by pairwise coupling. Ann. Stat. **26**(2), 451–471 (1998)
3. Jakaite, L., Schetinin, V.: Feature selection for Bayesian evaluation of trauma death risk. In: Katashev, A., Dekhtyar, Y., Spigulis, J. (eds.) 14th Nordic-Baltic Conference on Biomedical Engineering and Medical Physics. IFMBE Proceedings, vol. 20, pp. 123–126. Springer, Heidelberg (2008). https://doi.org/10.1007/978-3-540-69367-3_33
4. Jakaite, L., Schetinin, V., Maple, C.: Bayesian assessment of newborn brain maturity from two-channel sleep electroencephalograms. Comput. Math. Methods Med. 1–7 (2012)
5. Jakaite, L., Schetinin, V., Schult, J.: Feature extraction from electroencephalograms for Bayesian assessment of newborn brain maturity. In: 2011 24th International Symposium on Computer-Based Medical Systems (CBMS), pp. 1–6 (2011)
6. Kirby, M., Sirovich, L.: Application of the karhunen-loeve procedure for the characterization of human faces. IEEE Trans. Pattern Anal. Mach. Intell. **12**(1), 103–108 (1990)
7. Krizhevsky, A., Sutskever, I., Hinton, G.E.: ImageNet classification with deep convolutional neural networks. In: Advances in Neural Information Processing Systems, pp. 1097–1105 (2012)
8. Moghaddam, B., Jebara, T., Pentland, A.: Bayesian face recognition. Pattern Recogn. **33**(11), 1771–1782 (2000)
9. Moon, H., Phillips, P.J.: Computational and performance aspects of PCA-based face-recognition algorithms. Perception **30**(3), 303–321 (2001)

10. Nebauer, C.: Evaluation of convolutional neural networks for visual recognition. IEEE Trans. Neural Netw. **9**(4), 685–696 (1998)
11. Nyah, N., Jakaite, L., Schetinin, V., Sant, P., Aggoun, A.: Learning polynomial neural networks of a near-optimal connectivity for detecting abnormal patterns in biometric data. In: 2016 SAI Computing Conference (SAI), pp. 409–413 (2016)
12. Phung, S.L., Bouzerdoum, A.: A pyramidal neural network for visual pattern recognition. IEEE Trans. Neural Netw. **18**(2), 329–343 (2007)
13. Price, D., Knerr, S., Personnaz, L., Dreyfus, G.: Pairwise neural network classifiers with probabilistic outputs. In: Neural Information Processing Systems, pp. 1109–1116 (1994)
14. Samaria, F.S., Harter, A.C.: Parameterisation of a stochastic model for human face identification. In: IEEE Workshop on Applications of Computer Vision, pp. 138–142 (1994)
15. Schetinin, V., et al.: Comparison of the Bayesian and randomized decision tree ensembles within an uncertainty envelope technique. J. Math. Model. Algorithms **5**, 397–416 (2006)
16. Schetinin, V., Jakaite, L.: Classification of newborn EEG maturity with Bayesian averaging over decision trees. Expert Syst. Appl. **39**(10), 9340–9347 (2012)
17. Schetinin, V., Jakaite, L.: Extraction of features from sleep EEG for Bayesian assessment of brain development. PLOS ONE **12**(3), 1–13 (2017)
18. Schetinin, V., Jakaite, L., Krzanowski, W.J.: Prediction of survival probabilities with Bayesian decision trees. Expert Syst. Appl. **40**(14), 5466–5476 (2013)
19. Schetinin, V., Jakaite, L., Nyah, N., Novakovic, D., Krzanowski, W.: Feature extraction with GMDH-type neural networks for EEG-based person identification. Int. J. Neural Syst. (2018)
20. Schmidhuber, J.: Deep learning in neural networks: an overview. Neural Networks **61**, 85–117 (2015)
21. Schönborn, S., Egger, B., Morel-Forster, A., Vetter, T.: Markov chain monte carlo for automated face image analysis. Int. J. Comput. Vis. **123**(2), 160–183 (2017)
22. Soares, S., Antunes, C.H., Arajo, R.: Comparison of a genetic algorithm and simulated annealing for automatic neural network ensemble development. Neurocomputing **121**, 498–511 (2013). Advances in Artificial Neural Networks
23. Tivive, F., Bouzerdoum, A.: A face detection system using shunting inhibitory convolutional neural networks. In: IEEE International Joint Conference on Neural Networks, pp. 2571–2575 (2004)
24. Turk, M.A., Pentland, A.P.: Face recognition using eigenfaces. In: IEEE Conference on Computer Vision and Pattern Recognition, pp. 586–591 (1991)
25. Uglov, J., Jakaite, L., Schetinin, V., Maple, C.: Comparing robustness of pairwise and multiclass neural-network systems for face recognition. EURASIP J. Adv. Sign. Process. (2008)
26. Valenti, R., Sebe, N., Gevers, T., Cohen, I.: Machine learning techniques for face analysis. In: Cord, M., Cunningham, P. (eds.) Machine Learning Techniques for Multimedia. Cognitive Technologies, pp. 159–187. Springer, Heidelberg (2008). https://doi.org/10.1007/978-3-540-75171-7_7
27. Yawichai, K., Kitjaidure, Y.: Multiview invariant shape recognition based on neural networks. In: 3rd IEEE Conference on Industrial Electronics and Applications, vol. 3, pp. 1538–1542 (2008)

Mobile Application with Image Recognition for Persons with Aphasia

Jan Gonera, Krzysztof Szklanny$^{(\boxtimes)}$, Marcin Wichrowski,
and Alicja Wieczorkowska

Polish-Japanese Academy of Information Technology,
Koszykowa 86, 02-008 Warsaw, Poland
{s14197,kszklanny,mati}@pjwstk.edu.pl, alicja@poljap.edu.pl

Abstract. A person with aphasia, caused by a damage to a brain, loses (partially or completely) the ability to use speech or writing. Rehabilitation is based on mental or motor exercises, stimulating the brain areas responsible for communication. The aim of our work was to implement an application (app) for smartphones that could be used for this rehabilitation. We used Google's Cloud Vision for photo analysis. The initial prototype of the app was modified according to comments from a specialist. The usability tests with the target group of users prove the effectiveness of the app, and suggest the directions of further development.

Keywords: Smartphone application · Interactive storytelling
Aphasia

1 Introduction

Our linguistic competence is expressed by producing and understanding speech; disorders of language skills result in both mental and cultural disability. Aphasia is a disorder of speech understanding and production, reading, writing and other communication skills. It is caused by a damage to the areas of the left hemisphere of the brain. Most often (75% of cases) it is caused by a stroke; also, by brain cancer, epilepsy, and Alzheimer's disease. There is no one best rehabilitation method, and each patient needs an individual approach, taking into account his or her age, education, profession and interests, to make exercises useful and interesting. It can be based on speech therapy, or on digital approach [3].

Speech therapy methods aim at the treatment of speech disorders, and shaping the proper attitude. If the patient does not believe in the effectiveness of the therapy, this can lead to his or her isolation, and apathy. Digital methods, i.e. using digital devices, can also help aphasiac persons in returning to a normal life. Contact with other people is needed [5], but digital methods can be continued by patients at home, systematically, without involvement of others.

© Springer Nature Switzerland AG 2018
M. Ceci et al. (Eds.): ISMIS 2018, LNAI 11177, pp. 111–119, 2018.
https://doi.org/10.1007/978-3-030-01851-1_11

1.1 Software for the Rehabilitation of Aphasiac Patients

There are digital tools available for the rehabilitation of patients with aphasia [1,2,8]. *PhotoTalk* focuses on image capture. *Proloquo2Go* is a symbol-based communication app available for English, French, Spanish and Dutch. *Postboard* is a communication system consisting of a pair of whiteboards connected over the Internet. *Aphasia Create* for mobile devices provides a set of icons, photos from the built-in camera, and drawing with basic colors. The user can create images to share stories and present a problematic situation, thus suggesting the therapist what type of exercises to focus on. *Storytelling* app [6] can also create short stories using pictures, colors, photos, and audio recordings. In tests, aphasiac persons preferred using pictures and short descriptions, rather than keyboard and voice recording. Speech specialists pointed out that brain damage causing aphasia often causes the paresis of the right side of the body, and persons in the early stages of the disease cannot cope with complicated or colorful interface, so its interactive elements were moved to the left, and the color range was limited.

For Polish patients, *AfaSystem* and *Multimedialna Rehabilitacja Afazji* allow practicing language skills and overcome speech defects, using a desktop computer. Exercises consist in the selection of the words describing objects in the pictures, or building words using the available letters.

Our Approach. The described digital rehabilitation methods do not allow choosing words/expressions to practice, nor provide an individual approach like speech therapy methods. Our goal was to implement the app allowing individual approach, rehabilitation through contact with others, and self-rehabilitation. This resulted in *Słówka* (*Words*) app, prepared in Polish for Polish patients, but for the purpose of this work also prepared in English.

2 Design of *Words* App

The goal of *Words* app for mobile devices is to support aphasiac persons in the practising with the most needed words and expressions, and minimizing the involvement of others. The loss of communication skills and constant need of help negatively influence self-esteem and often cause depression. *Words* app allows patients preparing a set of words that they want to practice with, and facilitates rehabilitation, combined with sharing their progress with the loved ones.

2.1 Prototype

Our app supports rehabilitation of the patient in cooperation with a speech therapist or other specialist. The user can practice at home, and return faster to normal life. The app allows creating a database of photos and their descriptions, divided into categories (Fig. 1). The list of all categories is displayed on the main screen. Pictures are taken using the smartphone's camera. The app recognizes the object in the picture, using Cloud Vision, and assigns a label. This way, an entry is created, and marked as waiting for approval by a person supervising the progress; thus, errors can be corrected. The user can browse entries, and listen to words or expressions. The progress is checked in tests, and the mastered words are moved to the archive. The database of words, images and categories can be exported to other apps (to create games, crossword puzzles etc.).

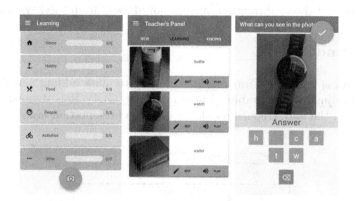

Fig. 1. Exemplary screens of *Words* app

2.2 Modification of the Project Design

The initial verification of the app indicated that adjustments are needed:

- the action button, located in the lower right corner (according to Google Material Design standards), could cause problems for persons with partial paralysis of the right side of the body, so this button was enlarged and moved to the middle of the screen, to be more easily accessible with the left hand;
- standard keyboard is used; this could pose difficulties for users with damaged Broca area, involved in speech production, so the keyboard was replaced with a set of buttons with letters from a given word, and buttons enlarged;
- *Restore to learn* button was added, as users forget mastered words, and over 50% of aphasia cases require years of rehabilitation;
- the mastered words and words for learning have been placed in 2 separate lists in 2 different tabs.

2.3 Modification of the Application Prototype

The prototype was consulted with a specialist, who suggested that:

- orange applied for emphasis should be replaced with another color, as orange is associated with warnings;
- more general categories should be used, and the categories should include typical activities, situations and places;

– all words in the same category should be listed in one list, without dividing into mastered and to practice words, as the patient may forget the word;
– *teacher* should be used instead of *caregiver*, as this suggests lack of independence due to the illness. Also, the user should be informed immediately if the answer was correct, and retrying should be allowed;
– since aphasiac persons may have motor problems, buttons in tests should be enlarged, and the confirmation button moved away from the keyboard area.

3   Implementation

Our program is based on *Clean Architecture* [7], to allow adding new functionalities and adjusting the graphical user interface (GUI). It is divided into the domain layer, the data (repository) layer, and the presentation layer, see Fig. 2.

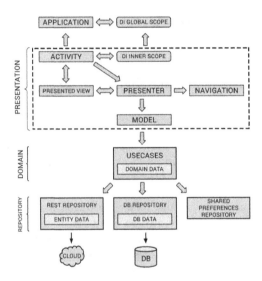

Fig. 2. The architecture of *Words* app

3.1   Architecture of the Application

The presentation layer implements displaying in Android GUI the screens:

– Category list - a list containing all available categories of words;
– Single category - a list containing words within one category;
– Camera - adding a new entry always starts with taking a picture;
– Crop - a tool for removing unnecessary areas of the photo;
– Recognition - recognizing an object in the image using Google Vision;
– Test - a list with all categories available for tests;
– End of test - a summary of the completed test in the selected category;

– Admin - lists of words to be added, in learning, and mastered;
– New entry details - description and enlarged photo, for edition by the admin.

The domain layer contains all data, the interfaces of data repositories, and the app logic. This layer also implements the following use cases: display all categories of words; save a new category; save a new photo with a description; update the photo and the description; display entries within one category, approved by the admin; display entries approved by the admin, marked as unmastered; display all entries in all categories; display all new unapproved entries; display all new entries marked as mastered; start the camera; display the current snapshot from the camera; save the photo; turn off the camera; ask to name the object in the picture; start the test in a given category; name the object in the photo in the displayed entry; display the next entry in a given category; end the test.

In the repository layer, data storage and reading is implemented:

– Camera repository - to turn on, turn off or load photos from the camera,
– Category - SQLite table with categories to which the user can assign words,
– Dictionary - SQLite table with the words added by the user,
– File repository - to save and read files from the device disk,
– Photo recognition - uses Google Vision to recognize an object in a photo.

3.2   Cloud Vision

Cloud Vision by Google is an image analysis tool, already applied in apps for users with disabilities [9]. It can recognize objects in a photo, find faces, and determine whether they are in the foreground or background. The mobile app sends the photo, the number of results expected, and a result list ordered according to decreasing accuracy is returned. Our app uses this cloud service to recognize the most important elements of the photo, to facilitate creating the dictionary. This list can be verified by the caregiver.

4   Usability Tests

In moderated usability tests, users performed tasks, and the sessions were recorded. First, the users were familiar with mobile apps on smartphones, and use them at least once a week. Next, testers were aphasiac patients and their partners. Finally, testers completed a System Usability Scale (SUS) questionnaire.

4.1   Phase 1

The testers were informed that this app is addressed to aphasiac users, and the disease and rehabilitation methods were described. Since the app was designed for both patients and their caregivers, the testers were requested to perform both roles, to find design errors. Each tester was given the following 4 tasks, requiring the use of all screens and key functionalities of the app:

1. Add 2 new words by taking pictures of 2 objects,
2. Review the set of added words, check their correctness and approve them,
3. Verify the dictionary with approved words,
4. Run and complete the exercise with words added earlier.

The first task required using the list of all categories, camera and cropping. The second task required using navigation screen, teacher panel, and new word edition. The third task used the single category screen. The fourth task required using screens of available categories, single word exercises, and summary.

All testers stated that they understood the app screens quickly, or in some cases understood but not entirely; task 2 was the most problematic.

The following problems to be addressed were discovered in Phase 1:

- After all new words were approved in the teacher's panel, "new" column became empty. An informative icon was added (no words awaiting approval);
- The tabs in the teacher's panel were marked as "I learn", "I can", referring to the student. This was reported by one tester, and was not adjusted;
- 3 users accidentally closed the app using the back button, when they wanted to return to the student screen. This bug required fixing;
- The option to change categories when approving a new word was difficult to find. Therefore, a small arrow was added to the right of the button for category selection, and a distinctive background was added to this button;
- The tasks in student screen should be expressed more friendly. Therefore, "give answer" was replaced with "name the object in the picture";
- The app did not allow adding a photo for a selected category. To allow it, an interactive icon with a brief description was added;
- The screens "I'm learning" and "Exercises" had similar titles, but different content. Only 1 tester reported this problem, so this was not changed;
- The icon for confirming photo cropping is unclear. This icon was replaced with a general approval icon.

4.2   Phase 2

These tests were conducted with 8 users with sensory or motor aphasia (Table 1). Two of them took the test in the company a relative; the use of the app in pairs was one of our assumptions. The remaining 6 users examined only student screens. Our goal was to test if the app is intuitive and easy to use, if users can learn the app quickly, even if they are not familiar with smartphone apps. We also wanted to check if the users with paresis of the right side of the body will manage to use the app, and if the division of the functionalities for the student and teacher is clear. The testers were given the following tasks:

1. Familiarize yourself with the app, check the available functions.
2. Find the menu of the app and tell us what functions are available here.

3. Add 2 new words by taking photos of 2 items.
4. Review the set of words added by the student, check their correctness and approve them (task for the relative, if the user was accompanied).
5. Review the dictionary of the approved words.
6. Run and complete the exercise with words added earlier.

Table 1. Testers in Phase 2 of usability tests.

Used ID	Age	Gender	Education	Uses smartphone apps	Aphasia	Rehabilitation
User1	36	M	Secondary	Yes	Sensory	Week
User2	24	F	Secondary	Yes	Sensory	Week
User3	70	M	Primary	No	Sensory	Month
User4	61	M	Vocational	No	Motor	Year
User5	47	F	Higher	No	Motor	6 months
User6	70	F	Higher	No	Motor	3 months
User7	80	M	Secondary	No	Motor	Month
User8	64	F	Secondary	No	Sensory	3 months

Table 2. Reactions of testers in Phase 2 - the user: 1. does not understand the app screen, 2. needed a long time to understand, or understood it after seeing other screens, 3. understood the screen but not entirely, 4. understood the screens quickly

Task	User1	User2	User3	User4	User5	User6	User7	User8
Task1	3	3	1	1	3	3	1	1
Task2	3	4	1	1	3	3	1	1
Task3	2	3	2	2	4	3	2	1
Task4	-	2	-	-	4	-	-	-
Task5	4	4	3	4	4	3	1	1
Task6	4	4	3	3	4	3	2	1

The aphasiac persons accompanied with relatives performed best (User2 and User5, see Table 2). The following problems to be addressed were discovered:

– Using touch screen and smartphone by patients over-60s, not familiar with standard icons, or Android menu (User6 needed more time).
– Confusing lack of initial content. Adding exemplary words in each category could help, and adding a button to launch the image capture screen.
– The first steps in the app pose a problem; again, initial words can help.
– *Play* button located on the right - problem for User7 (right side paresis). It should be enlarged, or the word on the screen should work as *Play* button.
– The learning and test screen differ only slightly (also learning and exercising screens are very similar, as reported in Phase 1).

The users also suggested adding instructions, and choosing a word/sentence from a list. Instructions could be helpful in the first use of the app; they should be designed in consultation with specialists.

4.3 System Usability Scale for the Application

At the end of each test session, users completed SUS questionnaire, to measure the perceived usability of the app [4]. The average score was 49.68 (F), so the app should be significantly improved. This is because User8 (with serious aphasia) had difficulties to complete SUS, and the moderator completed it, assigning the lowest scores. Without User8 the score was 56.78 (D), but it is still below average.

5 Summary and Future Work

We prepared an app for aphasiac persons (and those helping them), in consultation with a specialist, and performed usability tests. The users, not familiar with mobile apps, assessed the app as easy to use. To facilitate using the app for the first time, we plan adding instructions, a button to launch the image capture screen, and exemplary words in each category. *Play* button should be enlarged, or words work as *Play* button. Learning and test screen should be redesigned to look differently, or tests available to run for each category in the learning screen. We plan to combine all words in each category into one list, without dividing into mastered/to practice words, to allow continuous practicing with all words. Suggesting words was not available in Polish when making this app, and translation errors discouraged us from this option.

Our assumption of the parallel use by an aphasiac person and someone who helps in rehabilitation was accurate. The app seemed to be easier for users working with a partner, at least to start with. We consider connecting the users to online system, to collect data on the progress of the aphasiac users for therapists.

Acknowledgments. The authors would like to express thanks to Dr. Aleksandra Bala, a specialist in aphasiac patients, for her comments and help in experiments. This work was partially supported by the Research Center of PJAIT.

References

1. Allen, M., McGrenere, J., Purves, B.: The field evaluation of a mobile digital image communication application designed for people with aphasia. ACM Trans. Access. Comput. (TACCESS) **1**(1), 1–26 (2008)
2. Al Mahmud, A., Dijkhuis, S.Q., Blummel, L. Elberse, I.: Postboard: free-form tangible messaging for people with aphasia (and other people). In: CHI 2012, Austin, Texas, USA, pp. 1475–1480. ACM (2012)
3. Beukelman, D., Hux, K., Dietz, A., McKelvey, M., Weissling, K.: Using visual scene displays as communication support options for people with chronic, severe aphasia: a summary of AAC research and future research directions. Augment. Altern. Commun. **31**(3), 234–45 (2015)

4. Brooke, J.: SUS: a retrospective. J. Usability Stud. **8**(2), 29–40 (2013)
5. Chapey, R.: Language Intervention Strategies in Aphasia and Related Neurogenic Communication Disorders. Kluwer-Lippincott Williams & Wilkins, Philadelphia (2008)
6. Daemen, E., et al.: Designing a free style, indirect, and interactive storytelling application for people with aphasia. In: Baranauskas, C., Palanque, P., Abascal, J., Barbosa, S.D.J. (eds.) INTERACT 2007. LNCS, vol. 4662, pp. 221–234. Springer, Heidelberg (2007). https://doi.org/10.1007/978-3-540-74796-3_21
7. Martin, R.C.: Clean Code. Prentice Hall, Upper Saddle River (2009)
8. Menger, F., Morris, J., Salis, C.: Aphasia in an Internet age: wider perspectives on digital inclusion. Aphasiology **30**(2–3), 112–132 (2015)
9. Mulfari, D., Celesti, A., Fazio, M., Villari, M., Puliafito, A.: Using Google Cloud Vision in assistive technology scenarios. In: IEEE ISCC, Messina, Italy (2016)

A Comparative Study on Soft Biometric Approaches to Be Used in Retail Stores

Berardina De Carolis, Nicola Macchiarulo, and Giuseppe Palestra[⊠]

Department of Computer Science, University of Bari, Bari, Italy
berardina.decarolis@uniba.it, giuseppepalestra@gmail.com

Abstract. Soft biometric analysis aims at recognizing personal traits that provide some information about the individual. In this paper, we implemented and compared several approaches for soft biometric analysis in order to analyze humans soft biometric traits: age, gender, presence of eyeglasses and beard. Convolutional Neural Netoworks can be successfully used to understand soft biometric traits of passers-by looking at public displays and at shop windows.

Keywords: Soft biometrics · Retail · CNN

1 Introduction

Soft biometric analysis is a fast-growing research field and deep analysis already has several fields of application: security, health-care, marketing, and retail. In the retail and marketing field, biometric technology is becoming a part of daily lives [5]. In the context of the shop window and retail, the monitoring of passers-by soft biometrics is important for marketing strategies. Today face recognition technology for tracking the face of shoppers is starting to be adopted especially by fashion retailers [1,2]. This technology will help retailers to better understand who are their clients and build user's profiles based on biometric traits.

Soft biometric traits analysis aims at recognizing some characteristics that provide some information about the individual, but does not allow to sufficiently differentiate any two individuals [4] however it can also be used to aid or effect person recognition. Soft biometrics are human characteristics providing categorical information about people such as age, beard, gender, eyeglasses, ethnicity, eye/hair color, length of arms and legs, height, weight, skin/hair color, etc. In contrast to "hard" biometrics, soft biometrics provide some vague physical or behavioral information which is not necessarily permanent or distinctive. Such soft biometric traits are usually easier to capture from a distance and do not require cooperation from the subjects.

Few studies have investigated soft biometric and facial expression recognition during interaction with a shop window [5]. Moreover, none of them considers more than one or two soft biometrics traits at the same time. In this paper, we implemented and compared several approaches for soft biometric analysis in

M. Ceci et al. (Eds.): ISMIS 2018, LNAI 11177, pp. 120–129, 2018.
https://doi.org/10.1007/978-3-030-01851-1_12

order to analyze more than one humans soft biometric traits at the same time: age, gender, presence of eyeglasses and beard. Looking at results we decided to implement a system based on deep learning approaches in order to analyze several soft biometric traits of passers-by in front of a shop window. The system has been tested during preliminary experiments in a test environment.

This paper is structured as follows: Sect. 2 illustrates the software modules used for soft biometric analysis, Sect. 3 reports results, and Sect. 4 gives conclusions.

2   Soft Biometrics Analysis

This section describes software modules used for soft biometrics analysis. Soft biometric traits refer to physical and behavioral traits, such as gender, height and weight, which are not unique to a specific subject, but are useful for identification, verification, and description of human subjects. The proposed system is able to recognize age, gender, eyeglasses and beard presence. For each soft biometric trait, a specific software module has been implemented and described below.

2.1   Age and Gender Recognition

For age and gender estimation, our system has been based on the work of Levi and Hassner [8]. Their work focuses on automatic gender and age classification using deep Convolutional Neural Networks (CNNs). The authors use a CNN with three convolutional layers. This work provides age and gender classification outperforming the state-of-the-art on both tasks using unconstrained image dataset.

2.2   Eyeglasses Detection

Eyeglass detection is a two-class classification problem. For classification, we implemented and compared three different approaches: one non-learning based and two learning based. The first one is based on an algorithm described in [9] that is able to detect the presence or absence of eyeglasses by image processing elaboration. The second approach uses a Support Vector Machines (SVM) classifier (WEKA [3] implementation with default parameters and a polynomial kernel) as described in [9]. The third approach is based on deep learning classification. For Deep Learning we used TensorFlow and Keras in order to create a CNNs architecture.

The first step, shared by all methods, is the identification of the region of interest (ROI). In this case, the ROI represents a region where eyeglasses are certainly present. The ROI identified is where the eyeglass bridge is located, as shown in Fig. 1a. Using the Dlib machine-learning library [7], 68 face landmark points are locate [6]. The ROI is comprised in the rectangle of width equal to the distance between points 21 and 22 and height equal to the distance between points 21 and 28, as shown in Fig. 1b. The used dataset for training and test was the Color FERET v2 dataset [10]. For this particular task, 2712 images have been selected: 2366 people without eyeglasses and 346 people with eyeglasses.

Fig. 1. (a) Eyeglasses detection ROI; (b) ROI based on Dlib 68 face landmark.

Non-learning Approach. The non-learning approach is based on the study of [9]. After that ROI is detected as descripted above, preprocessing steps are necessary. Firstly, the Gaussian filter is applied to slightly blur the image and reduce noise. Subsequently, the image is converted to gray-scale and the Sobel operator is applied to detect the contours. For each pixel of coordinates (x, y) of the image obtained after preprocessing, the magnitude and orientation of the gradient is calculated, according to the formula:

$$G = \sqrt{G_x^2 + G_y^2} \tag{1}$$

where G_x and G_y are the horizontal and vertical derivatives of the pixel (x, y). Finally, if the number of pixels with a vertical direction (calculated according to formula 2) exceeds a predetermined threshold, then the presence of eyeglasses is detected. The threshold was set at 3% of the total number of pixels present in the region of interest.

$$\theta = arctan\left(\frac{G_y}{G_x}\right) \tag{2}$$

In Fig. 2, the pipeline of the non-learning approach is depicted. Two execution examples of the pipeline are given. The first one the eyeglasses are present in the second one the eyeglasses are not present.

SVM. Starting from the idea described in [9], the training of the SVM classifier is performed using the ROI identified in the images of the dataset. First, as described above the ROI is detected and it is resized to 32×48 pixels. Then, 1720 features are extracted from each image. Local Binary Pattern (LBP) is used to extract 1180 features and Histogram of Oriented Gradient (HOG) is used to

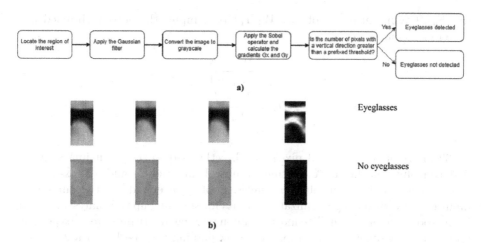

Fig. 2. Non-learning approach: (a) eyeglasses detection pipeline; (b) application to dataset images with and without eyeglasses.

extract the remaining 540 features. For this purpose, we used the SVM implementation of the WEKA [3] (Sequential Minimal Optimization) with default parameters and polynomial kernel as they yield the best results. The pipeline of the proposed approach is depicted in Fig. 3.

Fig. 3. Eyeglasses detection pipeline of the SVM approach.

CNNs. Deep learning algorithms are machine learning algorithms that use neural networks. In particular, in this case, convolutional neural networks (CNNs) were used. CNNs assume that the received inputs are images, each of them expressed in 3 dimensions [width, height, channels]. The proposed CNN architecture, depicted in Fig. 5, is based on reduced number of layers to reduce overfitting problem that occurring when training data are small, as in this case. The inputs of the network were ROI images, detected and resized as above described. The dimension of each input was $32 \times 48 \times 3$. This dimension is used to define the first layer of the network, which is the input layer. Then, two convolutional layers were defined as follows. Convolutional layer: divides the image into various overlapping fragments, which are then analyzed to identify their characteristics. The parameters of this layer are the number of filters (F), the size of the kernel (K_x, K_y), and the displacement factor along the width and height of the matrix

(S_x, S_y). Giving an image of size $[W_1 H_1 D_1]$ as input, the output obtained will have dimensions:

$$W_2 = \frac{W_1 - K_x}{S_x + 1} \tag{3}$$

$$H_2 = \frac{H_1 - K_y}{S_y + 1} \tag{4}$$

$$D_2 = F \tag{5}$$

32 filters of size $3 \times 3 \times 3$ pixels are directly used to the input image in the first convolutional layer. A Rectified Linear Unit (ReLU) and a max-pooling layer follow this first convolutional layer. Pooling layer reduces the number of input received through generalization, useful for speeding up the analysis without losing too much precision. The most common pooling algorithms are max-pooling and average-pooling. In this approach, max-pooling was used with a 2×2 size kernel in order to select the highest value among 4 neighboring elements of the input matrix, as shown in Fig. 4.

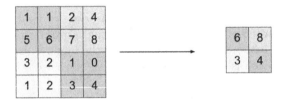

Fig. 4. Eyeglasses detection pipeline of the SVM approach.

The second convolutional layer processes the output of the previous layer. The second layer contained 64 filters of size $3 \times 3 \times 3$. As the first one, ReLU and max pooling layer defined with the same parameters as before follow it. A first fully connected layer works on the received output of the second convolutional layer and contains 64 neurons. A ReLU and a dropout layer follow it. A last fully connected layer maps to the final classes (with or without eyeglasses).

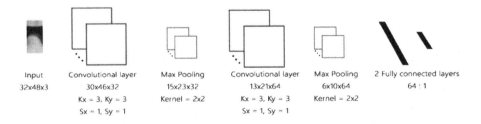

Fig. 5. Proposed eyeglasses detection CNN architecture.

2.3   Beard Detection

Beard detection is a two-class classification problem. As in the eyeglasses detection we implemented and compared three different approaches non learning based, SVM classifier, and CNN. In the case of the beard detection, the ROI to detect the beard is based on the rectangle of width equal to the distance of points 7 and 9 and height equal to the distance between points 6 and 7, as shown in Fig. 6. This region has been chosen because it is assumed that if the beard is present on a person's face, it will surely be present on this region. The used dataset for training and test was the Color FERET v2 dataset. For beard detection task, 2712 images have been selected: 2554 people without the beard and 158 bearded people. Deep learning approach works well and it reaches better accuracy performance.

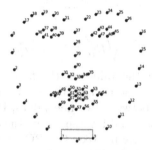

Fig. 6. Beard detection ROI based on Dlib 68 face landmark.

Non-learning Approach. Non-Learning approach for beard detection is based on the assumption that there are regions of the face with beard and there are regions without beard. Four ROIs of the face were selected and they are depicted in Fig. 7. ROIs without beard are located below eyes and on the forehead (1, 2, and 3), while bearded ROI is located in the area of the chin (part 4). Image processing function converts ROIs (1, 2, 3, and 4) to HSV color space. The proposed approach is based on the comparison of the ROI (4) with the without beard ROIs (1, 2 and 3), where the skin of the person is present.

For each region, k-means is performed to obtain a triple HSV corresponding to the centroid of the largest cluster. This step is necessary to eliminate the noise in each ROI. Once these values have been obtained, the distance of HSV triples between ROI 4 and ROIs (1, 2, and 3) was calculated, according to the following formulas:

$$D_H = \frac{min(|H_1 - H_0|, 360 - |H_1 - H_0|)}{180} \tag{6}$$

$$D_S = |S_1 - S_0| \tag{7}$$

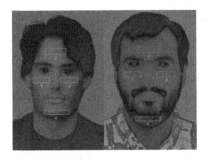

Fig. 7. Beard detection ROIs: 1, 2, 3 no beard regions; 4 beard region.

$$D_V = \frac{|V_1 - V_0|}{255} \tag{8}$$

$$D = \sqrt{D_H^2 + D_S^2 + D_V^2} \tag{9}$$

where H_0 and H_1 represent the values of the hue (H) of two chosen regions. S_0 and S_1 are the saturation values and V_0 and V_1 are brightness values (V). DH represents the distance between two hue values, DS represents the distance between two saturation values, and DV represents the distance of two brightness values. D represents the distance between two ROI in the HSV space. If at least two distances exceed a threshold (0.33 in the range 0–1), then the beard is detected.

SVM. The training of the SVM classifier is performed using the ROI (4) as described above. For each image in the dataset, the ROI (4) was extracted and resized to 64×16 pixels. Two versions of the classifier have been trained. The first one is based on the ROI in the RGB color space, while the second one is based on the ROI converted to the HSV color space. For both versions 1078 features were extracted, of which 826 correspond to the Local Binary Pattern (LBP) and 252 correspond to the Histogram of Gradient (HOG). For evaluation purpose, we used the SVM implementation of the WEKA [3] (Sequential Minimal Optimization) with default parameters and polynomial kernel as they yield the best results. The pipeline of the proposed approach is the same as for the eyeglasses detection.

CNNs. The CNN architecture described for the eyeglass detection task was also used for beard detection. The only change occurred in the parameters of the input layer in order to accept input images of different size $64 \times 16 \times 3$.

CNN training was performed twice. The first with regions of interest in the RGB color space. The second aimed to train the CNN using ROI in the HSV color space.

3 Results

We formulated beard and eyeglasses detection as two-class classification problem. For classification, we compared three different methods: non-learning method, Support Vector Machines, and Deep Learning. For non-learning method we used Dlib for image processing. For SVM, we used default parameters of the WEKA implementation. For Deep Learning we used TensorFlow and Keras Convolutional Neural Networks with proposed architecture in the previous sections. The training and the test of the two learning approaches and the test of the non learning approach was performed on the Color FERET v2 dataset. Frontal face pictures were extracted from it.

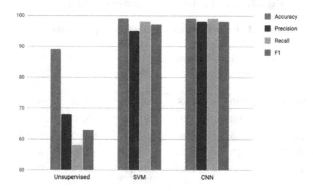

Fig. 8. Eyeglasses detection classifiers comparison.

In Table 1, we present classification results of the eyeglasses detection using non-learning method, SVM, and CNNs.

Table 1. Eyeglasses detection classifiers performance.

Approach	Accuracy	Precision NO	Precision YES	Recall NO	Recall YES	F-Measure NO	F-Measure YES
Non-Learning	89.64	92.80	67.90	95.19	58.02	63.99	62.58
SVM	99.19	99.75	95.36	99.33	98.21	99.54	96.77
CNNs	99.56	99.83	97.69	99.66	98.83	99.75	98.26

The graph in Fig. 8 depicts the performance comparison of the different classifiers for the eyeglasses detection task. The best performances are obtained from CNNs that reach an average accuracy of 99.56%.

In Table 2, we present the classification results of the beard detection using non-learning method, SVM, and CNNs.

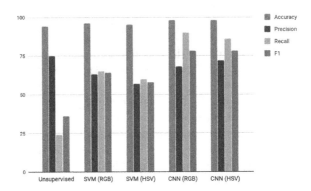

Fig. 9. Beard detection classifiers comparison.

Table 2. Beard detection classifiers performance.

Approach	Accuracy	Precision NO	Precision YES	Recall NO	Recall YES	F-Measure NO	F-Measure YES
Non-Learning	94.17	85.40	74.68	98.20	24.03	91.35	36.36
SVM (RGB)	95.87	97.93	62.66	97.70	65.13	97.87	63.87
SVM (RGB)	95.28	97.65	56.96	97.35	60.00	97.50	58.44
CNNs (RGB)	97.71	99.53	68.35	98.07	90.00	98.80	77.69
CNNs (HSV)	97.68	99.30	71.52	98.26	86.26	98.77	78.20

The graph in Fig. 9 illustrate the performance of the different classifiers for the beard detection task. The best performances are obtained by CNN classifiers. The difference in HSV or RGB images used during training did not have much impact on the final performance.

A shop window scenario was deployed in our university HCI laboratory. The system has been tested by 10 undergraduate students. In this preliminary test, the system was able to well recognize soft biometric traits (age, gender, beard and eyeglasses presence) of each participant.

4   Conclusions

In this paper, we presented an approach to soft biometrics traits analysis in which we compared several algorithms in order to recognize: gender, age, and the presence of beard and eyeglasses. Moreover, our approach considers more than one or two soft biometrics traits at the same time. CNNs can be successfully used to understand soft biometric traits of passers-by looking at public displays and at shop windows. In this way soft biometric traits can be collected and used in order to adapt advertising, recommendations or simply performing marketing analysis. In future work, we would like to analyze the whole body in order to detect non-verbal communication in front of a shop window.

References

1. De Carolis, B., Palestra, G.: Gaze-based interaction with a shop window. In: Proceedings of the International Working Conference on Advanced Visual Interfaces, pp. 304–305. ACM (2016)
2. De Carolis, B., Palestra, G.: Evaluating natural interaction with a shop window. In: Proceedings of the 12th Biannual Conference on Italian SIGCHI Chapter, p. 9. ACM (2017)
3. Hall, M., Frank, E., Holmes, G., Pfahringer, B., Reutemann, P., Witten, I.H.: The weka data mining software: an update. ACM SIGKDD Explor. Newslett. **11**(1), 10–18 (2009)
4. Jain, A.K., Dass, S.C., Nandakumar, K.: Soft biometric traits for personal recognition systems. In: Zhang, D., Jain, A.K. (eds.) ICBA 2004. LNCS, vol. 3072, pp. 731–738. Springer, Heidelberg (2004). https://doi.org/10.1007/978-3-540-25948-0_99
5. Jones, P., Williams, P., Hillier, D., Comfort, D.: Biometrics in retailing. Int. J. Retail Distrib. Manag. **35**(3), 217–222 (2007)
6. Kazemi, V., Josephine, S.: One millisecond face alignment with an ensemble of regression trees. In: 27th IEEE Conference on Computer Vision and Pattern Recognition, CVPR 2014, Columbus, United States, 23 June 2014–28 June 2014, pp. 1867–1874. IEEE Computer Society (2014)
7. King, D.E.: Dlib-ml: a machine learning toolkit. J. Mach. Learn. Res. **10**, 1755–1758 (2009)
8. Levi, G., Hassner, T.: Age and gender classification using convolutional neural networks. In: Proceedings of the IEEE Conference on Computer Vision and Pattern Recognition Workshops, pp. 34–42 (2015)
9. Mohammad, A.S., Rattani, A., Derahkshani, R.: Eyeglasses detection based on learning and non-learning based classification schemes. In: 2017 IEEE International Symposium on Technologies for Homeland Security (HST), pp. 1–5. IEEE (2017)
10. Phillips, P.J., Wechsler, H., Huang, J., Rauss, P.J.: The feret database and evaluation procedure for face-recognition algorithms. Image Vis. Comput. **16**(5), 295–306 (1998)

Low Cost Intelligent System for the 2D Biomechanical Analysis of Road Cyclists

Camilo Salguero[1], Sandra P. Mosquera[1], Andrés F. Barco[1(✉)] (iD),
and Élise Vareilles[2] (iD)

[1] LIDIS Research Group, Universidad de San Buenaventura Cali,
Santiago de Cali, Colombia
anfelbar@usbcali.edu.co
[2] Industrial Engineering Laboratory, Université de Toulouse, Mines Albi,
Albi, France

Abstract. This paper introduces an intelligent system focused on the biomechanical analysis of road bicycle cyclists. This type of analysis is carried out in specialized medical centers that operate using costly resources and are employed, mainly, for studies on athletes of high performance. The proposed system contrasts with these centers in that it provides the rookie cyclist with an accessible and affordable biomechanical analysis, although not as accurate. The architecture of the system rests in the advances in motion capture and augmented reality libraries. In the paper are discussed the motivations of the research, the internal design of the proposed system and the differences with various systems.

Keywords: Augmented reality · Biomechanical analysis
Intelligent systems · Motion capture · Road bicycle
Software architecture

1 Introduction

Biomechanics is defined as the study of the movement of living beings using the mechanics analysis, which involves forces as creators of movement [9]. The discipline studies the capture, measurement and analysis of human body movement with the aim of avoiding injuries and improve sports performance. Our research is focused on the movement performed by cyclists in the bicycle route category [12,15], one of the most important sports world-wide.

Frequently beginner cyclists, particularly those of bicycle route category, have a bad or uncomfortable posture when using their bicycles which generally has a negative impact on their performance [15]. This situation is due to unknowing of the correct positioning on the bicycle that may lead to an optimal performance. Normally, beginner cyclists come to good postures only through practice and imitation of positioning of professional or high performance athletes. This imitation of expert cyclists is not straightforward as differences on positioning cannot be observed with the naked eye and much less if the cyclist is riding.

© Springer Nature Switzerland AG 2018
M. Ceci et al. (Eds.): ISMIS 2018, LNAI 11177, pp. 130–138, 2018.
https://doi.org/10.1007/978-3-030-01851-1_13

Biomechanical studies with traditional systems carried out in specialized centers are usually done in controlled environments, mainly with controlled illumination, with several cameras and specialized sensors and with non versatile proprietary software [8]. In addition to this, access to a biomechanical evaluation is impossible for many cyclists in developing countries as it is an expensive service. In Colombia, for instance, where road cycling is one of the national sports, it exists only two biomechanical centers (visit [4,14] for details in Spanish) that performs an evaluation that cost around 250 US Dollars (around 750.000 COP), which is not affordable for most beginners cyclists.

Our goal is to develop a versatile and portable biomechanical analysis intelligent systems that contrasts with specialized center systems in that it does not require expensive devices or complex features. Given the low cost characteristic, the system does not have the same precision in its bicycle and cyclist measurements and is not able to implement all the biomechanical analysis techniques that standard systems can perform. However, it does strengthen the versatility with which athletes can improve their positioning on their bicycles.

The rest of the document presents the following topics. Related work of systems that currently use low resources are presented in Sect. 2. The design of an intelligent system for the biomechanical analysis is discussed in Sect. 3. Afterwards, the advantages and disadvantages of our proposal are shown in Sect. 4. Finally, conclusions and future work are drawn in Sect. 5.

2   Related Work

In this section we present related work of systems for the biomechanical analysis that use of low resources but that are not fully automatic systems, meaning that the human intervention is required, given opportunity to configuration failures and complication.

The first related work is called *Kinematic analysis of cyclist pedaling using systems of motion capture in 2D and 3D* [5]. This research focused on checking the reliability of the data thrown by a low cost analysis system that works in 2D and compares it with a system that works in 3D that makes use of different expensive resources. The study was carried out through the simultaneous use of both systems (2D and 3D) over a group of 12 cyclists, and the subsequent comparison of the data offered by both systems. Among the results and conclusions, authors did not find important differences between the sensitivity of both systems. On the other hand, as a drawback in the 2D measurement systems, the data obtained for the evaluation was not captured in real time, a characteristic of the 3D systems. The intelligent system that we propose works as well in 2D but it does operates real time, which addresses the disadvantage of the 2D system.

A second work on the subject is found in [11] and it is called *Biomechanical analysis for cyclists*. The research focuses on the analysis of the position of the lower joints of the rider. The authors developed an application for Android systems using the Java programming language and the OpenCV library [13]. To make use of their system, the cyclist must locate LED bulbs in his lower joints and record a video of the cyclist performing the pedaling movement on the bicycle. Finally, the system presents graphs with the data of the angles of the cyclist's joints movement. In contrast with this system, our system does not force the uses of LED bulbs but whatever color marker that facilitates the motion capture process. Additionally, the capture is done in real time without the need of an intermediate video.

Another work on biomechanical analysis is found under the name *Mobile tracking system and optical tracking integration for mobile mixed reality* [3]. This work aims at developing a system that combines optical tracking and augmented reality, operating in a local area network as it communicates with a server responsible for processing all tracking data. The idea behind it is to reduce the responsibilities of the mobile device that have limited processing and storage capabilities. The system is possible thanks to the implementation of a fast communication channel between mobile device and servers.

Apart from the described works, there are commercial systems that allow cyclists to perform a biomechanical analysis of low cost. To the best of the authors knowledge, these applications do not have related academic publications in which the underlying models and techniques are described.

Bike Fast Fit [1] is an application available for iOS devices, which allows to capture in video the pedaling of the cyclist to subsequently measure the angles presented in the recording. This measurement is done to a certain degree manually, meaning that the user has to choose the frames of the video where it is potentially correct to take the measurement. Also the user must choose the body angle to measure, and then locate the measurement points in the cyclist's joints. The advantage of this system is that it has a good precision in the measurements. Nonetheless, the precision may be affected by the intervention of the user as it can make mistakes when measuring and configuring the system.

The Roadie Bike Fit [2] is an application available for Android devices, which offers the user a guide and recommendations for the adjustment of the bicycle. The guide begins with the capture of two images, one of them of the bicycle and another of the cyclist. With these two images the application executed an object recognition algorithm with which it identifies body angles and bicycle measurements. The automation and guidance feature make of this a simple and intuitive application for the user. Nevertheless, the measurements are obtained from static positions of the cyclist, which does not give reliable data of the movements that can only be measured while the cyclist pedals.

3   System's Underlying Design

3.1   Solution Scheme

In the face of the problem, the proposed system is able to offer the route cyclist a kinematic analysis and recommendations to adapt the bicycle and body to their correct positions. This is done through the measurement of the body angles involved in the movements. The measurements are performed mostly in the sagittal plane (refer to [7] for information about different positions to capture movement data). Thus, our system uses a camera perpendicular to the direction of the bicycle, focusing on the cyclist's right profile, as shown in Fig. 1.a.

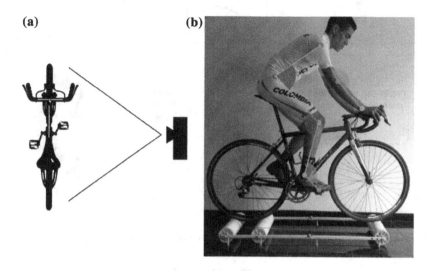

Fig. 1. (a) Camera captures right profile (sagittal plane). (b) Angles in the sagittal plane involved in the movement.

Having this configuration, the system captures the data necessary to execute the biomechanical analysis, i.e., the angles formed by the body parts. As presented in Fig. 1.b, in order to measure the angles, the cyclist must locate markers of any color that are distinctive in the image (high contrast to improve image recognition). The system interaction is as follows. Once the capturing device (e.g., smart-phone camera) is in front of the cyclist, it starts marker recognition on the video stream. Using the frames of the video stream it computes angles (as we will see further in Sect. 3.2). Then the angles are rendered over a display and finally a recommendation based on an analysis is given.

3.2 Architecture

The intelligent system implements the architecture presented in Fig. 2. Essentially, the application executes the biomechanical analysis on a device (such as a tablet, mobile device or laptop) and statistics are stored on a server.

Client's Side. The client's components are responsible for the most important tasks of the system, the core of the system; motion capture, data processing (biomechanical analysis) and visualization of results.

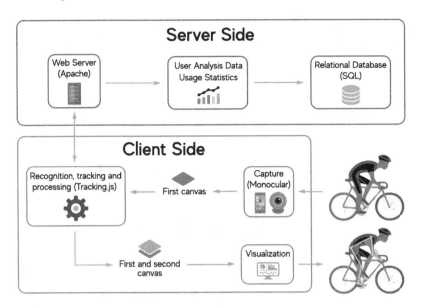

Fig. 2. System's architectural design.

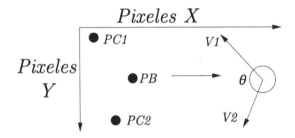

Fig. 3. Obtaining vectors from point in vector space R^2.

- Capture: This component captures the movement from the right profile of the cyclist. It uses only one video capturing device as a smart-phone camera and no sensors, making it low cost. The video stream is presented in a first HTML5 canvas.
- Processing: The processing of the captured images are performed on the client's side, making use of the library `tracking.js` [6]. The library integrates functionalities of image recognition and tracking of colors thus allowing to capture in real-time the position of the color markers that the cyclist is wearing. Having captured the positions of the markers, which are points in the vector space R^2, the system proceeds with the computation of angles. To do so two vectors are need to be computed (as presented in Fig. 3) by using three pixels coordinates (from the markers) and applying basic vector geometry equations.

Once the vectors $\overrightarrow{V1}$ and $\overrightarrow{V2}$ are computed, the system computes the vectors' magnitudes (Eqs. 1 and 2), computes the dot product using its algebraic notation (Eq. 3) and finally it computes the angle using the dot product geometry notation (isolating θ in Eq. 4).

$$|\overrightarrow{V1}| = \sqrt{V1x^2 + V1y^2} \tag{1}$$

$$|\overrightarrow{V2}| = \sqrt{V2x^2 + V2y^2} \tag{2}$$

$$\overrightarrow{V1} \cdot \overrightarrow{V2} = V1x \times V2x + V1y \times V2y \tag{3}$$

$$\overrightarrow{V1} \cdot \overrightarrow{V2} = |\overrightarrow{V1}| \times |\overrightarrow{V2}| \times \cos\theta \tag{4}$$

Finally, the biomechanical analysis of the cyclist is done by using cyclist and bicycle measures as well as the computed angles. This analysis is under implementation in the system as it requires the knowledge from experts in the field to be extracted.

- Visualization: The visualization is done in a fully augmented reality fashion. This technology integrates information captured from the real world with digital information, matching the two of them to build a new coherent complemented and enriched world [10]. The system does so by rendering information related to the positioning of the body and presenting recommendations of modification tailored to the bicycle. The data resulting from the analysis together with the bicycle adjustment recommendations are rendered in real time in a second HTML5 canvas.

Server's Side. In the server' side are implemented the persistent storage for statistics on the systems' use.

- Storage: The server is in charge of storing information about the biomechanical analysis made by cyclists. So far, the system does not manage users as it can be executed without sessions but it is planned to add such functionality. Nevertheless, it stores users data and results for future statistical analysis.
- Web Server: The server processes the collected data from users and analysis results in order to obtain use statistics. These statistics refer to the most common positioning errors made by novice cyclist, the number of times a given cyclist uses the tool, measurements of bicycles and execution time. This data is planned to be exploited with data mining algorithms in order to improve recommendations.

4 Key Differences with Other Systems

There exist some clear differences with respect to the systems in specialized centers and also with commercial systems available for mobile devices. We expose these as advantages and disadvantages.

1. One advantage w.r.t. systems developed as part of research or as a commercial product, is that out system requires very little configuration work on the part of the user. This is because the motion capture of the joints is highly automated, requiring only to select the appropriated color for the markers that the cyclist is wearing.
2. By having augmented reality as a visualization technology, the information related to the analysis is clearly displayed and easily understandable by the cyclist. This contrasts with the results in specialized centers that must be interpreted by experts.
3. The intelligent system offers the cyclist graphic recommendations, with the use of augmented reality, so that it can autonomously adjust the bicycle and thus implement a proper posture. No mechanics or expert intervention.
4. The system is implemented using a library that works in web environments, providing flexibility and a multi-platform development. Other systems either have too complex environment or are tied with a given platform.
5. The system uses only a video capture device and a processing unit (e.g. a smart-phone). Ergo, it is low cost unlike specialized centers.

Our system, on the other hand, presents the following disadvantages.

1. First, the system is affected in its measurements by the processing speed because it is designed to perform real-time analysis on a mobile devices that have limited processing and storage capabilities. It can therefore be slower than traditional systems.
2. Secondly, the angle and position of the camera may influence the reliability of the captured data which may generate inaccurate measurements.
3. Since it is a system that identifies the joints from markers, these markers are recognized by the system using an specific color. This process, made by the library `tracking.js`, is error prune and depends mostly on the color composition of the image (cyclist, bicycle and surroundings). Not being able to identify the marker implies not joints movement capture.
4. Finally, adequate lighting can be a determinant factor when capturing data. Unlike the specialized centers where there is a controlled environment, the lighting when using the proposed application may vary vastly in different executions making the analysis difficult or inaccurate.

5   Concluding Remarks

In this paper we have presented the motivations and internal design of an intelligent system focused on the biomechanical analysis of road bicycle cyclists. In particular, the paper briefly discussed the context of cycling, the intrinsic need of novice athletes for analysis of positioning and movement and related work found in the literature and Web. Our research is possible given the advances in mobile devices (cameras), their access, high-speed WI-FI networks and advances in libraries for motion capture and augmented reality. As such, the system is an important starting point for the novice cyclists as it allows to; do motion capture (thanks to the library `tracking.js`); show the cyclist movement over the display in real time (thanks to the augmented reality technology), do the biomechanical analysis in real time using the limited processing and storage capabilities of a mobile device, store usage statistics for further analysis and be used in a friendly and intuitive way.

Future Work. We have identified three strategic directions for future work.

- First, increase the accuracy of the analysis with the use of sensors. This step is fundamental for improving the analysis and is possible thanks to the inclusion of sensors in the latest mobile devices (mainly smart-phones).
- Second, it is important to dwell more on usage statistics. Here, it is planned to make a analysis of the stored data (how many times a user executed the system, measurements, etc.) in the aim of providing good recommendations for the novice cyclist. In short, it is planned to apply data mining.
- Lastly, it is planned to involve a knowledge base that helps the positioning of cyclist either by applying expertise knowledge or case-based reasoning.

References

1. Bike fast fit: App Store. https://goo.gl/AQP9vM. Accessed 26 Apr 2018
2. The roadie bike fit. Play Store. https://goo.gl/cNbPxE. Accessed 26 Apr 2018
3. Bajana, J., Francia, D., Liverani, A., Krajčovič, M.: Mobile tracking system and optical tracking integration for mobile mixed reality. Int. J. Comput. Appl. Technol. **53**(1), 13–22 (2016). https://doi.org/10.1504/IJCAT.2016.073606
4. Biomec: Bikefitting 3D. http://www.biomec.com.co/bikefitting.html. Accessed 26 Apr 2018
5. Abal del Blanco, P.: Análisis cinemático del pedaleo ciclista mediante sistemas de captura de movimiento en 2d y 3d. Universidad de León (2016)
6. tracking.js Community: Tracking.js. https://trackingjs.com/. Accessed 26 Apr 2018
7. Ferrer Roca, B.: Comparación de diferentes métodos de ajuste de la bicicleta en ciclistas entrenados. influencia de factores biomecánicos y energéticos. Universidad de León (2015)
8. John, S., Andrew, Z.: The economic impact of sports facilities, teams and mega-events. Aust. Econ. Rev. **39**(4), 420–427. https://doi.org/10.1111/j.1467-8462.2006.00431.x
9. Knudson, D.: Fundamentals of Biomechanics. Springer, New York (2007). https://doi.org/10.1007/978-0-387-49312-1
10. van Krevelen, D.W.F.R., Poelman, R.: A survey of augmented reality technologies, applications and limitations. Int. J. Virtual Reality **9**(2), 1–20 (2010)
11. Márquez, C.J., Pérez, L., Tocino, D.: Análisis biomecánico para ciclistas. Universidad Complutense de Madrid (2017)
12. Mota, L.N., Junior, A.J., Neto, G.A., Mota, E.N.: Influence of bike fit in lower limb biomechanics in road cyclists. Phys. Ther. Sport. **31**, e7 (2018). https://doi.org/10.1016/j.ptsp.2017.11.031

13. Pulli, K., Baksheev, A., Kornyakov, K., Eruhimov, V.: Real-time computer vision with OpenCV. Commun. ACM **55**(6), 61–69 (2012). https://doi.org/10.1145/2184319.2184337
14. aethos sport science: http://www.aethosport.com/laboratorio. Accessed 26 Apr 2018
15. Too, D.: Biomechanics of cycling and factors affecting performance. Sport. Med. **10**(5), 286–302 (1990). https://doi.org/10.2165/00007256-199010050-00002

Intelligent Systems

An Approach for the Police Districting Problem Using Artificial Intelligence

José Manuel Rodríguez-Jiménez[1,2]([✉]) [iD]

[1] Universidad de Malaga, Malaga, Spain
jmrodriguez@ctima.uma.es
[2] Mijas Police Department, Mijas, Spain

Abstract. Police patrols are usually assigned to a restricted zone where they have to serve and protect the law. This feature not only results in routine tasks, such imposing traffic tickets, but also there are other important tasks, like assisting in accidents or riot control, that need to be covered.

An efficient traffic Police patrol location and a schedule assignment across the streets of a city or in a road network ensure that the traffic Police comply with their functions.

How to distribute these patrols in the city is a complicated task that needs experience and a deep analysis of traffic and Police data. In this work, we present a method that uses artificial intelligence to analyse these data and propose how to distribute the Police patrols reacting to events that are monitored in real-time for a better service to the citizens.

Keywords: Traffic data · Police · Artificial intelligence

1 Introduction

Citizens request a fast and efficient service from Police patrols in their cities. These claims need that the strategic planning takes into account a deep analysis of Police data, that covers a broad framework of tasks.

The aim of a Police patrol, in particular, a traffic Police patrol, is to provide a service to the public mainly by enforcing laws and assisting road users who have an accident or other traffic issue which could be notified in the Police station.

Different Police forces and governments agree on making an effort to offer a predictive, more than reactive, Police activity based on a decision-making protocol.

Research on Police decision-making relies on official records or qualitative interviews and participant observation data. Also, social observation is a research field that simultaneously gather quantitative and qualitative data in a natural setting [1].

It is well known that, investment programs with big budgets, achieve great results. Program MOBESE [2], with the support of Istanbul's Governorship that includes approximately 600 modern cameras installed throughout the city,

© Springer Nature Switzerland AG 2018
M. Ceci et al. (Eds.): ISMIS 2018, LNAI 11177, pp. 141–150, 2018.
https://doi.org/10.1007/978-3-030-01851-1_14

achieves that crime rate in Istanbul has been considerably lower in recent years. This is an example of a Big Data project that also needs researchers focusing on the main ideas that need to be developed. Currently, Police forces have these researchers that know how to deal with Police problems.

To address the problem of achieving a good predictive model, the main task is to obtain an efficient distribution of Police patrols along the different areas of the city. In literature, there exist some works that examine this task, also known as the Police Districting Problem (PDP). In an effective distribution, it will be difficult, that a Police patrol come across any other patrol in its work if they are not required to do it. Different points of view and methods appear to analyse traffic and Police data proposing solutions for this problem:

- There is a review about **multicriteria methods** in Public Security [3]. Different criteria are presented and how they are used to define algorithms that valuate and offers detailed solutions to the problems. In particular, there exist works in Crime Prevention [4,5] where an optimization model for patrol sectors in the Spanish National Police is developed. This is part of the Decision Aid System for the implementation of predictive policing patrols in Madrid based on a multicriteria method.
- In [6], a **multiple-objective** linear program is developed with three distinct objectives. The objective functions maximize the following: traffic Police presence, Police presence at blackspots where frequent traffic offences occur and the time available for proactive work. In the second stage of their work, they consider also distance and time, that produce a detailed, daily shift schedule.
- **Geographical Information Systems** (GIS) are also studied to provide support for crime analysis in different ways [7], and crime-mapping became a breath of fresh air in crime analysis. Some of these analyses collected data in representative maps with criminal activity [8].
- **Maximal Covering models** represent an interesting point of view that proposes a solution for covering the complete studied area determining efficient spatial distributions of Police patrol areas. In [9], they propose to maximize the formula that sums all the incidents that are covered by at least one nearby located Police patrol.

These methods, combined with recent works related to mining traffic data to detect criminal patterns [10–15], are an interesting starting point for new studies. Adding dynamic data extracted with data mining algorithms, the related methodology could be improved.

In this paper, we will propose a new approach, analysing traffic and Police data. In Sect. 2, an analysis of the PDP in traffic Police patrols will be done. In Sect. 3, our proposal will be presented, with the results and discussion in Sect. 4. We will finish with a section with our conclusions and future work.

2   PDP in Traffic Police Patrols

Since the PDP have been studied for Police patrols without distinctions, traffic Police patrols could be included into the study, but, particular conditions of these patrols deserve an additional study.

Multicriteria methods take into account different attributes: area, isolation, risk and diameter [4,5]. With these attributes, it have been proposed different maps with different schedules related with an hour of the day and the type of day (holidays or weekends). The studied attributes need to be adapted to the traffic Police for a good measure. The potential risk of a vehicle which is parked illegally compared with a bank robbery is much lower.

These methods propose that an initial distribution of the workload has to be done to reduce the crime, based on a preventive disposition of the patrols. It is assumed that there are always the same number of patrols and that they are always capable of acting immediately in the event of any incident.

In real life, this perfect situation never exists and we have to consider that the number of active patrols are not always enough to cover all the areas of the city. When a patrol is working to solve an incident, this patrol could not be considered as an active patrol, unless an emergency requires its assistance and the patrol is allowed to move from the current incident to the new one.

It is required a hierarchy for the different Police tasks regarding their relevances. The hierarchy is not exhaustive, but it is rich enough to cover Traffic Police objectives. Traffic patrols act as collaborators in case of criminal activities and they are only available if requested, focusing on events related to traffic problems. An incident is said to be at a lower relevance level than other if a patrol can leave it to assist the other one. Table 1 shows examples of proposed levels.

Table 1. Relevance levels for Police incidents.

Level	Example
1	Traffic ticket which does not need the vehicle to be towed
2	Traffic ticket which needs the vehicle to be towed
3	Non-injury accident
4	Serious injury accident
5	Important criminal activity

The levels of the incidents need a human assesment. The real value of an incident could get values in an interval, including fuzzy values.

The current methods establish the limits of the working zones analysing Police databases, measuring different attributes and balancing the potential workloads. This analysis could consider the evolution of data along time. Also, to fix the period of time for the study is an open problem.

Since the response to an incident by Police patrols are part of the collected and used data for the analysis, the experience of acting patrols has a relative relevance. This experience has influence in response time and how the patrols solve the incidents, but not in how the zones are distributed.

The predetermined working zone for each patrol cannot have fixed limits, and it needs to be adapted to the number of patrols. Also these zones could have overlapped limits for assistance of a second patrol in case of necessity.

With these handicaps, we need to consider a flexible method that checks the status of each patrol and modify the distribution accordingly.

3   Proposed Methodology

Every Police force produces a big amount of data that needs to be analysed properly to solve different problems. There exist some of them that are recurrent and could be solved by removing their cause, such that additional signals in dangerous crossings that reduce the number of accidents. But, there are other problems that require Police action to avoid them, such as illegally parked vehicles or criminal activities.

Analysing multicriteria methods, we reduce those criteria that are not necessary in our study. This reduction simplifies the algorithm and the patrol distribution process. Also, it has a dynamic and fast analysis of the environment and the data using artificial intelligence (AI) agents.

Background data include events that are relevant to Police activity. These events are classified by the levels showed in Table 1 and the coordinates where the incidents were located are included. This information, adding actual events, composes the theoretical workload for Police patrols. The amount of theoretical workload will define the area and its diameter.

The AI agent reacts to every change in the initial data. Therefore this AI agent is incorporated in the database management systems and propose an action that retrofit the Police patrol distribution as soon as possible, covering efficiently the complete area.

In particular, the agent that we propose reacts to these events:

- Phone calls or advice that communicates a new incident: This incident is added to the database and assigned to a Police patrol. The working area of this patrol is often reduced due to the increase of workload.
- Police patrol that is not active, busy or out of service: The area covered by this deactivated patrol has to be reassigned to closer patrols. The theoretical workload increases and the extension of the working zone for each patrol grows, so this situation is not desirable. Increasing the number of active patrols when there is a prediction of high workload could minimize this problem.
- Police patrol that begins their service or is active again: The new Police patrol is assigned to a new working zone that reduces all the areas modifying their limits. Consequently, the workload for each patrol is reduced.

– Switch among several databases (morning, evening, night, holidays, weekend): The active database with the different incidents registered previously needs to be changed when the type of scenario is modified, for example: the restricted zones for parking that are only available in the morning, increased traffic at certain periods in the weekend, the reduction of traffic at night, etc. These data are not useful out of their own schedule so the incidents that we need to consider changes and also, the patrol distribution.

We have an array of patrols N, n_i is the i-th patrol, where $i = [1, |N|]$. We have also an array of incidents X, x_j is the j-th incident, where $j = [1, |X|]$. We have also a matrix $D = |N|x|X|$ of distances, having patrols over the rows and incidents over the columns, so that d_{ij} is the distance between the patrol n_i and the incident x_j. Furthermore, on the base of a predefined table (human supervised), we can assign a severity level to an accident, so we define the function $l(.)$, such that $l(x_j)$ will return the severity level of the incident x_j accordingly to the criterion mentioned above. Each incident from X is related to an unique patrol, so we define $X_i = \{x \in X | x$ is related to patrol $i\}$.

Assume that there are N active patrols and each one has n_i incidents related to them by zone that are registered in the considered database. Some of these incidents are solved, but are needed to know where the incidents usually occurred. Others are active incidents that need to be solved as soon as possible.

The objetive function to be minimized is given by:

$$f(N, X) = \sum_{i=1}^{N} (\sum_{j=1}^{|X_i|}(d_{ij} * l(x_j))^2) \qquad (1)$$

In Eq. 1, we take into account the theoretical workload for each Police patrol, so we have a sum of squares better than a simple sum. The distances are also multiplied by the level of the incident and squared. This is motivated because, in a straight line, the sum of the distances between each point of the line and the respective extremes is the same, so we need to enforce the location of the medium point in the formula.

The original position of each patrol is designed in the centre of the designed area. The formula is evaluated in 9 coordinates for each point. These 9 coordinates are the result of adding a specified distance in the four cardinal (N., E., S. and W.), the four semi-cardinal compass points (NE., SE., SW. and NW.) and the original position. At each step, we consider the values that minimize Eq. 1 for evaluating the new coordinates. The precision of the method is given by the number of iterations of the process. The specified distance is halved at each iteration.

A JAVA application has been developed to show graphically how the proposed methodology works. This visual representation offers an easy way to verify how the patrols could be distributed and the required modifications when an incident is added, a patrol turns to inactive or returns to the service (see Fig. 1).

On a map, the registered incidents appear as circles that have a proportional size related to their relevance. The colour of these points and the circumference

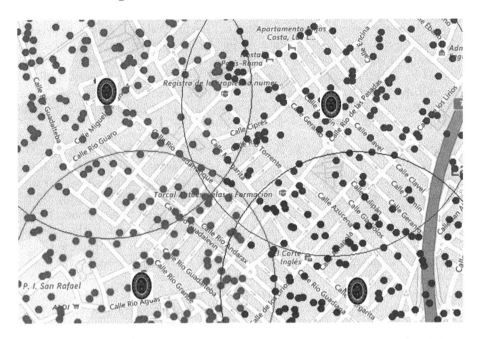

Fig. 1. JAVA Police-Patrol-Distribution program. Example of 4 patrols distribution with 500 random level 1 incidents.

around them indicates the patrol that is designated to the area where the incident is located. Also, a Police symbol is drawn in the centre of each circumference. These are the theoretical best positions for each patrol to cover, as soon as possible, all the incidents located in its area.

4   Results and Discussion

Some experiments were developed. It is not possible to compare properly the results with previous methodology due to different databases. Also there exists a human component that valuates all the considered attributes in previous works.

An adequate research needs that the theoretical results were used in experiments with real scenarios. According to these considerations, our experiments were divided into 2 phases.

First, we need to know the reaction time of the JAVA program when a new incident is added, a patrol is out of service or a patrol returns to the work. The case in which the database of incidents change is not studied specifically because it is included in the previous cases, such a change in the number of registered incidents.

Second, we need to check if the patrols in the work zones could give a fast response to citizens requirements according to the proposed work zones that the algorithm provides.

The studied area has been reflected on a map that has a size of 1.200×750 metres (600×500 pixels) and, in this case, we only iterate 8 times the algorithm. More precision is possible but, for this example, is not considered because the visual position of the patrol on the map is not distinguishable, and the patrol division is not affected, so a better accuracy is not required. Also the time of computation arises (12%–27%) with no relevant benefits.

4.1 Program Analysis

All the simulations were computed in a Dell Intel Core i3-2310M with 2.10 GHz and a RAM memory of 4GB with Windows 10. The program was developed in JAVA using Eclipse Luna.

Datasets are composed by a specified amount of random points with a fixed level 1. These points are located in the whole area, but there is no difference in computing time if the points are in a restricted area.

Table 2. Response time (ms) when successive incidents are added (4 patrols).

Initial incidents	Incidents added				
	1	2	3	4	5
100	3047	3031	3016	3063	3078
200	5906	5969	5969	6047	6079
300	8828	8829	8875	8906	8954
400	11753	11772	11860	11885	12200

In our first experiment (See Table 2), we use a fixed number of 4 patrols to study how the algorithm works when we add a new incident to different datasets (with 100, 200, 300 and 400 incidents respectively). Computation time is quite similar with respect to the number of incidents. There is an average time of 29.48 ms per incident in this case.

Table 3. Response time (ms) when the number of patrols change.

Incidents	Change in patrols								
	$0 \to 1$	$1 \to 2$	$2 \to 3$	$3 \to 4$	$4 \to 5$	$5 \to 4$	$4 \to 3$	$3 \to 2$	$2 \to 1$
100	0	47	266	2938	34142	2954	265	16	0
200	15	78	516	5891	65160	5906	515	47	0
300	31	109	782	8828	97710	8828	781	78	15
400	32	140	1031	11784	135882	12002	1031	93	0

In our second example (See Table 3), we work with the same number of incidents (100, 200, 300 and 400 incidents respectively) changing the number of patrols involved. This change is applied, adding or substracting patrols to check if the algorithm is sensitive to changes in the number of patrols in both directions. With 3 patrols or more, the direction did not affect the computation time of the algorithm, but with 1 or 2 patrols, the comparison of values points to better results when the patrols are substracted.

With this experiment, we can observe that the computing time for each incident is quite similar fixing the number of patrols. We can calculate the estimated time of computation in these cases. Nevertheless, it is quite different when we keep the number of incidents and the number of patrols changes. In these cases, predictions are not possible.

In spare cases, one random point is being assigned to a Police patrol that is not the nearest one due to the workload of this patrol. This singularity shows that the order of the incidences in the database can affect the distribution of the patrols when there are a big amount of incidents. This affection is quite small and, in real cases, this fact does not affect the distribution process.

4.2   Study of Real Case

For over a week we have observed the proposed Police patrol distribution and compare with the actual distribution in the morning schedule of working days in Mijas PD (Spain) in March 2018.

Analysing the dataset of incidents with the JAVA program, we propose a distribution of the studied zone of the city in 3 patrolling areas. The number of registered incidences that we use are 133 with levels from 1 to 3. A non-scheduled patrol was located in each of these areas for over an hour responding to the requirements. Scheduled patrols with usual distribution have not been informed about the experiment.

Since the number of patrolling areas and number of incidents are fixed, we only want to check responding time and efficiency of the non-scheduled patrol.

The percentage of covered requirement in less time than other patrols was 70.59% (12 of 17 cases). 4 of the successful requirements were located in nearby areas and, in 2 of them, the patrol was located just in an adjacent street. The successful requirement was covered in an average time under 4 min.

These results have to be carefully analised because of some of the incidents, there was at least one patrol out of service or busy, so the acting patrol was the nearest one. On the other hand, the non-scheduled patrol was covering some incidents and cannot cover other nearest requirements.

The amount of data used for the analysis could be improved. For this study, we use the data stored in a week. This is enough to obtain the main zones with incident activity. This experiment only considers response time related to the extension of the areas.

5   Conclusions and Future Work

The related methodology proposed the use of artificial intelligence agents that modify the distribution of Police patrols processing real-time events.

This methodology exploits the benefits of Multicriteria and Maximal Covering models, adapting them to a problem that has been thoroughly studied and analysed by different researchers.

A bigger amount of data makes the algorithm slower and the modifications are not easily distinguishable, except when a patrol is out of service or returns to the service. Since the number of patrols, and also the extension of the patrolling areas, depend on each Police force, we need to determine the desirable amount of data to be considered in each scenario.

Also, composition of Police patrols has to be considered. There are areas where the traffic is so dense and distances not so long that the best is to move in motorbikes. This is the fastest option to reach the destination, but are not suitable if there is an arrested person, that needs to be transferred to the Police headquarter into a car. These kind of patrols could be combined considering the commented restrictions.

High level incidences are unusual and they have an enormous effect on the distribution of Police patrols due to their relevance in the workload and how many patrols are out of service. For this reason, we consider that incidences with level 4 or 5 may not be used in Police patrol distribution, but they are really effective when they are studied separately for detecting criminal hotspots or black spots for accidents.

As future work, this methodology will be improved, using the real distance between Police patrols and incidents, instead of map distance.

Acknowledgments. Author wants to thank Police Inspector Lucas López and Police Officer Fernando Arribas from Mijas PD, for their comments about the methodology.
Special thanks to Dr. I.P. Cabrera from University of Málaga for her suggestions.

References

1. Schulenberg, J.L.: Systematic social observation of police decision-making: the process, logistics, and challenges in a Canadian context. Qual. Quant. **48**(1), 297–315 (2014)
2. Kula, S., Guler, A.: Smart public safety: application of mobile electronic system integration (MOBESE) in Istanbul. In: Gil-Garcia, J.R., Pardo, T.A., Nam, T. (eds.) Smarter as the New Urban Agenda. PAIT, vol. 11, pp. 243–258. Springer, Cham (2016). https://doi.org/10.1007/978-3-319-17620-8_13
3. Pereira Basilio, M., Pereira, V., Gomes Costa, H.: Review of the literature on multicriteria methods applied in the field of public security. Univers. J. Manag. **5**(12), 549–562 (2017)
4. Camacho-Collados, M., Liberatore, F., Angulo, J.M.: A multi-criteria Police Districting Problem for the efficient and effective design of patrol sector. Eur. J. Oper. Res. **246**(2), 674–684 (2015)
5. Liberatore, F., Camacho-Collados, M.: A comparison of local search methods for the multicriteria police districting problem on graph. Math. Probl. Eng. **3**, 1–13 (2016)
6. Adler, N., Hakkert, A.S., Raviv, T., Sher, M.: The traffic police location and schedule assignment problem. J. Multi-Criteria Decis. Anal. **21**, 315–333 (2014)

7. Daglar, M., Argun, U.: Crime mapping and geographical information systems in crime analysis. Int. J. Hum. Sci. **13**(1), 2208–2221 (2016)
8. Radhakrishnan, S., Milani, V.: Implementing geographical information system to provide evident support for crime analysis. Procedia Comput. Sci. **48**, 537–540 (2015)
9. Curtin, K.M., Hayslett-McCall, K., Qui, F.: Determining optimal police patrol areas with maximal covering and backup covering location models. Netw. Spat. Econ. **10**(1), 125–145 (2010)
10. Leuzzi, F., Del Signore, E., Ferranti, R.: Towards a pervasive and predictive traffic police. In: Leuzzi, F., Ferilli, S. (eds.) TRAP 2017. AISC, vol. 728, pp. 19–35. Springer, Cham (2018). https://doi.org/10.1007/978-3-319-75608-0_3
11. Ferilli, S., Redavid, D.: A process mining approach to the identification of normal and suspect traffic behavior. In: Leuzzi, F., Ferilli, S. (eds.) TRAP 2017. AISC, vol. 728, pp. 37–56. Springer, Cham (2018). https://doi.org/10.1007/978-3-319-75608-0_4
12. Rodriguez-Jimenez, J.M.: Detecting criminal behaviour patterns in Spain and Italy using formal concept analysis. In: Leuzzi, F., Ferilli, S. (eds.) TRAP 2017. AISC, vol. 728, pp. 57–68. Springer, Cham (2018). https://doi.org/10.1007/978-3-319-75608-0_5
13. Bernaschi, M., Celestini, A., Guarino, S., Lombardi, F., Mastrostefano, E.: Unsupervised classification of routes and plates from the *Trap-2017* dataset. In: Leuzzi, F., Ferilli, S. (eds.) TRAP 2017. AISC, vol. 728, pp. 97–114. Springer, Cham (2018). https://doi.org/10.1007/978-3-319-75608-0_8
14. Bernaschi, M., Celestini, A., Guarino, S., Lombardi, F., Mastrostefano, E.: Traffic data: exploratory data analysis with Apache Accumulo. In: Leuzzi, F., Ferilli, S. (eds.) TRAP 2017. AISC, vol. 728, pp. 129–143. Springer, Cham (2018). https://doi.org/10.1007/978-3-319-75608-0_10
15. Fumarola, F., Lanotte, P.F.: Exploiting recurrent neural networks for gate traffic prediction. In: Leuzzi, F., Ferilli, S. (eds.) TRAP 2017. AISC, vol. 728, pp. 145–153. Springer, Cham (2018). https://doi.org/10.1007/978-3-319-75608-0_11

Unsupervised Vehicle Recognition Using Incremental Reseeding of Acoustic Signatures

Justin Sunu[1(✉)], Allon G. Percus[1(✉)], and Blake Hunter[2(✉)]

[1] Claremont Graduate University, Claremont, CA 91711, USA
justinsunu@gmail.com, allon.percus@cgu.edu
[2] Microsoft, Redmond, WA 98052, USA
blake.hunter@microsoft.com

Abstract. Vehicle recognition and classification have broad applications, ranging from traffic flow management to military target identification. We demonstrate an unsupervised method for automated identification of moving vehicles from roadside audio sensors. Using a short-time Fourier transform to decompose audio signals, we treat the frequency signature in each time window as an individual data point. We then use a spectral embedding for dimensionality reduction. Based on the leading eigenvectors, we relate the performance of an incremental reseeding algorithm to that of spectral clustering. We find that incremental reseeding accurately identifies individual vehicles using their acoustic signatures.

Keywords: Spectral clustering · Machine learning · Vehicle audio

1 Introduction

Recognizing and distinguishing moving vehicles based on their audio signals are problems of broad interest. Applications range from traffic analysis and urban planning to military vehicle recognition. Audio data sets are small compared to video data, and multiple audio sensors can be placed easily and inexpensively. However, challenges arise due to equipment as well as to the underlying physics. Microphone sensitivity can result in disruption from wind and ambient noise. The Doppler shift can make a vehicle's acoustic signature differ according to its position.

In order to interpret acoustic signatures, one must extract information contained within the raw audio data. A natural feature extraction method is the short-time Fourier transform (STFT), using time windows large enough to carry sufficient frequency information but small enough to localize vehicle events. STFT has been used previously for classifying cars vs. motorcycles with principle component analysis [13], for characterizing ϵ-neighborhoods in vehicle frequency signatures [14], for vehicle classification based on power spectral density [1], and for vehicle engine identification [8,12]. Other related feature extraction

© Springer Nature Switzerland AG 2018
M. Ceci et al. (Eds.): ISMIS 2018, LNAI 11177, pp. 151–160, 2018.
https://doi.org/10.1007/978-3-030-01851-1_15

approaches have included the wavelet transform [1,2] and the one-third-octave filter bands [10].

Our study is motivated by previous work that uses a spectral embedding approach to identify different individual vehicles [11]. Representing each time window as an individual data point, we define a similarity measure between two points based on the cosine distance between their sets of Fourier coefficients, and then cluster according to the symmetric normalized graph Laplacian [9]. In this paper, we relate the eigenvectors of the Laplacian to a recently proposed clustering method, incremental reseeding (INCRES) [3], that iteratively propagates cluster labels across a graph. We compare the performance of INCRES with spectral clustering on the vehicle audio data. We find that both are promising unsupervised methods for vehicle identification, with INCRES correctly clustering 91.7% of the data points in a sequence of passages of three different vehicles.

2   Algorithms

Our clustering algorithms are based on the use of spectral embedding for dimensionality reduction [11]. Consider a signal of length n, with feature vector $\mathbf{x_i} \in \mathbb{R}^m$ associated with data point $i \in \{1, \ldots, n\}$. A spectral embedding represents data as vertices on a weighted graph, with edge weights S_{ij} expressing a similarity measure between data points i and j. The graph is encoded using the symmetric normalized graph Laplacian matrix [9]

$$\mathbf{L_s} = \mathbf{I} - \mathbf{D}^{-1/2}\mathbf{S}\mathbf{D}^{-1/2} \tag{1}$$

where $\mathbf{D}$ is a diagonal matrix with $D_{ii} = \sum_j S_{ij}$.

We use this embedding for vehicle identification with two related clustering methods: spectral clustering, and a recently developed incremental reseeding approach [3].

2.1   Spectral Clustering

The eigenvectors of $\mathbf{L_s}$ associated with the leading nontrivial eigenvalues $\lambda_2, \ldots \lambda_k$ form a $(k-1)$-dimensional approximation to $\mathbf{x_i}$. The approximation is justified when the spectral gap $|\lambda_{k+1} - \lambda_k|$ is large, which occurs when the data naturally form k clusters [9]. Spectral clustering uses k-means to cluster this $\mathbb{R}^{k-1}$ projection of the data.

2.2   Incremental Reseeding (INCRES) Algorithm

The INCRES algorithm [3] is a diffusive method that propagates cluster labels across the graph specified by $\mathbf{L_s}$. The approach (Algorithm 1) is incremental: it plants cluster seeds among nodes, grows clusters from these seeds, and then reseeds among the growing clusters.

Algorithm 1. INCRES

1: **Input** Similarity matrix S, number of clusters k, number of iterations s
2: **Initialize** Randomly partition data points into k clusters
3: **for** $i = 1$ to s **do**
4:    PLANT: Randomly choose certain points in each cluster as *seeds*
5:    GROW: Apply Laplacian $\mathbf{L_s}$ to matrix of seed labels
6:    HARVEST: Assign data points to cluster with largest label value
7: **end for**

To elaborate on the algorithm, an indicator matrix of seed labels is constructed in the PLANT step: $U_{ij} = 1$ if point i is picked as a seed for cluster j, and 0 otherwise. By applying the Laplacian in the GROW step, $\mathbf{L_s}U$ becomes a real-valued matrix representing the seeds propagating to neighbors according to similarity. Finally, the HARVEST step updates the assignment of data points to clusters, by finding for each point i the cluster j maximizing the propagated label matrix element $(\mathbf{L_s}U)_{ij}$. The process then repeats, with new seeds drawn from the updated clusters.

Since the process of seed propagation is governed by the graph Laplacian, INCRES is closely connected with spectral clustering. Eigenvectors of $\mathbf{L_s}$ are organized hierarchically: the second eigenvector separates data into two clusters at the coarsest resolution, the third eigenvector identifies a third cluster at a finer resolution, and so on. An application of INCRES with parameter k propagates seeds with k different labels through the graph, resulting in k clusters governed by the spectral properties of $\mathbf{L_s}$.

To illustrate the relation between the two algorithms, consider a similarity matrix S_{ij} given in block matrix form, with added "salt and pepper" noise. This is shown in Fig. 1, with lighter colors representing greater similarities. Figure 2 shows the second and third eigenvectors of $\mathbf{L_s}$, along with the results of INCRES using $k = 2$ and $k = 3$. The binary clustering results of both methods split the data into the same two larger classes, using only the second eigenvector for spectral clustering and using $k = 2$ for INCRES. When the third eigenvector is also used for spectral clustering and when $k = 3$ is used for INCRES, the two methods find the same subdivision of one of these two larger classes.

In less straightforward clustering examples, the reseeding process can allow INCRES to learn partitions that are not apparent to spectral clustering. Furthermore, the formulation of INCRES allows it to be applied even in cases of larger datasets where eigenpairs cannot be easily computed.

3   Data and Feature Extraction

Our audio data consists of recordings, provided by the US Naval Air Systems Command, of different vehicles moving multiple times through a parking lot at approximately 15mph. The original dataset consists of MP4 videos taken from a roadside camera; we extract the dual channel audio signal, and average the

Fig. 1. Synthetic similarity matrix with salt and pepper noise. White represents a similarity value of 1; black represents a similarity value of 0.

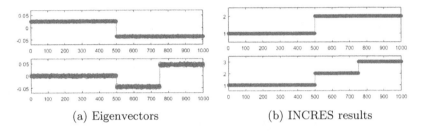

(a) Eigenvectors   (b) INCRES results

Fig. 2. (a) 2nd and 3rd eigenvectors of $\mathbf{L_s}$, and (b) INCRES results with $k = 2$ and $k = 3$, using the synthetic similarity matrix in Fig. 1. Note similar separation into two and three clusters.

channels together into a single channel. The audio signal has a sampling rate of 48,000 samples per second. Video information is used only to ascertain ground truth (vehicle identification) for training data.

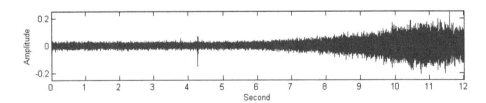

Fig. 3. Raw audio signal for a vehicle passage.

Each extracted audio signal is a sequence of a vehicle approaching from a distance, becoming audible after 5 or 6 s, passing the microphone after 10 s, and then leaving. An example of the raw audio signal is shown in Fig. 3. We form a composite sequence, shown in Fig. 4, from multiple passages of three different vehicles (a white truck, black truck, and jeep), selecting the two seconds where the vehicle is closest to the camera. The goal is to test the clustering algorithm's ability to differentiate the vehicles.

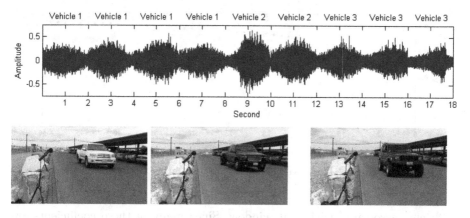

Fig. 4. Raw audio signal for composite data. Images show the three different vehicles, as seen in accompanying video (not used for analysis).

We preprocess the data by grouping audio samples into larger windows. With windows of 1/8 of a second, or 6000 samples per window, we find both a sufficient number of windows and sufficient information per window. While there is no clear standard in the literature, this window size is comparable to those used in other studies [13]. Discontinuities between successive windows can in some cases be reduced by applying a Hamming window as a filter, or by allowing overlap between windows [13]. However, in our study we found no conclusive benefit from either of these, and therefore used standard box windows with no overlap.

Relevant features are extracted from the raw audio signal using the short-time Fourier transform (STFT). The Fourier decomposition contains 6000 symmetric coefficients, leaving 3000 usable coefficients. Figure 5 shows the first 1000 Fourier coefficients, using a moving mean of 5 samples, for a time window representing a truck passing and a time window representing a sedan passing, both in similar positions. Note that a clear frequency signature is apparent for each vehicle, with much of the signal concentrated within the first 250 coefficients or 2000 Hz.

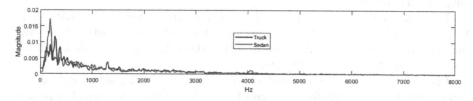

Fig. 5. First 1000 Fourier coefficients for a truck and a car, after applying a moving mean of size 5.

Each time window in the audio signal is taken as an independent data point to be clustered: we define the feature vector $\mathbf{x_i} \in \mathbb{R}^m$ as the set of m Fourier

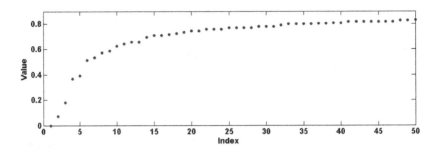

Fig. 6. Spectrum of $\mathbf{L_s}$ for vehicle data. Largest gap is after third eigenvalue.

coefficients associated with that window. Since many of these coefficients are relatively insignificant, we consider the cosine distance measure between data points

$$d_{ij} = 1 - \frac{\mathbf{x_i} \cdot \mathbf{x_j}}{\|\mathbf{x_i}\| \, \|\mathbf{x_j}\|}. \tag{2}$$

We then construct an M-nearest neighbor graph, where the edge $\{i, j\}$ is present if j is among the M closest neighbors of i or vice-versa, for a fixed value of M. Following standard methods [9], the similarity S_{ij} is taken to be a Gaussian function of distance,

$$S_{ij} = e^{-d_{ij}^2 / \sigma_i^2}, \tag{3}$$

where σ_i is defined adaptively [15] as the distance to vertex i's Mth neighbor.

4   Results

Our composite vehicle dataset contained the 18 s of raw audio shown in Fig. 4, resulting in $n = 144$ data points each representing 1/8-second time windows. We used only the first $m = 1500$ Fourier coefficients, reducing the risk of artifacts from window discontinuities. We set $M = 15$ for the M-nearest neighbor graph, so that neighborhoods contain the 16 data points used in the 2-second clips of a single vehicle passage.

Figure 6 shows the eigenvalues of the Laplacian for the vehicle data. The largest gap follows the third eigenvalue, consistent with three clusters representing the three vehicles actually present in the data. We therefore set $k = 3$ for both spectral clustering and INCRES.

Figure 7 shows the second and third eigenvectors of $\mathbf{L_s}$ and Fig. 8 shows typical results of INCRES for $k = 2$ and $k = 3$ (INCRES is stochastic, but results vary little from run to run). As in our earlier synthetic example, the second eigenvector and $k = 2$ INCRES result provide comparable binary separations of the data. Thresholding the eigenvector just above zero would place all of the vehicle 1 data in one cluster, and most of the vehicle 2 and 3 data in the other cluster (the exceptions are primarily data points at the beginning and end of a vehicle passage, where the signal is weakest). The third eigenvector

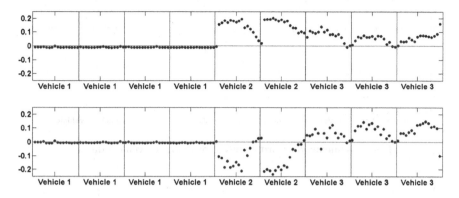

Fig. 7. 2nd and 3rd eigenvectors of $\mathbf{L_s}$ for vehicle data.

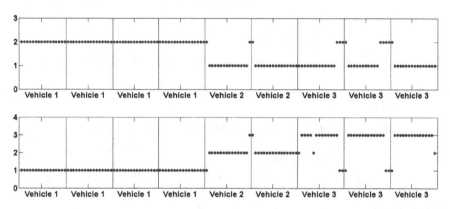

Fig. 8. INCRES results for vehicle data.

mostly distinguishes vehicle 2 (negative values) and vehicle 3 (positive values). The $k = 3$ INCRES result recognizes the three vehicles very accurately, and is discussed below.

Note that unlike in the straightforward synthetic data problem, the third eigenvector is not by itself sufficient to separate the three clusters. Figure 9 shows the results of k-means clustering, with $k = 3$, on the third eigenvector alone. While all vehicle 1 data points are clustered together, a significant fraction of vehicle 2 and 3 data points are incorrectly placed in that cluster as well.

Figure 10 shows results of the more conventional spectral clustering method, using k-means on the $\mathbb{R}^2$ projection of the data given by the 2nd and 3rd eigenvectors. The inclusion of the 2nd eigenvector is sufficient to cluster the vast majority of vehicle 2 and 3 data points correctly.

Tables 1 and 2 interpret the spectral clustering results of Fig. 10 and the INCRES $k = 3$ results of Fig. 8 as classifications. Both methods classify all of vehicle 1 correctly. but INCRES performs noticeably better than spectral clustering on vehicle 2, and they perform comparably on vehicle 3. Overall purity

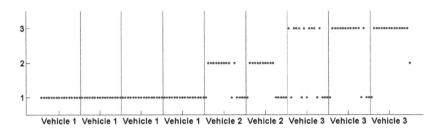

Fig. 9. k-means on third eigenvector of $\mathbf{L_s}$ for vehicle data.

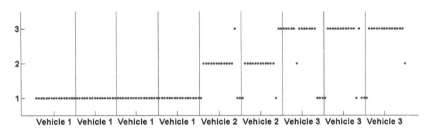

Fig. 10. k-means on second and third eigenvectors of $\mathbf{L_s}$ (standard spectral clustering) for vehicle data.

Table 1. Vehicle clustering results using spectral clustering.

True	Obtained cluster		
	Vehicle 1 (white truck)	Vehicle 2 (black truck)	Vehicle 3 (jeep)
Vehicle 1 (white truck)	64	0	0
Vehicle 2 (black truck)	5	24	3
Vehicle 3 (jeep)	8	2	38

Table 2. Vehicle clustering results using INCRES with $k = 3$.

True	Obtained cluster		
	Vehicle 1 (white truck)	Vehicle 2 (black truck)	Vehicle 3 (jeep)
Vehicle 1 (white truck)	64	0	0
Vehicle 2 (black truck)	1	29	2
Vehicle 3 (jeep)	6	3	39

scores are 87.5% for spectral clustering, and 91.7% for INCRES, with misclassifications again occurring primarily at the beginning or end of a vehicle passage.

The runtimes for both algorithms in Matlab were both under one second, insignificant compared to the minutes needed for processing the initial data.

5   Conclusions

We have presented a method to identify moving vehicles from audio recordings, by clustering their frequency signatures with an incremental reseeding method (INCRES) [3]. Motivated by the approach in [11], we decompose the audio signal with a short-time Fourier transform (STFT), and treat each 1/8-second time window as an individual data point. We then apply a spectral embedding and consider the symmetric normalized graph Laplacian. We find that spectral clustering, which uses the leading eigenvectors of the Laplacian, correctly clusters 87.5% of the data points. INCRES, which directly uses the Laplacian to construct a random walk on the graph, correctly clusters 91.7% of the data points. Almost all incorrectly clustered points lie at the very beginning or very end of a vehicle passage, when the vehicle is furthest from the recording device. The vast majority of data points result in correct vehicle recognition.

We observe that there is a close relation between the kth eigenvector and the INCRES output for k clusters. This suggests that clustering results might be improved by simultaneously taking the INCRES output for 2 through k clusters, and then using k-means on this $\mathbb{R}^{k-1}$ projection of the data just as spectral clustering does on the 2nd through kth eigenvectors. While doing so does not noticeably change our INCRES $k = 3$ results, the difference could be significant for larger values of k. This could be tested, using a dataset with a larger number of vehicles.

Finally, we note that, since time windows are treated as independent data points, our approach ignores most temporal information. Explicitly taking advantage of the time-series nature of our data in the clustering algorithm could improve results, by clustering data points according not only to their own frequency signatures but also to those of preceding or subsequent time windows. Furthermore, while the STFT is a standard method for processing audio signals, it suffers from two drawbacks: the use of time windows imposes a specific time scale for resolving the signal that may not always be the appropriate one, and vehicle sounds may contain too many distinct frequencies for the Fourier decomposition to yield easily learned signatures. These difficulties may best be addressed by using multiscale techniques such as wavelet decompositions that have been proposed for vehicle detection and classification [1,2], as well as more recently developed sparse decomposition methods that learn a set of basis functions from the data [4–7].

Acknowledgments. The authors are grateful to Dr. Arjuna Flenner at the US Navy's Naval Air Systems Command (NAVAIR) for having supplied the vehicle data. We also wish to thank the anonymous reviewers for their comments and suggestions, which helped improve the clarity of the paper.

References

1. Aljaafreh, A., Dong, L.: An evaluation of feature extraction methods for vehicle classification based on acoustic signals. In: IEEE International Conference on Networking, Sensing and Control (2010)
2. Averbuch, A., Zheludev, V.A., Rabin, N., Schclar, A.: Wavelet-based acoustic detection of moving vehicles. Multidimension. Syst. Signal Process. **20**, 55–80 (2009)
3. Bresson, X., Hu, H., Laurent, T., Szlam, A., von Brecht, J.: An incremental reseeding strategy for clustering. Mathematics and Visualization (2018, to appear)
4. Chui, C.K., Mhaskar, H.: Signal decomposition and analysis via extraction of frequencies. Appl. Comput. Harmonic Anal. **40**(1), 97–136 (2016)
5. Daubechies, I., Lu, J., Wu, H.T.: Synchrosqueezed wavelet transforms: an empirical mode decomposition-like tool. Appl. Comput. Harmonic Anal. **30**(2), 243–261 (2011)
6. Gilles, J.: Empirical wavelet transform. IEEE Trans. Sign. Process. **61**(16), 3999–4010 (2013)
7. Hou, T.Y., Shi, Z.: Adaptive data analysis via sparse time-frequency representation. Adv. Adapt. Data Anal. **03**(01n02), 1–28 (2011)
8. Kozhisseri, S., Bikdash, M.: Spectral features for the classification of civilian vehicles using acoustic sensors. In: IEEE Workshop on Computational Intelligence in Vehicles and Vehicular Systems (2009)
9. Luxburg, U.V.: A tutorial on spectral clustering. Stat. Comput. **17**(4), 395–416 (2007)
10. Rahim, N.A., Paulraj, P.M., Adom, A.H., Kumar, S.S.: Moving vehicle noise classification using multiple classifiers. In: IEEE Student Conference on Research and Development (2011)
11. Sunu, J., Percus, A.G.: Dimensionality reduction for acoustic vehicle classification with spectral embedding. In: IEEE International Conference on Networking, Sensing and Control, pp. 129–133 (2018)
12. Wieczorkowska, A., Kubera, E., Słowik, T., Skrzypiec, K.: Spectral features for audio based vehicle and engine classification. J. Intell. Inf. Syst. **50**(2), 265–290 (2018)
13. Wu, H., Siegel, M., Khosla, P.: Vehicle sound signature recognition by frequency vector principle component analysis. IEEE Trans. Instrum. Meas. **48**, 1005–1009 (1999)
14. Yang, S.S., Kim, Y.G., Choi, H.: Vehicle identification using wireless sensor networks. In: IEEE SoutheastCon (2007)
15. Zelnik-Manor, L., Perona, P.: Self-tuning spectral clustering. Adv. Neural Inf. Process. Syst. **17**, 1601–1608 (2004)

A Big Data Framework for Analysis of Traffic Data in Italian Highways

Claudia Diamantini, Domenico Potena, and Emanuele Storti[(✉)]

DII, Polytechnic University of Marche, via Brecce Bianche, 60131 Ancona, Italy
{c.diamantini,d.potena,e.storti}@univpm.it

Abstract. The analysis of traffic data can provide decision-makers with invaluable information. Despite the availability of methodologies specifically oriented to processing this kind of data and extract knowledge from them, few tools provide a rich set of functionalities tailored to traffic analysis in large-scale, stream-like contexts. In this paper we aim to fill this gap, by introducing an exploratory framework supporting the analysis of massive stream traffic data by either OLAP-like exploration or by resorting to advanced data mining techniques.

1 Introduction

The analysis of data generated by daily activities and transactional systems has been widely recognized as the new oil[1], fostering evidence-based decision and management practices. It is not difficulty- and error-prone though: in order to correctly obtain valid, correct, and useful knowledge which can drive decision makers, many operations on data must be performed, and results correctly interpreted by decision makers, who typically do not have the necessary skills. The recent emergence of Big Data even worsens the issue, since huge volumes of data, possibly different in format and content, and produced in a continuous, streaming fashion, require adequate computational architectures and proper methodologies [3]. Traffic data fall into this category: although possibly generated by a unique system (as considered in this paper) and hence homogeneous in terms of format and content, they are continuously produced in huge quantity, and each vehicle typically leaves a trace of successive spatial points crossed during its trip. Traffic data can be analyzed from many perspectives and for different goals: spatial distribution at a certain point in time can provide valuable knowledge about busiest places, while time distribution at a certain place allows to evaluate peak traffic. The two perspectives can also be combined. Traces can be analyzed to discover common travel habits or anomalies [2,5], and so on. For an overview of challenges and possible applications of analysis techniques to traffic data we refer the reader to [6].

---

[1] Originally coined by Clive Humby in 2006, the analogy has been used since then by many others, among which Peter Sondergaard, SVP Gartner.

© Springer Nature Switzerland AG 2018
M. Ceci et al. (Eds.): ISMIS 2018, LNAI 11177, pp. 161–168, 2018.
https://doi.org/10.1007/978-3-030-01851-1_16

Among the different solutions available to discover knowledge from data, Exploratory Analysis systems [4] have the advantage to provide a range of alternatives to view information and dig into it in an interactive way, guiding the analyst towards a progressive discovery of the knowledge hidden into data by either OLAP-like exploration [7] or by resorting to more advanced techniques like data mining [3]. From a user perspective, support functionalities can be guided by a fixed logic (implemented by predefined queries/analyses and Graphical User Interfaces), which satisfies the requirements of unexperienced users by providing a set of predefined analysis functionalities, or by languages that can be exploited by experienced users to perform more sophisticated and non-predefined analysis, in response to unpredictable needs.

The aims of this paper are (1) to propose an architecture for large-scale exploratory analysis of traffic data which is able to satisfy both kinds of requirements, and (2) to demonstrate its ability to deal both with OLAP-like combined with advanced analytic techniques like clustering and trace/trajectory-like analyses in a stream context, providing also pre-processing capabilites. The proposal is in the mainstream of recent work building user-oriented functionalities on top of high performance distributed architectures [1]. With respect to the latter, this work is more oriented towards the development of a full-fledged exploratory tool for the benefit of the end user.

The paper is organized as follows: Sect. 2 introduces the architecture of the developed system. Section 3 describes the real traffic data used to demonstrate pre-processing and analysis capabilities, then Sect. 4 presents some of the functionalities implemented at the application layer. Finally, Sect. 5 draws some conclusions and discusses possible extensions of the present work.

2   Architecture

This Section is aimed to discuss the architecture we have developed to allow the management and analysis of traffic data, ensuring scalability, availability and usability. The architecture scheme is shown in Fig. 1.

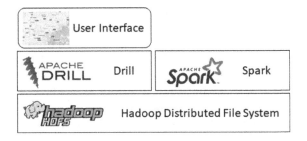

Fig. 1. The scheme of developed architecture.

The lowest layer of the architecture is used for storing and managing data. Noteworthy that traffic data are characterized by large volumes and are produced very frequently (see Sect. 3 for details). In order to deal with these characteristics, the Apache Hadoop Distributed File System (HDFS) has been chosen. HDFS allows to manage high-volume of data, ensuring high-availability and scalability. In the middle layer we find modules devoted to data processing and querying, as well as tools for the discovery of knowledge. In particular, the layer is implemented through Apache Spark (https://spark.apache.org/) and Apache Drill (https://drill.apache.org/). Spark is a state-of-the-art system for big data analysis. It enables parallel and distributed processing of large datasets, which are partitioned over several nodes and processed by them, reducing the processing time due to data transfer. Spark provides a number of coarse-grained operators that are used in cascade to transform input data. These operators are applied only when an action is required, that is only when the analyst requests to visualize the results. Delaying the application of the required transformations allows Spark to optimize the execution plan. On the top of core functionalities, Spark provides powerful libraries for data processing. In particular, we use SparkSQL for the preprocessing of traffic data and MLlib for extraction of data models (see Sect. 4 for details). Moreover, by exploiting Spark Streaming library, the proposed architecture is ready to handle streams of traffic data. Finally, Apache Drill is used to access data through SQL. Drill has been chosen as it allows to access several data sources, whatever data are structured and wherever they are stored (e.g. structured text files, JSON files, HIVE tables, HBASE dataset, MongoDB documents, and so on) by a unique query language (i.e., SQL). Drill has an optimizer and a parallelizer that ensure high-performances and scalable query execution plans. Furthermore, Drill provides REST APIs to query data sources hosted on the Hadoop ecosystem from external tools over the Web.

The system is designed to support two kinds of users, namely decision-makers as well as advanced data analysts. The former can access traffic data through a Web application, which allows to build up a query through a set of pre-defined selection tools. Drill REST APIs are used to connect the Web application with Drill. Results are rendered over a map[2], showing both traffic data (e.g., position of vehicles, number of vehicles per road section) and behavioral models (e.g., speed-based clustering, traffic trends). The latter kind of users can access both Drill to build ad-hoc SQL queries and Spark to use libraries like Mlib for the induction of new models.

The infrastructure implementing the architecture described in this Section is formed by 3 nodes with a 2.3 GHz quad-core processor and 8 GB RAM each, hosted at the Daisy lab of Marche Polytechnic University[3].

3   Dataset

In this work, we elaborate upon the TRAP2017 dataset, which contains traffic data of an Italian motorway. The dataset was provided by the Italian National

---

[2] The Web application is based on OpenStreetMap (www.openstreetmap.org).

[3] Data Analytics, Artificial Intelligence and Cybersecurity laboratory - http://daisy.dii.univpm.it/.

Police and was used for the contest organized during the First European Conference on Traffic Mining Applied to Police Activities[4].

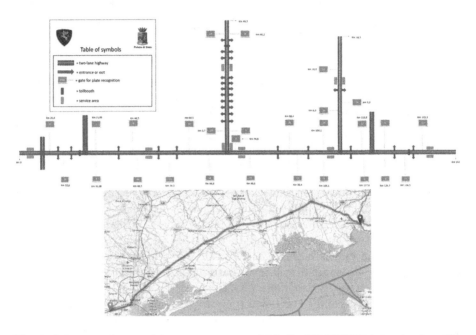

Fig. 2. Scheme and map of the motorways to which the TRAP2017 dataset refers. The map represents the main route in the scheme, i.e. the horizontal part.

The dataset contains traffic information collected through automatic Number Plate Reading Systems (NPRS), which are placed in fixed positions on motorways A4 and A57. In Fig. 2 we show both the map of the sections of motorways, to which the dataset refers to, and a schematic description of the routes with ID and position of each detection gate (boxes) as well as other useful information like entrances/exits (vertical arrows), tollbooths and service areas. The TRAP2017 dataset is formed by several CSV files, each related to a day of 2016. On average we have around 425100 records per day, ranging from 152 (when some problems occurred) to 698710, for a total of 155585944 records concerning 10 million of different vehicles. For each vehicle, the plate number (anonymised for privacy reasons), the gate where the NPRS is installed, the detection timestamp, the lane traveled by the vehicle and the nationality of the plate (this information is missing in 23% of cases) are stored.

4   Analysis of Traffic Data

The main analysis goal is to understand the behavior of vehicles in different road sections, namely the road between two detection gates.

---

[4] https://trap2017.poliziadistato.it/.

4.1   Preprocessing

A preprocessing procedure has been defined in order to clean data and to transform them from information on the passage under a gate to information related to travel a road section. The first step of the procedure is devoted to clean the dataset by removing errors due to malfunctions at NPRS sensors level or at system level. In this step duplicate records are removed, returning a new dataset with 155368907 records.

It is to be noted that not all pairs of gates define a valid road section. We define a *road section* as the pair of gates $<g_1, g_2>$ if:

1. they are in the same direction of travel in the scheme of Fig. 2;
2. g_1 occurs before g_2, and
3. there are no other gates between them.

For instance, the pair $<26, 10>$ represents a road section, while $<26, 18>$ is not valid as the gate 10 is between them; $<26, 16>$ refers to gates in different directions and in $<10, 26>$ the two gates occur in reverse order.

Given a road section $<g_1, g_2>$, a *travel* is defined as a vehicle moving from gate g_1 to gate g_2. In order to obtain travels , the TRAP2017 dataset has been joined with itself pivoting on the plate number and selecting only pairs of gates defining a valid road section. Then, on the basis of distances between gates, the average speed to each travel has been computed and added to the dataset.

From an initial analysis of the average speed, highly unlikely values (e.g., average speed greater than 300 km/h) have been recognized and removed by the dataset. They are certainly due to detection errors and could not be identified only by detections under the gates. Finally, we have 81567101 travels after the pre-processing phase. The whole pre-processing, running on the infrastructure available at Daisy lab, takes 9.1 min in total. Noteworthy, the pre-processing time depends on data to be analyzed and the used infrastructure. It can be considerably reduced by adding nodes to Hadoop/Spark; furthermore, in real scenario, traffic data are updated with a given frequency (even as data stream), thus reducing the data to be processed each time.

4.2   Application

In order to support non-IT expert decision-makers, on the top of the architecture, a Web application offers the possibility to graphically perform various OLAP-like analyses on traffic data, specifying a motorway, a certain level of aggregation of the temporal dimension (e.g., day, month, year) and the nationality of the vehicle. As shown in Fig. 3, the interface provides general statistical information on each road section of the chosen motorway (in the example, the A4), for the specified time level and value (e.g., at Day level). As a result, the selected motorway is highlighted and colored according to the selected indicator (e.g., average speed). The relative slow speed in the whole section visible in the Figure can be explained by considering that the chosen day, namely 6th August 2016, is known to have been the day with the heaviest traffic of the whole year.

Fig. 3. User interface for OLAP-like traffic analysis. (Color figure online)

The pop-up panel, corresponding to the selected section, enables also the possibility to show the statistical distribution of both hourly traffic, i.e. the number of vehicles per hour between the specified gates of the section, and distribution of speed. The extraction of records for a single section and a day can be practically done at run-time, while the extraction of data for a section and a week requires around 5.1 s on average.

Figure 4 shows histograms representing the distribution of speed values aggregated for the whole year 2016, for travels in two different road sections of the A4 motorway, namely <10, 18> and <26, 10>. As for the first case, a bimodal distribution can be easily recognized, meaning that vehicles mostly follow two patterns. By focusing the analysis on the typology of vehicle, such a trend can be explained by considering that heavy vehicles have stricter speed limits, namely around 80 km/h, while for the others it is around 130 km/h. On the other hand, Fig. 4b considerably differs from the former as it shows a multi-modal distribution including a further peak. In this case, however, this trend can be explained by the fact that a service area is located within the gates 26 and 10. Hence, such a peak is likely to represent the situation of those vehicles that passed through gate 26, then stopped for a while at the service area, thus increasing the overall time to reach gate 10 and, as a consequence, reducing their computed average speed in the section. Running times for the generation of histograms of speed values between two gates for a time period of a year is between 1 and 3 s. By conducting a clustering analysis on these data through the algorithm k-Means, it is possible to recognize three clusters centered respectively in 21.97 km/h, 91.1 km/h and 125.66 km/h. The execution of 100 iterations of k-Means to generate the clusters required around 20 s.

A more detailed route analysis can be carried out at vehicle level. By considering all records in the log that refer to the same plate number and highway, and can be temporally ordered in a given time window (i.e., traces), it is possible to track the specific path of a certain vehicle on a motorway, as shown in Fig. 5.

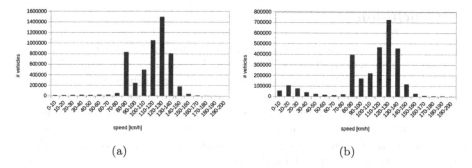

Fig. 4. Distribution of speeds for year 2016 in two road sections of motorway A4, respectively (a) <10, 18> and (b) <26, 10>, which includes a service area.

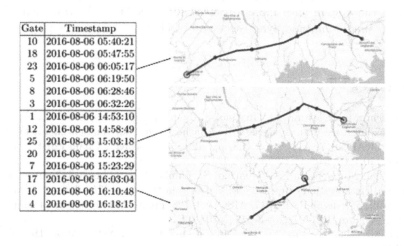

Fig. 5. Route analysis on motorway A4, for a specific day. For each trace, the log and the corresponding path on the map is shown. The bigger circle represent the gate that recorded the first event in the trace.

Further typologies of analyses can be supported by the approach, and will be investigated in future work, among which:

- frequent pattern analysis, that is devoted to recognize which are the most frequent paths for a certain vehicle (or for a class of vehicles or for all vehicles) within a motorway or considering the whole motorway network;
- outlier analysis, which involves focusing on those vehicles that follow infrequent patterns, e.g. a vehicle driving at an infrequent average speed, or that follows a rare pattern like stopping at several service areas in a row;
- traffic flux analysis, that computes the percentage of vehicles choosing each route on a motorway junction. This model can be also useful for predictive analyses to evaluate the likelihood for a car to turn towards a specific direction on the junction.

5   Conclusion

The paper introduced the proposal of an Exploratory Analysis system supporting large-scale analysis of traffic data. The system is based on a well established suite of tools to manage and process huge amounts of stream data. On top of these, we designed and implemented a number of functionalities, ranging from pre-processing data, visualization and cluster analysis. The system is designed to support both unexperienced decision-makers and advanced data analysts. The former access traffic data through a Web application, which allows to build up a query through a set of pre-defined selection tools, while the latter can submit their own queries through Apache Drill and Spark.

At present, we implemented a subset of the advanced data analytics functionalities actually supported by the architecture. Future work will be devoted to enrich the system with more advanced functionalities like trace clustering, trajectory prediction, and synthesis of user behaviors by process mining techniques.

References

1. Bernaschi, M., Celestini, A., Guarino, S., Lombardi, F., Mastrostefano, E.: Traffic data: exploratory data analysis with Apache Accumulo. In: Leuzzi, F., Ferilli, S. (eds.) TRAP 2017. AISC, vol. 728, pp. 129–143. Springer, Cham (2018). https://doi.org/10.1007/978-3-319-75608-0_10
2. Bernaschi, M., Celestini, A., Guarino, S., Lombardi, F., Mastrostefano, E.: Unsupervised classification of routes and plates from the *Trap-2017* dataset. In: Leuzzi, F., Ferilli, S. (eds.) TRAP 2017. AISC, vol. 728, pp. 97–114. Springer, Cham (2018). https://doi.org/10.1007/978-3-319-75608-0_8
3. Chen, M., Mao, S., Liu, Y.: Big data: a survey. Mob. Netw. Appl. **19**, 171–209 (2014)
4. Di Blas, N., Mazuran, M., Paolini, P., Quintarelli, E., Tanca, L.: Exploratory computing: a challenge for visual interaction. In: Proceedings of the 2014 International Working Conference on Advanced Visual Interfaces, AVI 2014, pp. 361–362. ACM, New York (2014)
5. Giannotti, F., et al.: Unveiling the complexity of human mobility by querying and mining massive trajectory data. VLDB J. **20**(5), 695–719 (2011)
6. Leuzzi, F., Del Signore, E., Ferranti, R.: Towards a pervasive and predictive traffic police. In: Leuzzi, F., Ferilli, S. (eds.) TRAP 2017. AISC, vol. 728, pp. 19–35. Springer, Cham (2018). https://doi.org/10.1007/978-3-319-75608-0_3
7. Deshpande, P.M., Ramasamy, K.: Data warehousing, multi-dimensional data models and OLAP. In: Rivero, L.C., Doorn, J.H., Ferraggine, V. (eds.) Encyclopedia of Database Technologies and Applications. IGI Global, Hershey (2005)

Traffic Data Classification for Police Activity

Stefano Guarino[1], Fabio Leuzzi[2], Flavio Lombardi[1(✉)],
and Enrico Mastrostefano[1]

[1] Institute for Applied Mathematics (IAC-CNR), Rome, Italy
{s.guarino,e.mastrostefano}@iac.cnr.it, flavio.lombardi@cnr.it
[2] Italian National Police, Rome, Italy
fabio.leuzzi@poliziadistato.it

Abstract. Traffic data, automatically collected *en masse* every day, can be mined to discover information or patterns to support police investigations. Leveraging on domain expertise, in this paper we show how unsupervised clustering techniques can be used to infer trending behaviors for road-users and thus classify both routes and vehicles. We describe a tool devised and implemented upon openly-available scientific libraries and we present a new set of experiments involving three years worth data. Our classification results show robustness to noise and have high potential for detecting anomalies possibly connected to criminal activity.

1 Introduction

The advances and diffusion of automatic Number Plate Reading Systems (NPRS) make available to Law Enforcement Agencies (LEAs) large amounts of traffic data having great potential for supporting crime detection and police investigations at large, and that must be sent to a police force to be compliant with italian privacy regulations. In this context, data sent to police systems are affected by several open issues [7], and are large and noisy, justifying the need for effective and scalable techniques and tools, capable of extracting meaningful traffic patterns and identifying anomalous and criminal behaviors. To foster contributions by the research community, the Italian National Police (INP) provided a dataset containing anonymized transits to promote their advanced analysis.

Following on from previous work on the TRAP dataset [2], we report on a new set of experiments using a revised and extended classifier for traffic data. Our classifier allows inferring many characteristics of the gate network (without help from the NPRS owner), is suitable for extension to other NPRS, and is reasonably robust to noise. Notably, the produced classification is informative enough to allow *a posteriori* understanding and labelling of the clusters by police officers. The contributions of the present paper include: (i) a revision of the tool and a deeper validation of its performance upon a much larger dataset, spanning over three years (2014–2016); (ii) a two-tier analysis of the 2016 dataset, based on recursively applying vehicles clustering/classification to gain a more

© Springer Nature Switzerland AG 2018
M. Ceci et al. (Eds.): ISMIS 2018, LNAI 11177, pp. 169–178, 2018.
https://doi.org/10.1007/978-3-030-01851-1_17

detailed description of the distinctive features of road-users; (iii) an insight into the benefits of leveraging domain expertise to interpret the results brought by the classifier, thanks to the collaboration between IAC-CNR (Institute for Applied Mathematics) and the INP.

2   Related Work

Extracting meaningful information from the large datasets collected by cameras, GPS tracking, and other means, is not a trivial task. In the following we provide a brief overview of main related work.

A broad overview of both problems and solutions is given by Sivaraman's et al. [11], detailing advances in vehicle detection, and also characterizing on-road behavior and introducing some performance metrics and benchmarks. Our work differs in that we cannot control or tune data acquisition mechanisms but simply infer information based on the classification of the given raw data. Frequent road traffic patterns are analyzed by Jindal et al. [4], mining spatio-temporal periodic patterns as representatives of the road network. Jindal uses density-based clustering to cluster the sensors on the road network based on the similarities between their periodic behavior as well as their geographical distance, thus combining similar nodes to form a road network with larger but fewer nodes. Our present approach is more general as it can be applied to heterogeneous data with little or no prerequisites. Also Kiran et al. [5] aim at discovering partial periodic itemsets in temporal databases. They introduce a new measure (periodic-frequency) to determine the periodic interestingness of itemsets by taking into account their number of cyclic repetitions in the entire data. Kiran's contributions is interesting and can be deployed for a given analysis following our approach described here. A statistical analysis to identify significant contiguous set of road segments and time intervals was performed by Necula [9], mining vehicle traces to extract outlier traffic patterns. One similarity to what we did is that he organized the road infrastructure as segments in a graph to tracks the visits for each vehicle. Finally, Grossi et al. [3] aim at clustering data using a machine learning and an optimization-finding approach. They present a constraint programming model for a centroid based clustering and one for a density based clustering. In particular, they show that density-based clustering by constraint programming is similar to a label propagation problem. We plan to further investigate benefits and pitfalls of our approach in comparison to different approaches such as Grossi's.

3   Unsupervised Classification of Road Usage

In the absence of a complete ontology of road users, unsupervised learning is one feasible option for detecting and classifying highly variable and unpredictable/unknown criminal activities (*i.e.*, not just straightforward examples, such as cloned plates). Along this line, we designed a classifier that produces a clear and informative output to be fed to officers for interpretation and labeling, relying on the assumption that similar road users exhibit similar travel behaviors on similar routes.

In [1,2] we presented a statistical analysis of the TRAP dataset that highlighted the presence of apparently incoherent transit records (*e.g.*, supersonic speeds and vehicles crossing 300+ gates in a day). With no control upon the data collection process, preprocessing the available data without risk of losing relevant information is however problematic. We only cut out police vehicles and vehicles whose nationality was unknown, which is symptomatic of an OCR error. Notably, this yields significantly fewer vehicles in total for the 2016 dataset when compared with results in [2]. Anomalies can be hardly distinguished from errors, to the point that inferring the actual itinerary of a vehicle and breaking down its transits sequence into a series of journeys is not always possible. Without imposing restrictions on *feasible* itineraries, our approach is based on a very general definition of *route* as "any pair of gates traveled at least once by at least one vehicle". Practically speaking, for each vehicle $\mathcal{V}$ we consider the time-ordered series of all transit records $\{(\mathcal{V}, \mathcal{G}_i, \mathcal{T}_i)\}_{i \geq 0}$, and we say that $\mathcal{V}$ traveled over the route $(\mathcal{G}_i, \mathcal{G}_{i+1})$ in time $\mathcal{T}_{i+1} - \mathcal{T}_i$. Notably, in our model $\mathcal{G}_i$ and $\mathcal{G}_{i+1}$ are *not* necessarily adjacent on the highway – indeed, this information might not even be available. Simply, there must be no other record $(\mathcal{V}, \mathcal{G}', \mathcal{T}')$ with $\mathcal{T}' \in [\mathcal{T}_i, \mathcal{T}_{i+1}]$.

For each route, we extract *global* statistics by considering all travel times for journeys along that route, irrespective of the vehicle. Routes are then classified based on the similarity of their global statistics, whereas vehicles are clustered according to how their individual statistics fit the global statistics for each class of routes. The rationale is that the global statistics of a route are correlated to its unique characteristics (*e.g.*, length, number of lanes, traffic patterns, number of service areas), and that the individual statistics of a vehicle can be interpreted as indicators of its driver's travel habits (*e.g.*, having a preferred class of routes or travelling below/above the average speed). To tune the parameters of our classifier we used the map released with the TRAP dataset (see [1]), concluding that the classifier benefits from taking the logarithm of all travel times, so that time differences reflect the proportion minute : second = hour : minute. As detailed in [2], this choice maximizes the accuracy of our route classifier in detecting two types of routes: (i) those compliant with the direction of travel, and (ii) those delimited by two adjacent gates. Some technical details on our classifier follow:

Routes Clustering. We apply Gaussian Mixture Modeling (GMM) [8] to the set of all travel times of a fixed route. The obtained Gaussian Mixture (GM) $\{(G_i, w_i)\}_{i=0}^{M-1}$ can be interpreted as a set of weighted driving trends for that route. For instance, Gaussian G_0 models driving on the route when free, G_1 models driving when congested, G_2 models stopping at a service area, and so on. Since the inferred GMs are *de facto* probability distributions, we can compare them using the Earth Mover's Distance (EMD) [10]. In particular, we use the Kullback-Leibler divergence [6] as the *ground distance* between any two Gaussians taken from the two mixtures (KL has a closed form for Gaussians). We apply Hierarchical Agglomerative Clustering [12] to find the desired routes classification.

Vehicles Clustering. Using each route's GM as a generative model, we define a $N \times M$ matrix for each vehicle, where N is the number of routes clusters and M the number of Gaussians composing each route's GM. The element (i, j) of the matrix tells us how frequently that vehicle's travels can be associated with Gaussian j of (a route of) cluster i, so that the matrix summarizes the behavior of that vehicle. We then use the norm of the difference of two such matrices to measure the distance between the corresponding vehicles, in such a way that two vehicles are similar if they have two similar travel times patterns. Finally, we use K-means clustering to classify plates, setting K based on a silhouette score. Notably, M and N need to be fixed for all routes to allow comparing matrices (a fixed M is also consistent with EMD-based metrics). We experimentally found $M = N = 5$ to be a good choice (see [2]).

4   Experimental Findings

This section summarizes experimental results over data spanning three years.

4.1   The 2016 Dataset

Figure 1 shows a table that summarizes the results of our routes classifier over the 2016 dataset. Specifically, in Fig. 1a for each route cluster we report the cluster size and the GM of the most representative route (the *medoid*), which is also showed graphically in Fig. 1b. From Fig. 1a it is apparent that our classifier clusters all *reasonable* routes into Routes Cluster 3 (RC 3). RC 3 is composed of routes that are conformant to the direction of travel. In fact, it is characterized by 88% of travels of duration ~ 10 min, 7% of duration ~ 25 min, 3% of duration ~ 53 min, plus a tail that weighs 2% in total. By isolating this set of routes, we could, in principle, identify the most reliable source of data for comprehending and classifying driving behaviors, other than for finding evidence of standard violations (*e.g.*, travelling above the speed limit). In addition, RC 3 can be used as a base model to understand novel data collected from the same highway and to detect infrastructural anomalies (*e.g.*, switching on/off of a gate, opening/closing of a service area, a parking area, a ramp to enter/exit the highway). In a more general view, being able to classify routes means gaining a deeper understanding of the road network and its dynamics, and allowing the detection of outliers or patterns (also identifying the vehicles if needed), thus facilitating prompt interventions and preventing risks/crimes.

   If we switch to vehicles, in Fig. 2 we describe each vehicles cluster highlighting the size of the cluster and the three most frequent behaviors. A behavior is expressed in terms of: (i) routes cluster; (ii) gaussian and its mean time; (iii) weight (*i.e.*, frequency). For instance, the first quadruplet of Vehicles Cluster 2 (VC 2) in Fig. 2a says that vehicles belonging to that cluster travel for 71% of their time along routes of RC 3 with an average travel time of ~ 10 min (associable with Gaussian 0 of RC 3). Based on the classification reported in Fig. 2a we can filter out uninteresting vehicles and identify those clusters that

RC	Size	G_0		G_1		G_2		G_3		G_4	
		μ	w	μ	w	μ	w	μ	w	μ	w
0	5	21h 15m 10s	0.42	23h 59m 54s	0.16	48h 15m 41s	0.17	154h 35m 43s	0.15	597h 59m 24s	0.1
1	655	1h 32m 14s	0.17	4h 35m 31s	0.13	23h 33m 16s	0.27	118h 53m 15s	0.27	921h 37m 39s	0.16
2	1	1h 7m 12s	0.01	35h 33m 54s	0.08	193h 30m 28s	0.27	1395h 46m 13s	0.32	3636h 41m 17s	0.32
3	67	0h 10m 26s	0.88	0h 25m 20s	0.07	0h 53m 36s	0.03	11h 39m 29s	0.01	349h 1m 7s	0.01
4	1	0h 38m 51s	0.18	0h 52m 30s	0.39	1h 13m 57s	0.19	13h 33m 11s	0.15	543h 25m 46s	0.09

(a) For each routes cluster: size and GM of the most representative route (the *medoid*) of that cluster. The GM is summarized reporting only the mean μ and the weight w of each constituent Gaussian.

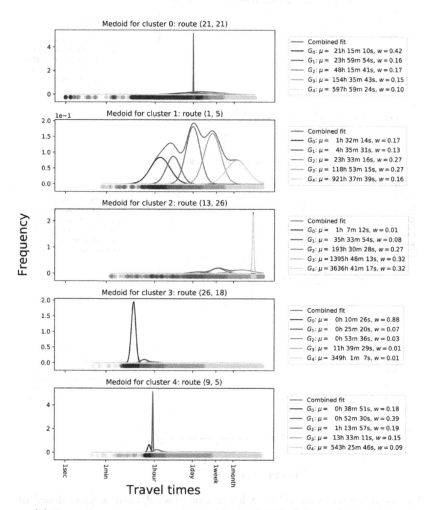

(b) For each routes cluster: scatter plot and GM of the medoid route.

Fig. 1. Routes clusters for the 2016 dataset.

VC	Size	Behavior 1				Behavior 2				Behavior 3			
		RC	G	μ	w	RC	G	μ	w	RC	G	μ	w
0	943154	3	1	0h 25m 20s	0.2	3	0	0h 10m 26s	0.16	1	1	4h 35m 31s	0.13
1	263718	1	4	921h 37m 39s	0.5	1	0	1h 32m 14s	0.22	1	1	4h 35m 31s	0.07
2	2008398	3	0	0h 10m 26s	0.71	3	1	0h 25m 20s	0.08	1	3	118h 53m 15s	0.04

(a) For each vehicles cluster: size and top 3 behaviors for vehicles of that cluster. A behavior is defined in terms of: a RC, a Gaussians of that RC characterized by an average travel time μ, and a weight w measuring the frequency of that behavior.

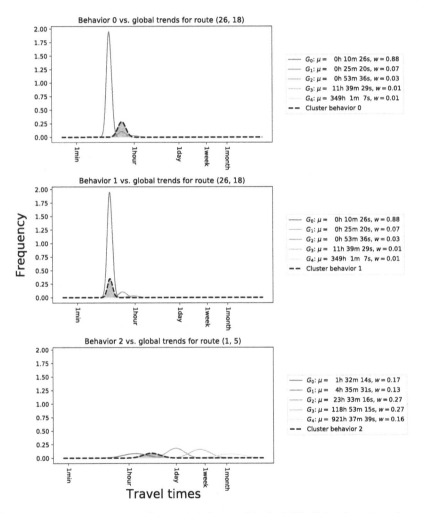

(b) The top 3 behaviors for VC 0: each behavior (dashed/filled) is plotted against the GM of the RC where it takes place to highlight divergences with typical drivers.

Fig. 2. Vehicles clusters for the 2016 dataset.

VC	Size	Behavior 1				Behavior 2				Behavior 3			
		RC	G	μ	w	RC	G	μ	w	RC	G	μ	w
0	262803	3	1	0h 25m 20s	0.43	3	0	0h 10m 26s	0.19	3	2	0h 53m 36s	0.12
1	247836	3	0	0h 10m 26s	0.33	1	1	4h 35m 31s	0.1	3	1	0h 25m 20s	0.1
2	51099	1	1	4h 35m 31s	0.97	1	3	118h 53m 15s	0.01	1	2	23h 33m 16s	0.01
3	42232	1	3	118h 53m 15s	0.96	1	2	23h 33m 16s	0.01	1	4	921h 37m 39s	0.01
4	41406	3	1	0h 25m 20s	0.89	3	0	0h 10m 26s	0.03	1	1	4h 35m 31s	0.02
5	29799	3	2	0h 53m 36s	0.51	3	4	349h 1m 7s	0.27	3	1	0h 25m 20s	0.05
6	48045	0	4	597h 59m 24s	0.37	0	3	154h 35m 43s	0.24	0	0	21h 15m 10s	0.12
7	182387	1	3	118h 53m 15s	0.22	1	2	23h 33m 16s	0.18	1	1	4h 35m 31s	0.17
8	37547	1	2	23h 33m 16s	0.96	1	3	118h 53m 15s	0.02	1	1	4h 35m 31s	0.01

(a) For each vehicles cluster: size and top 3 behaviors for vehicles of that cluster. A behavior is defined in terms of: a RC, a Gaussians of that RC characterized by an average travel time μ, and a weight w measuring the frequency of that behavior.

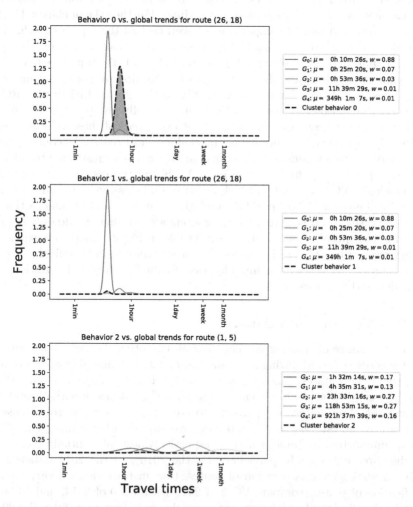

(b) The top 3 behaviors for VC 0.4: each behavior (dashed/filled) is plotted against the GM of the RC where it takes place to highlight divergences with typical drivers.

Fig. 3. Vehicles subclusters for VC 0 of the 2016 dataset.

may be of particular interest for LEAs. Specifically, we opt for discarding VC 1 and VC 2 and focusing on VC 0. The vehicles in VC 1, in fact, exhibit a very odd set of behaviors: most of their travels occur over anomalous routes, and 50% of their travels last more than one month. A possible explanation is that these vehicles have been traveling on surrounding roads and only occasionally entering the highway section under study. More generally, we infer that the available data miss crucial information which would be needed to properly characterize these vehicles. Conversely, VC 2 is the most numerous and (apparently) the most reasonable cluster, characterized by a significant preference for *rational* travel times (*e.g.*, 71% of travels lasting $\sim$ 10 min). This probably means that VC 2 is mostly composed by *standard* road users.

Differently from VC 1 and VC 2, the vehicles in VC 0 exhibit a reasonable yet somewhat unexpected pattern of travel times. For the sake of clarity, Fig. 2b proposes a direct graphical comparison between each of the top three behaviors for VC 0 and the global travel habits for the same route. Two aspects of the travel times distribution of VC 0 stand out: it is very balanced (the top three behaviors only weigh 49% in total), and travel times of $\sim$ 25 min are more frequent than travel times of $\sim$ 10 min for RC 3, contrarily to the global statistics for RC 3. These two observations suggest that VC 0 may actually group together vehicles with very different travel habits, and that (at least) part of these vehicles spend suspiciously more time than needed to travel along short routes. To come up with a more fine-grained classification, we ran again the classifier restricted to vehicles of VC 0, reporting our findings in Fig. 3. Analyzing the results, we concluded that vehicles in VC 0.4 – that is, VC 4 of this second classification round – could hide criminal intents. VC 0.4 is in fact a relatively small cluster of vehicles that in 89% of the cases spend $\sim$ 25 min to travel along a route that should only require $\sim$ 10 min. This peculiar behavior, clearly visible in Fig. 3b, might be explained by the presence in VC 0.4 of vehicles driven by criminals who quickly stop at all service areas to select a victim. This calls for further targeted investigation, that is deferred to future work.

4.2   The 2014 and 2015 Datasets

We got the chance of running our classifier on two other datasets, consisting of transit records collected during 2015 and 2014 over the same highway section of the 2016 TRAP dataset. Since the gate network is not consistent among these three dataset (due to inevitable changes in NPRS' positions over the years), we analyzed the three datasets separately to concurrently test the robustness of our classifier and establish if the datasets have any remarkable divergence. The results, summarized in Tables 1 and 2, respectively, are substantially consistent over the three datasets. In particular, for the 2016 and the 2014 datasets we obtain an analogous classification of routes, that in turn yields a very similar classification of plates (compare VC 2 of 2016 with VC 0 of 2014, and VC 0 of 2016 with VC 3 of 2014), although one more cluster is found for 2014. For 2015, the classifier is not able to identify reasonable routes, and further investigation of this dataset is needed to understand the reasons. As a consequence, the obtained classification of plates seems weaker, albeit still informative. A first hypothesis is that in 2015 some gates have been switched on/off (or moved) during the year, without leaving tracks of these events in the dataset. Similar changes were not reported on purpose in the 2016 dataset (for adding a further mining challenge to the task). Nevertheless, this will require further investigation.

Table 1. Classification results for the 2015 dataset.

RC	Size	G_0		G_1		G_2		G_3		G_4	
		μ	w	μ	w	μ	w	μ	w	μ	w
0	2	0h 7m 38s	0.0	2h 13m 20s	0.07	29h 4m 38s	0.3	114h 55m 29s	0.4	501h 27m 27s	0.23
1	3	0h 38m 51s	0.0	201h 33m 31s	0.11	689h 12m 3s	0.19	1921h 40m 29s	0.28	4401h 50m 14s	0.41
2	1605	0h 21m 32s	0.14	2h 17m 51s	0.35	11h 24m 30s	0.24	78h 28m 33s	0.19	491h 10m 4s	0.08
3	1	0h 0m 1s	0.02	0h 7m 13s	0.03	2h 27m 7s	0.06	43h 0m 11s	0.39	330h 2m 43s	0.5
4	1	0h 0m 1s	0.01	10h 41m 18s	0.15	23h 59m 50s	0.1	91h 7m 15s	0.53	604h 28m 30s	0.21

(a) Routes clusters: for each cluster, mean time (μ) and weight (w) of each Gaussian (G_j) of the medoid route.

VC	Size	Behavior 1				Behavior 2				Behavior 3			
		RC	G	μ	w	RC	G	μ	w	RC	G	μ	w
0	525368	2	4	491h 10m 4s	0.38	2	3	43h 0m 11s	0.24	2	2	11h 24m 30s	0.18
1	2323156	2	0	0h 21m 32s	0.74	2	1	2h 17m 51s	0.09	2	2	11h 24m 30s	0.07
2	614309	2	1	2h 17m 51s	0.54	2	0	0h 21m 32s	0.22	2	2	11h 24m 30s	0.11

(b) Vehicles clusters: for each cluster, summary of the top three behaviors, in terms of routes cluster (RC), Gaussian (G), mean time (μ) and weight (w).

Table 2. Classification results for the 2014 dataset.

RC	Size	G_0		G_1		G_2		G_3		G_4	
		μ	w	μ	w	μ	w	μ	w	μ	w
0	2	0h 7m 38s	0.0	2h 13m 20s	0.07	29h 4m 38s	0.3	114h 55m 29s	0.4	501h 27m 27s	0.23
1	3	0h 38m 51s	0.0	201h 33m 31s	0.11	689h 12m 3s	0.19	1921h 40m 29s	0.28	4401h 50m 14s	0.41
2	1605	0h 21m 32s	0.14	2h 17m 51s	0.35	11h 24m 30s	0.24	78h 28m 33s	0.19	491h 10m 4s	0.08
3	1	0h 0m 1s	0.02	0h 7m 13s	0.03	2h 27m 7s	0.06	43h 0m 11s	0.39	330h 2m 43s	0.5
4	1	0h 0m 1s	0.01	10h 41m 18s	0.15	23h 59m 50s	0.1	91h 7m 15s	0.53	604h 28m 30s	0.21

(a) Routes clusters: for each cluster, mean time (μ) and weight (w) of each Gaussian (G_j) of the medoid route.

VC	Size	Behavior 1				Behavior 2				Behavior 3			
		RC	G	μ	w	RC	G	μ	w	RC	G	μ	w
0	525368	2	4	491h 10m 4s	0.38	2	3	43h 0m 11s	0.24	2	2	11h 24m 30s	0.18
1	2323156	2	0	0h 21m 32s	0.74	2	1	2h 17m 51s	0.09	2	2	11h 24m 30s	0.07
2	614309	2	1	2h 17m 51s	0.54	2	0	0h 21m 32s	0.22	2	2	11h 24m 30s	0.11

(b) Vehicles clusters: for each cluster, summary of the top three behaviors, in terms of routes cluster (RC), Gaussian (G), mean time (μ) and weight (w).

5   Conclusions and Future Work

In this paper we applied a purely automated classification/clustering method to analyze a traffic dataset extended over three years. All work was performed with self-written python scripts only leveraging on state of the art open source scientific libraries. Our classification mechanism provides significant information about the obtained clusters of routes/vehicles and it is inherently robust towards many quality issues that are unavoidable with real datasets. Our findings show that unsupervised clustering can extract usable knowledge building upon a formal and descriptive definition of behavior of a vehicle, only relying on a type of information (*who* was driving *where* and *when*) that is usually easily available to police forces. Finally, domain expertise allowed us to further understand/interpret the obtained results, paving the way for the deployment of a tool that police officers can practically use in their daily activities.

References

1. Bernaschi, M., Celestini, A., Guarino, S., Lombardi, F., Mastrostefano, E.: Traffic data: exploratory data analysis with Apache Accumulo. In: Leuzzi, F., Ferilli, S. (eds.) TRAP 2017. AISC, vol. 728, pp. 129–143. Springer, Cham (2018). https://doi.org/10.1007/978-3-319-75608-0_10

2. Bernaschi, M., Celestini, A., Guarino, S., Lombardi, F., Mastrostefano, E.: Unsupervised classification of routes and plates from the *Trap-2017* dataset. In: Leuzzi, F., Ferilli, S. (eds.) TRAP 2017. AISC, vol. 728, pp. 97–114. Springer, Cham (2018). https://doi.org/10.1007/978-3-319-75608-0_8

3. Grossi, V., Monreale, A., Nanni, M., Pedreschi, D., Turini, F.: Clustering formulation using constraint optimization. In: Bianculli, D., Calinescu, R., Rumpe, B. (eds.) SEFM 2015. LNCS, vol. 9509, pp. 93–107. Springer, Heidelberg (2015). https://doi.org/10.1007/978-3-662-49224-6_9

4. Jindal, T., Giridhar, P., Tang, L.A., Li, J., Han, J.: Spatiotemporal periodical pattern mining in traffic data. In: UrbComp 2013 (2013)

5. Kiran, R.U., Shang, H., Toyoda, M., Kitsuregawa, M.: Discovering partial periodic itemsets in temporal databases. In: Proceedings of SSDBM 2017, pp. 30:1–30:6, New York, NY, USA. ACM (2017)

6. Kullback, S., Leibler, R.A.: On information and sufficiency. Ann. Math. Stat. **22**(1), 79–86 (1951)

7. Leuzzi, F., Del Signore, E., Ferranti, R.: Towards a pervasive and predictive traffic police. In: Leuzzi, F., Ferilli, S. (eds.) TRAP 2017. AISC, vol. 728, pp. 19–35. Springer, Cham (2018). https://doi.org/10.1007/978-3-319-75608-0_3

8. McLachlan, G., Peel, D.: Finite Mixture Models. Wiley, New York (2004)

9. Necula, E.: Analyzing traffic patterns on street segments based on GPS data using R. Transp. Res. Procedia, **10**, 276–285 (2015). EWGT 2015, Delft

10. Rubner, Y., Tomasi, C., Guibas, L.J.: The Earth Mover's Distance as a metric for image retrieval. Intl. J. Comput. Vis. **40**(2), 99–121 (2000)

11. Sivaraman, S., Trivedi, M.M.: Looking at vehicles on the road: a survey of vision-based vehicle detection, tracking, and behavior analysis. IEEE Trans. Intell. Transp. Syst. **14**(4), 1773–1795 (2013)

12. Ward Jr., J.H.: Hierarchical grouping to optimize an objective function. J. Am. Stat. Assoc. **58**(301), 236–244 (1963)

Multipurpose Web-Platform for Labeling Audio Segments Efficiently and Effectively

Ayman Hajja[1]([⊠]), Griffin P. Hiers[1], Pierre Arbajian[2], Zbigniew W. Raś[1,3], and Alicja A. Wieczorkowska[3]

[1] Department of Computer Science, College of Charleston, 66 George Street, Charleston, SC 29424, USA
hajjaa@cofc.edu, hiersgp@g.cofc.edu, ras@uncc.edu
[2] Department of Computer Science, University of North Carolina, 9201 University City Blvd., Charlotte, NC 28223, USA
arbajian@uncc.edu
[3] Polish-Japanese Academy of Information Technology, Koszykowa 86, 02-008 Warsaw, Poland
alicja@poljap.edu.pl

Abstract. One of the principal reasons for the success of machine learning discoveries can be attributed to the utilization of large sums of labeled datasets used to train various learning models. The availabilities of annotated data depend, to a large extent, on the nature of the domain, and how easy it is to obtain labeled data-points. One of the areas that we believe still lacks substantial labeled data is audio. This is not surprising, since labeling audio segments can be rather tedious and time-consuming, mainly due to the temporal nature of it. In this paper, we present a free and open-source web-based platform that we developed, which allows individuals and research teams to crowdsource large sums of labeled audio segments efficiently and effectively. Once an individual or a team signs up to use the platform as researchers, they will be granted administrative access that will enable them to upload their own audio files, and customize the labeling and data collection process according to their study needs. Examples of customizing the study include listing the different labels of interest, specifying the duration of audio segments and how they should be extracted from the audio file(s), and dictating how labelers should be prompted with the audio segments based on a set of pre-determined user-defined rules. Our system will automatically handle generating the audio segments from the audio files, presenting labelers with an intuitive interface using the rules specified by the study administrators, and finally recording the labelers' responses and providing them to the administrators of the study in a readable and easy-to-access format.

Keywords: Crowdsourcing · Labeling · Human computation
Web application · Speech analysis

© Springer Nature Switzerland AG 2018
M. Ceci et al. (Eds.): ISMIS 2018, LNAI 11177, pp. 179–188, 2018.
https://doi.org/10.1007/978-3-030-01851-1_18

1 Introduction and Background

Voice-based analysis solutions fall into two groups: the lexical- and the semantic-based analysis, and the prosodic-based. In the lexical analysis category, the speaker's voice is first transcribed into words and sentences, then text analysis is performed [6,7]. Whereas in the prosodic variant, researchers are interested in exploiting the tonality of voice such as pitch, pauses, rhythms, relative emphases, and certain forms of stutter [2,3]; for instance, when analyzing prosodic aspects of a voice recording, one could detect fear or anger, even if the semantic meaning is void of any indications for such emotional state.

Prosodic analysis of voice is a broad and rich field with interesting implementation possibilities [1,5], it encompasses important audio traits such as voice quality, which is fundamental to human communication for conveying meaning, but also for expressing feeling and disposition. A speaker's tone of speech can be telling as to his emotional state, a human listener could tell if a certain utterance is joyful or sad. One can easily argue that humans (and many animals for that matter) are particularly superior to machines at detecting emotional states of other beings; that being said, with the help of large sums of labeled data, we believe that a well-trained machine learning model would be able to make some of these vocal-based distinctions, somewhat successfully.

In this work, we present a web-based platform that we developed, which will allows individuals and research teams to crowdsource large sums of labeled audio segments efficiently and effectively. We believe that we are filling a growing need for an easy-to-use, open-source and freely-available tool that can be used by the machine learning research community; the source code for this project is hosted on GitHub (https://github.com/grifffin/Web-Platform-for-Labeling-Audio).

We envision that the rich set of possibilities for the detection of soft speech content (i.e. fear, conviction, anger, etc.) would be of particular interest to our targeted audience; as a result, we expect that potential users of our system will consist of researchers and research teams who are particularly interested in the development of automatic intelligent speech emotion detection systems, many of whom we believe will be attracted from the social science fields such as psychology.

2 Proposed System

In this section, we will describe the functionalities that our proposed web platform provides, along with screenshots for the most important user interfaces of it. Through the use of our system, study administrators will be able to create a customized crowdsourcing study that will enable them to collect data quickly and efficiently according to a set of user-defined rules, which will be further used by the system to optimize the data-collection process. The developed web platform will allow users to upload audio files from which the segments will be extracted, specify the duration of segments and the potential labels for each segment, and define other study-related features that will be covered in the following subsections.

Once an administrator creates a study according to his or her desired rules, he or she will be provided with a randomly generated string code that can be consequently shared with other users of the website to label the audio segments; we will refer to users that label audio segments by the term *workers*. A *worker*, who could potentially be any individual such as a lab member or a non-affiliated contributor, can contribute to the labeling process by joining a study after creating a profile on our website and entering the corresponding code provided by the study administrator(s).

One of the main novelties of our proposed system stems from the fact that our system dynamically assigns audio segment to workers based on the previous labels provided by other workers for that particular segment, and in accordance with the rules specified by the study administrators. As will be shown below, by dynamically assigning workers audio segments, our system optimizes the data-collection process in such a way that will result in maximum usable labeled audio segments. Finally, once a study is completed, administrators will be able to download comma-separated values (CSV) files to be used for further analysis and model building. In addition to being able to download CSV files, administrators are also able to download a zip file containing folders of the labeled audio segments (grouped by label), appropriately-named for convenient mapping with the CSV files; a thorough explanation of the results page will be provided in Subsect. 2.3.

2.1 Creating a Crowdsourcing Study

In this subsection, we will provide a summary of all the pieces of information that a study administrator needs to provide our system with to create a crowdsourcing study; we will also discuss the impact that each value has on the resulting study. Here is the list of information with their description:

File(s): Every crowdsourcing study needs to be linked to one (or more) audio files that the study administrator would need to upload prior to creating the study. Currently, our system allows study administrators to upload up to 20 audio files (in .mp3 or .wav formats) with no limit on file size. The uploaded files can be further used with any number of studies.

Segment Size: As mentioned in Sect. 1, one of the motives that led us to develop this system is to facilitate the process of labeling short segments of audio files. The segment duration (in seconds) needs to be provided by the study administrators when the study is being created, and our system will automatically present workers with audio segments accordingly. The choice of this value will highly depend on the domain of the study. For example, if an administrator wants to label audio segments for stutter detection, then a reasonable segment size would be between .5 and 2 s [8]; if on the other hand, the goal of the study is to label audio segments for the purpose of emotions-in-speech detection, then a much bigger segment size would be needed [4].

Stride Size: The stride size can be defined as the distance between (the beginning of) any two consecutive segments, and will consequently determine whether there will be an overlap in the segments or not, and what the exact ratio of the overlap is. In addition to having a direct impact on the overlap ratio, the stride size (in conjunction with the audio file(s) duration) will also determine the number of candidate segments generated for labeling. Next, we will demonstrate the effects of the stride size on hypothetical study.

Let us assume that we have a single audio file that has the duration of 18 s; let us also assume that the user-defined stride size in our study is 2 s, which means that the beginning of each segment that will be presented to workers will be a multiple of 2 (0, 2, 4, ..., 16). Now, if the duration of each segment that needs to be labeled happens to be also 2 s (equal to the stride size), then the generated segments will exhibit no overlap between them. Figure 1 shows the segments that the workers will be prompted to label.

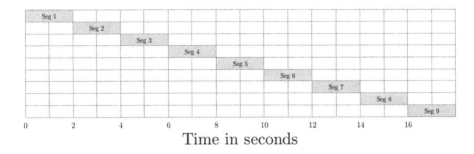

Fig. 1. Resulting segments when segment size and stride size are both 2 s

If for example, the segment duration is greater than the stride size (say segment size is equal to 3, and stride size is equal to 2), then the resulting segments will contain an overlap in the segments that the workers will be prompted to label. Figure 2 shows the segments that the workers will be prompted to label.

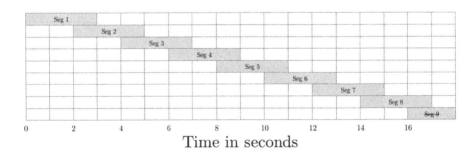

Fig. 2. Resulting segments when segment size is 3 s and stride size is 2 s

As can be shown in Fig. 2, we notice here that there is a 1/3 overlap between each two consecutive segments. This can be useful, and rather necessary, when the start and end times of a label needs to be precise, such as in the case of stutter detection. The ratio of overlap p can be defined as:

$$max\left(0, 1 - \frac{\text{stride size}}{\text{segment size}}\right)$$

We needed to introduce a upper bound (0) to prevent our ratio of overlap p from resulting in a negative value when the stride size is greater than the segment size.

The number of unique segments n can be defined as:

$$\sum_{i=1}^{m} \left\lfloor \frac{\text{duration of file i}}{\text{stride size}} \right\rfloor$$

where m is the number of files associated with the given study.

Labels: There is no limit on the number of different labels that workers can be prompted with for a given study; our system allows study administrators to define as many labels as their study requires.

Maximum Number of Overall Responses: Study administrators can set a limit on the total number of responses gathered from workers; this value can be useful to define when a study would end, or to limit expenses in the case when workers are being compensated. When this value is not specified, there will be no limit on the number of responses, and the study will remain open indefinitely.

Maximum (and Minimum) Number of Responses for a Segment: These two values are crucial to the success of our platform, and they play a major role (jointly with the stop threshold value, which will be explained next) in the optimization and adaptable data-collection process being developed in our website. Our system will determine the exact number of responses (within the minimum and maximum range) that will satisfy the stop threshold value depending on the previous responses for every unique segment, as will be explained next.

Stop Threshold for a Single Segment: The stop threshold can be defined as the minimum percentage of agreeability that study administrators require for any segment to be marked useful (or credible) for further analysis. As soon as a particular segment becomes *credible*, and as long as the number of responses for that segment is within the minimum and maximum range, then our system will not present this segment for additional labeling.

Let us demonstrate how the stop threshold, the minimum and maximum responses, and the number of labels affect the labeling process when they are used altogether in a single study. Let us assume that the number of unique labels for a given study is 4, the minimum and maximum number of responses for a single segment is 3 and 10, respectively, and that the stop threshold value is 70%.

Since any segment needs to be labeled by at least 3 (minimum responses) unique workers to become *credible*, our system will start by prompting 3 unique workers to label some random segment s, which will result in three different potential responses:

1. The three workers will agree on one label (highest percentage is 100%)
2. Only two workers will agree on one label (highest percentage is 66.67%)
3. Each worker will choose a different label (highest percentage is 33.33%)

Notice here that the third scenario is only possible because the number of labels in this study is greater than two.

If the first case scenario occurs, then our system would mark the segment s as *credible*, and no additional labeling will be needed for that segment. If on the other hand, either scenario 2 or 3 occur, then our system will need to ask a fourth worker to label segment s, which would result in the three following potential responses:

1. Three (out of four) workers agree on one label (highest percentage is 75%)
2. Two (out of four) workers agree on one label (highest percentage is 50%)
3. The four workers choose a different label each (highest percentage is 25%)

Again, notice here that the third scenario is only possible because the number of labels in our study is greater than three (4). Also, we notice here that we cannot have all four workers agree on a single label, since for this to occur, the three original workers would have agreed on the same label initially, which would have caused our system to mark segment s as credible, and stop presenting for further labeling.

Based on the three possible scenarios listed above, only if the first one occurs, our system would mark segment s as credible. If either the second or third scenario occurs, then our system will prompt two more workers to label segment s. The reason why our system will ask for two additional responses instead of one is because in this case, it would be impossible for segment s to become *credible* by obtaining only one other response, due to the value of the stop threshold specified (70%); to elaborate, here are the two scenarios that we would end up with if we ask for only one additional response:

1. Three (out of five) workers agree on one label (highest percentage is 60%)
2. Only two (out of five) workers agree on one label (highest percentage is 40%)

For some x number of responses, a segment can potentially be *credible* only if the minimum number of responses (out of the x total) that satisfies the stop threshold is equal to the minimum number of responses that satisfies the stop threshold for $x - 1$ responses.

For example, when the total number of responses is 3, the minimum number of responses that satisfies the stop threshold would be 3 (since 2/3 is less than 70%), and since the minimum number of responses required to satisfy the stop threshold in a total of 4 responses is also 3 (3/4 is greater than 70%), then that would imply that we can have a *credible* segment using 4 responses. In the case of five responses, that will not be the case since, and we cannot obtain a *credible* segment. Table 1 shows the number of responses that we can obtain *credible* segments in.

Table 1. The number of responses that can yield credible segments per our example

# of total responses for a given segment	3	4	5	6	7	8	9	10
Min. # of responses needed to satisfy stop threshold (70%)	3	3	4	5	5	6	7	7
Can segment potentially be marked *credible*?	✓	✓	✗	✗	✓	✗	✗	✓

Notice here that the minimum number of responses that satisfies the stop threshold also depends on the number of labels in our study. Our website will dynamically (and on-the-fly) show whether a segment can be *credible* or not for all the number of responses between the minimum and maximum values provided. Figure 3 shows a screenshot of a filled study.

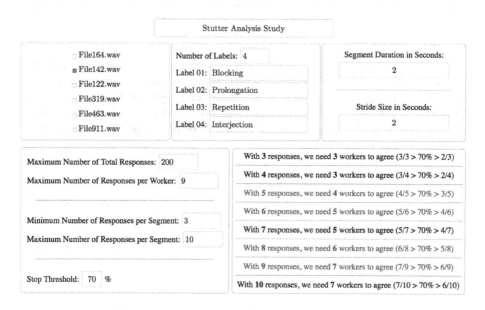

Fig. 3. Interface for creating a new crowdsourcing study

2.2   Worker Page

Once the study is created by an administrator, he or she will be provided with a random string code that can be shared with other users of the system, primarily workers. For workers to contribute to the labeling process, they would need to create an account on our website, and enroll in the study by entering the code associated with it.

Workers who join a study can request audio segments (by visiting the study page), which would be provided by our platform according to the study rules and specifications, and in accordance with the algorithm provided above. After a worker listens to the audio segment, he or she can select one of the pre-defined labels by clicking on the corresponding button; a worker also has the option of re-playing the segment if desired. Figure 4 shows a screenshot of the worker interface for a stutter analysis study with four labels.

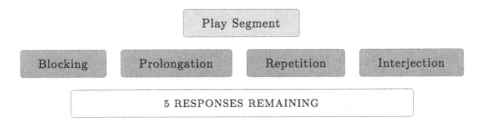

Fig. 4. Worker interface that shows available functions for a stutter analysis study

2.3   Results Page

At any point during the lifetime of a study, administrators can terminate the study or view its results. The study's responses are made available in a CSV file format, containing the filename, start and end segment time, distribution of workers' responses, and percentage of most-agreed-upon label; administrators can also play the audio segment from the results page. Figure 5 shows a screenshot of the results page.

ID	Filename	Start Time	End Time	Blocking	Prolongation	Repetition	Interjection	Final Label	Percent	Play
1	F142.wav	00.00	02.00	7	1	1	1	Blocking	70%	▶
2	F142.wav	02.00	04.00	2	2	0	2	Inactive	NA	▶
3	F142.wav	04.00	06.00	0	0	3	0	Repetition	100%	▶
4	F142.wav	06.00	08.00	0	0	5	2	Repetition	71%	▶
5	F142.wav	08.00	10.00	1	3	0	0	Prolongation	75%	▶
6	F142.wav	10.00	12.00	3	3	1	0	Inactive	NA	▶
7	F142.wav	12.00	14.00	1	1	5	0	Repetition	71%	▶
8	F142.wav	14.00	16.00	5	1	1	0	Blocking	71%	▶
9	F142.wav	16.00	18.00	1	0	0	3	Interjection	75%	▶

Download CSV   Download Audio Segments

Fig. 5. Screenshot of sample results from a stutter detection analysis study

In addition to being able to download a CSV file with the study results, study administrators can download the set of all labeled audio segments as WAV files for further analysis. Our system will group audio segments in folders based on the *final label* value of the files, and will compress all the folders in a single downloadable zip file. For convenience, audio segments are named based on the identification numbers of the segments provided in the table (or CSV file).

2.4   Logic Behind Segment-Worker Label Assignment

In this section, we will provide a thorough explanation of the logic that our system employs to optimize the data-collection process; before we present the algorithm however, we would like to define a few terms. Each audio segment has a start time, end time, and file associated with it; in addition, every segment exhibits one of these four statuses:

1. **Credible**: a credible segment is a segment that: (a) has been labeled x number of times, where x is between the min. and max. number of responses provided by the study administrator(s), and (b) the percentage of the most-agreed-upon label by the workers satisfies the stop threshold provided
2. **Inactive**: as soon as our system determines that a particular segment cannot reach the stop threshold defined, even if that segment has not reached its max. number of responses, that segment would be marked as *Inactive*, and our system will no longer present that segment for further labeling
3. **High Priority**: high priority segments refers to segments that have been labeled at least once, and they still have the potential of becoming *credible*
4. **Low Priority**: a low priority segment is a segment that has not been labeled by anyone yet. A random *low priority* segment will be assigned to a worker only if there are no *high priority* segments available for labeling for that particular worker

In Algorithm 1, we provide a summary of the segment-worker assignment algorithm deployed by our platform.

Algorithm 1. Summary of our implemented algorithm

---

start by creating the audio segments based on the stride size and the files associated with the study, and mark all of them as *low priority*;
for *every worker w requesting an audio segment for labeling* **do**
   populate the set H_w with all the *high priority* segments, excluding the ones that have been already labeled by worker w;
   if $|H_w|$ *is greater than or equal to 1* **then**
      randomly choose segment $h : h \in H_w$ and return to worker w;
      after worker w labels segment h, update the status of segment s to either *Credible* or *Inactive* if necessary, according to the label provided by worker w;
   else
      randomly choose and return to worker w segment $h : h \in L$, where L is the set of segments that exhibit status *low priority*;
      after worker w labels segment h, update the status of segment h to either *High prioriy* or *Credible* (in the case where the min. responses is equal to 1)
   end
end

3   Conclusion and Future Work

In this paper, we described a web-platform designed to provide researchers with the ability to efficiently and effectively gather pre-labeled audio information for further analysis. The system we have designed allows streamlined data management with easy scope extension. We have developed this system in support of an on-going research in the area of stuttered speech detection; therefore, our initial implementation was limited to audio data recordings. Future releases will support video segment labeling hence expanding the scope of possible applications.

Acknowledgments. This work was partially supported by the Research Center of the Polish-Japanese Academy of Information Technology, supported by the Ministry of Science and Higher Education in Poland.

References

1. El Ayadi, M., Kamel, M.S., Karray, F.: Survey on speech emotion recognition: features, classification schemes, and databases. Pattern Recognit. **44**(3), 572–587 (2011)
2. Ferguson, J., Durrett, G., Klein, D.: Disfluency detection with a semi-Markov model and prosodic features. In: Proceedings of the 2015 Conference of the North American Chapter of the Association for Computational Linguistics: Human Language Technologies, pp. 257–262 (2015)
3. Huang, Z., Chen, L., Harper, M.: An open source prosodic feature extraction tool. In: Proceedings of the Language Resources and Evaluation Conference (LREC) (2006)
4. Kim, S., Georgiou, P.G., Lee, S., Narayanan, S.: Real-time emotion detection system using speech: multi-modal fusion of different timescale features. In: IEEE 9th Workshop on Multimedia Signal Processing, MMSP 2007, pp. 48–51. IEEE (2007)
5. Scherer, S., Siegert, I., Bigalke, L., Meudt, S.: Developing an expressive speech labeling tool incorporating the temporal characteristics of emotion. In: LREC (2010)
6. Snover, M., Dorr, B., Schwartz, R.: A lexically-driven algorithm for disfluency detection. In: Proceedings of HLT-NAACL 2004: Short Papers, HLT-NAACL-Short 2004, pp. 157–160. Association for Computational Linguistics, Stroudsburg (2004). http://dl.acm.org/citation.cfm?id=1613984.1614024
7. Taboada, M., Brooke, J., Tofiloski, M., Voll, K., Stede, M.: Lexicon-based methods for sentiment analysis. Comput. Linguist. **37**(2), 267–307 (2011)
8. Tumanova, V., Zebrowski, P.M., Throneburg, R.N., Kayikci, M.E.K.: Articulation rate and its relationship to disfluency type, duration, and temperament in preschool children who stutter. J. Commun. Disord. **44**(1), 116–129 (2011)

A Description Logic of Typicality
for Conceptual Combination

Antonio Lieto[1,2] and Gian Luca Pozzato[1(✉)]

[1] Dipartimento di Informatica, Università di Torino, Torino, Italy
{antonio.lieto,gianluca.pozzato}@unito.it
[2] ICAR-CNR, Palermo, Italy

Abstract. We propose a nonmonotonic Description Logic of typicality able to account for the phenomenon of combining prototypical concepts, an open problem in the fields of AI and cognitive modelling. Our logic extends the logic of typicality $\mathcal{ALC} + \mathbf{T_R}$, based on the notion of rational closure, by inclusions $p :: \mathbf{T}(C) \sqsubseteq D$ ("we have probability p that typical Cs are Ds"), coming from the distributed semantics of probabilistic Description Logics. Additionally, it embeds a set of cognitive heuristics for concept combination. We show that the complexity of reasoning in our logic is ExpTime-complete as in $\mathcal{ALC}$.

1 Introduction

Inventing novel concepts by combining the typical knowledge of pre-existing ones is an important human creative ability. Dealing with this problem requires, from an AI perspective, the harmonization of two conflicting requirements that are hardly accommodated in symbolic systems: the need of a syntactic compositionality (typical of logical systems) and that one concerning the exhibition of typicality effects [1]. According to a well-known argument [2], in fact, prototypical concepts are not compositional. The argument runs as follows: consider a concept like *pet fish*. It results from the composition of the concept *pet* and of the concept *fish*. However, the prototype of *pet fish* cannot result from the composition of the prototypes of a pet and a fish: e.g. a typical pet is furry and warm, a typical fish is grayish, but a typical pet fish is neither furry and warm nor grayish (typically, it is red).

In this work we provide a framework able to account for this type of human-like concept combination. We propose a nonmonotonic Description Logic (from now on DL) of typicality called $\mathbf{T^{CL}}$ (typical compositional logic). This logic combines two main ingredients. The first one relies on the DL of typicality $\mathcal{ALC} + \mathbf{T_R}$ introduced in [3]. In this logic, "typical" properties can be directly specified by means of a "typicality" operator $\mathbf{T}$ enriching the underlying DL, and a TBox can contain inclusions of the form $\mathbf{T}(C) \sqsubseteq D$ to represent that "typical Cs are also Ds". As a difference with standard DLs, in the logic $\mathcal{ALC} + \mathbf{T_R}$ one can consistently express exceptions and reason about defeasible inheritance as well. For instance, a knowledge base can consistently express that "normally, athletes

© Springer Nature Switzerland AG 2018
M. Ceci et al. (Eds.): ISMIS 2018, LNAI 11177, pp. 189–199, 2018.
https://doi.org/10.1007/978-3-030-01851-1_19

are in fit", whereas "sumo wrestlers usually are not in fit" by $\mathbf{T}(Athlete) \sqsubseteq$ $InFit$ and $\mathbf{T}(SumoWrestler) \sqsubseteq \neg InFit$, given that $SumoWreslter \sqsubseteq Athlete$. The semantics of the $\mathbf{T}$ operator is characterized by the properties of *rational logic* [4], recognized as the core properties of nonmonotonic reasoning. $\mathcal{ALC} + \mathbf{T_R}$ is characterized by a minimal model semantics corresponding to an extension to DLs of a notion of *rational closure* as defined in [4] for propositional logic: the idea is to adopt a preference relation among $\mathcal{ALC} + \mathbf{T_R}$ models, where intuitively a model is preferred to another one if it contains less exceptional elements, as well as a notion of *minimal entailment* restricted to models that are minimal with respect to such preference relation. As a consequence, $\mathbf{T}$ inherits well-established properties like *specificity* and *irrelevance*: in the example, the logic $\mathcal{ALC} + \mathbf{T_R}$ allows us to infer $\mathbf{T}(Athlete \sqcap Bald) \sqsubseteq InFit$ (being bald is irrelevant with respect to being in fit) and, if one knows that Hiroyuki is a typical sumo wrestler, to infer that he is not in fit, giving preference to the most specific information.

As a second ingredient, we consider a distributed semantics similar to the one of probabilistic DLs known as DISPONTE [5], allowing to label axioms with degrees representing probabilities, but restricted to typicality inclusions. The basic idea is to label inclusions $\mathbf{T}(C) \sqsubseteq D$ with a real number between 0.5 and 1, representing its probability[1], assuming that each axiom is independent from each others. The resulting knowledge base defines a probability distribution over *scenarios*: roughly speaking, a scenario is obtained by choosing, for each typicality inclusion, whether it is considered as true or false. In a slight extension of the above example, we could have the need of representing that both the typicality inclusions about athletes and sumo wrestlers have a probability of 80%, whereas we also believe that athletes are usually young with a higher probability of 95%, with the following KB: (1) $SumoWrestler \sqsubseteq Athlete$; (2) $0.8 :: \mathbf{T}(Athlete) \sqsubseteq InFit$; (3) $0.8 :: \mathbf{T}(SumoWrestler) \sqsubseteq \neg InFit$; (4) $0.95 :: \mathbf{T}(Athlete) \sqsubseteq YoungPerson$. We consider eight different scenarios, representing all possible combinations of typicality inclusion: as an example, $\{((2), 1), ((3), 0), ((4), 1)\}$ represents the scenario in which (2) and (4) hold, whereas (3) does not. We equip each scenario with a probability depending on those of the involved inclusions, then we restrict reasoning to scenarios whose probabilities belong to a given and fixed range.

We show that the proposed logic $\mathbf{T}^{CL}$ is able to tackle the problem of composing prototypical concepts. As an additional element of the proposed formalization we employ a method inspired by cognitive semantics [6] for the identification of a dominance effect between the concepts to be combined: for every combination, we distinguish a HEAD, representing the stronger element of the combination, and a MODIFIER. The basic idea is: given a KB and two concepts C_H (HEAD) and C_M (MODIFIER) occurring in it, we consider only *some* scenarios in order to define a revised knowledge base, enriched by typical properties of the combined concept $C \sqsubseteq C_H \sqcap C_M$ (the heuristics for the scenario selections are detailed in Sect. 2, Definition 7).

---

[1]   Here, we focus on the proposal of the formalism itself, therefore the machinery for obtaining probabilities from an application domain will not be discussed.

In this paper we exploit $\mathbf{T}^{\mathsf{CL}}$ by showing how it is able to capture well established examples in the literature of cognitive science concerning concept combination. On the other hand, we have also shown [7] how $\mathbf{T}^{\mathsf{CL}}$ can be also used as a tool for the generation of novel creative concepts, that could be useful in many applicative scenarios. We also show that the reasoning complexity of $\mathbf{T}^{\mathsf{CL}}$ is ExpTime-complete, as in the standard $\mathcal{ALC}$ logic.

2   A Logic for Concept Combination

The nonmonotonic Description Logic $\mathbf{T}^{\mathsf{CL}}$ combines the semantics based on the rational closure of $\mathcal{ALC} + \mathbf{T_R}$ [3] with the probabilistic DISPONTE semantics [5]. By taking inspiration from [8], we consider two types of properties associated to a given concept: rigid and typical. Rigid properties are those defining a concept, e.g. $C \sqsubseteq D$ (all Cs are Ds). Typical properties are represented by inclusions equipped by a probability. Additionally, we employ a cognitive heuristic for the identification of a dominance effect between the concepts to be combined, distinguishing between HEAD and MODIFIER[2].

The language of $\mathbf{T}^{\mathsf{CL}}$ extends the basic DL $\mathcal{ALC}$ by *typicality inclusions* of the form $\mathbf{T}(C) \sqsubseteq D$ equipped by a real number $p \in (0.5, 1)$ representing its probability, whose meaning is that "normally, Cs are also D with probability p".

Definition 1 (Language of $\mathbf{T}^{\mathsf{CL}}$). *We consider an alphabet of concept names* C, *of role names* R, *and of individual constants* O. *Given $A \in$ C *and $R \in$ R, we define:*

$$C, D := A \mid \top \mid \bot \mid \neg C \mid C \sqcap C \mid C \sqcup C \mid \forall R.C \mid \exists R.C$$

We define a knowledge base $\mathcal{K} = \langle \mathcal{R}, \mathcal{T}, \mathcal{A} \rangle$ where: • $\mathcal{R}$ is a finite set of rigid properties of the form $C \sqsubseteq D$; • $\mathcal{T}$ is a finite set of typicality properties of the form $p :: \mathbf{T}(C) \sqsubseteq D$, where $p \in (0.5, 1) \subseteq \mathbb{R}$ is the probability of the inclusion; • $\mathcal{A}$ is the ABox, i.e. a finite set of formulas of the form $C(a)$ where $a \in$ O.

It is worth noticing that we avoid typicality inclusions with degree 1. Indeed, an inclusion $1 :: \mathbf{T}(C) \sqsubseteq D$ would mean that it is a certain property, that we represent with $C \sqsubseteq D \in \mathcal{R}$. Also, observe that we only allow typicality inclusions equipped with probabilities $p > 0.5$. Indeed, the very notion of typicality derives from that one of probability distribution, in particular typical properties attributed to entities are those characterizing the majority of instances involved. Moreover, in our effort of integrating two different semantics – DISPONTE and typicality logic – the choice of having probabilities higher than 0.5 for typicality inclusions seems to be the only one compliant with both the formalisms. In fact, despite the DISPONTE semantics allows to assign also low probabilities/degrees of belief to standard inclusions, in the logic $\mathbf{T}^{\mathsf{CL}}$ it would be misleading to also allow low probabilities for typicality inclusions. Please, note that

---

[2] Here we assume that some methods for the automatic assignment of the HEAD/MODIFER pairs are/may be available and focus on the discussion of the reasoning part.

this is not a limitation of the expressivity of the logic $\mathbf{T}^{\text{CL}}$: we can in fact represent properties not holding for typical members of a category, for instance if one need to represent that typical students are not married, we can have that $0.8 :: \mathbf{T}(Student) \sqsubseteq \neg Married$.

Following from the DISPONTE semantics, each axiom is independent from each others. This avoids the problem of dealing with probabilities of inconsistent inclusions.

A model $\mathcal{M}$ of $\mathbf{T}^{\text{CL}}$ extends standard $\mathcal{ALC}$ models by a preference relation among domain elements as in the logic of typicality [3]. In this respect, $x < y$ means that x is "more normal" than y, and that the typical members of a concept C are the minimal elements of C with respect to this relation. An element $x \in \Delta^{\mathcal{I}}$ is a *typical instance* of some concept C if $x \in C^{\mathcal{I}}$ and there is no C-element in $\Delta^{\mathcal{I}}$ more normal than x.

Definition 2 (Model of $\mathbf{T}^{\text{CL}}$). *A model $\mathcal{M}$ is any structure $\langle \Delta^{\mathcal{I}}, <, .^{\mathcal{I}} \rangle$ where: (i) $\Delta^{\mathcal{I}}$ is a non empty set of items called the domain; (ii) $<$ is an irreflexive, transitive, well-founded and modular (for all x, y, z in $\Delta^{\mathcal{I}}$, if $x < y$ then either $x < z$ or $z < y$) relation over $\Delta^{\mathcal{I}}$; (iii) $.^{\mathcal{I}}$ is the extension function that maps each concept C to $C^{\mathcal{I}} \subseteq \Delta^{\mathcal{I}}$, and each role R to $R^{\mathcal{I}} \subseteq \Delta^{\mathcal{I}} \times \Delta^{\mathcal{I}}$. For concepts of $\mathcal{ALC}$, $C^{\mathcal{I}}$ is defined as usual. For the $\mathbf{T}$ operator, we have $(\mathbf{T}(C))^{\mathcal{I}} = Min_<(C^{\mathcal{I}})$, where $Min_<(C^{\mathcal{I}}) = \{x \in C^{\mathcal{I}} \mid \nexists y \in C^{\mathcal{I}} \text{ s.t. } y < x\}$.*

A model $\mathcal{M}$ can be equivalently defined by postulating the existence of a function $k_{\mathcal{M}} : \Delta^{\mathcal{I}} \longmapsto \mathbb{N}$, where $k_{\mathcal{M}}$ assigns a finite rank to each domain element [3]: the rank of x is the length of the longest chain $x_0 < \cdots < x$ from x to a minimal x_0, i.e. such that there is no x' such that $x' < x_0$. The rank function $k_{\mathcal{M}}$ and $<$ can be defined from each other by letting $x < y$ if and only if $k_{\mathcal{M}}(x) < k_{\mathcal{M}}(y)$.

Definition 3 (Model satisfying a KB). *Let $\mathcal{K} = \langle \mathcal{R}, \mathcal{T}, \mathcal{A} \rangle$ be a KB. Given a model $\mathcal{M} = \langle \Delta^{\mathcal{I}}, <, .^{\mathcal{I}} \rangle$, we assume that $.^{\mathcal{I}}$ is extended to assign a domain element $a^{\mathcal{I}}$ of $\Delta^{\mathcal{I}}$ to each individual constant a of $\mathcal{O}$. We say that: (i) $\mathcal{M}$ satisfies $\mathcal{R}$ if, for all $C \sqsubseteq D \in \mathcal{R}$, we have $C^{\mathcal{I}} \subseteq D^{\mathcal{I}}$; (ii) $\mathcal{M}$ satisfies $\mathcal{T}$ if, for all $q :: \mathbf{T}(C) \sqsubseteq D \in \mathcal{T}$, we have that $\mathbf{T}(C)^{\mathcal{I}} \subseteq D^{\mathcal{I}}$, i.e. $Min_<(C^{\mathcal{I}}) \subseteq D^{\mathcal{I}}$; (iii) $\mathcal{M}$ satisfies $\mathcal{A}$ if, for all assertion $F \in \mathcal{A}$, if $F = C(a)$ then $a^{\mathcal{I}} \in C^{\mathcal{I}}$, otherwise if $F = R(a, b)$ then $(a^{\mathcal{I}}, b^{\mathcal{I}}) \in R^{\mathcal{I}}$.*

Even if the typicality operator $\mathbf{T}$ itself is nonmonotonic (i.e. $\mathbf{T}(C) \sqsubseteq E$ does not imply $\mathbf{T}(C \sqcap D) \sqsubseteq E$), what is inferred from a KB can still be inferred from any KB' with KB $\subseteq$ KB', i.e. the resulting logic is monotonic. In order to perform useful nonmonotonic inferences, in [3] the authors have strengthened the above semantics by restricting entailment to a class of minimal models. Intuitively, the idea is to restrict entailment to models that *minimize the untypical instances of a concept*. The resulting logic corresponds to a notion of *rational closure* on top of $\mathcal{ALC} + \mathbf{T_R}$. Such a notion is a natural extension of the rational closure construction provided in [4] for the propositional logic. This nonmonotonic semantics relies on minimal rational models that minimize the *rank of domain elements*. Informally, given two models of KB, one in which a given domain element x has rank 2 (because for instance $z < y < x$), and another in which it has rank 1 (because only $y < x$), we prefer the latter, as in this model the element x is assumed to be "more typical" than in the former. Query entailment, i.e. checking whether a formula holds in all models satisfying a KB, is then restricted to minimal *canonical models*. The intuition is that a canonical model contains all the individuals that enjoy properties that are consistent with KB. This is needed when reasoning about the rank of the concepts: it is important to have them all represented. For details, see [3].

A query F is minimally entailed from a KB if it holds in all minimal canonical models of KB. In [3] it is shown that query entailment in the nonmonotonic $\mathcal{ALC} + \mathbf{T_R}$ is in ExpTime.

Definition 4 (Entailment in $\mathbf{T}^{CL}$). *Let $\mathcal{K} = \langle \mathcal{R}, \mathcal{T}, \mathcal{A} \rangle$ be a KB and let F be either $C \sqsubseteq D$ (C could be $\mathbf{T}(C')$) or $C(a)$ or $R(a,b)$. We say that F follows from $\mathcal{K}$ if, for all minimal $\mathcal{M}$ satisfying $\mathcal{K}$, then $\mathcal{M}$ also satisfies F.*

Let us now define the notion of *scenario* of the composition of concepts. Intuitively, a scenario is a knowledge base obtained by adding to all rigid properties in $\mathcal{R}$ and to all ABox facts in $\mathcal{A}$ only *some* typicality properties. More in detail, we define an *atomic choice* on each typicality inclusion, then we define a *selection* as a set of atomic choices in order to select which typicality inclusions have to be considered in a scenario.

Definition 5 (Atomic choice). *Given $\mathcal{K} = \langle \mathcal{R}, \mathcal{T}, \mathcal{A} \rangle$, where $\mathcal{T} = \{E_1 = q_1 :: \mathbf{T}(C_1) \sqsubseteq D_1, \ldots, E_n = q_n :: \mathbf{T}(C_n) \sqsubseteq D_n\}$ we define (E_i, k_i) an atomic choice, where $k_i \in \{0, 1\}$.*

Definition 6 (Selection). *Given $\mathcal{K} = \langle \mathcal{R}, \mathcal{T}, \mathcal{A} \rangle$, where $\mathcal{T} = \{E_1 = q_1 :: \mathbf{T}(C_1) \sqsubseteq D_1, \ldots, E_n = q_n :: \mathbf{T}(C_n) \sqsubseteq D_n\}$ and a set of atomic choices ν, we say that ν is a selection if, for each E_i, one decision is taken, i.e. either $(E_i, 0) \in \nu$ and $(E_i, 1) \notin \nu$ or $(E_i, 1) \in \nu$ and $(E_i, 0) \notin \nu$ for $i = 1, 2, \ldots, n$. The probability of ν is $P(\nu) = \prod\limits_{(E_i,1) \in \nu} q_i \prod\limits_{(E_i,0) \in \nu} (1 - q_i)$.*

Definition 7 (Scenario). *Given $\mathcal{K} = \langle \mathcal{R}, \mathcal{T}, \mathcal{A} \rangle$, where $\mathcal{T} = \{E_1 = q_1 :: \mathbf{T}(C_1) \sqsubseteq D_1, \ldots, E_n = q_n :: \mathbf{T}(C_n) \sqsubseteq D_n\}$ and given a selection σ, we define a scenario $w_\sigma = \langle \mathcal{R}, \{E_i \mid (E_i, 1) \in \sigma\}, \mathcal{A} \rangle$. We also define the probability of a scenario w_σ as the probability of the corresponding selection, i.e. $P(w_\sigma) = P(\sigma)$. Last, we say that a scenario is consistent with respect to $\mathcal{K}$ when it admits a model in the logic $\mathbf{T}^{cl}$ satisfying $\mathcal{K}$.*

We denote with $\mathcal{W}_\mathcal{K}$ the set of all scenarios. It follows that the probability of a scenario $P(w_\sigma)$ is a probability distribution over scenarios, that is to say $\sum\limits_{w \in \mathcal{W}_\mathcal{K}} P(w) = 1$.

Given a KB $\mathcal{K} = \langle \mathcal{R}, \mathcal{T}, \mathcal{A} \rangle$ and given two concepts C_H and C_M occurring in $\mathcal{K}$, our logic allows defining the compound concept C as the combination of the HEAD C_H and the MODIFIER C_M, where $C \sqsubseteq C_H \sqcap C_M$ and the typical properties of the form $\mathbf{T}(C) \sqsubseteq D$ to ascribe to the concept C are obtained in the set of scenarios that: i) are consistent with respect to $\mathcal{K}$; ii) are not trivial, i.e. those with the highest probability, in the sense that the scenarios considering *all* properties that can be consistently ascribed to C are discarded; iii) are those giving preference to the typical properties of the HEAD C_H (with respect to those of the MODIFIER C_M) with the highest probability.

Lastly, we define the ultimate output of our mechanism: a knowledge base in the logic $\mathbf{T}^{\mathrm{CL}}$ whose set of typicality properties is enriched by those of the compound concept C. Given a scenario w satisfying the above properties, we define the properties of C as the set of inclusions $p :: \mathbf{T}(C) \sqsubseteq D$, for all $\mathbf{T}(C) \sqsubseteq D$ that are entailed (Definition 4) from w in the logic $\mathbf{T}^{\mathrm{CL}}$. The probability p is such that: (i) if D is a property inherited either from the HEAD or from both the HEAD and the MODIFIER, that is to say $\mathbf{T}(C_H) \sqsubseteq D$ is entailed from w, then p corresponds to the probability of such inclusion in the initial knowledge base, i.e. $p :: \mathbf{T}(C_H) \sqsubseteq D \in \mathcal{T}$; (ii) otherwise, i.e. $\mathbf{T}(C_M) \sqsubseteq D$ is entailed from w, then p corresponds to the probability of such inclusion in the initial knowledge base, i.e. $p :: \mathbf{T}(C_M) \sqsubseteq D \in \mathcal{T}$.

The knowledge base obtained as the result of combining concepts C_H and C_M into the compound concept C is called *C-revised* knowledge base:

$$\mathcal{K}_C = \langle \mathcal{R}, \mathcal{T} \cup \{p :: \mathbf{T}(C) \sqsubseteq D\}, \mathcal{A} \rangle,$$

for all D such that $\mathbf{T}(C) \sqsubseteq D$ is entailed in w and p defined as above. Notice that, since the *C-revised* knowledge base is still in the language of the $\mathbf{T}^{\mathrm{CL}}$ logic, we can iteratively repeat the same procedure in order to combine not only atomic concepts, but also compound concepts. We leave a detailed analysis of this topic for future works.

Let us now define the probability that a query is entailed from a C-revised knowledge base. We restrict our concern to ABox facts. The idea is that, given a query $A(a)$ and its probability p, the probability of $A(a)$ is the product of p and the probability of the inclusion in the C-revised knowledge base which is responsible for that.

Definition 8 (Probability of query entailment). *Given a knowledge base* $\mathcal{K} = \langle \mathcal{R}, \mathcal{T}, \mathcal{A} \rangle$, *the C-revised knowledge base* $\mathcal{K}_C$, *a query* $A(a)$ *and its probability* $p \in (0, 1]$, *we define the probability of the entailment of* $A(a)$ *from* $\mathcal{K}_C$, *denoted as* $\mathbb{P}(A(a), p)$ *as follows:*

- $\mathbb{P}(A(a), p) = 0$, *if* $A(a)$ *is* not *entailed from* $\mathcal{K}_C$;
- $\mathbb{P}(A(a), p) = p \times q$, *where either* $q :: \mathbf{T}(C) \sqsubseteq A$ *belongs to* $\mathcal{K}_C$ *or* $q :: \mathbf{T}(C) \sqsubseteq D$ *belongs to* $\mathcal{K}_C$ *and* $D \sqsubseteq A$ *is entailed from* $\mathcal{R}$ *in standard* $\mathcal{ALC}$, *otherwise.*

We conclude by showing that reasoning in $\mathbf{T}^{\mathrm{CL}}$ remains in the same complexity class of standard $\mathcal{ALC}$. For the completeness, let n be the size of KB, then the number of typicality inclusions is $O(n)$. It is straightforward to observe that we have an exponential number of different scenarios, for each one we need to check whether the resulting KB is consistent in $\mathcal{ALC} + \mathbf{T_R}$ which is EXPTIME-complete. Hardness immediately follows form the fact that $\mathbf{T}^{\mathrm{CL}}$ extends $\mathcal{ALC} +$

$\mathbf{T_R}$. Reasoning in the revised knowledge base relies on reasoning in $\mathcal{ALC} + \mathbf{T_R}$, therefore we can conclude:

Theorem 1. *Reasoning in* $\mathbf{T}^{CL}$ *is* ExpTime-*complete.*

3   Applications of the Logic $\mathbf{T}^{CL}$

We propose two different types of examples adopting the logic $\mathbf{T}^{CL}$ to model the phenomenon of typicality-based conceptual combination, in order to show how our logic is able to handle two paradigmatic examples of typicality-based concept composition coming from the Cognitive Science. Such cases have been historically problematic to model by adopting other kinds of logics (for example fuzzy logic, [2,6,9]) and have been recently considered as a testbed for computational models aiming at dealing with this type of problem [10]. The exploitation of $\mathbf{T}^{CL}$ as a possible application in the field of computational creativity for the generation of new characters is proposed in [7].

3.1   Pet Fish

Let $\mathcal{K} = \langle \mathcal{R}, \mathcal{T}, \emptyset \rangle$ be a KB, where $\mathcal{R} = \{Fish \sqsubseteq \exists livesIn.Water\}$ and $\mathcal{T}$ is:

1. $0.9 :: \mathbf{T}(Pet) \sqsubseteq \exists livesIn.(\neg Water)$
2. $0.8 :: \mathbf{T}(Pet) \sqsubseteq Affectionate$
3. $0.7 :: \mathbf{T}(Fish) \sqsubseteq \neg Affectionate$
4. $0.8 :: \mathbf{T}(Pet) \sqsubseteq Warm$
5. $0.6 :: \mathbf{T}(Fish) \sqsubseteq Greyish$
6. $0.9 :: \mathbf{T}(Fish) \sqsubseteq Scaly$
7. $0.8 :: \mathbf{T}(Fish) \sqsubseteq \neg Warm$

By the properties of the typicality operator $\mathbf{T}$, we have that $(*)$ $\mathbf{T}(Pet \sqcap Fish) \sqsubseteq \exists livesIn.Water$. Indeed, $Fish \sqsubseteq \exists livesIn.Water$ is a rigid property, which is always preferred to a typical one: in this case, additionally, the rigid property is also associated to the HEAD element fish. Therefore, this element is reinforced.

Since $\mid \mathcal{T} \mid = 7$, we have $2^7 = 128$ different scenarios. We can observe that some of them are not consistent, more precisely those (i) containing the inclusion 1 by $(*)$; (ii) containing inclusions 2 and 3; (iii) containing inclusions 4 and 7. The scenarios with the highest probabilities (up to 17%) are both trivial and inconsistent: indeed, probabilities p_i equipping typicality inclusions are such that $p_i > 0.5$, therefore the higher is the number of inclusions belonging to a scenario the higher is the associated probability. Since typicality inclusions introduce properties that are pairwise inconsistent, it follows that such scenarios must be discarded.

Consistent scenarios with the highest probabilities (from 0.14% to 0.21%) contain 2 and do not contain 3, namely they privilege the MODIFIER with respect to the corresponding negation in the HEAD, obtaining that being affectionate is a typical property of a pet fish. These scenarios are consistent, but they are trivial, since they include either all the properties associated to the MODIFIER or those with the highest probability. In both cases, in these scenarios we pay the price of discarding some properties of the HEAD. For the same reason we discard the scenarios with probabilities 0.12% containing 4 and not 7. The remaining scenarios privilege the properties associated to the HEAD, namely they contain 3 and do not contain 2 and 7 and not 4. Again, these scenarios range from the most trivial to the most surprising ones. It turns out that the scenarios able to account for the phenomenon of conceptual composition are not those with the highest probability, but those belonging to the probability range immediately lower. In the example, most trivial scenarios where we have that typical pet fishes are greyish (probability 0.12%)The not trivial scenario defining prototypical properties of a pet fish comes from the selection $\sigma = \{(1,0),(2,0),(3,1),(4,0),(5,0),(6,1),(7,1)\}$, and contains 3, 6, and 7:

$0.8 :: \mathbf{T}(Pet \sqcap Fish) \sqsubseteq \neg Affectionate$
$0.6 :: \mathbf{T}(Pet \sqcap Fish) \sqsubseteq Scaly$
$0.8 :: \mathbf{T}(Pet \sqcap Fish) \sqsubseteq \neg Warm$

Notice that in $\mathbf{T}^{CL}$, adding a new inclusion $\mathbf{T}(Pet \sqcap Fish) \sqsubseteq Red$, would not be problematic (i.e. our logic tackles the phenomenon of prototypical *attributes emergence* [6]).

3.2   Linda the Feminist Bank Teller

We exploit the logic $\mathbf{T}^{CL}$ in order to tackle the so called conjunction fallacy problem [11]: Linda is 31 years old, single, outspoken, and bright. She majored in philosophy and was concerned with issues of discrimination and social justice, and also participated in anti-nuclear demonstrations. When asked to rank the probability of the statements 1) "Linda is a bank teller" and 2)"Linda is a bank teller and is active in the feminist movement", the majority of people rank 2) as more probable than 1), violating the classic probability rules. In $\mathbf{T}^{CL}$, let $\mathcal{K} = \langle \mathcal{R}, \mathcal{T}, \mathcal{A} \rangle$ be a KB, where $\mathcal{A} = \emptyset$, $\mathcal{T}$ is:

$0.8 :: \mathbf{T}(Feminist) \sqsubseteq Bright$
$0.9 :: \mathbf{T}(Feminist) \sqsubseteq OutSpoken$
$0.8 :: \mathbf{T}(Feminist) \sqsubseteq \exists.fightsForSocialJustice$
$0.9 :: \mathbf{T}(Feminist) \sqsubseteq Environmentalist$
$0.6 :: \mathbf{T}(BankTeller) \sqsubseteq \neg\exists.fightsForSocialJustice$
$0.8 :: \mathbf{T}(BankTeller) \sqsubseteq Calm$

and $\mathcal{R}$ is as follows: $\{BankTeller \sqsubseteq \exists isEmployed.Bank$, $Feminist \sqsubseteq \exists believesIn.Feminism$, $Feminist \sqsubseteq Female$, $Environmentalist \sqsubseteq \exists isAgainst.NuclearEnergyDevelopment\}$. Let us consider the compound concept $Feminist \sqcap BankTeller$. It can be obtained in two different ways, namely by choosing *Feminist* as the HEAD and *BankTeller* as the MODIFIER, or vice versa. In $\mathbf{T}^{CL}$, the compound concept inherits all the rigid properties, that is to say $Feminist \sqcap BankTeller$ is included in $\exists isEmployed.Bank$, in $\exists believesIn.Feminism$ and in *Female*. Concerning the typical properties, two of them are in contrast, namely typical feminists fight for social justice, whereas typical bank tellers do not. All the scenarios including both $\mathbf{T}(Feminist) \sqsubseteq \exists fightsFor.SocialJustice$

and $\mathbf{T}(BankTeller) \sqsubseteq \neg\exists fightsFor.SocialJustice$ are then inconsistent. Concerning the remaining ones, if we consider *Feminist* as the HEAD, then scenarios including $\mathbf{T}(BankTeller) \sqsubseteq \neg\exists fightsFor.SocialJustice$ are discarded, in favor of scenarios including $\mathbf{T}(Feminist) \sqsubseteq \exists fightsFor.SocialJustice$. The most obvious scenario, with the highest probability, corresponds to the one including all the typicality inclusions related to the HEAD. In the logic $\mathbf{T}^{CL}$ we discard it and we focus on the remaining ones. Among them, one of the scenarios having the highest probability is the one not including $\mathbf{T}(Feminist) \sqsubseteq Bright$. This scenario defines the following *Feminist* $\sqcap$ *BankTeller*-revised knowledge base:

$0.9 :: \mathbf{T}(Feminist \sqcap BankTeller) \sqsubseteq OutSpoken$
$0.8 :: \mathbf{T}(Feminist \sqcap BankTeller) \sqsubseteq \exists fightsFor.SocialJustice$
$0.9 :: \mathbf{T}(Feminist \sqcap BankTeller) \sqsubseteq Environmentalist$
$0.8 :: \mathbf{T}(Feminist \sqcap BankTeller) \sqsubseteq Calm$

Let us now consider the case of the instance Linda, described as: $YoungWoman(linda)$, $\exists graduatedIn.Philosophy(linda)$, $Outspoken(linda)$, $Bright(linda)$, $Single(linda)$, $\exists fightsFor.SocialJustice(linda)$, $\exists isAgainst.$ $NuclearEnergyDevelopment(linda)$. In our logic, solving the conjunction fallacy problem means that we have to find the most appropriate category for Linda. In our case the choice is between *BankTeller* and *Feminist* $\sqcap$ *BankTeller*. We can assume that, in absence of any other information, the described properties that are explicitly assigned to the instance Linda can be set to a default probability value of 0.6. Let us first consider the *Feminist* $\sqcap$ *BankTeller*-revised knowledge base, with an ABox asserting that Linda is a bank teller, i.e. $\mathcal{A}_1 = \{BankTeller(linda)\}$, and let us consider each property of the instance Linda and the associated probability of entailment. None of such properties are entailed by the *Feminist* $\sqcap$ *BankTeller*-revised knowledge base with $\mathcal{A}_1$, therefore, for each property of the form $D(linda)$ we have $\mathbb{P}(D(linda), 0.6) = 0$ (by Definition 8). On the other hand, let us consider an ABox asserting that Linda is a feminist bank teller, i.e. $\mathcal{A}_2 = \{(Feminist \sqcap BankTeller)(linda)\}$. We have: (i) $YoungWoman(linda)$ is not entailed from the *Feminist* $\sqcap$ *BankTeller*-revised knowledge base, therefore $\mathbb{P}(YoungWoman(linda), 0.6) = 0$; the same for $\exists graduatedIn.Philosophy(linda)$ and $Single(linda)$; (ii) $Outspoken(linda)$ is entailed from the *Feminist* $\sqcap$ *BankTeller*-revised KB with $\mathcal{A}_2$, then, by Definition 8, we have $\mathbb{P}(Outspoken(linda), 0.6) = 0.6 \times 0.9 = 0.54$, where 0.9 is the probability of $\mathbf{T}(Feminist \sqcap BankTeller) \sqsubseteq OutSpoken$ in the *Feminist* $\sqcap$ *BankTeller*-revised KB; (iii) the same for $\exists fightsFor.SocialJustice(linda)$ entailed by using $\mathcal{A}_2$: in this case, we have $\mathbb{P}(\exists fightsFor.SocialJustice(linda), 0.6) = 0.6 \times 0.8 = 0.48$; (iv) $\exists isAgainst.NuclearEnergyDevelopment(linda)$ is entailed by using $\mathcal{A}_2$. Observe that $Environmentalist \sqsubseteq NuclearEnergyDevelopment$ follows from $\mathcal{R}$ in standard $\mathcal{ALC}$, then $\mathbb{P}(\exists isAgainst.NuclearEnergyDevelopment (linda), 0.6) = 0.6 \times 0.9 = 0.54$ by Definition 8, where 0.9 is the probability of $\mathbf{T}(Feminist \sqcap BankTeller) \sqsubseteq Environmentalist$ in the *Feminist* $\sqcap$ *BankTeller*-revised KB. Computing the sum of the probabilities of the queries of all facts about Linda, we obtain $0.54 + 0.48 + 0.54 = 1.56$, to witness that the choice of $\mathcal{A}_2$ is more appropriate w.r.t. the choice of $\mathcal{A}_1$ where the sum is 0. This means that, in our logic, the human choice of classifying Linda as a feminist bank teller sounds perfectly plausible and has to be preferred to the alternative one of classifying her as a bank teller.

4 Related and Future Works

Lewis and Lawry [10] present a detailed analysis of the limits of the AI approaches (i.e. set-theory, fuzzy logics, vector-space models and quantum probability) proposed to model the phenomenon of prototypical concept composition. In addition, they show how hierarchical conceptual spaces allow to model this phenomenon. While we agree with the authors about the comments moved to the described approaches, we showed that our logic can equally model, in a cognitively compliant-way, the composition of prototypes by using a computationally effective nonmonotonic DL formalism. The logic $\mathbf{T}^{CL}$ also extends the works of [12,13] in that it provides a set of mechanisms allowing to block inheritance of prototypical properties in concept combination.

Other attempts similar to the one proposed here, and employing the $\mathcal{EL}^{++}$ [14], concerns the conceptual blending: a task where the obtained concept is *entirely novel* and has no strong association with the two base concepts (while, in concept combination, the compound concept is always a subset of the base concepts, for the differences see [15]). A technical difference of our work w.r.t [14] is in that what the authors call prototypes are expressed in a monotonic DL, which does not allow reasoning about typicality.

More recently, a different approach is proposed in [16], where the authors see the problem of concept blending as a nonmonotonic search problem and proposed to use Answer Set Programming (ASP) to deal with this search problem in a nonmonotonic way. Also in this case the authors propose a framework for the conceptual blending task and not for the concept combination.

As future research we aim at extending our approach to more expressive DLs. Starting from the work of [17], applying the logic with the typicality operator and the rational closure to $\mathcal{SHIQ}$, we intend to study whether and how $\mathbf{T}^{CL}$ could provide an alternative solution to the problem of the "all or nothing" behavior of rational closure w.r.t. property inheritance. We also aim at implementing efficient reasoners for our logic, relying on the prover RAT-OWL [18] which allows to reason in the underlying $\mathcal{ALC} + \mathbf{T_R}$.

Acknowledgements. This work is partially supported by the project "ExceptionOWL", Università di Torino and Compagnia di San Paolo, call 2014 "Excellent (young) PI", http://di.unito.it/exceptionowl.

References

1. Frixione, M., Lieto, A.: Representing concepts in formal ontologies: compositionality vs. typicality effects. Log. Log. Philos. **21**(4), 391–414 (2012)
2. Osherson, D.N., Smith, E.E.: On the adequacy of prototype theory as a theory of concepts. Cognition **9**(1), 35–58 (1981)

3. Giordano, L., Gliozzi, V., Olivetti, N., Pozzato, G.L.: Semantic characterization of Rational Closure: from Propositional Logic to Description Logics. Artif. Intell. **226**, 1–33 (2015)
4. Lehmann, D., Magidor, M.: What does a conditional knowledge base entail? Artif. Intell. **55**(1), 1–60 (1992)
5. Riguzzi, F., Bellodi, E., Lamma, E., Zese, R.: Reasoning with probabilistic ontologies. In: Yang, Q., Wooldridge, M. (eds.) Proceedings of the Twenty-Fourth International Joint Conference on Artificial Intelligence, IJCAI 2015, Buenos Aires, Argentina, 25–31 July 2015, pp. 4310–4316 AAAI Press (2015)
6. Hampton, J.A.: Inheritance of attributes in natural concept conjunctions. Mem. Cogn. **15**(1), 55–71 (1987)
7. Lieto, A., Pozzato, G.L.: Creative concept generation by combining description logic of typicality, probabilities and cognitive heuristics. In: Proceedings of the 17th International Conference of the Italian Association for Artificial Intelligence AI*IA 2018 (2018)
8. Lieto, A., Minieri, A., Piana, A., Radicioni, D.P.: A knowledge-based system for prototypical reasoning. Connect. Sci. **27**, 137–152 (2015)
9. Hampton, J.A.: Conceptual combinations and fuzzy logic. Concepts Fuzzy Log. **209**, 209–232 (2011)
10. Lewis, M., Lawry, J.: Hierarchical conceptual spaces for concept combination. Artif. Intell. **237**, 204–227 (2016)
11. Tversky, A., Kahneman, D.: Extensional versus intuitive reasoning: the conjunction fallacy in probability judgment. Psychol. Rev. **90**(4), 293 (1983)
12. Pozzato, G.L.: Reasoning in description logics with typicalities and probabilities of exceptions. In: Antonucci, A., Cholvy, L., Papini, O. (eds.) ECSQARU 2017. LNCS (LNAI), vol. 10369, pp. 409–420. Springer, Cham (2017). https://doi.org/10.1007/978-3-319-61581-3_37
13. Pozzato, G.L.: Reasoning about plausible scenarios in description logics of typicality. Intelligenza Artificiale **11**(1), 25–45 (2017)
14. Confalonieri, R., Schorlemmer, M., Kutz, O., Peñaloza, R., Plaza, E., Eppe, M.: Conceptual blending in EL++. In: Lenzerini, M., Peñaloza, R. (eds.) Proceedings of the 29th International Workshop on Description Logics, Cape Town, South Africa, 22–25 April 2016, vol. 1577 of CEUR Workshop Proceedings (2016). CEUR-WS.org
15. Nagai, Y., Taura, T.: Formal description of concept-synthesizing process for creative design. In: Gero, J.S. (ed.) Design Computing and Cognition '06. Springer, Dordrecht (2006). https://doi.org/10.1007/978-1-4020-5131-9_23
16. Eppe, M., et al.: A computational framework for conceptual blending. Artif. Intell. **256**, 105–129 (2018)
17. Giordano, L., Gliozzi, V., Olivetti, N., Pozzato, G.L.: Rational closure in $\mathcal{SHIQ}$. In: DL 2014, 27th International Workshop on Description Logics, vol. 1193 of CEUR Workshop Proceedings, pp. 543–555 (2014). CEUR-WS.org
18. Giordano, L., Gliozzi, V., Pozzato, G.L., Renzulli, R.: An efficient reasoner for description logics of typicality and rational closure. In: Artale, A., Glimm, B., Kontchakov, R. (eds.) Proceedings of the 30th International Workshop on Description Logics, Montpellier, France, 18–21 July 2017, vol. 1879 of CEUR Workshop Proceedings (2017). CEUR-WS.org

Mining Complex Patterns

Sparse Multi-label Bilinear Embedding on Stiefel Manifolds

Yang Liu[1,2(✉)], Guohua Dong[3], and Zhonglei Gu[1]

[1] Department of Computer Science, Hong Kong Baptist University,
Kowloon, Hong Kong SAR
{csygliu,cszlgu}@comp.hkbu.edu.hk
[2] HKBU Institute for Research and Continuing Education,
Shenzhen, People's Republic of China
[3] Songshan Lake TechX Institute,
Dongguan, People's Republic of China
dong.gh@xbotpark.com

Abstract. Dimensionality reduction plays an important role in various machine learning tasks. In this paper, we propose a novel method dubbed Sparse Multi-label bILinear Embedding (SMILE) on Stiefel manifolds for supervised dimensionality reduction on multi-label data. Unlike the traditional multi-label dimensionality reduction algorithms that work on the vectorized data, the proposed SMILE directly takes the second-order tensor data as the input, and thus characterizes the spatial structure of the tensor data in an efficient way. Differentiating from the existing tensor-based dimensionality reduction methods that perform the eigen-decomposition in each iteration, SMILE utilizes a gradient ascent strategy to optimize the objective function in each iteration, and thus is more efficient. Moreover, we introduce column-orthonormal constraints to transformation matrices to eliminate the redundancy between the projection directions of the learned subspace and add an L_1-norm regularization term to the objective function to enhance the interpretability of the learned subspace. Experiments on a standard image dataset validate the effectiveness of the proposed method.

Keywords: Sparse multi-label bilinear embedding
Dimensionality reduction · Second-order tensor
Column-orthonormal constraints · Stiefel manifolds
L_1-norm regularization

1 Introduction

Dimensionality reduction (DR), generally speaking, aims to generate a compact representation of high-dimensional observations according to some criteria governing the representation's quality. Due to the capacity of DR in finding the intrinsic and informative features for data with huge dimensions, it has been

© Springer Nature Switzerland AG 2018
M. Ceci et al. (Eds.): ISMIS 2018, LNAI 11177, pp. 203–213, 2018.
https://doi.org/10.1007/978-3-030-01851-1_20

widely applied to many real-world applications, especially the supervised learning tasks such as information retrieval [16], image classification [5], and emotion recognition [9].

In traditional supervised DR models, each labeled data sample is generally assumed belonging to only one class, i.e., associated with a single label. In many real-world applications, however, each object might be associated with multiple semantic meanings, i.e., multiple labels [9,12,15]. For instance, in image annotation, an image can be associated with several concepts such as sunset, sea, and boat; in movie categorization, a movie can belong to more than one genres, such as drama, comedy, and animation. To characterize multiple semantic meanings of data in the low-dimensional subspace, multi-label DR algorithms have been proposed to model multiple labels simultaneously in the learning procedure [13,17–19,22,24].

Classical multi-label DR algorithms generally require the input in vector form. However, in various learning tasks, such as image recognition and video retrieval, the data are naturally represented as second-order tensors (i.e., matrices) or higher-order tensors. Therefore, vectorizing these tensor data into vectors not only dramatically increases the computational complexity but also destroys the natural tensor structure of the original data. In order to overcome these problems, some multi-label tensor-based DR algorithms have been developed. Panagakis *et al.* proposed a multi-label multilinear approach to learn the intrinsic features for music genre classification [11]. Panagiotis *et al.* utilized the higher order singular value decomposition to find the low-dimensional representations for social tag recommendation [14]. Liu *et al.* presented the multi-label tensor-based DR algorithms, which aim to preserve the label similarities in the learned subspace for music emotion recognition [8,9].

However, the aforementioned methods performed the eigen-decomposition to find the transformation matrices for DR, which is time-consuming and may encounter the ill posed problem, especially when the dimension of the data is very high. To address this challenging issue, in this paper, we propose a novel method dubbed **S**parse **M**ulti-label b**IL**inear **E**mbedding (SMILE) for supervised DR on second-order tensor data. Unlike the existing multi-label tensor-based DR algorithms that perform the eigen-decomposition in each iteration, SMILE utilizes a gradient ascent strategy to optimize the objective function in each iteration, and thus is more efficient. Moreover, we introduce column-orthonormal constraints to transformation matrices to eliminate the redundancy between the projection directions of the learned subspace and add an L_1-norm regularization term to the objective function to enhance the interpretability of the learned subspace.

The rest of the paper is organized as follows. Section 2 introduces the proposed method, including the problem formulation, objective function, optimization procedure, and computational cost analysis. Section 3 then evaluates the performance of the proposed method on a real-world multi-label tensor dataset: the Natural Scene Image Dataset. Finally, the paper is concluded in Sect. 4 with future works.

2   Sparse Multi-label Bilinear Embedding

2.1   Problem Formulation

Given the dataset $\mathcal{X} = \{\mathbf{X}_1, ..., \mathbf{X}_n\}$, where $\mathbf{X}_i \in \mathbb{R}^{D_1 \times D_2}$ $(i = 1, ..., n)$ denotes the original feature representation of the i-th data sample, n is the number of data samples in the set, and $D_1 \times D_2$ denotes the dimension of the original feature space. Let $\mathcal{L}$ be the label set including C labels. The labels associated with the i-th data sample in $\mathcal{X}$ constitute a subset of $\mathcal{L}$, which can be represented as a C-dimensional binary vector $\mathbf{l}_i \in \{0,1\}^C$, with 1 indicating that the data point is belonging to the corresponding class and 0 otherwise.

Given the training dataset $\{(\mathbf{X}_1, \mathbf{l}_1), ..., (\mathbf{X}_n, \mathbf{l}_n)\}$, SMILE aims to learn two sparse transformation matrices $\mathbf{W}_1 \in \mathbb{R}^{D_1 \times d_1}$ and $\mathbf{W}_2 \in \mathbb{R}^{D_2 \times d_2}$ $(d_i \leq D_i, i = 1, 2)$, which are capable of projecting the original high dimensional data to an intrinsically low-dimensional subspace $\mathcal{Y} = \mathbb{R}^{d_1 \times d_2}$ by $\mathbf{Y}_i = \mathbf{W}_1^T \mathbf{X}_i \mathbf{W}_2$, where the second-order tensor structure of the data sample, the geometric distribution of the dataset, and the discriminant information in the label space can be well preserved.

2.2   Geometry Preserving Criterion

In order to preserve the geometric structure of the dataset, we take both locality and globality into consideration. To describe the local structure of the dataset, we define a neighborhood graph matrix $\mathbf{M}^L$, the (i, j)-th element of which is given as:

$$M_{ij}^L = e^{-\frac{||\mathbf{X}_i - \mathbf{X}_j||_F^2}{2\sigma}}, \tag{1}$$

where $||\cdot||_F$ denotes the Frobenius norm, and the bandwidth σ is empirically set by $\sigma = \sum_{i=1}^{n} ||\mathbf{X}_i - \mathbf{X}_{i_K}||_F^2 / n$ with $\mathbf{X}_{i_K}$ being the K-th nearest neighbor of $\mathbf{X}_i$ under the Frobenius norm [7]. Accordingly, we construct the locality preserving objective:

$$\{\mathbf{W}_1, \mathbf{W}_2\} = \arg \min_{\mathbf{W}_1, \mathbf{W}_2} J_L(\mathbf{W}_1, \mathbf{W}_2), \tag{2}$$

where

$$J_L(\mathbf{W}_1, \mathbf{W}_2) = \sum_{i,j=1}^{n} M_{ij}^L ||\mathbf{W}_1^T (\mathbf{X}_i - \mathbf{X}_j) \mathbf{W}_2||_F^2. \tag{3}$$

By minimizing $J_L(\mathbf{W}_1, \mathbf{W}_2)$ in Eq. (2), the nearby data points in the original space will also be close to each other in the learned subspace, and thus the local structure is expected to be well captured.

To represent the global structure of the dataset, we define the globality weight matrix $\mathbf{M}^G$, the (i, j)-th element of which is given as $M_{ij}^G = 1 - M_{ij}^L$. We then construct the globality preserving objective:

$$\{\mathbf{W}_1, \mathbf{W}_2\} = \arg \max_{\mathbf{W}_1, \mathbf{W}_2} J_G(\mathbf{W}_1, \mathbf{W}_2), \tag{4}$$

where

$$J_G(\mathbf{W}_1, \mathbf{W}_2) = \sum_{i,j=1}^{n} M_{ij}^G ||\mathbf{W}_1^T(\mathbf{X}_i - \mathbf{X}_j)\mathbf{W}_2||_F^2. \tag{5}$$

It is obvious that the larger the distance between $\mathbf{X}_i$ and $\mathbf{X}_j$, the smaller the M_{ij}^L, hence, the larger the M_{ij}^G. This indicates that $J_G(\mathbf{W}_1, \mathbf{W}_2)$ pays more attention to the relationship between faraway data points, which characterizes the structure of the dataset from the global perspective. By maximizing $J_G(\mathbf{W}_1, \mathbf{W}_2)$ in Eq. (4), the distant data points in the original space are still faraway from each other in the low-dimensional space, and thus the global structure of the dataset is preserved.

2.3 Discrimination Preserving Criterion

The idea behind discrimination preserving is straightforward: if two data samples possess similar labels, they should be close to each other in the subspace; otherwise, they should be faraway from each other. To describe the label similarity between data samples, we define the label similarity matrix $\mathbf{M}^S$ for the dataset, the (i, j)-th element of which is given as:

$$M_{ij}^S = \langle \mathbf{l}_i, \mathbf{l}_j \rangle, \tag{6}$$

where $\langle \cdot, \cdot \rangle$ denotes the inner product operation, and M_{ij}^S denotes the similarity between the label vector of the i-th data sample and that of the j-th data sample. The reason of using the inner product form as the similarity measure is that it is able to capture the information of correlations between different labels, which is of great importance in multi-label learning [9]. To normalize the similarity values into the interval $[0, 1]$, we define the normalized label similarity:

$$\hat{M}_{ij}^S = \langle \hat{\mathbf{l}}_i, \hat{\mathbf{l}}_j \rangle = \langle \mathbf{l}_i / ||\mathbf{l}_i||, \mathbf{l}_j / ||\mathbf{l}_j|| \rangle. \tag{7}$$

Then we can construct the similarity preserving objective:

$$\{\mathbf{W}_1, \mathbf{W}_2\} = \arg \min_{\mathbf{W}_1, \mathbf{W}_2} J_S(\mathbf{W}_1, \mathbf{W}_2), \tag{8}$$

where

$$J_S(\mathbf{W}_1, \mathbf{W}_2) = \sum_{i,j=1}^{n} \hat{M}_{ij}^S ||\mathbf{W}_1^T(\mathbf{X}_i - \mathbf{X}_j)\mathbf{W}_2||_F^2. \tag{9}$$

By minimizing $J_S(\mathbf{W}_1, \mathbf{W}_2)$ in Eq. (8), the data samples with similar or even the same label vectors will be close to each other in the learned subspace, and thus the label similarity is well maintained.

To characterize the relationship between data samples with dissimilar label vectors, we define the normalized label diversity matrix $\hat{\mathbf{M}}^D$, the (i, j)-th element of which is given as $\hat{M}_{ij}^D = 1 - \hat{M}_{ij}^S$. Then the diversity preserving objective is constructed as follows:

$$\{\mathbf{W}_1, \mathbf{W}_2\} = \arg \max_{\mathbf{W}_1, \mathbf{W}_2} J_D(\mathbf{W}_1, \mathbf{W}_2), \tag{10}$$

where

$$J_D(\mathbf{W}_1, \mathbf{W}_2) = \sum_{i,j=1}^{n} \hat{M}_{ij}^D \|\mathbf{W}_1^T(\mathbf{X}_i - \mathbf{X}_j)\mathbf{W}_2\|_F^2. \tag{11}$$

By maximizing $J_D(\mathbf{W}_1, \mathbf{W}_2)$ in Eq. (10), the data samples with diverse label vectors are expected to be faraway from each other in the low-dimensional space, and thus the discriminant information can be well preserved.

2.4   Objective Function with L_1 Regularization

To simultaneously model the geometry and discrimination in the proposed method, we integrate the aforementioned four objectives in Eqs. (2), (4), (8), and (10) to form a unified objective function:

$$\{\mathbf{W}_1, \mathbf{W}_2\} = \arg \max_{\mathbf{W}_1, \mathbf{W}_2} J_U(\mathbf{W}_1, \mathbf{W}_2)$$
$$= \arg \max_{\mathbf{W}_1, \mathbf{W}_2} \frac{J_G(\mathbf{W}_1, \mathbf{W}_2) + \lambda J_D(\mathbf{W}_1, \mathbf{W}_2)}{J_L(\mathbf{W}_1, \mathbf{W}_2) + \lambda J_S(\mathbf{W}_1, \mathbf{W}_2)}, \tag{12}$$

where $\lambda \in [0, +\infty)$ is the trade-off parameter to balance the weight of geometric structure preserving and that of discriminant information preserving. To enhance the interpretability of the learned subspace, we add an L_1-norm regularization term $J_R(\mathbf{W}_1, \mathbf{W}_2)$ into Eq. (12) to enforce $\mathbf{W}_1$ and $\mathbf{W}_2$ to be sparse, then we have

$$\{\mathbf{W}_1, \mathbf{W}_2\} = \arg \max_{\mathbf{W}_1, \mathbf{W}_2} J(\mathbf{W}_1, \mathbf{W}_2)$$
$$= \arg \max_{\mathbf{W}_1, \mathbf{W}_2} \left(J_U(\mathbf{W}_1, \mathbf{W}_2) - \mu J_R(\mathbf{W}_1, \mathbf{W}_2)\right), \tag{13}$$

where $J_R(\mathbf{W}_1, \mathbf{W}_2) = |vec(\mathbf{W}_1)|_1 + |vec(\mathbf{W}_2)|_1$, $vec(\cdot)$ converts a matrix to a vector by concatenating its columns, $|\cdot|_1$ denotes the L_1-norm of a vector, and $\mu \in [0, +\infty)$ is the regularization parameter.

To eliminate the redundancy between the projection directions of the learned subspace, we further introduce the column-orthonormal constraints on $\mathbf{W}_1$ and $\mathbf{W}_2$, i.e., $\mathbf{W}_1^T\mathbf{W}_1 = \mathbf{I}_{d_1}$ and $\mathbf{W}_2^T\mathbf{W}_2 = \mathbf{I}_{d_2}$, respectively. It is known that $\mathbf{W}_1$ and $\mathbf{W}_2$ lie on the so-called Stiefel manifolds $\mathcal{O}_1$ and $\mathcal{O}_2$, respectively [1,2]. Therefore, the objective function of SMILE could be formulated as:

$$\{\mathbf{W}_1, \mathbf{W}_2\} = \arg \max_{\substack{\mathbf{W}_1 \in \mathcal{O}_1 \\ \mathbf{W}_2 \in \mathcal{O}_2}} J(\mathbf{W}_1, \mathbf{W}_2). \tag{14}$$

2.5   Optimization Strategy

In this subsection, we propose a straightforward yet effective gradient ascent strategy to solve the objective function of SMILE. Since the regularization term $J_R(\mathbf{W}_1, \mathbf{W}_2)$ in Eq. (13) is non-smooth, we first smooth it by using bump function [6], which can polish the sharp corner of the absolute value function near the origin while preserving its shape away from the corner invariant.

For a one-dimensional variable $x \in \mathbb{R}$, its bump function $\Psi(x)$ is defined as:

$$\Psi(x) = \begin{cases} e^{-\frac{1}{1-x^2}}, & \text{for } |x| < 1; \\ 0, & \text{otherwise.} \end{cases} \tag{15}$$

Accordingly, $|x|$ is smoothened as

$$|x|_{1,\epsilon} = [1 - e\Psi(\frac{x}{\epsilon})] \cdot |x|, \tag{16}$$

where $\epsilon \to 0$ is a small positive constant. It is easy to prove that $|x|_{1,\epsilon}$ is differentiable at the origin. Then we smooth the regularization term $J_R(\mathbf{W}_1, \mathbf{W}_2)$ as

$$J_{\widetilde{R}}(\mathbf{W}_1, \mathbf{W}_2) = |vec(\mathbf{W}_1)|_{1,\epsilon} + |vec(\mathbf{W}_2)|_{1,\epsilon}, \tag{17}$$

where $|vec(\mathbf{W}_1)|_{1,\epsilon}$ and $|vec(\mathbf{W}_2)|_{1,\epsilon}$ denote the sum of $|\cdot|_{1,\epsilon}$ values of all the elements in $vec(\mathbf{W}_1)$ and $vec(\mathbf{W}_2)$, respectively. Now the objective function becomes

$$\begin{aligned} \{\mathbf{W}_1, \mathbf{W}_2\} &= \arg \max_{\substack{\mathbf{W}_1 \in \mathcal{O}_1 \\ \mathbf{W}_2 \in \mathcal{O}_2}} J(\mathbf{W}_1, \mathbf{W}_2) \\ &= \arg \max_{\substack{\mathbf{W}_1 \in \mathcal{O}_1 \\ \mathbf{W}_2 \in \mathcal{O}_2}} \left(J_U(\mathbf{W}_1, \mathbf{W}_2) - \mu J_{\widetilde{R}}(\mathbf{W}_1, \mathbf{W}_2) \right). \end{aligned} \tag{18}$$

In order to utilize the gradient ascent strategy to solve the problem, we need to find out the gradient of the above objective function. Let

$$\begin{aligned} A_1 &= J_G(\mathbf{W}_1, \mathbf{W}_2) + \lambda J_D(\mathbf{W}_1, \mathbf{W}_2), \\ A_2 &= J_L(\mathbf{W}_1, \mathbf{W}_2) + \lambda J_S(\mathbf{W}_1, \mathbf{W}_2), \end{aligned} \tag{19}$$

then we have

$$\frac{\partial A_1}{\partial \mathbf{W}_1} = 2 \sum_{i,j=1}^{n} (M_{ij}^G + \lambda \hat{M}_{ij}^D)(\mathbf{X}_i - \mathbf{X}_j)\mathbf{W}_2\mathbf{W}_2^T(\mathbf{X}_i - \mathbf{X}_j)^T\mathbf{W}_1,$$

$$\frac{\partial A_1}{\partial \mathbf{W}_2} = 2 \sum_{i,j=1}^{n} (M_{ij}^G + \lambda \hat{M}_{ij}^D)(\mathbf{X}_i - \mathbf{X}_j)^T\mathbf{W}_1\mathbf{W}_1^T(\mathbf{X}_i - \mathbf{X}_j)\mathbf{W}_2,$$

$$\frac{\partial A_2}{\partial \mathbf{W}_1} = 2 \sum_{i,j=1}^{n} (M_{ij}^L + \lambda \hat{M}_{ij}^S)(\mathbf{X}_i - \mathbf{X}_j)\mathbf{W}_2\mathbf{W}_2^T(\mathbf{X}_i - \mathbf{X}_j)^T\mathbf{W}_1,$$

$$\frac{\partial A_2}{\partial \mathbf{W}_2} = 2 \sum_{i,j=1}^{n} (M_{ij}^L + \lambda \hat{M}_{ij}^S)(\mathbf{X}_i - \mathbf{X}_j)^T\mathbf{W}_1\mathbf{W}_1^T(\mathbf{X}_i - \mathbf{X}_j)\mathbf{W}_2. \tag{20}$$

Furthermore, the derivative of $|x|_{1,\epsilon}$ with respect to x is

$$sgn_\epsilon(x) = [1 - e\Psi(\frac{x}{\epsilon})] \cdot sgn(x) + 2e\epsilon^2 x|x|\Psi(\frac{x}{\epsilon})\left[1 - (\frac{x}{\epsilon})^2\right]^{-2}, \tag{21}$$

Algorithm 1. Sparse Multi-label bILinear Embedding (SMILE)

Input: The training dataset $\{(\mathbf{X}_1, \mathbf{l}_1), ..., (\mathbf{X}_n, \mathbf{l}_n)\}$; the reduced dimensions d_1 and d_2; the balancing parameters λ and μ; the step size δ; and the stopping threshold ζ

Output: Transformation matrices $\mathbf{W}_1$ and $\mathbf{W}_2$

1 $t \leftarrow 0$;
2 Initialize $\mathbf{W}_1(t)$ and $\mathbf{W}_2(t)$ as column orthogonal matrices;
3 $\mathbf{W}_1(t+1) = \mathbf{W}_1(t) + \delta \nabla_{\mathbf{W}_1} J(t)$;
4 $\mathbf{W}_2(t+1) = \mathbf{W}_2(t) + \delta \nabla_{\mathbf{W}_2} J(t)$;
5 **while** $J(t+1) - J(t) > \zeta$ **do**
6 $\quad$ $t \leftarrow t+1$;
7 $\quad$ $\mathbf{W}_1(t+1) = \mathbf{W}_1(t) + \delta \nabla_{\mathbf{W}_1} J(t)$;
8 $\quad$ $\mathbf{W}_2(t+1) = \mathbf{W}_2(t) + \delta \nabla_{\mathbf{W}_2} J(t)$;
9 **end**

---

where $sgn(x)$ denotes the signum function. Accordingly, we have

$$
\begin{aligned}
\frac{\partial J}{\partial \mathbf{W}_1} &= \frac{1}{A_2^2}(A_2 \frac{\partial A_1}{\partial \mathbf{W}_1} - A_1 \frac{\partial A_2}{\partial \mathbf{W}_1}) - \mu \cdot sgn_\epsilon(\mathbf{W}_1), \\
\frac{\partial J}{\partial \mathbf{W}_2} &= \frac{1}{A_2^2}(A_2 \frac{\partial A_1}{\partial \mathbf{W}_2} - A_1 \frac{\partial A_2}{\partial \mathbf{W}_2}) - \mu \cdot sgn_\epsilon(\mathbf{W}_2),
\end{aligned}
\tag{22}
$$

where J is the abbreviation of $J(\mathbf{W}_1, \mathbf{W}_2)$, $sgn_\epsilon(\mathbf{W})$ is a matrix with the same dimension as $\mathbf{W}$, obtained by taking $sgn_\epsilon(\cdot)$ on each element of $\mathbf{W}$. By projecting the gradients $\frac{\partial J}{\partial \mathbf{W}_1}$ and $\frac{\partial J}{\partial \mathbf{W}_2}$ onto the corresponding Stiefel manifolds $\mathcal{O}_1$ and $\mathcal{O}_2$, respectively, we can obtain the gradient of the objective function in Eq. (18):

$$
\begin{aligned}
\nabla_{\mathbf{W}_1} J &= (\mathbf{I}_{D_1} - \frac{1}{2}\mathbf{W}_1 \mathbf{W}_1^T)\frac{\partial J}{\partial \mathbf{W}_1} - \frac{1}{2}\mathbf{W}_1(\frac{\partial J}{\partial \mathbf{W}_1})^T \mathbf{W}_1, \\
\nabla_{\mathbf{W}_2} J &= (\mathbf{I}_{D_2} - \frac{1}{2}\mathbf{W}_2 \mathbf{W}_2^T)\frac{\partial J}{\partial \mathbf{W}_2} - \frac{1}{2}\mathbf{W}_2(\frac{\partial J}{\partial \mathbf{W}_2})^T \mathbf{W}_2.
\end{aligned}
\tag{23}
$$

After having obtained the above gradient, we can update $\mathbf{W}_1$ and $\mathbf{W}_2$ alternately until convergence. The detailed procedure of SMILE is summarized in Algorithm 1[1].

2.6 Computational Complexity Analysis

In this subsection, we analyze the computational cost of the proposed method as well as some representative DR algorithms. For simplicity, we assume that $D_1 = D_2 = D$ and $d_1 = d_2 = d$. For SMILE, the most demanding step is the calculation in Eq. (20), and the training time cost of SMILE is $O(tn^2 D^2 d)$, where t is the number of iterations. For the traditional vector-based DR methods such

---

[1] In Algorithm 1, $\mathbf{W}_1(t)$, $\mathbf{W}_2(t)$, $\nabla_{\mathbf{W}_1} J(t)$, $\nabla_{\mathbf{W}_2} J(t)$, and $J(t)$ denote the values of $\mathbf{W}_1$, $\mathbf{W}_2$, $\nabla_{\mathbf{W}_1} J$, $\nabla_{\mathbf{W}_2} J$, and J after the t-th iteration, respectively.

as multi-label linear discriminant analysis (ML-LDA) [17], the most demanding step is performing the eigen-decomposition on a $D^2 \times D^2$ matrix and thus the training time cost is $O(D^6)$. For the tensor-based DR methods such as bilinear multi-emotion similarity preserving embedding (BME-SPE) [9], the training cost is $O(Cn^2 + tnD^2d + tD^3)$, where C is the number of labels.

3   Experimental Results

In this section, we evaluate the performance of the proposed algorithm on a standard multi-label tensor dataset: the Natural Scene Image Dataset [10,23]. This dataset consists of 2000 natural scene images, each of which may belong to one or several of the following five classes: *desert*, *mountains*, *sea*, *sunset*, and *trees*. Figure 1 shows two examples of multi-labeled images from the dataset. Figure 1(a) is an image associated with two labels: *mountains* and *trees*; while Fig. 1(b) is associated with other two labels: *sea* and *sunset*. Averagely, each image in Natural Scene Image Dataset is associated with 1.24 class labels. For each image, it is represented as a bag of nine 15-dimensional instances generated by the SBN method [10]. Therefore, the feature dimension of the dataset is 9×15 (for the tensor-based methods) or 135 (for the vector-based methods). In our experiments, the original data vectors/tensors are centralized by subtracting the mean vector/tensor before DR.

(a)   (b)

Fig. 1. Examples of multi-label images from the Natural Scene Image Dataset. (a) Image associated with labels *mountains* and *trees*; (b) Image associated with labels *sea* and *sunset*.

To evaluate the performance of the proposed method, we compare it with four representative DR algorithms: principal component analysis (PCA) [4], locality preserving projections (LPP) [3], multi-label linear discriminant analysis (ML-LDA) [17], and bilinear multi-emotion similarity preserving embedding (BME-SPE) [9], where PCA and LPP are classical DR approaches, ML-LDA is a typical multi-label DR algorithm, and BME-SPE is a representative multi-label tensor-based DR method. We also include the classification result on the raw data (i.e., without DR) as the baseline.

Two groups of criteria are used to evaluate the performance. The first group is the label-based metrics, including the Macro/Micro average precision and F1 score [21]. The nearest neighbor classifier is used for final classification after dimensionality reduction. For these four criteria, larger value indicates better performance. The second group is the example-based metrics, including average precision, Hamming loss, one-error, and ranking loss [21]. The multi-label nearest neighbor classifier [20] is used for the final classification. For average precision, larger value indicates better performance. For Hamming loss, one-error, and ranking loss, smaller value indicates better performance. In our experiments, we randomly select 90% of the data for training and use the remaining 10% for test. For the proposed method, we randomize the initialization process to generate a total of 10 sets of $\mathbf{W}_1$ and $\mathbf{W}_2$, and choose the set which yields the best result. Moreover, we set $\lambda = 0.1$, $\mu = 0.05$, $\delta = 1$, $\zeta = 0.01$, and $\epsilon = 0.01$ for our method.

Table 1 lists the results of all the methods. All the DR methods' performances are comparable to or better than that of the raw data, which shows the necessity of dimensionality reduction. In addition, by taking the multi-label correlation and tensor structure into consideration, BME-SPE and SMILE perform better than the other DR methods. With further constraints on sparse structure and column orthonormality, the proposed SMILE achieves the best results.

Table 1. Performance of PCA, LPP, ML-LDA, BME-SPE, and SMILE on the Natural Scene Image Dataset. Eight standard criteria are used for evaluation: Macro average precision, Micro average precision, Macro F1 score, Micro F1 score, average precision, Hamming loss, one-error, and ranking loss.

	Raw	PCA	LPP	ML-LDA	BME-SPE	SMILE
Macro Aver. Prec.	0.8407	0.8426	0.8395	0.8431	0.8523	**0.8635**
Micro Aver. Prec.	0.8436	0.8450	0.8417	0.8462	0.8551	**0.8665**
Macro F1	0.8350	0.8393	0.8383	0.8412	0.8452	**0.8592**
Micro F1	0.8407	0.8415	0.8406	0.8501	0.8522	**0.8624**
Aver. Prec.	0.6657	0.6678	0.6812	0.6723	0.6833	**0.6906**
Hamming Loss	0.2511	0.2398	0.2401	0.2451	0.2356	**0.2329**
One-Error	0.5131	0.4974	0.4869	0.4902	0.4871	**0.4823**
Ranking Loss	0.3024	0.3050	0.2928	0.2912	0.2891	**0.2882**

4 Conclusions

In this paper, we proposed a novel dimensionality reduction algorithm called Sparse Multi-label bILinear Embedding (SMILE) to extract discriminant features for second-order tensor data with multiple labels. By integrating the sparse and column-orthonormal constraints into the learning framework, the proposed method achieved better performance than other representative DR algorithms on the Natural Scene Image Dataset.

Future work will be explored from the following two aspects. Firstly, unlike the linear DR methods in which the learned low dimensional representations carries clear and explainable physical meaning, the physical meaning of the learned low dimensional representations using the proposed bilinear method is rather unclear, which affects the interpretability of our method. Further investigation will be engaged emphasizing on analyzing the physical meanings of the learned low dimensional representations, so as to improve the explanation power of our proposed method. Moreover, performance of the proposed method relies on quality of the initialization process. This is undesirable as it limits the generality of the proposed method. Future works are planned to develop suitable initialization strategy for the proposed method, to ensure it can be applied to a variety of scenarios with a stable performance.

Acknowledgment. This work was supported in part by the National Natural Science Foundation of China (NSFC) under Grant 61503317, in part by the General Research Fund (GRF) from the Research Grant Council (RGC) of Hong Kong SAR under Project HKBU12202417, and in part by the SZSTI Grant with the Project Code JCYJ20170307161544087.

References

1. Boothby, W.M.: An Introduction to Differentiable Manifolds and Riemannian Geometry. Academic Press, New York (2002)
2. Cunningham, J.P., Ghahramani, Z.: Linear dimensionality reduction: survey, insights, and generalizations. J. Mach. Learn. Res. **16**, 2859–2900 (2015)
3. He, X., Yan, S., Hu, Y., Niyogi, P., Zhang, H.J.: Face recognition using laplacianfaces. IEEE Trans. Pattern Anal. Mach. Intell. **27**(3), 328–340 (2005)
4. Hotelling, H.: Analysis of a complex of statistical variables into principal components. J. Edu. Psychol. **24**(417–441), 498–520 (1933)
5. Huang, J., Nie, F., Huang, H., Ding, C.: Supervised and projected sparse coding for image classification. In: Proceedings of 27th AAAI, pp. 438–444 (2013)
6. Lee, J.M.: Introduction to Smooth Manifolds. Graduate Texts in Mathematics, 2nd edn. Springer, New York (2012). https://doi.org/10.1007/978-1-4419-9982-5
7. Liu, Y., Liu, Y., Chan, K.C.C., Hua, K.A.: Hybrid manifold embedding. IEEE Trans. Neural Netw. Learn. Syst. **25**(12), 2295–2302 (2014)
8. Wang, C., et al.: What strikes the strings of your heart? - multi-label dimensionality reduction for music emotion analysis via brain imaging. IEEE Trans. Auton. Mental Develop. **7**(3), 176–188 (2015)
9. Liu, Y., Liu, Y., Zhao, Y., Hua, K.A.: What strikes the strings of your heart? - feature mining for music emotion analysis. IEEE Trans. Affect. Comput. **6**(3), 247–260 (2015)
10. Maron, O., Ratan, A.L.: Multiple-instance learning for natural scene classification. In: Proceedings of 15th ICML, pp. 341–349 (1998)
11. Panagakis, I., Benetos, E., Kotropoulos, C.: Music genre classification: a multilinear approach. In: Proceedings of 9th ISMIR, pp. 583–588 (2008)

12. Qi, G.J., Hua, X.S., Rui, Y., Tang, J., Zhang, H.J.: Two-dimensional multilabel active learning with an efficient online adaptation model for image classification. IEEE Trans. Pattern Anal. Mach. Intell. **31**(10), 1880–1897 (2009)
13. Sun, L., Ji, S., Ye, J.: Multi-Label Dimensionality Reduction. Chapman and Hall/CRC Machine Learning & Pattern Recognition, Chapman & Hall/CRC (2013)
14. Symeonidis, P., Nanopoulos, A., Manolopoulos, Y.: Tag recommendations based on tensor dimensionality reduction. In: Proceedings of 2nd ACM RecSys, pp. 43–50 (2008)
15. Ueda, N., Saito, K.: Parametric mixture models for multi-labeled text. In: NIPS, vol. 15, pp. 737–744 (2003)
16. Venna, J., Peltonen, J., Nybo, K., Aidos, H., Kaski, S.: Information retrieval perspective to nonlinear dimensionality reduction for data visualization. J. Mach. Learn. Res. **11**, 451–490 (2010)
17. Wang, H., Ding, C., Huang, H.: Multi-label linear discriminant analysis. In: Daniilidis, K., Maragos, P., Paragios, N. (eds.) ECCV 2010. LNCS, vol. 6316, pp. 126–139. Springer, Heidelberg (2010). https://doi.org/10.1007/978-3-642-15567-3_10
18. Yu, K., Yu, S., Tresp, V.: Multi-label informed latent semantic indexing. In: Proceedings of 28th ACM SIGIR, pp. 258–265 (2005)
19. Zeng, J., Liu, Y., Leng, B., Xiong, Z., Cheung, Y.: Dimensionality reduction in multiple ordinal regression. IEEE Trans. Neural Netw. Learn. Syst. **29**(9), 4088–4101 (2018). https://doi.org/10.1109/TNNLS.2017.2752003
20. Zhang, M.L., Zhou, Z.H.: Ml-knn: A lazy learning approach to multi-label learning. Pattern Recogn. **40**(7), 2038–2048 (2007)
21. Zhang, M.L., Zhou, Z.H.: A review on multi-label learning algorithms. IEEE Trans. Knowl. Data Eng. **26**(8), 1819–1837 (2014)
22. Zhang, Y., Zhou, Z.H.: Multilabel dimensionality reduction via dependence maximization. ACM Trans. Knowl. Discov. Data **4**(3), 14:1–14:21 (2010)
23. Zhou, Z.H., Zhang, M.L.: Multi-instance multi-label learning with application to scene classification. In: Schölkopf, B., Platt, J.C., Hoffman, T. (eds.) NIPS 19, pp. 1609–1616 (2007)
24. Zhu, X., Li, X., Zhang, S.: Block-row sparse multiview multilabel learning for image classification. IEEE Trans. Cybern. **46**(2), 450–461 (2016)

Learning Latent Factors in Linked Multi-modality Data

Tiantian He$^{(\boxtimes)}$ ⓘ and Keith C. C. Chan ⓘ

The Hong Kong Polytechnic University, Hung Hom, Kowloon, Hong Kong SAR
{csthe, cskcchan}@comp.polyu.edu.hk

Abstract. Many real-world data can be represented as networks in which the vertices and edges represent data entities and the interrelationship between them, respectively. The discovery of network clusters, which are typical latent structures, is one of the most significant tasks of network analytics. Currently, there are no effective approaches that are able to deal with linked data with features from multimodality. To address it, we propose an effective model for learning latent factors in linked multimodality data, named as LFLMD. Given the link structure and multimodality features associated with vertices, LFLMD formulates a constrained optimization problem to learn corresponding latent spaces representing the strength that each vertex belongs to the latent components w.r.t. link structure and multimodality features. Besides, LFLMD further adopts an effective method to model the affinity between pairwise vertices so that the cluster membership for each vertex can be revealed by grouping vertices sharing more similar latent structures. For model inference, a series of iterative algorithms for updating the variables in the latent spaces are derived. LFLMD has been tested on several sets of networked data with different modalities of features and it is found LFLMD is very effective.

Keywords: Linked multi-modality data · Complex network
Community detection · Network clustering · Graph clustering
Latent factor analysis

1 Introduction

Many real-world data are linked type in which the data entities and the interrelationship between them are represented as vertices (nodes) and links (edges), respectively. Always, they are treated as networks. Different from networks that are randomly generated, real-world networks always possess some latent features beneath the chaotic structure. Therefore, how to learn such latent features from linked data has drawn much attention in the recent [1].

Amongst all the latent features, clusters are one of the most significant as discovering them may bring us better understandings of the network data and it relates to several real-world applications, e.g., social community detection [2], document classification [3], and biological graph analytics [4]. To discover these clusters effectively, several approaches so-called network clustering algorithms are proposed, by formulating the problem taking into the consideration different information carried in the network.

© Springer Nature Switzerland AG 2018
M. Ceci et al. (Eds.): ISMIS 2018, LNAI 11177, pp. 214–224, 2018.
https://doi.org/10.1007/978-3-030-01851-1_21

When discovering clusters in the network, many approaches take into the consideration network topology. Either these approaches are model based, or heuristic based, they are able to discover clusters in networks by grouping together those vertices that are connected more. For example, a density-based algorithm presented in [5] can discover network clusters by searching densely connected sub-regions. In [6], a clustering algorithm called affinity propagation (AP) is proposed to detect clusters based on the similarities between candidate cluster centers and other vertices. Besides, spectral clustering is also very effective in network clustering. For example, in [7], spectral based methods are used in network clustering by utilizing normalized cuts [8]. In [22], another spectral clustering based method, LM is proposed. LM is able to discover network communities by analyzing the spectral features of the transition matrix which is constructed based on the local connectivity of the network. In [9], a framework is proposed for community detection based on structural correlation. In [10], a generative model called CoDa is proposed to detect clusters in the network by maximizing the posterior probability that a pair of vertices are connected in the same cluster.

As other types of information can be collected, and they can be treated as attributes characterizing the linked data entities, there are several approaches proposed to cluster network data utilizing attributes rather than topology. For example, k-means clustering [11] has been used to discover clusters in network data by assigning vertices with higher similarity of attributes into the same groups.

Although those mentioned approaches are effective to some extent, they may not discover more meaningful clusters in the linked data as they take more emphasis either on topology, or attributes associated with the vertices. To discover more meaningful clusters in the network, there are several approaches proposed to perform the task considering both network topology and attributes. In [12], attributed random network kernel is proposed to construct the transition matrix combining both structural and attribute similarity. Then the clustering task can be performed by spectral clustering. In [13], CESNA is proposed to detect network clusters using a generative process. The cluster membership for each vertex can be revealed by maximizing the posterior probability representing the edge density and attribute similarity between each pair of linked vertices. In [14], the authors propose a diffusion model for network analysis and the cluster membership can be inferred by an EM algorithm.

There are some other approaches proposed to make use of different objective functions to measure the overall quality of discovered clusters and the cluster membership of each vertex is obtained by optimizing the objective function. For examples, MISAGA [15] and FSPGA [16], are two effective approaches that are able to perform the task by maximizing the objective function measuring the overall edge density and attribute similarity in all the clusters.

We find that almost all the previous approaches to clustering in linked data, consider neither the attributes may come from different modalities, nor how the attributes may affect the interrelationship between pairwise vertices. Examples of such attributes from multi-modality are easily found in real-world linked data: an online social network user can be characterized by its user profile, hobbies, topics concerned, etc. Such attributes from different modalities may also take effect on both the relationship between vertices, and the cluster membership. To address these mentioned challenges, we propose a novel latent factor model for learning in linked multimodality data

(LFLMD). LFLMD is able to discover clusters in linked data containing multiple modalities of attributes. To perform the learning task, LFLMD considers the cluster membership for each vertex is determined by both link structure and attributes from each modality. By adopting an effective objective function that evaluates the overall clustering quality related to edge structure, attributes from each modality, and the interrelationship between pairwise vertices, LFLMD is able to identify the optimal cluster membership for each vertex through an effective updating algorithm. LFLMD has been tested with several sets of real-world network data and compared with the state-of-the-art approaches to network clustering. The experimental results show that LFLMD outperforms those compared baselines in most testing datasets, which means that learning latent factors in linked multi-modality data may lead to a better understanding in network data, and LFLMD is a promising approach to the given task.

2   Mathematical Preliminaries and Notations

In this section, the mathematical preliminaries and notations used in this paper are introduced. As attributes characterizing the data entities may come from more than one modality, the definition of linked data in this manuscript is different from the previous. Given a set of linked data containing n vertices and $|E|$ edges, it can be represented as $N = \{V, E, \Lambda\}$, where V, E, and Λ represent the vertex, edge, and attribute set in the network data, respectively. For the vertex set, it is defined as $V = \{v_i \,|\, 1 \leq i \leq n\}$. The edge set, is defined as $E = \{e_{ij} = 1 \,|\, v_i$ and v_j are connected$\}$. As for the attribute set, it is defined as $\Lambda = \left\{ \Lambda_j^i | 1 \leq i \leq d, \, 1 \leq j \leq m^i \right\}$, where d and m^i represent the total number of attribute modalities, and total number of attributes in modality i, respectively. It should be noted that d is always larger than 1 and $m^1 + \ldots + m^d = m$, where m is the total number of attributes. LFLMD uses $\mathbf{G}$ to represent the link structure between pairwise vertices in N. Apparently, $\mathbf{G}$ is an n-by-n symmetric matrix each element of which, say $\mathbf{G}_{ij}$, is equal to 1 if vertex i and vertex j are connected, and vice versa. LFLMD uses a m^i-by-n binary matrix, $\mathbf{F}^i$ to represent whether a vertex is labeled with an attribute from Λ^i, i.e., $\mathbf{F}_{jk}^i$, is equal to 1 if vertex k is labeled with attribute j from Λ^i, and vice versa. For notations, we use a subscript, e.g., $\mathbf{G}_i$, to represent the ith column of a given matrix, say $\mathbf{G}$. We use $\mathbf{G}_{ij}$, to represent the entry of $\mathbf{G}$, in ith row, jth column. We use a superscript, e.g., $\mathbf{U}^i$, to represent a matrix related to ith modality. $tr(\cdot)$ represents the matrix trace. $\|\cdot\|_F$ represents the matrix Frobenius norm. All these mentioned mathematical preliminaries and notations are used by LFLMD to model the problem of learning latent factors in linked multimodality data.

3   The Proposed Model

In this section, how LFLMD formulates the problem of learning latent factors in linked multi-modality data, and how LFLMD infers the parameters are introduced.

3.1 Modeling in Linked Multimodality Data

As mentioned in Sect. 1, LFLMD aims at searching for optimal cluster membership for each vertex taking into the consideration link structure, attributes of each modality, and the interrelationship between pairwise vertices. To build the learning model, LFLMD firstly assume that the linked data are generated by the corresponding low-dimensional latent spaces, then the model can be fitted by optimizing the difference between the original data and the generated.

To model the link structure in N, LFLMD constructs an n-by-k latent matrix $\mathbf{V}$, to represent the strength that each vertex belongs to one of the k latent clusters. Obviously, a higher value of an entry in $\mathbf{V}$, say $\mathbf{V}_{ij}$, means vertex v_i is more probable to belong to jth cluster. LFLMD assumes that each link in N is generated by $\mathbf{V}$, which in other words each element of $\mathbf{G}$ is generated by $\mathbf{V}$, so that the following function can be used to evaluate the difference between original link structure and the generated

$$\left\| \mathbf{G} - \mathbf{V}\mathbf{V}^T \right\|_F^2 \tag{1}$$

It can be seen from (1), LFLMD assumes that $\mathbf{G}_{ij}$ is generated by $\mathbf{V}\mathbf{V}_{ij}^T$, which means the product of ith and jth row in $\mathbf{V}$. When (1) is optimized, the link density within each cluster can be maximized.

Similar to model the link structure, LFLMD constructs a m^i-by-k latent space, $\mathbf{U}^i$, and an n-by-k matrix, $\mathbf{A}$ to represent the weight factors that each attribute from Λ^i contributes to each of k latent components, and the strength that each vertex belongs to each of the k latent components, respectively. LFLMD assumes that whether or not each attribute in Λ^i, is associated with a vertex, is generated by $\mathbf{U}^i$ and $\mathbf{A}$, so that the following function can be used to evaluate the difference between the original attributes associated with each vertex and the generated in all d modalities

$$\sum_{i=1}^d \left\| \mathbf{F}^i - \mathbf{U}^i \mathbf{A}^T \right\|_F^2 \tag{2}$$

When (2) is optimized, $\mathbf{F}^i$ is best approximated by $\mathbf{U}^i$ and $\mathbf{A}$, and $\mathbf{A}$ may reveal the strength that each vertex belongs to each of k latent components, in terms of attributes from all the modalities.

As mentioned above, both link structure and attributes may affect the interrelationship between pairwise vertices and the cluster membership of each vertex in the network. Such interrelationship can be seen as affinity between pairwise vertices, which stands for how similar/related two vertices are. For LFLMD, we propose to use the following method to model the affinity between pairwise vertices. Let $\mathbf{X}$ be a matrix representing the degree of affinity between pairwise vertices in N. For each entry in $\mathbf{X}$, LFLMD assumes that it can be generated by $\mathbf{V}$ and $\mathbf{A}$. Hence, we may obtain the following function evaluating the difference between $\mathbf{X}$ and the one generated by $\mathbf{V}$ and $\mathbf{A}$

$$\left\|\mathbf{X} - \mathbf{A}\mathbf{V}^T\right\|_F^2 \tag{3}$$

Given (3), we find that how affinitive that two vertices are is determined by both $\mathbf{V}$ and $\mathbf{A}$, which are the latent spaces learned from (1) and (2). This means both structural and attribute features of the linked data take effect on the affinity between each pair of vertices. If (3) can be optimized, latent spaces $\mathbf{V}$ and $\mathbf{A}$ are regularized to assign those closely affinitive vertices into the same latent components. As for $\mathbf{X}$, it can be obtained by computing pairwise similarity between vertices, using either link or attribute information in N. In this manuscript, $\mathbf{X}$ is obtained by computing the cosine similarity of local neighbors between pairwise vertices.

Given (1), (2) and (3), we proposed LFLMD to make use of the following objective function to evaluate the learning in linked multimodality data:

minimize
$$O = \left\|\mathbf{G} - \mathbf{V}\mathbf{V}^T\right\|_F^2 + \sum_{i=1}^{d} \left\|\mathbf{F}^i - \mathbf{U}^i\mathbf{A}^T\right\|_F^2 + \alpha\left\|\mathbf{X} - \mathbf{A}\mathbf{V}^T\right\|_F^2 \tag{4}$$
subject to $\mathbf{V} \geq 0$, $\mathbf{A} \geq 0$, $\mathbf{U}^i \geq 0$

where α is a positive real which is used to control the effect of (3) in the learning process. Given (4), we may find out LFLMD has the following peculiarities when it is learning latent factors in linked network data. First, it considers the effect from multimodality of attributes, which is not concerned by almost all the previous works. Second, LFLMD considers modeling the affinity between pairwise vertices using both topological and attribute properties of the linked data, so that vertices sharing with similar latent structural and attribute features are more possible to be assigned into the same clusters. For such modelling method, this is also the first attempt, compared with previous works. Third, LFLMD is very flexible to be extended or simplified. As it shows in (4), the number of modalities of attributes can be easily configured according to the data to learn. Therefore, LFLMD can be seen as a generic model for learning latent factors in linked data. When (4) is optimized, the cluster membership for each vertex can be easily obtained by LFLMD.

3.2   Learning as Optimization

How to search for the optimal latent factors is essential for a learning model. As the objective function (4) used by LFLMD is a constrained optimization problem, we may derive a series of iterative rules for updating the corresponding parameters in LFLMD, based on the KKT condition.

Let β_{jk} be the Lagrange multiplier for the constraint $\mathbf{V}_{jk} \geq 0$. The Lagrange function L for $\mathbf{V}$ is

$$L(\mathbf{V}, \boldsymbol{\beta}) = O - tr(\boldsymbol{\beta}^T\mathbf{V}) \tag{5}$$

where $\boldsymbol{\beta} = \left[\beta_{jk}\right]$ is the matrix of Lagrange multipliers for the non-negativity of $\mathbf{V}$. Based on the KKT condition, we have the following element-wise equation system

$$\frac{\partial L}{\partial \mathbf{V}_{jk}} = \left[-4\mathbf{GV} + 4\mathbf{VV}^T\mathbf{V} - 2\alpha\mathbf{XA} + 2\alpha\mathbf{VA}^T\mathbf{A} - \boldsymbol{\beta} \right]_{jk} \tag{6}$$

$$\boldsymbol{\beta}_{jk}\mathbf{V}_{jk} = 0, \quad \boldsymbol{\beta} \geq 0$$

Based on (6), we may obtain the following rule for updating the variables in $\mathbf{V}$

$$\mathbf{V}_{jk} \leftarrow \mathbf{V}_{jk} \frac{\sqrt{\sqrt{\Delta_{jk}} - \left[\alpha\mathbf{VA}^T\mathbf{A} \right]_{jk}}}{\sqrt{4\left[\mathbf{VV}^T\mathbf{V} \right]_{ij}}} \tag{7}$$

$$\Delta_{jk} = \left[\alpha\mathbf{VA}^T\mathbf{A} \right]_{jk}^2 + 8\left[\mathbf{VV}^T\mathbf{V} \right]_{jk} [2\mathbf{GV} + \alpha\mathbf{XA}]_{jk}$$

Similarly, we may derive the updating rules for $\mathbf{A}$ and $\mathbf{U}^i$. Here we directly present the updating rules for $\mathbf{A}$ and $\mathbf{U}^i$ due to the space limitation. Based on the KKT condition, we have the following iterative updating rule for the inference of $\mathbf{A}$

$$\mathbf{A}_{jk} \leftarrow \mathbf{A}_{jk} \frac{\alpha[\mathbf{XV}]_{jk} + \sum\limits_{i=1}^{d} \left[\mathbf{F}^{iT}\mathbf{U}^i \right]_{jk}}{\alpha\left[\mathbf{AV}^T\mathbf{V} \right]_{jk} + \sum\limits_{i=1}^{d} \left[\mathbf{AU}^{iT}\mathbf{U}^i \right]_{jk}} \tag{8}$$

And similarly, we have the following iterative updating rule for the inference of $\mathbf{U}^i$

$$\mathbf{U}_{jk}^i \leftarrow \mathbf{U}_{jk}^i \frac{\left[\mathbf{F}^i\mathbf{A} \right]_{jk}}{\left[\mathbf{U}^i\mathbf{A}^T\mathbf{A} \right]_{jk}} \tag{9}$$

Based on (7)–(9), LFLMD is able to search for the local optima when updating the variables in $\mathbf{V}$, $\mathbf{A}$ and all the $\mathbf{U}^i$s in each iteration of model optimization. Following the methods proposed in [17, 18], the convergence of the proposed updating rules can be verified.

The complexity of these updating rules may determine the efficiency of the learning process. Based on (7), updating the variables in $\mathbf{V}$ in each iteration follows the order of $O(4nk^2 + 2n^2k)$. Based on (8), updating the latent factors in $\mathbf{A}$ in each iteration follows the order of $O(n(1+d)k^2 + n^2k + mnk)$. Based on (9), updating the latent factors in each $\mathbf{U}^i$ follows the order of $O(m^ink + m^ik^2)$. Thus, LFLMD is a model having the computational complexity that is approximate to $O(n^2)$, as k is much smaller than n and m is near to n. One may improve the efficiency of LFLMD by using some high-performance computation techniques.

3.3 Summary of the Model

Based on the descriptions from Sects. 3.1 and 3.2, the learning process of LFLMD can be summarized as the pseudo codes shown in Fig. 1. After the regulation parameter α, maximum number of iteration *max_iteration*, tolerance for improvement τ, and the number of clusters k are determined, LFLMD will iteratively search for the optimal settings for the variables in **V**, **A**, and all the **U**^{i} till the variation of **V** between each two iterations is less than τ or the objective function converges to the local minima. After inferring the parameters, LFLMD obtains the matrix **V**, which contains the optimal cluster membership between each vertex and k clusters.

LFLMD	
Input:	$N, \alpha, max\_iteration, \tau, k$

Randomly initialize **V**, **A**, **U**^{i};
for count=1: *max_iteration*
    Fixing **A** and **U**^{i}
    update **V** according to (7);
    Fixing **V** and **U**^{i}
    update **A** according to (8);
    Fixing **A**
    update each **U**^{i} according to (9);
    if ($|\mathbf{V}^i - \mathbf{V}^{i-1}|_F < \tau$)
        compute objective value according to (4); break;
    end if
end for
return **V**, **A**, **U**^{i} ($1 \leq i \leq d$);

Fig. 1. Pseudo codes of LFLMD

4   Experimental Results and Analysis

In this section, we present the details on how we test the effectiveness of LFLMD using series of experiments.

Baselines for Comparison. As the main task performed by LFLMD is to detect clusters in linked multi-modality data, we selected six prevalent approaches to network clustering as compared baselines. Specifically, Spectral clustering (SC) [7], and CoDa [10] are two approaches to clustering based on link structure. k-means [11] can discover network clusters based on node attributes. While, Collective Non-negative Matrix Factorization (CollNMF) [19], MISAGA [15], and FSPGA [16] are three methods for network clustering utilizing both network topology and attributes. It should be noted that currently there are not many effective methods for analyzing networks with multi-modality features. Hence, we select CollNMF, which is a model for segmenting multi-modality image data, as one of the baselines.

Experimental Set-Up. For performance comparison, we either used the source code, or executables of the baselines which are online available. All experiments were conducted under the same environment which included a workstation with 4-core 3.4 GHz CPU and 16 GB RAM. For the parameter settings of different approaches, we set the number of clusters, k to be the number of ground truth clusters of each testing dataset before these approaches run. For other parameters that are required by the baselines, we used the best settings recommended. For LFLMD, we set α to be 1, the maximum number of iteration to be 300. All the algorithms, including LFLMD, were executed 10 times to obtain statistical averages for the performance measures.

Evaluation Metric. For performance evaluation, we used Normalized Mutual Information (*NMI*), which is a widely-used measure for verifying the clustering quality in network data. The detailed definition for *NMI* are presented in [15, 16].

Dataset Descriptions. For performance testing, we used five real-world datasets with ground truth. These network data are with different sizes and contain different number of modalities of node attributes. They include Football (FB) [20], PoliticsUK (PUK) [20], Olympics (Olym) [20], Cora [21], and Googleplus (Gplus) [10]. All these datasets have been widely used for testing the effectiveness of network clustering. The detailed information on these datasets have been summarized in Table 1.

Table 1. Statistics of the testing datasets

	FB	PUK	Olym	Cora	Gplus
N	248	419	464	2708	8725
$|E|$	3222	23003	9749	5278	972899
m	15407	22747	21552	1432	5913
d	2	2	2	1	5
Ground truth k	20	5	28	7	130

4.1   The Experimental Results

For performance comparison, we used LFLMD and other baselines to discover clusters in the mentioned network data. And the discovered clusters are evaluated by *NMI*. The experimental results of *NMI* obtained in the testing datasets are summarized in Table 2.

Table 2. Clustering performance evaluated by *NMI* (%)

	FB	PUK	Olym	Cora	Gplus
SC	54.386	47.38	46.905	7.841	12.915
CoDa	30.4	21.654	25.253	9.411	18.9
k-means	38.201	27.527	48.257	14.065	9.735
FSPGA	82.338	51.314	80.227	12.202	27.557
CollNMF	54.998	39.659	66.683	16.806	36.458
MISAGA	81.636	54.643	82.437	14.536	21.553
LFLMD	**88.276**	**59.892**	82.129	**32.467**	**43.671**
Improvement[#]	**7.212**	**9.606**	–	**93.187**	**19.784**

#: The percentage of improvement when LFLMD is compared with the second-best approach in each dataset.

As Table 2 shows, LFLMD performs robustly in all the datasets, when *NMI* is considered. In 5 datasets except of Olym, LFLMD outperforms any other compared baselines when the discovered clusters are evaluated by *NMI*. LFLMD outperforms FSPGA in FB by 7.212%, MISAGA in PUK by 9.606%, CollNMF in Cora, and Gplus by 93.187% and %19.784, respectively. Given the experimental results shown in Table 2, it is said that LFLMD is a very effective model for learning latent factors as clusters in linked data, and the clusters discovered by LFLMD are better matched with the ground truth.

4.2 Sensitivity Test of the Model

As mentioned in Sect. 3, α is used to control the effect of affinity between pairwise vertices within the optimization process. By varying the settings of α, the clustering performance of LFLMD might be changed. To investigate how parameter α affects the clustering performance of LFLMD, we let LFLMD learn latent factors in all the testing datasets, using α ranging from 1 to 10, with a changing step of 1. The discovered clusters are then evaluated by *NMI*. The corresponding experimental results are shown in Fig. 2. As the table shows, LFLMD performs relatively steadily in most datasets when α changes. The clustering performance evaluated by *NMI* has relatively evident variations in datasets PUK and FB. Obtaining such results indicates that the modeling of pairwise affinity used by LFLMD leads it to obtain an optimal cluster membership that better matches the ground truth. In most datasets, the clustering performance is better when the pairwise affinity generated jointly by **V** and **A**, approximates more to the original. This also satisfies the common sense that those vertices are more similar are more probable to be in the same cluster. Modeling pairwise affinity is one of the significant reasons that LFLMD is able to be effective in clustering in different types of linked data. Based on the results shown, we recommend α to be set to 1 by default, when one is using LFLMD to learn latent factors in the linked data.

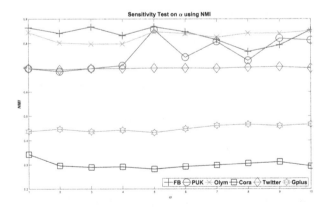

Fig. 2. Sensitivity test on α

5 Conclusion

In this paper, a novel model for learning latent factors in linked data, LFLMD is proposed. LFLMD models link structure, multimodality attributes, and affinity between pairwise vertices, when it formulates the problem of unsupervised learning in the linked data. To evaluate its effectiveness, LFLMD has been used to discover clusters in 5 sets of networked data and compared with the state-of-the-art approaches to network clustering. The experimental results show that LFLMD performs more robustly in different types of networked data, when it is compared with other baselines. LFLMD is a very promising model for learning in linked multimodality data. In future, we will further improve the effectiveness and efficiency of LFLMD by searching for more effective methods for affinity modeling and developing parallel version of LFLMD.

References

1. Fortunato, S.: Community detection in graphs. Phys. Rep. **486**(3–5), 75–174 (2010)
2. He, T., Chan, K.C.C.: Evolutionary community detection in social networks. In: Proceedings of CEC 2014, Beijing, pp. 1496–1503 (2014)
3. Hu, L., Chan, K.C.C.: Fuzzy clustering in a complex network based on content relevance and link structures. IEEE TFS **24**(2), 456–470 (2016)
4. He, T., Chan, K.C.C.: Evolutionary graph clustering for protein complex identification. IEEE/ACM TCBB **15**(3), 892–904 (2018)
5. Chang, L., Li, W., Qin, L., Zhang, W., Yang, S.: pSCAN: fast and exact structural graph clustering. IEEE TKDE **29**(2), 387–401 (2017)
6. Frey, B.J., Dueck, D.: Clustering by passing messages between data points. Science **16**, 972–976 (2007)
7. Luxburg, U.: A tutorial on spectral clustering. Stat. Comput. **17**(4), 395–426 (2007)
8. Shi, J., Malik, J.: Normalized cuts and image segmentation. IEEE TPAMI **22**(8), 888–905 (2000)
9. Veldt, N., Gleich, D.F., Wirth, A.: A correlation clustering framework for community detection. In: Proceedings of WWW 2018, Lyon, pp. 439–448 (2018)
10. Yang, J., McAuley, J., Leskovec, J.: Detecting cohesive and 2-mode communities in directed and undirected networks. In: Proceedings of WSDM 2014, New York, pp. 323–332 (2014)
11. MacKay, D.J.C.: Information Theory, Inference, and Learning Algorithms. Cambridge University Press, Cambridge (2003)
12. Guo, T., Wu, J., Zhu, X., Zhang, C.: Combining structured node content and topology information for networked graph clustering. ACM Trans. Knowl. Disc. Data (TKDD) **11**(3), 29 (2017)
13. Yang, J., McAuley, J., Leskovec, J.: Community detection in networks with node attributes. In: Proceedings of ICDM 2013, Dallas, pp. 1151–1156 (2013)
14. Bao, Q., Cheung, W.K., Zhang, Y., Liu, J.: A component-based diffusion model with structural diversity for social networks. IEEE TCYB **47**(4), 1078–1089 (2017)
15. He, T., Chan, K.C.C.: MISAGA: an algorithm for mining interesting sub-graphs in attributed graphs. IEEE TCYB **48**(5), 1369–1382 (2018)
16. He, T., Chan, K.C.C.: Discovering fuzzy structural patterns for graph analytics. IEEE TFS (2018). https://doi.org/10.1109/TFUZZ.2018.2791951

17. Dempster, A., Laird, N., Rubin, D.: Maximum likelihood from incomplete data via the EM algorithm. J. R. Stat. Soc. B **39**, 1–38 (1977)
18. Wang, F., Li, T., Wang, X., Zhu, S., Ding, C.: Community discovery using nonnegative matrix factorization. Data Min. Knowl. Disc. **22**, 493–521 (2011)
19. Akata, Z., Thurau, C., Bauckhage, C.: Non-negative matrix factorization in multimodality data for segmentation and label prediction. In: Proceedings of CVWW 2011, Mitterberg, pp. 1–8 (2011)
20. Greene, D., Cunningham, P.: Producing a unified graph representation from multiple social network views. In: Proceedings of WebSci 2013, Paris, pp. 1–4 (2013)
21. Lu, Q., Getoor, L.: Link-based classification. In: Proceedings of ICML 2003, Washington DC, pp. 189–207 (2003)
22. Yang, B., Liu, J., Feng, J.: On the spectral characterization and scalable mining of network communities. IEEE TKDE **24**(2), 326–337 (2012)

Researcher Name Disambiguation: Feature Learning and Affinity Propagation Clustering

Zhizhi Yu[1,2] and Bo Yang[1,2(✉)]

[1] College of Computer Science and Technology,
Jilin University, Changchun 130012, China
yuzz16@mails.jlu.edu.cn, ybo@jlu.edu.cn
[2] Key Laboratory of Symbol Computation and Knowledge Engineering,
(Jilin University), Ministry of Education, Changchun 130012, China

Abstract. Name ambiguity has been considered as a challenging task in the field of information retrieval. When we want to query all the papers of a researcher in the current literature integration system, we will find that many irrelevant papers written by the same researcher name appear in the retrieval results, which seriously affect the quality of retrieval. To tackle this problem, name disambiguation task was proposed to correctly distinguish the papers, thus making papers contained in each part belongs to a unique researcher. Certain information sources can help disambiguate researchers, e.g., CoResearcher, affiliation, homepages and paper titles. However, such information sources may be costly to obtain or unavailable. Therefore, it is necessary to solve name disambiguation task under the condition of insufficient information sources. Another challenge is how to accomplish the task without knowing the number of distinct researchers. In this paper, we sufficiently use the relational network between papers. Our proposed method learns the feature representations of papers and then uses affinity propagation clustering to solve name disambiguation task. The experimental results show that our proposed method can obtain better accuracy at solving name disambiguation task comparing to existing methods.

Keywords: Name disambiguation · Feature learning
Affinity propagation clustering

1 Introduction

With the rapid development of information technology, how to help us find the desired paper quickly and accurately is a challenging task to be solved in the current literature retrieval system. However, most literature retrieval systems only stay in keyword retrieval stage. We can only spend our time and energy to evaluate whether the retrieval results meet our need. If we want to predict or estimate the quality of a paper, we often need to know extra information of the researcher, such as what other papers written by the researcher and academic position held by the researcher.

Those information are useful for us to determine the quality of a paper, but there is no such information in the general literature database. In this context, the literature retrieval system with researcher as the core was born. However, when we use the

M. Ceci et al. (Eds.): ISMIS 2018, LNAI 11177, pp. 225–235, 2018.
https://doi.org/10.1007/978-3-030-01851-1_22

system, we face another problem, when querying all the papers of an researcher, we will find that many irrelevant papers written by the same researcher name appear in the retrieval results, causing us to spend a lot of time to filter papers of the researcher we are interested in, which seriously affects the search quality, so name disambiguation task was designed to solve this problem in the field of information retrieval.

Name disambiguation task [2, 8] aims to eliminate the ambiguity of names in cross-document situations and classify the same names into different entities in the real world, so as to provide information to users after effective information clustering. Researcher name disambiguation is a kind of multiple document name disambiguation, which can be used to determine papers clustering process by judging whether the same researcher name in different papers points to a unique researcher in real life.

Many existing studies has been proposed to solve name disambiguation task [1, 8, 10], among them, most methods [14] use extra information sources to help identify researchers, e.g., CoResearcher, affiliation, homepages and paper titles, However, some information sources are difficult to obtain, e.g., home pages. Therefore, it is necessary to solve name disambiguation task under the condition of insufficient information sources. In addition, another challenge in name disambiguation task is, given a researcher name and a series of papers containing that researcher, how to accomplish name disambiguation task without knowing the number of distinct researchers?

In our work, we sufficiently use relationship information between papers to solve name disambiguation task. For a given researcher name, we first construct the relationship network between papers which written by the researcher, then learn the feature representations of papers and finally use affinity propagation clustering to cluster the papers so that the papers contained in each part belong to a unique researcher.

In this paper, we systematically study name disambiguation task and make the following contributions:

1. We sufficiently use the relationship network between papers to propose the feature learning model, and then use the generated feature representations of papers to solve name disambiguation task.
2. We propose a novel method to cluster the feature representations of papers by using affinity propagation clustering, which is effectively for solving name disambiguation task without knowing the number of distinct researchers.
3. We evaluated our proposed method on a real dataset. The experimental results demonstrate that our proposed method is superior to some existing methods in the field of name disambiguation.

2   Related Work

Researchers have done a lot of work on name disambiguation task [1, 7, 13]. In general, there are three ways to solve name disambiguation task, supervised, unsupervised and semi-supervised. The supervised disambiguation method [15] needs to mark the training data, then obtain the classification model on the basis of a large number of training data, finally use the model to judge whether new researcher and the researcher in the sample belong to same researcher. The unsupervised disambiguation method [11, 16] uses extra information calculate the similarity between papers, and then solve name disambiguation task by using a given clustering algorithm. The semi-supervised disambiguation method [17, 18] uses a small amount of annotation data to guide a large amount of unlabeled data, thereby improving accuracy of name disambiguation task.

In recent years, researchers have begun to use other methods to tackle name disambiguation task, such as author graphs, and extra information extracted from other sources. Authors in [2] presented a graph-based approach to solve name disambiguation task by using self-citation and Co-Researcher, Authors in [6] proposed a method which calculates the similarity between papers by using graph kernels and then use SVM to solve name disambiguation task. Authors in [4] propose an ADANA method which solves name disambiguation task by using active user interactions. Authors in [5] use deep neural-network to automatically learn feature representation of papers for solving name disambiguation task. Authors in [9] only use a time-stamped graph topology in an anonymous network and solve name disambiguation tasks by using link information in collaboration network.

Our proposed method first learns feature representations of papers in Paper-Paper network, and then uses affinity propagation clustering to solve name disambiguation task, which is inspired by word embedding model in natural language processing, such as DeepWalk [12] and Node2Vec [3].

3 Problem Formalization

In this section, we first give some basic definitions, and then give the definition of name disambiguation task.

3.1 Definitions

Definition 1 (Principle Researcher and Secondary Researcher): For a given paper p_i, We denote $A_i = \{r_1, r_2, \ldots, r_n\}$ to be a set of researchers, where the disambiguation researcher is recorded as the principle researcher r_1, the rest $R_i = \{r_2, \ldots, r_n\}$ as secondary researcher.

Definition 2 (CoResearcher): For two given papers p_i and p_j, if $R_i \cap R_j \neq \emptyset$, where R_i denotes the secondary researcher of paper p_i, R_j denotes the secondary researcher of paper p_j, then we believe the two papers have a CoResearcher relation.

Definition 3 (r-CoResearcher): r-CoResearcher represents r-extension collaborato-rs relation. For a given paper dataset, we first construct a CoResearcher network $C(V, E)$, where V represents a set of researchers, E represents a set of CoResearcher relation. For two given papers p_i and p_j, we can obtain their corresponding coresearcher set R_i and R_j. If the intersection of R_i and R_j is not empty, then we consider the two papers have 1-CoResearcher relation. For judging whether the two papers have 2-CoResearcher relation, we constructed the collaborator set R_i^2 and R_j^2 according to CoResearcher network $C(V, E)$. R_i^2 is the set of researchers by extending R_i^1 with all neighbors of the researchers in R_i^1, represents as $R_i^2 = R_i^1 \cup \{NB(r)\}$, where $r \in R_i^1$, $NB(r)$ is the set of neighbors of paper p_i, If the two sets of R_i^2 and R_j^2 have intersection, we think there is 2-CoResearcher relation between two papers, Therefore, in order to judge whether two papers have r-CoResearcher relation, we further extend R_i^{r-1} to obtain the collaborators set R_i^r. If two sets of collaborators have intersection, we consider that there are r-CoResearcher relation between two papers.

3.2   Name Disambiguation Task

For a given researcher r, we denote the set of papers associated with researcher r as P and use the Paper-Paper network to represent the relationship between papers, where each node represent a paper. For two given papers p_i and p_j, if r-CoResearcher relation of two papers is intersecting, then there is a certain connection between two papers, we build an edge between them, so we can define the Paper-Paper network as follows:

Definition 4 (Paper-Paper Network): Let P (V,E) represents a Paper-Paper network, Where V represents a set of papers, E represents a set of paper relations. The weights between papers are simply defined as the intersection of r-CoResearcher between papers, represent as $w_{ij} = \left| R_i^r \cap R_j^r \right|$.

Suppose that the set of papers associated with researcher r is denote as $P\{p_1, p_2 \ldots p_n\}$, Name disambiguation task is to divide the paper set P into M (the number of distinct researcher) disjoint sets, so that papers contained in each part belongs to a unique researcher r. although it may consider as a simple clustering task, name disambiguation task is difficult to tackle in real life. Since in most cases the number of distinct researcher is unknown, most traditional clustering methods are performing badly on this task. In order to tackle this task, we use the following steps deal with name disambiguation task without knowing the number of distinct researchers.

1. Feature Learning: We obtain the feature representations of papers by using the feature learning model combines structural information from Paper-Paper Network.
2. Affinity Propagation Clustering: We solve name disambiguation task by using affinity propagation clustering to implement iterative clustering between paper feature representations, and divide the paper set into several disjoint parts such that each part belongs to a unique researcher.

4   Model

In this section, we will introduce our method to solve name disambiguation problem. Our approach is first to learn the feature representations of papers, and then cluster feature representations by using affinity propagation clustering to implement name disambiguation task.

4.1   Feature Learning

We believe that neighboring nodes in the network should have more similar feature representation than non-neighboring nodes. In Paper-Paper Network P, the similarity between two nodes p_i and p_j can be obtained by using the inner product of their corresponding feature representation, represented as $S_{ij} = p_i^T p_j$. So for a given triad $(p_i, p_j, p_k) \in P$, where $p_i,\ p_j$ are neighboring nodes, $p_j,\ p_k$ are non-neighboring node, then $S_{ij} > S_{ik}$. For the convenience of calculation, we use logic function to preserve the probability of neighboring nodes being larger than non-neighboring nodes, which is defined as follows:

$$P(S_{ij} > S_{ik}) = \frac{1}{1 + \exp(-(S_{ij} - S_{ik}))} \tag{1}$$

where $S_{ij} - S_{ik}$ is defined as below:

$$S_{ij} - S_{ik} = p_i^T p_j - p_i^T p_k \tag{2}$$

From Eq. 1 we found: the larger difference in similarity, the larger probability of $S_{ij} > S_{ik}$. For a Paper-Paper network E, let P_{N*K} represents feature representations matrix of papers. By assuming that the ranking order of triad $(p_i, p_j, p_k) \in P$ is mutually independent, the feature representations of papers can be obtained by maximizing the probability that S_{ij} is larger than S_{ik}, the objective function is defined as follows:

$$L = max \prod_{(p_i, p_j, p_k) \in P} P(S_{ij} > S_{it} | p_i, p_j, p_k) \tag{3}$$

For the convenience of computation, we use the method of minimizing the negative log-likelihood function to calculate, so the objective function is abbreviated to:

$$L = \min_{P} - \log P(S_{ij} > S_{it} | p_i, p_j, p_k) + \lambda ||P||_F^2 \tag{4}$$

where λ is the norm set to prevent overfitting of the model. $(p_i,\ p_j,\ p_k)$ represents a sampling triad. for a given paper node p_i, we use neighbor sampling strategy to sample its neighbor node p_j, and use a uniform sampling strategy to sample its non-neighbor node p_k.

Therefore, For a given sampling triad $(p_i, p_j, p_k) \in P$, we use stochastic gradient descent algorithm to optimize the objective function so as to learn feature representations of papers.

4.2 Affinity Propagation Clustering

After obtaining the feature representations of papers, we use the affinity propagation clustering to realize the iterative clustering between papers. The goal of clustering is to minimize the distance between the sample paper and its representative paper.

The affinity propagation clustering regards each paper as a potential cluster center, and its goal is to find the optimal Cluster center, which maximizes the similarity of all papers to their class representative paper. For two given papers p_i and p_j, The Similarity between them can be calculated by Euclidean distance corresponding to their feature representations, denotes as $S(i,j) = -|p_i - p_j|^2$, Where S represents an $N \times N$ similarity matrix and the larger $S(k,k)$, the greater probability that the paper k becomes Cluster Center.

The Affinity Propagation Clustering mainly contains two information parameters: Responsibility and Availability. Let $R = [r(i,j)]_{nxn}$ represents the responsibility matrix, $A = [a(i,j)]_{n \times n}$ represents the Availability matrix, where $r(i,j)$ represents the Responsibility of paper p_i as the representative paper of paper p_j, and $a(i,j)$ represents the Availability of paper p_j as the representative paper of paper p_i. In the initial stage, $a(i,j) = r(i,j) = 0$, $r(i,j)$ and $a(i,j)$ are updated iteratively as follows:

$$r(i,j) = s(i,j) - \max_{j' \neq j}\{a(i,j') + s(i,j')\} \tag{5}$$

$$If\ i \neq j, a(i,j) = \min\{0, r(i,j)\} + \sum_{i' \neq j}\max\{0, r(i',j)\}\} \tag{6}$$

$$a(j,j) = \sum_{i' \neq j}\max\{0, r(i',j)\}\} \tag{7}$$

Therefore, the objective function for obtaining the class representative paper p_j of the paper p_i can be defined as follows:

$$\underset{j}{\mathrm{argmax}}(a(i,j) + r(i,j)) \tag{8}$$

5 Experiment

We conducted some experiments to verify the ability of our proposed method on name disambiguation task without knowing the number of distinct researchers, Besides, we compare our method with several existing methods to show that is better than those methods.

5.1 Datasets

A major challenge in name disambiguation task is to find data sets with real tags to evaluate the performance of the proposed method. In recent years, the bibliographic database website Arnetminer provides a set of data with researcher names and disambiguation results. In our experiment, we selected 11 researcher names on Arnetminer and test the performance of our proposed method on these researcher names. The statistical information of researchers is shown in Table 1. For each researcher name, we also give corresponding number of papers and the number of distinct researcher.

Table 1. Arnetminer name disambiguation dataset.

Researcher name	No. of papers	No. of actual
Bin Li	181	60
Bo Liu	124	47
Cheng Chang	27	5
David Brown	61	25
Feng Liu	149	32
Gang Luo	47	9
Hao Wang	178	48
Jing Zhang	231	85
Lei Wang	308	112
Ning Zhang	127	33
Wei Xu	153	48

5.2 Comparison Methods

In order to test the name disambiguation ability of our proposed method, we compare it with several existing methods, which can be divided into two classes: Traditional disambiguation methods; Feature Learning method. For all methods, we set the embedded space dimension to be 20.

1. Rand: Rand randomly assigns papers to one class. We use the number of real researchers as the predefined class number.
2. DeepWalk: DeepWalk encodes relations of the paper node in a continuous vector space by using skip-gram method.
3. Node2vec: Node2Vec obtains the feature representations of papers by designing a biased random walk and maximizing the likelihood of flexible network neighborhoods.

5.3 Experimental Setting

We use Macro-*F1* value to compare with previously proposed method. The Macro-*F1* value is defined as follows:

$$Macro - F1 = \frac{\sum_{i=1}^{L} F1(i)}{L} \tag{9}$$

where L represents the number of clusters, $F1$ represents the harmonic mean of accuracy and recall, denote as:

$$F1(i) = \frac{2 \times recall(i) \times pecision(i)}{recall(i) \times precision(i)} \tag{10}$$

There are some other parameters in our model, in the feature learning stage, we set collaborator extension length to be 2 and embedding space dimension to be 20, and perform a grid search over the parameters, setting $\lambda \in \{0.01, 0.02, 0.1\}$, Besides, we fix the learning rate $a = 0.001$.

5.4   Comparisons with Name Disambiguation Methods

For all 11 researcher names, we conducted disambiguation experiments on them, The data in Table 2 shows the results of comparing our method with other methods.

Table 2. Compare our method with other methods.

Researcher name	Our method	Rand	DeepWalk	Node2Vec
Bin Li	**0.841**	0.542	0.234	0.399
Bo Liu	**0.822**	0.471	0.765	0.749
Cheng Chang	**0.831**	0.520	0.742	0.496
David Brown	**0.885**	0.544	0.765	0.837
Feng Liu	**0.705**	0.584	0.530	0.425
Gang Luo	**0.776**	0.309	0.202	0.626
Hao Wang	**0.749**	0.590	0.599	0.605
Jing Zhang	**0.783**	0.609	0.465	0.663
Lei Wang	**0.838**	0.570	0.503	0.351
Ning Zhang	**0.713**	0.437	0.436	0.507
Wei Xu	**0.741**	0.464	0.319	0.639

As we can observe from the Table 2, our proposed method performs better than other methods and the overall improvement of our method is much higher than that of the second method. Among the competing methods, The Rand does not perform well because it randomly distributes the paper to a class. Feature Learning methods such as DeepWalk performs badly, The possible reason is that DeepWalk does not consider the weight of the edges between the papers when calculating the similarity of the papers. Besides, due to inappropriate selection of parameters p and q, the Node2Vec method also performs poorly.

5.5   Parameter Sensitivity

We conduct experiments to evaluate how the collaborator extension length r and the embedding dimension k affect the disambiguation effect of our proposed method. We set $r \in \{1, 2, 3, 4, 5\}$, $k \in \{10, 20, 30, 40, 50\}$. To save space, we show the average results of the 11 researcher names, which are shown in Figs. 1 and 2. Figure 1 shows the result of Macro-*F1* changing with r, as the collaborator extension length r increases, the value of Macro-*F1* increases first and then becomes stable. The probable reason is that when r is small, the network of papers constituted by collaborator relations is relatively sparse and we may lose some information. However, as the collaborator extension length r become larger, the Paper-Paper network is gradually fixed and the value of Macro-*F1* tends to be stable. Figure 2 shows the result of Macro-*F1* changing with k, as the embedded dimension k increases, the value of Macro-*F1* first increases and then decreases. The reason for this phenomenon is that when k is small, the feature representations ability of papers is weak and we may lose some feature information. However, as the embedding dimension k becomes larger, the model we propose is too complex and may lead to over-fitting of data.

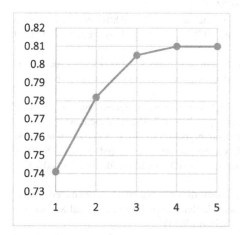

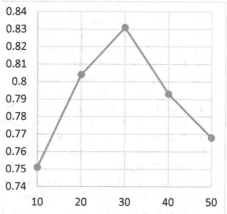

Fig. 1. The influence of collaborator extension length r.

Fig. 2. The influence of embedding dimension k.

6   Conclusion

In this paper, we proposed a combination method of feature learning and affinity propagation clustering to solve name disambiguation task without knowing the number of distinct researchers. For a given researcher name, we propose building a Paper-Paper network based on the Principle Researcher, Secondary Researcher and r-CoReasearch relation, then use Paper-Paper network to learn feature representations of papers, and finally completing name disambiguation task by using affinity propagation clustering. In addition, we use the disambiguation dataset on Arntminer for testing. The experimental results show that our proposed method can obtain better accuracy at solving name disambiguation comparing to existing methods.

Our proposed model provides a simple and effective way to solve the name disambiguation task, but the model still needs to be improved in prediction of the number of distinct researcher. Therefore, in the future work, we will use some other relevant information to improve prediction of the number of distinct researcher, e.g., paper title, keywords, etc.

Acknowledgments. This work was supported in part by National Natural Science Foundation of China under grant 61572226, and Jilin Province Key Scientific and Technological Research and Development project under grants 20180201067GX and 20180201044GX.

References

1. Hasan, M.A., Baichuan, Z.: Name disambiguation in anonymized graphs using network embedding. In: Proceedings of the 2017 ACM on Conference on Information and Knowledge Management, pp. 1239–1248. ACM, Singapore (2017)
2. Shadbolt, N.R., Mcrae-Spencer, D.M.: Also by the same author: Aktiveauthor, a citation graph approach to name disambiguation. In: ACM/IEEE Joint Conference on Digital Libraries, pp. 53–54. ACM (2006)
3. Grover, A., Leskovec, J.: node2vec: Scalable feature learning for networks. In: Proceedings of the 22nd ACM SIGKDD International Conference on Knowledge Discovery and DataMining, pp. 855–864. ACM (2016)
4. Wang, X., Tang, J., Cheng, H., Yu, P.S.: ADANA: active name disambiguation. In: 11th IEEE International Conference on Data Mining, pp. 794–803. IEEE, Vancouver (2011)
5. Tran, H.N., Huynh, T., Do, T.: Author name disambiguation by using deep neural network. In: Nguyen, N.T., Attachoo, B., Trawiński, B., Somboonviwat, K. (eds.) ACIIDS 2014. LNCS (LNAI), vol. 8397, pp. 123–132. Springer, Cham (2014). https://doi.org/10.1007/978-3-319-05476-6_13
6. Hermansson, L., Kerola, T., Johansson, F.: Entity disambiguation in anonymized graphs using graph kernels. In: 22nd ACM International Conference on Information and Knowledge Management, pp. 1037–1046. ACM, San Francisco (2013)
7. Zhu, J., Wu, X., Lin, X.: A novel multiple layers name disambiguation framework for digital libraries using dynamic clustering. Scientometrics **114**(3), 781–794 (2018)
8. Zhang, B., Choudhury, S., Hasan, M.A., Ning, X., Agarwal, K., Purohit, S.: Trust from the past: Bayesian personalized ranking based link prediction in knowledge graphs. CoRR (2016)
9. Zhang, B., Saha, T.K., Hasan, M.A.: Name disambiguation from link data in a collaboration graph. Soc. Netw. Anal. Min. **5**, 11 (2014)
10. On, B., Lee, I., Lee, D.: Scalable clustering methods for the name disambiguation problem. Knowl. Inf. Syst. **31**(1), 129–151 (2012)
11. Mann, G.S., Yarowsky, D.: Unsupervised personal name disambiguation. In: Proceedings of the Seventh Conference on Natural Language Learning, pp. 33–40. ACL, Edmonton (2003)
12. Perozzi, B., Al-Rfou, R., Skiena, S.: Deepwalk: online learning of social representations. In: The 20th ACM SIGKDD International Conference on Knowledge Discovery and Data Mining, pp. 701–710. ACM, New York (2014)
13. Chen, C., Hu, J., Wang, H.: Clustering technique in multi-document personal name disambiguation. In: Proceedings of the 47th Annual Meeting of the Association for Computational Linguistics and the 4th International Joint Conference on Natural Language Processing of the AFNLP, pp. 88–95, Singapore (2009)

14. Saha, T.K., Zhang, B., Hasan, M.A.: Name disambiguation from link data in a collaboration graph using temporal and topological features. Soc. Netw. Anal. Min. **5**(1), 1–14 (2015)
15. Han, H., Giles, L., Zha, H., Li, C., Tsioutsiouliklis, K.: Two supervised learning approaches for name disambiguation in author citations. In: ACM/IEEE Joint Conference on Digital Libraries, pp. 296–305. ACM, Tucson (2004)
16. Cen, L., Dragut, E.C., Si, L., Ouzzani, M.: Author disambiguation by hierarchical agglomerative clustering with adaptive stopping criterion. In: The 36th International ACM SIGIR Conference on Research and Development in Information Retrieval, pp. 741–744. ACM, Dublin (2013)
17. Tang, J., Fong, A.C.M., Wang, B., Zhang, J.: A unified probabilistic framework for name disambiguation in digital library. IEEE Trans. Knowl. Data Eng. **24**(6), 975–987 (2012)
18. Zhang, D., Tang, J., Li, J., Wang, K.: A constraint based probabilistic framework for name disambiguation. In: Proceedings of the Sixteenth ACM Conference on Information and Knowledge Management, pp. 1019–1022. ACM, Lisbon (2007)

Hierarchical Clustering of High-Dimensional Data Without Global Dimensionality Reduction

Ilari Kampman[1] and Tapio Elomaa[2(✉)]

[1] Wapice Ltd., Tampere, Finland
ilari.kampman@gmail.com
[2] Tampere University of Technology, Tampere, Finland
tapio.elomaa@tut.fi

Abstract. Very few clustering algorithms can cope with a high number of dimensions. Many problems arise when dimensions grow to the order of hundreds. Dimensionality reduction and feature selection are simple remedies to these problems. In addition to being somewhat intellectually disappointing approaches, both also lead to loss of information.

Furthermore, many elaborated clustering algorithms become unintuitive for the user because she is required to set the values of several hyperparameters without clear understanding of their meaning and effect.

We develop PCA-based hierarchical clustering algorithms that are particularly geared for high-dimensional data. Technically the novelty is to describe data vectors iteratively by the their angles with eigenvectors in the orthogonal basis of the input space. The new algorithms avoid the major curse of dimensionality problems that affect cluster analysis.

We aim at expressive algorithms that are easily applicable. This entails that the user only needs to set few intuitive hyperparameters. Moreover, exploring the effects of tuning parameters is simple since they are directly (or inversely) proportional to the clustering resolution. Also, the clustering hierarchy has a comprehensible interpretation and, therefore, moving between nodes in the hierarchy tree has an intuitive meaning.

1 Introduction

Clustering is the predominant unsupervised learning technique in which the data is grouped into subsets with similar members. Traditional clustering algorithms, however, cannot cope with data that has a very high number of dimensions (in hundreds or thousands) that may have clusters in arbitrarily oriented subspaces. This is common situation in document clustering [1] and in cluster analysis of gene expressions [2]. The problems arise mainly because the distance measures cannot handle the sparse instance space unavoidably induced by the high-dimensional data. Therefore, *feature selection* or *dimensionality reduction* becomes an almost necessary (preprocessing) task in clustering.

Hence, there is a demand for new algorithms specifically designed to handle high-dimensional data. The most promising algorithms are based on recognizing

© Springer Nature Switzerland AG 2018
M. Ceci et al. (Eds.): ISMIS 2018, LNAI 11177, pp. 236–246, 2018.
https://doi.org/10.1007/978-3-030-01851-1_23

the arbitrarily oriented, most significant components in the input space to characterize the data. *Principal Component Analysis* (PCA) [3] provides the mature formal background of many of these algorithms.

The downside of such algorithms is often pretty cumbersome implementation requiring the user to set quite a few *hyperparameter* values. Being able to choose suitable values for the parameters would actually necessitate the user to have a good understanding of the internal structure of the data rather than explore its hidden structure. Also details of algorithm implementation or background theory often need to be understood. These requirements reduce the applicability as well as comprehensibility of the algorithm and its results.

An example of a PCA-based clustering algorithm producing excellent results, but having an excessive number of unintuitive hyperparameters, is ERiC [4]: The user must choose the number of nearest neighbors k and minimum cluster size minPts. Furthermore, α is the share of variance explained by the "strong" eigenvectors and δ specifies the threshold for approximate linear dependency of the strong eigenvectors. Finally, τ specifies the threshold for the maximum distance between the approximately linear dependent subspaces of two objects.

A large number of parameters allows the user to fine tune the clustering results to suit the situation at hand. However, the parameters of ERiC cannot be linearly adjusted to improve the accuracy. Our new algorithms have parameters that are directly or inversely proportional to the clustering resolution allowing easy exploration. To "clean" the clusters ERiC discards some of the data points. This exposes it to falsely rejecting instances (see *Problem 4* in Sect. 3). Anyhow, it is hard to determine on what basis does one set the values of α, δ, and τ correctly. In our algorithms an exceptional data point is identified through variations of eigenvectors indicating how significant the deviation is.

The levels of the clustering hierarchy produced by ERiC are ordered according to dimensionality: the root level stands for the whole data set with full dimensionality and it reduces when one proceeds to the children. The node labels do not have any connection with the hierarchy and, hence, do not bear any intuitive significance. In the algorithms that we present, the labels have a meaningful interpretation and each data point can in principle be reconstructed with some accuracy if the required information (factor loads) is retained.

We develop two new hierarchical correlation clustering algorithms for high-dimensional data, CHUNX and CRUSHES, both of which are firmly based on the background of PCA. We aim at ready-to-use clustering algorithms that do not require the user to provide her guesses on unintuitive hyperparameter values. We describe each data point individually using the principal components that best represent it. Hence, the cluster hierarchy retains individual special characteristics of data points. We particularly have sequential time series data in mind.

Kriegel et al. [5] conducted a systematic study of clustering algorithms for high-dimensional data and identified four major *curse of dimensionality* problems affecting the setting. We will take all of these problems into account in the algorithm design and devise means to circumnavigate the potential problems.

We deploy *cosine distance* to measure the (dis)similarity of data points. It quantifies the angle between two vectors in radians. We examine the correlation between each feature vector and (positive and negative) eigenvectors. Data points that share the same most significant component are clustered together.

2   Measuring Dissimilarity

Throughout this paper we denote by $n \in \mathbb{N}$ the number of data points to be clustered. In other words, the data set that we receive is $S = \{x_1, x_2, \ldots, x_n\}$. Each data point is a vector of $d \in \mathbb{N}$ components; $x_i = \begin{bmatrix} x_{i,1} & x_{i,2} & \cdots & x_{i,d} \end{bmatrix}$. We assume that the data is real-valued; i.e., $x_{i,j} \in \mathbb{R}$ and $x_i \in \mathbb{R}^d$ for all i and j.

In order to partition a set of data S a clustering algorithm must infer dissimilarity information about S from the *data matrix*

$$X = \begin{bmatrix} x_1 & \cdots & x_n \end{bmatrix}^\top \in \mathbb{R}^{n \times d}. \tag{1}$$

Each row in X represents one data point. A dissimilarity matrix can be computed from a data matrix using different measures of dissimilarity.

The most common measures of distance used in clustering are Euclidean distance and Manhattan distance [6]. Euclidean distance refers to the shortest distance between two points in the d-dimensional space. This is also known as the L_2-norm of vector $x_i - x_j$. Manhattan distance, on the other hand, is based on taking the L_1 norm of the difference of x_i and x_j. A further commonly used measure, *Chebyshev distance*, takes the L_∞-norm of vector $x_i - x_j$ [6].

Rather than measuring distance of vectors by the difference of a chosen norm, *cosine distance* or *correlation* measures the angle between vectors,

$$\delta_{\cos}(x_i, x_j) = \mathrm{corr}(x_i, x_j) = \frac{\langle x_i, x_j \rangle}{\|x_i\|_2 \|x_j\|_2} = \frac{\sum_{k=1}^{d} x_{i,k} x_{j,k}}{\|x_i\|_2 \|x_j\|_2}, \tag{2}$$

where $\mathrm{corr}(\cdot, \cdot)$ stands for the correlation of vectors [6]. Correlation between vectors x_i and x_j is also denoted by $\rho_{i,j}$. Observe that correlation of 1 represents the angle of 0 radians and correlation of -1 the angle of π radians. Let $\mathrm{corr}(X)$ be the correlation matrix of X with diagonal elements 1 and other elements $\rho_{i,j} \in [-1, 1]$ represent the correlations between row vectors of X. In order to use correlation distance data, the distances of vectors must be represented as the angles between vectors. This can be done by taking the arcus cosine of the correlation distance data,

$$D_{\mathrm{angles}} = \arccos\left(\mathrm{corr}(X)\right) = \arccos \left(\begin{bmatrix} 1 & \rho_{1,2} & \cdots & \rho_{1,n} \\ \rho_{2,1} & 1 & \cdots & \vdots \\ \vdots & \vdots & \ddots & \rho_{n-1,n} \\ \rho_{n,1} & \rho_{n,2} & \cdots & 1 \end{bmatrix} \right), \tag{3}$$

3   Problems of Clustering in High Dimensions

Even though there is a broad spectrum of clustering methods from different types of approaches, a significant portion of them (have been designed to) function only when the number of dimensions is low. When the number of dimensions in a clustering task is in hundreds or thousands, unintuitive problems start to arise and clustering often becomes cumbersome [5]. It is usual that only a small number of dimensions is relevant in terms of clustering—the remaining dimensions tend to produce a lot of noise and mask the relevant features.

High-dimensional data is often sparsely distributed and most of the data vectors tend to be mutually orthogonal. A well-known set of problems related to high-dimensional data in general is the *curse of dimensionality*. In clustering, it can be highlighted with four basic problems [5, pp. 43–46] described below.

- *Problem 1.* The more there are dimensions, the more there are possible value combinations. Clustering in general assumes that a data set can be generated by a finite number of functions, which become more complex as the number of dimensions, i.e., function parameters, increases. Visualization of high-dimensional data is another topic related to this particular problem.
- *Problem 2.* As the number of dimensions increases, the concepts of distance and proximity of two data points become almost useless. It is known that

$$\lim_{d \to \infty} (\delta_{\max} - \delta_{\min}) / \delta_{\min} = 0,$$

  where $\delta_{\max}$ and $\delta_{\min}$ are the maximum and minimum distance, respectively, between two data points in the d-dimensional data set [7]. Therefore, using distance as a measure of similarity in high-dimensional spaces, it can be hard to say whether two data points are similar to each other or not.
- *Problem 3.* The applications of clustering low-dimensional data aim at grouping data points whose attributes are most similar to each other. As the number of dimensions increases, the data points can be related to each other by only a number of relevant attributes. The other remaining irrelevant attributes are considered as noise which often masks the relevant features.
- *Problem 4.* Many high-dimensional clustering algorithms aim to extract the most significant attributes for clustering. It is possible that some attributes that are considered as noise are actual characteristic features of data. Therefore, care must be taken when reducing noise in high-dimensional data.

3.1   Related Work

Many high-dimensional algorithms cluster in axis-parallel subspaces [5]. Thus the infinite search space is limited to subspaces which are parallel to the dimensional axes. Even though the number of possible subspaces in this approach is finite, the number is still quite large in higher dimensions: For d dimensions the number of all k-dimensional subsets is $\sum_{k=1}^{d} \binom{d}{k} = 2^d - 1$. On the other hand, the finite search space can be very useful in applications where *Problem 3* is on display.

The algorithms can be categorized based on the assumptions they make about the underlying problem. *Projected clustering* algorithms assume that each data point can be assigned uniquely to one cluster or considered as noise [8]. These algorithms project the original data to clusters. Some assume that there is a fixed number of clusters to which the data can be optimally assigned. Such *soft projected clustering* algorithms [9], do not produce hard unique assignments, but optimized and approximate projections avoiding *Problem 4*.

Subspace clustering algorithms compute all possible subspaces where clusters can be identified [10]. *Hybrid clustering algorithms* combine methods from the other categories of axis-parallel clustering [11].

Clustering in axis-parallel subspaces sometimes collides with *Problem 2* by determining the similarity of data points using a measure of distance or density. In the next category of high-dimensional clustering—*pattern-based clustering*—the algorithms look for characteristic behavior of data points concerning certain attributes of data [12]. They also operate in axis-parallel subspaces.

Algorithms which cluster data based on patterns in the data matrix are often referred to as *biclustering* algorithms. They operate on both the set of row vectors X and the set of column vectors Y of the data matrix $\boldsymbol{X} \in \mathbb{R}^{n \times d}$. Since $x_{p,q} \in \boldsymbol{X}$ is an element on a row vector $\boldsymbol{x}_p \in X$ and a column vector $\boldsymbol{y}_q \in Y$, $\boldsymbol{X}$ can be denoted by a tuple (X, Y). In general, biclustering algorithms try to find a set of submatrices $\{(I_1, J_1), \ldots, (I_k, J_k)\}$ of $\boldsymbol{X}$, $k \in \mathbb{N}$. Each of the k submatrices, also known as biclusters, must satisfy a given homogeneity criterion [5].

Biclustering algorithms tend to find patterns which are quite special or even artificial. The main inconvenience is caused by *Problem 1*: The patterns produced often rely on simple mathematical models and cannot represent complex relationships between data points.

In *correlation clustering* [4,5,13,14] the clusters may appear in arbitrarily oriented subspaces of $\mathbb{R}^d$. These algorithms are able to find clusters in which data points have strong linear relationships: The data points can be located along a line or a hyperplane which are not axis-parallel. The algorithms differ from each other by the methods they use to determine the orientation of subspaces.

3.2   Principal Component Analysis

A large number of correlation clustering algorithms use PCA for subspace projection [5]. PCA [3] is a well-known statistical multivariate method for extracting important features from data by simplifying data representation and describing the structure of multivariate data. Let $\boldsymbol{X} \in \mathbb{R}^{n \times d}$ be an arbitrary data matrix (see Eq. 1). Furthermore, let $\mathrm{col}(\cdot)$ denote the set of column vectors of a matrix. The *sample covariance* of vectors $\boldsymbol{y}_i, \boldsymbol{y}_j \in \mathrm{col}(\boldsymbol{X})$ is defined as

$$\sigma_{i,j} = \mathrm{cov}(\boldsymbol{y}_i, \boldsymbol{y}_j) = \langle (\boldsymbol{y}_i - \mu_i \boldsymbol{1}_n), (\boldsymbol{y}_j - \mu_j \boldsymbol{1}_n) \rangle / (n-1), \tag{4}$$

where μ_i and μ_j are the mean values of ith and jth vectors of $\mathrm{col}(X)$, $i, j \in \{1, \ldots, n\}$, and $\mathbf{1}_n \in \mathbb{R}^n$ is a vector of ones [15]. The sample covariance of vector y_i with itself is called *sample variance*. It can be represented as [15]:

$$\sigma_i^2 = \mathrm{cov}(y_i, y_i) = \langle (y_i - \mu_i \mathbf{1}_n), (y_i - \mu_i \mathbf{1}_n) \rangle / (n-1) = \|y_i - \mu_i \mathbf{1}_n\|_2^2 / (n-1).$$

The *centered* data matrix X_C is defined as $X_C = X - \mathbf{1}_n \mu^\top$, where $\mu = \begin{bmatrix} \mu_1 & \mu_2 & \cdots & \mu_d \end{bmatrix}^\top$ is a vector containing mean values of each random variable. PCA starts by computing the sample covariance of matrix X whose columns are considered as random variables,

$$\mathrm{cov}(X) = \frac{1}{n-1} X_C^\top X_C = \begin{bmatrix} \sigma_1^2 & \sigma_{1,2} & \cdots & \sigma_{1,d} \\ \vdots & \vdots & \ddots & \vdots \\ \sigma_{d,1} & \sigma_{d,2} & \cdots & \sigma_d^2 \end{bmatrix} = \Sigma_X. \tag{5}$$

Covariance matrix Σ_X of Eq. (5) has the variances of each vector of $\mathrm{col}(X)$ as its diagonal elements and the covariances of all vectors in its other elements.

PCA proceeds by solving the *eigenvalues* $\lambda \in \mathbb{R}_+$, which are the roots of the *characteristic equation* [16]: $\det(\Sigma_X - \lambda I) = 0$, where I is the identity matrix. Thus, for each distinct eigenvalue λ there exists an *eigenvector* $v \in \mathbb{R}^n$ such that $\Sigma_X v = \lambda v$. If the roots $\lambda \in \mathbb{R}_+$ in the characteristic equation are distinct, the eigenvectors of matrix Σ_X diagonalize it:

$$\Sigma_X = V \Lambda V^{-1},$$

where V is the *eigenvector matrix*, V^{-1} its inverse, and Λ is the *eigenvalue matrix*. It has the eigenvalues as its diagonal elements in a decreasing order [16]. Vectors of V are mutually orthogonal [16]. Matrix Λ can be tought of as the presentation of Σ_X in a different coordinate system where the first eigenvector $v_1 \in \mathrm{col}(V)$ points to the direction of highest variance in X [3]. PCA-based correlation clustering algorithms differ in the way they determine the most interesting subspaces generated and in the similarity measures they use.

PCA-based correlation clustering algorithms often collide with *Problem 2*. Diagonalization is a powerful technique for revealing information about the internal structures of a matrix. The full potential of PCA-based correlation algorithms has not been reached yet. Next, we propose two new algorithms which hierarchically cluster the high-dimensional data based on the variations of the most significant eigenvectors of each data point.

4   Algorithms CHUNX and CRUSHES

PCA is a sophisticated method for overcoming several issues related to *Problem 1* and suits as the basis of CHUNX and CRUSHES. The new algorithms provide easy-to-use and mathematically expressive clustering algorithms for applications in high-dimensional data. Furthermore, they provide generic tools for exploring the internal structures of data exposed by matrix diagonalization.

First we compute the orthogonal basis V for the data matrix X and the angle between each data point and each eigenvector in V. The angles are labeled as $\pm\theta_i$, where $i = 1, \ldots, d$ and $0 \leq \theta_i \leq \pi/2$. Since the vectors in V are mutually orthogonal, it is essential to take both the positive and negative vector directions into account. The vector in $v \in \mathrm{col}(V)$ with the smallest angle between x is the *most significant component* of x. Initially, the data points are divided into tentative clusters $\{C_1, \ldots, C_k\}$ according to their most significant components. If a stopping criterion is not met for cluster C_i, the process continues by dividing its members according to their second most significant component, and so forth.

CHUNX and CRUSHES have different stopping criteria. Both construct a tree whose root corresponds to the whole data set X. Its children are those components that are the most significant ones for at least one data point. Since both positive and negative angles are taken into account, there are at most $2d$ candidates as children. Because the division can continue using the next most significant components, the tree defines a hierarchical division of the data points.

The stopping criterion in CHUNX is a hyperparameter *max cluster size* $n_{\max}$. A cluster is refined as long as its element count is above $n_{\max}$. Because an angle can appear only with one sign on a path from the root to a leaf, in the worst case d components are needed and there is still no guarantee of fulfilling the desired limit. Typically, after examining some $k < d$ most significant components, the recursion yields a set of leaf nodes that contain at most $n_{\max}$ elements.

In CRUSHES stopping is determined by the correlation of a data point and its representation by the most significant components. The level of required correlation is provided as hyperparameter ρ. Part of the data points that fall to a node of the tree may already have correlation greater than ρ w.r.t. their representation, while others do not yet fulfill the requirement. Those $x \in X$ that satisfy the criterion stay in the node, while those that still need to gather correlation are passed down to nodes corresponding to their next most significant components. Hence, the final clusters are in the leaves of tree and in some internal nodes—those in which the correlation limit ρ holds for part of the data points.

Using the component labeling information, both algorithms construct a tree in which the root node represents the entire data set X. Each child node of the root represent the set of sample vectors which have the smallest angle with the same vector in $\pm V$. Each cluster label in CHUNX and CRUSHES is a path—a sequence of node labels—between the root and some descendant node.

In a CHUNX tree the clusters appear only in its leaf nodes. An imaginary example is illustrated on left in Fig. 1 in which a data matrix $X \in \mathbb{R}^{500 \times d}$, where $d > 5$, is clustered into 8 clusters with $n_{\max} = 100$. The leaf nodes with grey fill and element counts represent the constructed clusters. The labels of the 8 clusters are: $[-\theta_1, -\theta_3]$, $[-\theta_1, +\theta_4, +\theta_2]$, $\ldots$, $[+\theta_3]$.

Figure 1 also illustrates an example of a CRUSHES tree (right), where the same data matrix X is clustered into 8 clusters with some value of ρ. The grey nodes and element counts represent the formed clusters with labels $[-\theta_1]$, $[-\theta_1, -\theta_3]$, $\ldots$, $[+\theta_3, +\theta_4, -\theta_5]$. The CRUSHES clusters are not necessarily leaf nodes.

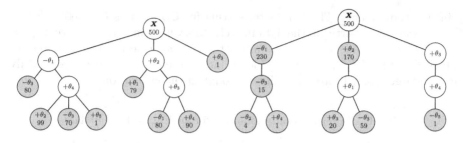

Fig. 1. An imaginary example of CHUNX (left) and CRUSHES (right) clustering hierarchy with $n = 500$ and $n_{max} = 100$. The number in each node represents the data point count. The nodes marked with grey fill are the formed clusters.

The clustering hierarchy trees of CHUNX and CRUSHES provide a mathematically expressive illustration of the complex relationships within the data matrix and they provide a data structure for noise reduction: As a tree is formed, it can be filtered with queries related to certain component variations and cluster sizes. Since both algorithms build their tree hierarchies based on the most significant components of data, they are not sensitive to *Problem 3*. Contrary to many PCA-based algorithms, CHUNX and CRUSHES also avoid *Problem 4* by not dismissing any component of the data sets orthogonal basis V.

Both algorithms operate on centered and normalized data and, hence, they dismiss clustering by data scale. Since the algorithms focus on approximate data vector directions by grouping data points by the angles between them and their most significant components, they operate very loosely on the Euclidean distances between data points. In other words, CHUNX and CRUSHES cluster shapes of high-dimensional data. Therefore, they practically avoid *Problem 2*.

5   Empirical Evaluation

Let us test CHUNX and CRUSHES using a real-life high-dimensional data set: *analcatdata authorship*[1] [17]. It contains word count distributions from different text sources labeled with their author's name. The clustering results are compared with a PCA-based correlation clustering algorithm COPAC [14], k-MEANS [18] clustering, and EM [19] clustering. Both k-MEANS and EM were tested with the original data and a version whose dimensionality was reduced using PCA.

The quality of each clustering is evaluated using common metrics for clustering performance: *Homogeneity* [20], *Completeness* [20], *V-measure* [20], *Adjusted Rand index* (ARI) [21], *Adjusted Mutual Information* (AMI) [22], *Normalized Mutual Information* (NMI) [23], and *Fowlkes-Mallows score* (FMI) [20].

The evaluations of clustering result are given in Table 1. Since the test data contained observations from four authors, k-MEANS and EM were set to find 4 clusters. The reduced data has 8 dimensions according to the most significant

---

[1] https://www.openml.org/d/458.

principal components. The parameter setting for COPAC was $k = 500$ in k-NN. CHUNX was set to partition the data into clusters with a max cluster size $n_{max} = 84$ and CRUSHES had correlation value of $\rho = 0.95$. Due to their hierarchical nature CHUNX and CRUSHES produce a large number of small clusters. Only clusters larger than 10 are taken into account in the evaluation.

Table 1. Clustering result evaluation for analcat_authorship data.

Algorithm	Homog.	Compl.	V-score	ARI	AMI	NMI	FMI
k-MEANS	0.770	0.719	0.744	0.728	0.718	0.744	0.809
EM	0.786	0.727	0.756	0.728	0.726	0.756	0.809
$k = 4$ Reduced k-MEANS	0.753	0.700	0.725	0.709	0.699	0.726	0.796
Reduced EM	0.825	0.774	0.799	0.766	0.773	0.799	0.836
COPAC	0.371	0.260	0.305	0.239	0.255	0.310	0.433
CHUNX	0.897	0.419	0.571	0.314	0.407	0.613	0.488
CRUSHES	1.000	0.677	0.807	0.618	0.666	0.823	0.732

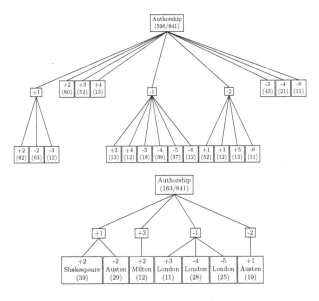

Fig. 2. Tree hierarchy of authorship data by CHUNX (top) and CRUSHES (bottom).

The cluster hierarchy is presented for CHUNX (top) and CRUSHES (bottom) in Fig. 2. The number in the root is the total number of data points used in the evaluation of the results. Similarly, in other nodes the number indicates the size of the cluster. Since CRUSHES produces a fully homogeneous clustering according to the labels in data, each cluster in Fig. 2 is labeled with the author name.

CHUNX and CRUSHES do not outperform the other algorithms on all areas. Nevertheless, they are able to produce informative results about relationships in high-dimensional test data. The other algorithms require prior knowledge about the structure of data and reduction of global dimensionality—CRUSHES is able to find quality clusters hierarchically with a simple hyperparameter interface.

6 Conclusion

High-dimensional data has become common in many application areas through the modern capabilities of automatic data acquisition and generation. Also clustering algorithms need to adjust to such a setting. We introduced two mathematically well-founded, though simple, correlation clustering algorithms for high-dimensional data. Both have only one easy-to-control parameter. Hence, it is easy to explore how changes of its value affect the clustering result.

References

1. Dhillon, I.S.: Co-clustering documents and words using bipartite spectral graph partitioning. In: Proceedings of the 7th ACM SIGKDD, pp. 269–274 (2001)
2. Bhattacharya, A., De, R.K.: Average correlation clustering algorithm (ACCA) for grouping of co-regulated genes with similar pattern of variation in their expression values. J. Biomed. Inform. **43**(4), 560–568 (2010)
3. Abdi, H., Williams, L.J.: Principal component analysis. Wiley Interdiscip. Rev. Comput. Stat. **2**(4), 433–459 (2010)
4. Achtert, E., Böhm, C., Kriegel, H.P., Kröger, P., Zimek, A.: On exploring complex relationships of correlation clusters. In: Proceedings of the 19th SSBDM, pp. 52–61. IEEE (2007)
5. Kriegel, H.P., Kröger, P., Zimek, A.: Clustering high-dimensional data: a survey on subspace clustering, pattern-based clustering, and correlation clustering. ACM TKDD **3**(1), 1–58 (2009)
6. Choi, S.S., Cha, S.H., Tappert, C.C.: A survey of binary similarity and distance measures. J. Syst. Cybern. Inform. **8**(1), 43–48 (2010)
7. Hinneburg, A., Aggarwal, C.C., Keim, D.A.: What is the nearest neighbor in high dimensional spaces? In: Proceedings of the 26th International Conference on VLDB, pp. 506–515 (2000)
8. Aggarwal, C.C., Procopiuc, C.M., Wolf, J.L., Yu, P.S., Park, J.S.: Fast algorithms for projected clustering. In: Proceedings of the ACM SIGMOD, pp. 61–72. ACM (1999)
9. Domeniconi, C., Papadopoulos, D., Gunopulos, D., Ma, S.: Subspace clustering of high dimensional data. In: Proceedings of the 4th SIAM SDM, pp. 517–521 (2004)
10. Agrawal, R., et al.: Automatic subspace clustering of high dimensional data for data mining applications. In: Proceedings of the ACM SIGMOD, pp. 94–105 (1998)
11. Procopiuc, C.M., Jones, M., Agarwal, P.K., Murali, T.M.: A Monte Carlo algorithm for fast projective clustering. In: Proceedings of the 2002 ACM SIGMOD, pp. 418–427 (2002)
12. Pei, J., Zhang, X., Cho, M., Wang, H., Yu, P.S.: MaPle: a fast algorithm for maximal pattern-based clustering. In: Proceedings of the 3rd IEEE ICDM, pp. 259–266 (2003)

13. Aggarwal, C.C., Yu, P.S.: Finding generalized projected clusters in high dimensional spaces. In: Proceedings of the 2000 ACM SIGMOD, pp. 70–81 (2000)

14. Achtert, E., Böhm, C., Kriegel, H.P., Kröger, P., Zimek, A.: Robust, complete, and efficient correlation clustering. In: Proceedings of the 2007 SIAM SDM, pp. 413–418 (2007)

15. Zhang, Y., Wu, H., Cheng, L.: Some new deformation formulas about variance and covariance. In: Proceedings of the 4th ICMIC, pp. 987–992. IEEE (2012)

16. Strang, G.: Linear Algebra and Its Applications. Thomson, Brooks/Cole, Pacific Grove (2006)

17. Simonoff, J.S.: Analyzing Categorical Data. Springer, New York (2003). https://doi.org/10.1007/978-0-387-21727-7

18. MacQueen, J.B.: On convergence of k-means and partitions with minimum average variance (abstract). Ann. Math. Stat. **36**, 1084 (1965)

19. Dempster, A.P., Laird, N.M., Rubin, D.B.: Maximum likelihood from incomplete data via the EM algorithm. J. R. Stat. Soc. Ser. B **39**, 1–38 (1977)

20. Rosenberg, A., Hirschberg, J.: V-Measure: a conditional entropy-based external cluster evaluation measure. In: Proceedings of the 2007 EMNLP-CoNLL (2007)

21. Hubert, L., Arabie, P.: Comparing partitions. J. Classification **2**(1), 193–218 (1985)

22. Vinh, N.X., Epps, J., Bailey, J.: Information theoretic measures for clusterings comparison. In: Proceedings of the 26th ICML, pp. 1073–1080. ACM (2009)

23. Strehl, A., Ghosh, J.: Cluster ensembles – a knowledge reuse framework for combining multiple partitions. JMLR **3**, 583–617 (2002)

Exploiting Order Information Embedded in Ordered Categories for Ordinal Data Clustering

Yiqun Zhang[1] and Yiu-ming Cheung[1,2(✉)]

[1] Department of Computer Science, Hong Kong Baptist University,
Kowloon Tong, Hong Kong SAR, China
{yqzhang,ymc}@comp.hkbu.edu.hk
[2] HKBU Institute of Research and Continuing Education, Shenzhen, China

Abstract. As a major type of categorical data, ordinal data are those with the attributes whose possible values (also called categories interchangeably) are naturally ordered. As far as we know, all the existing distance metrics proposed for categorical data do not take the underlying order information into account during the distance measurement. This will make the produced distance incorrect and will further influence the results of ordinal data clustering. We therefore propose a specially designed distance metric, which can exploit the order information embedded in the ordered categories for distance measurement. It quantifies the distance between two ordinal categories by accumulating the sub-entropies of all the categories ordered between them. Since the proposed distance metric takes the order information into account, distance produced by it will be more reasonable than the other metrics proposed for categorical data. Moreover, it is parameter-free and can be easily applied to different ordinal data clustering tasks. Experimental results show the promising advantages of the proposed distance metric.

Keywords: Distance metric · Ordinal data · Order information
Entropy · Clustering analysis · Categories

1 Introduction

Clustering analysis has been widely used in a variety of scientific areas [5,6,8,13]. In this paper, we will focus on the clustering of ordinal data, which, as a major type of categorical data, is common in the field of data analysis, machine learning, pattern recognition and knowledge discovery [2]. The categories of ordinal data are naturally ordered without a well-defined distance, and the categories are stored in a nominal scale [4]. Therefore, some commonly used distance metrics for numerical data, e.g., Euclidean distance [7] and Mahalanobis distance

This work was fully supported by Faculty Research Grant of Hong Kong Baptist University under Project FRG2/17-18/082.

M. Ceci et al. (Eds.): ISMIS 2018, LNAI 11177, pp. 247–257, 2018.
https://doi.org/10.1007/978-3-030-01851-1_24

[20] are inapplicable for ordinal data clustering. On the opposite, some distance metrics for categorical data, e.g., Hamming Distance Metric (HDM) [17], Ahmad Distance Metric (ADM) [3], Association-Based Distance Metric (ABDM) [14], Context-Based Distance Metric (CBDM) [9] and Frequency-and-Relationship-Based Distance Metric (FRBDM) [11], can be directly applied to measure the distance for ordinal datasets. Among the above mentioned five categorical data distance metrics, HDM is the most conventional and popular one. Although HDM is easy to use, it assigns the same distance to all the pairs of different categories, which is somewhat unreasonable. Later, ADM and ABDM adopt similar ideas that, the corresponding values from the other attributes of two similar categories will have similar conditional probability distributions, have been proposed to achieve better clustering performance. CBDM takes use of a more reasonable idea than ADM and ABDM that, not all the attributes offer valuable information, has also been proposed, which selects important attributes for distance measurement. All the ADM, ABDM and CBDM metrics have not considered the case that some attributes are independent of each other [18]. Therefore, FRBDM is proposed, which considers the independent case. However, all the existing distance metrics proposed for categorical data have not considered the case that the categories are naturally ordered. Therefore, a distance metric, which can reasonably exploit the underlying order information of ordinal data for clustering analysis is in urgent need.

In this paper, we therefore propose a distance metric for ordinal data clustering. From the perspective of information theory [10], each category will have its information amount, which can be viewed as its category width towards the direction of the order of its attribute. Based on this, distance between categories can be viewed as the information cost for moving from one category to another. Therefore, the distance between two categories can be quantified by adding the information amount of them and all the categories ordered between them up. We take use of entropy [15] to quantify the information amount for the distance measurement and the different distance scales of attributes caused by different number of categories are also unified in the proposed metric. To prove the effectiveness of the proposed distance metric, and to study the relationship between the exploiting degree of order information and the clustering performance, three experiments are conducted on six real ordinal datasets in this paper. The main contributions of this paper are summarized into three points: (1) A parameter-free distance metric for ordinal data clustering is proposed, which is the first attempt to consider the underlying order information. (2) The proposed metric appropriately exploits the order information embedded in the ordered categories to measure the distance. Further, we design a variant of entropy to indicate the distance between ordinal categories with considering their order. (3) Relationship between the exploiting degree of order information and clustering performance is studied to show the effectiveness of the proposed distance metric.

2   Related Work

HDM [17] is straight-forward and easy to use. It uniformly defines the distance
between different categories by "1", while the distances between identical categories are set at "0". However, it does not consider the statistical information
of categories. For this reason, ABDM [14] has been presented to measure the
distance between two categories of an attribute by considering the conditional
probability distributions of the categories from the other attributes given the
two target categories. Later, ADM proposed in [3] has adopted similar idea as
ABDM. The difference is that, ABDM considers the co-occurrence of the categories from the other attributes given two target categories, while ADM selects
one representative co-occurred category from each of the other attributes and
utilizes their corresponding conditional probabilities for distance measurement.
However, ABDM and ADM are unable to distinguish the contributions of different attributes. Under this circumstance, CBDM has been proposed in [9], which
measures the distance between two categories from one attribute according to
the conditional probability distributions of the selected relevant attributes. Since
the inter-attribute relationships have been taken into account by the three above-
mentioned metrics [3,9,14], their performance will decline when the attributes
are independent of each other. To avoid this, FRBDM [11] has been proposed,
which simultaneously takes the occurrence frequency of categories in their own
attributes and the relationship [18] between different attributes into account for
distance measurement.

3   The Proposed Metric

Since ordinal categories are naturally ordered, it is more reasonable to measure
the distance between two of them according to the cost for moving from one
ordinal category to another towards their order. For example, the cost can be
understand as the thinking cost for a paper reviewer to change his/her decision from "neutral" to "strong accept" for five naturally ordered choices, i.e.,
"strong accept","accept", "neutral","reject" and "strong reject". Because the
choice"accept" is ordered between "strong accept" and"neutral", it must be
considered by the reviewer during the process of changing mind from "neutral"
to "strong accept". Each category will have its own cost and we measure it using
entropy because entropy has been commonly used for indicating the information
amount of a category. "accept" choice with larger information amount will cost
more thinking for a reviewer to change his/her mind from "neutral" to "strong
accept". Therefore, the distance between two ordinal categories can be measured
by adding the entropy of them and all the categories ordered between them up.
Specifically, given an ordinal dataset $X = \{x_1, x_2, ..., x_n\}$ with n data objects.
A data object x_i consisted of d ordinal categories from d attributes, respectively.
The rth value of x_i is o_{r,i_r}, where the i_r stands for the order of this category

among all the categories from A_r. The distance between two different categories o_{r,i_r} and o_{r,j_r} from attribute A_r is defined as

$$D(o_{r,i_r}, o_{r,j_r}) = \sum_{s=min(i_r,j_r)}^{max(i_r,j_r)} -\frac{\sigma(A_r = o_{r,s})}{n} \log \frac{\sigma(A_r = o_{r,s})}{n}, \tag{1}$$

where the function $\sigma(A_r = o_{r,s})$ counts the number of objects in X with their rth value equals to $o_{r,s}$. Because it is impossible to change mind between two identical categories, the distance between two identical categories will always be 0.

Since different attributes will have different number of categories, the distances measured for different attributes according to Eq. (1) will have different scale. To solve this problem, the distance between two categories o_{r,i_r} and o_{r,j_r} from A_r should be divided by the maximum entropy of A_r. If A_r has v_r categories, the distance with unified scale can be written as

$$D(o_{r,i_r}, o_{r,j_r}) = \begin{cases} \frac{\sum_{s=min(i_r,j_r)}^{max(i_r,j_r)} -\frac{\sigma(A_r=o_{r,s})}{n} \log \frac{\sigma(A_r=o_{r,s})}{n}}{-log\frac{1}{v_r}} & , if\ i_r \neq j_r \\ 0 & , if\ i_r = j_r, \end{cases} \tag{2}$$

where the factor $-log\frac{1}{v_r}$ stands for the maximum entropy of A_r. Subsequently, the distance between two ordinal data objects x_i and x_j is defined as

$$D(x_i, x_j) = \sqrt{\sum_{r=1}^{d} D(o_{r,i_r}, o_{r,j_r})^2}. \tag{3}$$

The distance measurement procedures of the proposed distance metric is summarized in Algorithm 1. To make the distance measurement more efficient, we can calculate the distances between the categories of an attribute, and store the distances using a matrix. In this way, distance between two objects can be easily read off from the maintained distance matrices during clustering.

Algorithm 1. Distance Measurement Algorithm

Input: Data objects x_i and x_j
Output: $D(x_i, x_j)$
1: **for** $r = 1$ to d **do**
2:     **if** $i_r \neq j_r$ **then**
3:         calculate the distance between the rth value of x_i and x_j by $D(o_{r,i_r}, o_{r,j_r}) =$
4:         $\frac{\sum_{s=min(i_r,j_r)}^{max(i_r,j_r)} -\frac{\sigma(A_r=o_{r,s})}{n} \log \frac{\sigma(A_r=o_{r,s})}{n}}{-log\frac{1}{v_r}};$
5:     **else**
6:         $D(o_{r,i_r}, o_{r,j_r}) = 0;$
7:     **end if**
8: **end for**
9: calculate the distance between x_i and x_j by $D(x_i, x_j) = \sqrt{\sum_{r=1}^{d} D(o_{r,i_r}, o_{r,j_r})^2};$

---

4   Experiments

In this section, comparative experiments are conducted on six real ordinal datasets to prove the effectiveness of the proposed distance metric. Among the datasets, Internship Questionnaire (IQ) and Teaching Evaluation (TE) are collected from the College of International Exchange of an university. Employee Rejection/Acceptance (ER), Employee Selection (ES), Lecturer Evaluation (LE) and Social Workers (SW) are collected from the web site of Weka [1]. Statistical information of the six real world ordinal datasets are shown in Table 1. In the experiments, five counterparts, i.e., HDM, ADM, ABDM, CBDM and FRBDM, are compared with the proposed Entropy-Based Distance Metric (EBDM). Moreover, to illustrate that the performance of ordinal data clustering analysis depends on if the corresponding distance metric exploits order information during the distance measurement, we propose a new validity index to evaluate the exploiting degree of order information. Based on the new validity index, the relationship between the exploiting degree of order information and the clustering performance is also studied.

Table 1. Statistical information of the six ordinal datasets

Dataset	No. of Instances	No. of Attributes	No. of Classes
IQ	90	3	2
TE	66	5	3
ER	1,000	4	9
ES	488	4	9
LE	1,000	4	5
SW	1,000	10	4

4.1   Experiment 1: Comparative Studies

Two validity indices, i.e., accuracy and Rand Index (RI) [11,12,19], are utilized to evaluate the performance of all the compared distance metrics. Each of the metrics is embedded into the k-modes [16] clustering framework and the clustering performance is averaged on 10 runs for each of the datasets. Performance of all the six distance metrics in terms of accuracy and RI are compared in Tables 2 and 3, respectively. To evaluate the stability of each distance metric, standard deviation of their performance is also recorded in the two tables. Among the results of each dataset, the best and the second best results are highlighted by boldface and underline, respectively.

It can be observed from the results that the proposed EBDM metric outperforms the others on four of the six datasets, i.e., IQ, TE, ER and LE. It is because that EBDM exploits order information for ordinal data distance measurement, but the other five metrics do not. For the ES dataset, even EBDM's performance is the second best, the gap between it and the best one is very tiny, i.e., 0.01

Table 2. Clustering performance in terms of clustering accuracy.

Dataset	HDM	ADM	ABDM	CBDM	FRBDM	EBDM
IQ	0.620 ± 0.07	0.571 ± 0.06	0.524 ± 0.01	0.502 ± 0.00	0.573 ± 0.05	**0.734 ± 0.06**
TE	0.521 ± 0.07	0.550 ± 0.06	0.605 ± 0.07	0.624 ± 0.09	0.535 ± 0.08	**0.629 ± 0.08**
ER	0.188 ± 0.01	0.202 ± 0.01	0.203 ± 0.01	0.196 ± 0.01	0.186 ± 0.01	**0.206 ± 0.01**
ES	0.380 ± 0.04	0.396 ± 0.04	0.400 ± 0.05	**0.412 ± 0.03**	0.348 ± 0.04	0.402 ± 0.04
LE	0.330 ± 0.04	0.328 ± 0.02	0.316 ± 0.02	0.313 ± 0.02	0.320 ± 0.04	**0.367 ± 0.02**
SW	0.376 ± 0.03	**0.411 ± 0.02**	0.405 ± 0.02	0.378 ± 0.03	0.362 ± 0.03	0.396 ± 0.04

Table 3. Clustering performance in terms of RI.

Dataset	HDM	ADM	ABDM	CBDM	FRBDM	EBDM
IQ	0.620 ± 0.07	0.571 ± 0.06	0.524 ± 0.01	0.502 ± 0.00	0.573 ± 0.05	**0.734 ± 0.06**
TE	0.681 ± 0.05	0.700 ± 0.04	0.736 ± 0.05	0.750 ± 0.06	0.690 ± 0.06	**0.753 ± 0.05**
ER	0.819 ± 0.00	0.822 ± 0.00	0.823 ± 0.00	0.821 ± 0.00	0.819 ± 0.00	**0.824 ± 0.00**
ES	0.862 ± 0.01	0.866 ± 0.01	0.867 ± 0.01	**0.870 ± 0.01**	0.855 ± 0.01	0.867 ± 0.01
LE	0.732 ± 0.02	0.731 ± 0.01	0.726 ± 0.01	0.725 ± 0.01	0.728 ± 0.02	**0.747 ± 0.01**
SW	0.688 ± 0.01	**0.706 ± 0.01**	0.702 ± 0.01	0.689 ± 0.02	0.681 ± 0.02	0.698 ± 0.02

and 0.003 in terms of accuracy and RI, respectively. For the SW dataset, the gap between EBDM and the others is around 0.01, which is also small. Overall, EBDM is obvious competent in comparison with the other distance metrics. In Sects. 4.2 and 4.3, we will further analyze the reasons why EBDM cannot outperform some of the counterparts on the ES dataset and the SW dataset.

4.2   Experiment 2: Distance Matrices Demonstration

To intuitively observe the distance produced by different metrics, we demonstrate the distance matrices of all the attributes of TE dataset produced by all the compared metrics except FRBDM metric, because FRBDM cannot form distance matrix for an attribute. All the distance metrics are normalized into the interval [0, 1] and represented by grey-scale blocks as shown in Fig. 1.

In this figure, pure black colour indicates distance "0" between two categories while pure white colour indicates distance "1". Therefore, all the blocks on the diagonals from the left top corner to the right bottom corner are pure black because they indicate the distance from categories to themselves. If a distance matrix can reasonably indicate the natural distances between the categories of an ordinal attribute, the right top block should be pure white and the other blocks closer to the diagonal will be darker. This is because the distance between the two categories locate on two sides of the order should be the largest and vice versa. According to the above analysis, distance matrices produced by EBDM is completely consistent with the natural distance structure of an ordinal dataset. It can also be observed that the distance matrices produced by ADM, ABDM

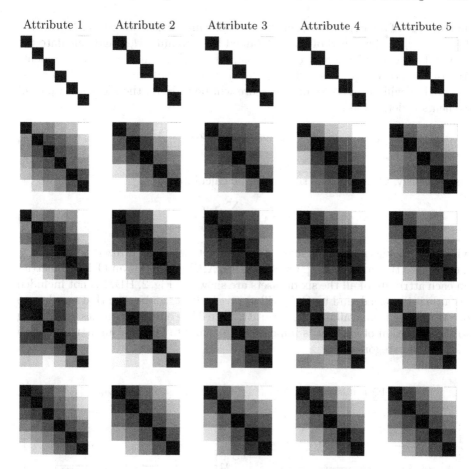

Fig. 1. Distance matrices of TE dataset. Row 1–5 shows the matrices produced by HDM, ADM, ABDM, CBDM and EBDM, respectively.

and CBDM can also indicate the natural ordinal data distance structure to a certain extent. Therefore, if we consider the matrices of these distance metrics in combination with the performance demonstrated in Tables 2 and 3, we can find that if the distance matrices produced by a distance metric are closer to the nature ordinal data distance structure, its clustering performance will be better. This observation will be further studied in Sect. 4.3.

4.3   Experiment 3: Study of the Order Information Exploiting

To study the relationship between exploiting degree of order information and clustering performance for the distance metrics, we propose a new validity index, called Ordinal Consistency Score (OCS), to quantify the consistent level between the distance matrices and the order of categories. Specifically, given a distance matrix M_{A_r} of an attribute A_r, elements of this matrix is denoted as $M_{A_r}(a,b) = Dist(o_{r,a}, o_{r,b})$. The OCS of A_r is obtained by searching M_{A_r} and calculate the accumulation for the scores of each pair of M_{A_r}'s elements. If two elements $M_{A_r}(a,b)$ and $M_{A_r}(a,c)$ satisfy $M_{A_r}(a,b) < M_{A_r}(a,c)$ with $a < b < c$ or $c < b < a$, they will get a score, or, the score will be 0. Here, the score for a pair of elements is defined as

$$S_{a,b,c} = \begin{cases} \frac{v_r - (c-b) - 1}{v_r - 2}, & if\ M_{A_r}(a,b) < M_{A_r}(a,c) \\ 0, & otherwise. \end{cases} \quad (4)$$

Subsequently, OCS of an attribute can be calculated by

$$OCS_{A_r} = \frac{\sum S_{a,b,c}}{T_{A_r}}, \quad (5)$$

where T_{A_r} is the maximum score M_{A_r} can get, which is utilized to normalize OCS_{A_r} into the interval $[0,1]$. OCS of ADM, ABDM, CBDM and EBDM metrics on each attribute of all the six datasets are shown in Fig. 2. HDM is not included because it is special, and its OCS will always be 0. It can be seen that attribute 6 of SW dataset is unavailable, because attribute 6 has only 2 categories. According to the definition of OCS, it is impossible to compute OCS for an attribute with less than 3 categories.

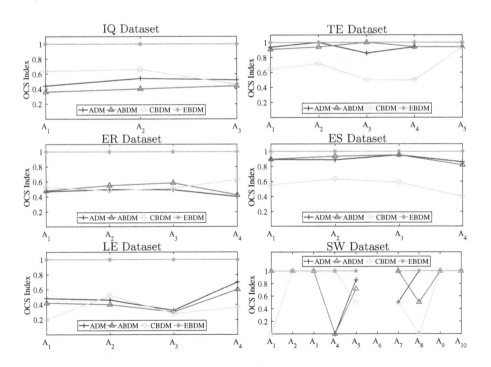

Fig. 2. OCS index on each attribute of the six datasets.

To study the relationship between order information exploiting and clustering performance, we demonstrate OCS averaged on all the attributes of each dataset, OCS-Accuracy correlation, and OCS-RI correlation in Fig. 3.

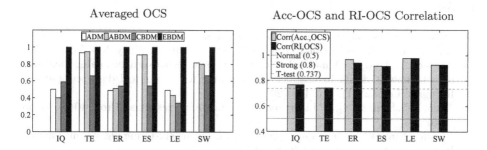

Fig. 3. Averaged OCS(left) and Acc-OCS, RI-OCS Correlations(right).

In the right part of Fig. 3, we use one-sided T-test at 90% confidence level with degree of freedom equals to 2 to prove the significance of the correlation. Moreover, two commonly used critical values of correlation, i.e. normal correlation (0.5) and strong correlation (0.8) are also shown in this figure for reference. It can also be observed that the correlations on all the six datasets pass the T-test, which means that the correlations are significant. In addition, four of them are above the strong correlation value and all of them are above the normal correlation value. To sum up, for a distance metric, its clustering performance on ordinal datasets is obvious in proportion to its exploiting level of order information. Therefore, EBDM outperforms its counterparts because it can better exploit the order information. In addition, because ADM, ABDM, CBDM and EBDM have relative high order information exploiting degree on the ES and SW datasets, their clustering performance shown in Tables 2 and 3 are close to each other. Besides, we can also find that although EBDM has obvious higher exploiting degree than the other metrics on ER dataset, its clustering performance is still close to the others. To explain this phenomenon, we carefully study the ER dataset and find that not all the attributes in ER dataset have obvious ordered categories. Categories of these attributes are viewed as ordered categories by EBDM and will influence the performance of EBDM. EBDM has high OCS on ER dataset because we force EBDM to treat each attribute of ER dataset as pure ordinal attributes.

5   Conclusion

This paper has presented an effective distance metric for ordinal data clustering, which quantifies the distance between ordinal categories with considering their order by using accumulated entropy. Furthermore, the relationship between order information exploiting and clustering performance has also been studied. Experimental results have shown that the proposed metric outperforms the existing metrics for ordinal data clustering.

References

1. https://www.cs.waikato.ac.nz/ml/weka/datasets.html
2. Agresti, A.: An Introduction to Categorical Data Analysis. Wiley, New York (1996)
3. Ahmad, A., Dey, L.: A method to compute distance between two categorical values of same attribute in unsupervised learning for categorical data set. Pattern Recognit. Lett. **28**(1), 110–118 (2007)
4. Allen, I.E., Seaman, C.A.: Likert scales and data analyses. Qual. Prog. **40**(7), 64–65 (2007)
5. Cheung, Y.m.: k*-means: a new generalized k-means clustering algorithm. Pattern Recognit. Lett. **24**(15), 2883–2893 (2003)
6. Cheung, Y.m.: Maximum weighted likelihood via rival penalized EM for density mixture clustering with automatic model selection. IEEE Trans. Knowl. Data Eng. **17**(6), 750–761 (2005)
7. Danielsson, P.E.: Euclidean distance mapping. Comput. Graph. Image Process. **14**(3), 227–248 (1980)
8. Fayyad, U.M., Piatetsky-Shapiro, G., Smyth, P., Uthurusamy, R.: Advances in Knowledge Discovery and Data Mining. MIT Press (1996)
9. Ienco, D., Pensa, R.G., Meo, R.: Context-based distance learning for categorical data clustering. In: Adams, N.M., Robardet, C., Siebes, A., Boulicaut, J.-F. (eds.) IDA 2009. LNCS, vol. 5772, pp. 83–94. Springer, Heidelberg (2009). https://doi.org/10.1007/978-3-642-03915-7_8
10. Jaynes, E.T.: Information theory and statistical mechanics. Phys. Rev. **106**(4), 620–630 (1957)
11. Jia, H., Cheung, Y.m., Liu, J.: A new distance metric for unsupervised learning of categorical data. IEEE Trans. Neural Netw. Learn. Syst. **27**(5), 1065–1079 (2016)
12. Krieger, A.M., Green, P.E.: A generalized rand-index method for consensus clustering of separate partitions of the same data base. J. Classification **16**(1), 63–89 (1999)
13. Law, L.t., Cheung, Y.m.: Color image segmentation using rival penalized controlled competitive learning. In: Proceedings of the International Conference on Joint Neural Networks, pp. 108–112 (2003)
14. Le, S.Q., Ho, T.B.: An association-based dissimilarity measure for categorical data. Pattern Recognit. Lett. **26**(16), 2549–2557 (2005)
15. Michalski, R.S., Carbonell, J.G., Mitchell, T.M.: Machine Learning: An Artificial Intelligence Approach. Springer Science & Business Media, Heidelberg (2013). https://doi.org/10.1007/978-3-662-12405-5
16. Ng, M.K., Li, M.J., Huang, J.Z., He, Z.: On the impact of dissimilarity measure in k-modes clustering algorithm. IEEE Trans. Pattern Anal. Mach. Intell. **29**(3), 503–507 (2007)
17. Norouzi, M., Fleet, D.J., Salakhutdinov, R.R.: Hamming distance metric learning. In: Proceedings of the Advances in Neural Information Processing Systems, pp. 1061–1069 (2012)
18. Peng, H., Long, F., Ding, C.: Feature selection based on mutual information criteria of max-dependency, max-relevance, and min-redundancy. IEEE Trans. Pattern Anal. Mach. Intell. **27**(8), 1226–1238 (2005)

19. Rand, W.M.: Objective criteria for the evaluation of clustering methods. J. Am. Stat. Assoc. **66**(336), 846–850 (1971)
20. Xiang, S., Nie, F., Zhang, C.: Learning a mahalanobis distance metric for data clustering and classification. Pattern Recognit. **41**(12), 3600–3612 (2008)

User-Emotion Detection Through Sentence-Based Classification Using Deep Learning: A Case-Study with Microblogs in Albanian

Marjana Prifti Skenduli[1]([✉]), Marenglen Biba[1], Corrado Loglisci[2],
Michelangelo Ceci[2], and Donato Malerba[2]

[1] University of New York Tirana, Tirana, Albania
{marjanaskenduli,marenglenbiba}@unyt.edu.al
[2] Universita' degli Studi di Bari, Bari, Italy
{corrado.loglisci,michelangelo.ceci,donato.malerba}@uniba.it

Abstract. Human emotion analysis has always stimulated studies in different disciplines, such as Cognitive Sciences, Psychology, and thanks to the diffusion of the social media, it is attracting the interests of computer scientists too. Particularly, the growing popularity of Microblogging platforms, has generated large amounts of information which in turn represent an attractive source of data to be further subjected to opinion mining and sentiment analysis. In our research, we leverage the analysis performed on micro-blogging texts and postings in Albanian language, which enables the use of technologies to monitor and follow the feelings and perception of the people with respect to products, issues, events, etc. Our approach to emotion analysis tackles the problem of classifying a text fragment into a set of pre-defined emotion categories and therefore aims at detecting the emotional state of the writer conveyed through the text. In order to achieve this goal, we perform a comparative analysis between different classifiers, using deep learning and other classical machine learning classification algorithms. We also adopt a domestic stemming tool for Albanian language in order to preprocess the datasets used in a second round of experiments. Experimental evaluation shows that deep learning produces overall better results compared with the other methods in terms of classification accuracy. We present also other findings related to the length of the texts being processed and the impact on the classifiers' accuracy.

1 Introduction

The use of social media has essentially brought two main novelties in the society, the facilitation of intertwining social relationships and the possibility to express and share feelings and emotions. Human emotion analysis has always stimulated studies in different disciplines, such as Cognitive sciences and Psychology, and, thanks to the diffusion of the social media, is attracting the interests of computer

© Springer Nature Switzerland AG 2018
M. Ceci et al. (Eds.): ISMIS 2018, LNAI 11177, pp. 258–267, 2018.
https://doi.org/10.1007/978-3-030-01851-1_25

scientists. The reason may be seek in the impact that the analysis of emotions may have on real-world applications and emerging challenges. A representative case is the analysis of micro-bloggings and postings, which enables the use of technologies to monitor and follow the feelings and perception of the people with respect to products, issues and events. Emotion analysis is chiefly concerned with the problem of classifying a text fragment into a set of pre-defined emotion categories and therefore aims at detecting the emotional state of the writer conveyed through the text. To define the categories, many works resort to cognitive-based emotion theory, for instance, the Ekman' model [7]. The recognition of emotions within a text fragment is more complex than other tasks of text classification for several reasons related. First, there are no grammatical or syntactic structures able to characterize a category and discriminate others. Second, the same emotion can be expressed with different lexical forms. Third, emotions can be attributed to different causes and can be induced by different behaviors of human beings or events.

The length and nature of the text distinguish the problem of emotion detection into sentence-level classification and document-level classification. The sentences are prone to the sparseness due to the limited content and quality of the writing. The typical case is represented by the short texts in social media, which are often characterized by abbreviations, emoticons and typos [10]. The documents are often formal, written in correct language but may contain impersonal content which does not express an emotional state [5]. Contrary to the documents, the sentential forms are often subjective, personal and represent manifestations of the personality of individuals. Moreover, the sentences are information units of the documents, therefore the analysis of emotions at the level of the sentences can support the analysis at the level of the documents.

For both problems, research works are essentially based on unsupervised learning and supervised learning [21]. The approaches of the first category largely rely on the availability of lexical resources, which often depend on the specific language of the resources and thus offer coverage of a specific language in emotive texts. The supervised learning approaches rely on the annotated datasets, which are generated especially for frequently spoken languages and are subject to manual labeling.

However, many linguistic resources and annotated texts have been generated for wide-coverage languages [2], such as English, Chinese and Arabic, while no attempt has been made for other rare Indo-European languages, such as the Albanian. Nowadays, Albanian is spoken by 7.6 million Albanians living in Albania, Kosovo, Montenegro, northwest of Macedonia, northwest of Greece, in some Western Europe countries and in North America where thousands of Albanians have migrated and currently live. Despite its late documentation, the Albanian language is of great interest to linguists and not only, due to its unique and archaic traits. Albanian has distinctive features which range from morphological to lexical viewpoints. It has a large alphabet (thirty-six letters), and constitutes a lot of words with particles in two units, which are difficult to label [16]. The words used to express the same idea can have different types of grammatical relations, they can be nouns and verbs. In addition, they can be associated to general semantic categories or specific categories [3]. Albanian is a very rich language in words that hold more than one meaning. They are called poly-semantic words, which may mislead an automatic classifier since the emotion being expressed might be related to only one of the meanings of the word [4]. The richness of the lexical characteristics and syntactic rules however are not enough to stimulate the generation of linguistic resources, such as thesaurus and dictionary, which drives us towards application of supervised learning to analyse emotive texts.

In this paper, we present a case study for the classification of emotion on micro-blogging sentences. Motivated by the growing relevance of the social media to mirror political information influenced by emotion states, we generated a dataset of Facebook statuses posted by Albanian politicians and manually selected the sentences and annotated them by using the Ekman's emotion categories [7]. Then, we performed extensive experiments with several state-of-art classifiers under different perspectives.

2   Related Work and Contribution

In recent research, there has been great attention in the study of various aspects of emotion analysis, such as emotion resource creation, emotion cause event analysis, reader emotion detection, and emotion detection for writers authoring long-texts (documents) and short-texts (messages), which is the focus of the current work.

In [14], the authors generated a lexicon of pairs word-emotion based on hashtags from an annotated Twitter dataset. Experiments show an improvement in accuracy by using SVM classifiers for the six basic emotion categories. Different lexical resources, such as, WordNetAffect, SentiWordnet and SenticNet have been used in [6] with a Conditional Random Field classifier. The algorithm relies on three scoring methods and is able to outperform many systems which use only one lexicon.

An approach which does not need lexicon has been described in [1]. The algorithm works on semantic and syntactic relatedness. The approach described in [10] does not resort to lexicons but exploit other features internal to the texts written in Chinese. It incorporates both the label dependence among the emotion labels and the context dependence among the contextual instances into a factor graph, where the label and context dependence is modeled as various factor functions.

In [21], the authors use intra-sentence based features to determine the emotion label set of a target sentence coarsely through the statistical information gained from the label sets of the most similar sentences in the training data. Then, they use the emotion probabilities between neighboring sentences to refine the emotion labels of the target sentences. The proposed algorithm is evaluated on Ren-CECps, a Chinese blog emotion corpus. The same corpus has been analyzed in [17] which use a polynomial kernel to compute similarities between sentences and basis sets of emotions. A different language is studied in [5], which consider a Bengali blog dataset annotated at the word-level and proposed a scoring technique for the constituents of the sentences.

The method reported in [19] identifies emotional patterns from part-of-speech tags of emotion triggered terms and its co-occurrence terms. Patterns are classified hierarchically into categories referred to positive and negative emotions. Sentences are categorized by capturing the degree of emotive content with respect to the semantics of patterns.

Recently, some studies overcome the constraint to require emotive words, common to many works, by considering emotion signals, such as polarity shifters, negations, emoticons and slangs [2]. The authors propose a rule-based classification framework which combines a pipeline of classifiers which learned from the emotion signals.

A research that is recently attracting attention is that of focusing the study on the emotions manifested by specific users [15,20]. This has a two-fold motivation. First, in social media platforms, the content of the messages is often expression of the reaction to or influence of the messages posted by specific users. A typical example is represented by Twitter users with many followers. Second, the distribution of the emotion categories could be not fair because some users could have never been expressed some emotions. In these cases, a user-centric study could be more reliable because we could better learn the lexical characteristics. The current work leverages upon these considerations and reports a case study focused on the emotive tweets of (influential) users. We propose a supervised learning approach based on a Deep learning architecture, which compared against some competitors, is able to accurately account for the specific distribution of the emotion categories over the posts of the selected politicians.

3   Construction of the Sentence-Level Datasets

The extant studies report sentiment classification at varying levels of granularity (document and sentence based), mainly for popular languages like English, German, Spanish, Chinese etc. While fully recognizing the scarcity of Albanian linguistic resources i.e. corpora, gazetteers and dictionaries, we decided to build our own corpuses using as a primary source public Facebook pages belonging to high rank Albanian politicians. It goes without saying that the finished corpuses represent an important contribution of this work. It is fitting therefore that we describe the steps followed for the construction of corpuses, which were time consuming, yet imperative to the aim of this work.

3.1   Data Collection and Assembly

For the purpose of collecting abundant quantities of micro blogging data, we used RestFB[1] - a simple and flexible Facebook Graph API client written in Java language. It is an open source software released under the terms of the

---

[1] https://restfb.com//.

MIT License. Expanding upon this step, specifically, we had to reuse the source code and setup a whole framework that allowed us to fetch posts out of public/community pages belonging to Albanian public figures. The framework we propose entails the following depicted modules as shown in Fig. 1. We fetched around 60000 posts belonging to 119 Albanian politicians, shortlisted among the ones who are pretty active in Facebook and popular to the general perception of the social media audience. The fetched posts where captured and stored in a local SLQ database, from where we could extract and build sample datasets for experimental purposes.

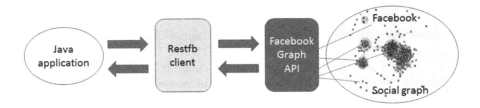

Fig. 1. Data collection framework.

3.2   Data Preprocessing

Prior to execution of sentence-based classification, we had to perform the preparation tasks on the raw text datasets, such as data cleaning and stemming. Initially, we exported six sample datasets from our SQL raw collection. The six datasets were manually cleaned and annotated using the Ekman' model [7] - a vocabulary consisting on seven emotions: JOY, ANGER, DISGUST, FEAR, SHAME, GUILT and SADNESS. As a result, we constructed six datasets, further referred to as per their respective ID numbers: D1, D2, D3, D4, D5 and D6. Dataset D1 is a multiuser dataset that assembles 2325 posts from 119 users, not longer than 200 characters distributed across the period of time January-March 2018. The remaining datasets D2, D3, D4, D5 and D6 are single-user datasets consisting of 159, 322, 1002, 1481 and 1069 posts respectively, not longer than 200 characters each and distributed across the period of time 2008–2018. Moreover, we manually annotated the sentiment polarity of each sentence in both training and testing corpuses. We chose a 70:30 split of our datasets into training and testing sets. The annotation for training corpuses is used to train the classifier, while the annotation for testing corpuses is used to test the accuracy of sentiment classification at sentence level.

4   Sentence-Based Classification

In this work, we address the categorization of the emotions expressed in the collected sentences with a Deep learning (DL) architecture, which revises a state of the art model originally designed for binary classification [8].

DL architectures have attracted much interest for their peculiarities to learn with small intervention on the data representation and feature engineering. In Natural Language Processing, we can enumerate many DL methods which have been proposed to investigate as many tasks, such as speech recognition and sequence labeling (e.g., POS tagging).

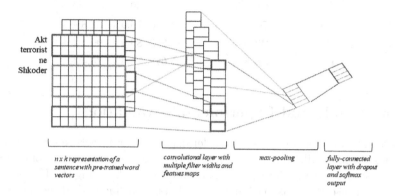

Fig. 2. Deep learning architecture for sentence-based classification of emotions.

In this paper, we extend the model proposed in [8] to the multi-class problem (the categories refer to the Ekman's basic emotions) and apply it to Albanian language. The model is sketched in Fig. 2 and summarized in the following. The Input layer embeds words into low-dimensional vectors. More precisely, a raw sentence is represented as a concatenation of the vectors, each referring a word of the sentence. All the vectors have length k and the sentence representation has length n, so all the sentences have n vectors. A sentence is padded if necessary.

The Convolutional layer performs filters over the embedded word vectors using multiple widths. More precisely, we apply convolution operations to windows of width h (that is, h words), in order to produce a long feature vector of length hk. Then, we apply a max-pool operation on the feature map, in order to select the most important feature. A dropout regularization is applied to the Convolutional layer, in order to prevent the co-adaptation of hidden units and force them to learn individually useful features. Finally, there is fully connected *softmax* layer whose output is the probability distribution over labels. Further details can be found in [8]. The experimental configuration in terms of the architecture parameters is reported in Sect. 5.

As to the word embeddings in Albanian, which is a specific characteristic of the current paper and it is beyond the original model, we consider the word vectors pre-trained on a Twitter corpus in English [9] and build the dictionary of the words by performing a translation of the words, which have no social tags (e.g., hashtag), into Albanian. Further details on this vocabulary are reported in Sect. 5.

Table 1. Experimental results in terms of accuracy (correctly classified unstemmed instances).

Dataset ID	Deep Learning	NB	IBK	SMO
D1	**91.2%**	75.5%	78.3%	81.8%
D2	**86.3%**	82.9%	75.3%	83.5%
D3	**81.9%**	79.1%	80.7%	80.7%
D4	70.2%	77.2%	79.9%	84.8%
D5	**90.5%**	81.2%	89.9%	90.1%
D6	**88.8%**	65.6%	75.4%	75.4%

5   Case Study and Experimental Evaluation

In this section we report experimental details on the architecture setup, vocabulary preparation in Albanian and performance of the learner in terms of accuracy.

The Deep learning architecture has been designed by using the Keras framework[2] and Tensorflow[3] as back-end. The hyper-parameters are configured as follows: sentence length $(n) = 33$, word embedding size $(k) = 300$, windows width $(h) = 3$, dropout rate $= 0.2$, number of neurons of the Convolutional layer $=100$, number of epochs $= 10$. The size of the vocabulary is almost 1M of words and the translation step has been performed with public API libraries[4].

We have performed experiments along two main perspectives: learning emotive posts regardless of authors (politicians) and learning emotive posts of specific individual authors. The results are discussed by following this distinction. As to the first perspective, we considered 2325 annotated posts (training set), equally distributed over the authors under examination. In the second perspective, we considered the posts posted by five specific authors[5] and built the annotated training sets with 159, 322, 1002, 1486 and 1069 posts. The performance has been evaluated in terms of classification accuracy on the testing sets.

The datasets are structured as follows: Dataset D1 is composed of 119 users whose posts have been labelled using the seven emotional tags while datasets D2-D6 are single-user datasets which represent the data of every single user over the period of reference. We decided to measure the classification accuracy for both cases: multi-user context and single-user context so that we could understand how well algorithms perform in both different scenarios.

In addition, we decided to investigate the effect of linguistic processing such stemming for Albanian, on the classification accuracy. For this purpose we have used the algorithm developed in [18], which uses a rule-based approach for stemming of texts in Albanian language. The experiments are divided into two groups, the first without stemming and the second with stemming.

---

[2] https://keras.io/.
[3] https://www.tensorflow.org/.
[4] https://github.com/matheuss/google-translate-api.
[5] We do not report their names due to privacy reasons.

Table 2. Experimental results in terms of accuracy (correctly classified stemmed instances).

Dataset ID	Deep Learning	NB	IBK	SMO
D1	**92.4%**	66.7%	73.6%	76.1%
D2	**84.5%**	81.6%	78.5%	82.3%
D3	**83.1%**	77.9%	81.0%	81.0%
D4	67.0%	78.1%	80.0%	84.8%
D5	**91.0%**	83.5%	90.3%	90.3%
D6	**85.3%**	65.4%	75.4%	75.3%

Experimental results are shown in Tables 1 and 2. We have used algorithms implemented in Weka for the other three classical machine learning classification algorithms, such as Naive Bayes (NB), Instance-based learner (IBK), and Support Vector Machines (SMO).

Table 1 shows the results on texts without being stemmed, with 10-fold cross-validation. As we can see, the DL approach has produced better results in terms of classification accuracy. Only in one case the DL approach has scored worse than the other algorithms. We suspect this worse result may be due to shorter sentences in this dataset in general. However, this case needs further investigation which we leave for future work as it requires extensive experiments by varying the length of the sentences to understand how this relates to the performance of the various algorithms.

Table 2 shows results on stemmed texts. As it can be seen, again the DL approach performs in general better than the other classical algorithms. Again only in one case the DL approach has performed worse which leaves again room for future exploration of the reasons of this specific result.

Another interesting finding of the experiments, is that using the stemming step, did not lead to relevant improvement in the overall performance of all the algorithms, with some worse results in some cases. We suspect this is due to short length of the sentences which after being stemmed carry even less information than before being stemmed. We believe this matter requires further investigation on the relationship between length of the texts being stemmed and the performance of the classifier.

6   Conclusions

In this paper we have presented an approach for analyzing micro-blogging texts and postings in Albanian language. Our approach to emotion analysis is by classifying a text fragment into a set of pre-defined emotion categories and therefore detecting the emotional state of the writer conveyed through the text. We have performed a comparative analysis between different classifiers, using deep learning and other classical machine learning classification algorithms. We have also used a stemming tool for Albanian language to perform a second round of experiments, preprocessing our datasets. Experimental evaluation shows that deep learning produces overall better results compared with the other methods in terms of classification accuracy. Interesting findings related to the length of the texts being processed and the impact on the classifiers' accuracy, show that this matter requires future investigation. As future work, we plan to investigate three research directions: *(i)* sentence-based classification by considering semantic relations among named entities [13], *(ii)* time-variability of the user emotion [12] and *(iii)* correlation between the emotive status of different users [11].

Acknowledgments. This work has been partially supported by the Apulia Regional Government through the project "Computer-mediated collaboration in creative projects" (8GPS5R0) collocated in "Intervento cofinanziato dal Fondo di Sviluppo e Coesione 2007–2013 – APQ Ricerca Regione Puglia - Programma regionale a sostegno della specializzazione intelligente e della sostenibilita' sociale ed ambientale - FutureIn-Research".

References

1. Agrawal, A., An, A.: Unsupervised emotion detection from text using semantic and syntactic relations. In: 2012 IEEE/WIC/ACM International Conferences on Web Intelligence, WI 2012, Macau, China, 4–7 December 2012, pp. 346–353 (2012). https://doi.org/10.1109/WI-IAT.2012.170
2. Asghar, M.Z., Khan, A., Bibi, A., Kundi, F.M., Ahmad, H.: Sentence-level emotion detection framework using rule-based classification. Cogn. Comput. **9**(6), 868–894 (2017). https://doi.org/10.1007/s12559-017-9503-3
3. Beci, B.: Gramatika e gjuhes shqipe. Publisher, Logos-A (2005)
4. Biba, M., Mane, M.: Sentiment analysis through machine learning: an experimental evaluation for Albanian. In: Proceedings of the Second International Symposium on Intelligent Informatics, ISI 2013, India, pp. 195–203 (2013). https://doi.org/10.1007/978-3-319-01778-5_20
5. Das, D., Bandyopadhyay, S.: Sentence to document level emotion tagging a coarse-grained study on Bengali blogs. In: Advances in Pattern Recognition - Second Mexican Conference on Pattern Recognition, Mexico, pp. 332–341 (2010)
6. Das, D., Bandyopadhyay, S.: Sentence-level emotion and valence tagging. Cogn. Comput. **4**(4), 420–435 (2012). https://doi.org/10.1007/s12559-012-9173-0
7. Ekman, P.: Facial expression and emotion. Am. Psychol. **48**(4), 384–392 (1993). https://doi.org/10.1037/0003-066X.48.4.384
8. Kim, Y.: Convolutional neural networks for sentence classification. In: Proceedings of the 2014 Conference on Empirical Methods in Natural Language Processing, EMNLP 2014, 25–29 October 2014, Doha, Qatar, pp. 1746–1751 (2014)
9. Li, Q., Shah, S., Liu, X., Nourbakhsh, A.: Data sets: word embeddings learned from tweets and general data. In: Proceedings of the Eleventh International Conference on Web and Social Media, ICWSM 2017, Canada, 2017, pp. 428–436 (2017)
10. Li, S., Huang, L., Wang, R., Zhou, G.: Sentence-level emotion classification with label and context dependence. In: Proceedings of the 53rd Annual Meeting of the Association for Computational Linguistics, ACL 2015, China, pp. 1045–1053 (2015)
11. Loglisci, C.: Time-based discovery in biomedical literature: mining temporal links. IJDATS **5**(2), 148–174 (2013). https://doi.org/10.1504/IJDATS.2013.053679

12. Loglisci, C., Ceci, M., Malerba, D.: Relational mining for discovering changes in evolving networks. Neurocomputing **150**, 265–288 (2015). https://doi.org/10. 1016/j.neucom.2014.08.079
13. Loglisci, C., Ienco, D., Roche, M., Teisseire, M., Malerba, D.: Toward geographic information harvesting: Extraction of spatial relational facts from web documents. In: 12th IEEE International Conference on Data Mining Workshops, ICDM Workshops, Brussels, Belgium, 10 December 2012, pp. 789–796 (2012). https://doi.org/ 10.1109/ICDMW.2012.20
14. Mohammad, S.M., Kiritchenko, S.: Using hashtags to capture fine emotion categories from tweets. Comput. Intell. **31**(2), 301–326 (2015). https://doi.org/10. 1111/coin.12024
15. Oleri, O., Karagoz, P.: Detecting user emotions in twitter through collective classification. In: Proceedings of the 8th International Joint Conference on Knowledge Discovery, Portugal, 2016, pp. 205–212 (2016). https://doi.org/10.5220/ 0006037502050212
16. Piton, O., Lagji, K., Përnaska, R.: Electronic dictionaries and transducers for automatic processing of the Albanian language. In: 12th International Conference on Applications of Natural Language to Information Systems, NLDB 2007, France, 2007, pp. 407–413 (2007). https://doi.org/10.1007/978-3-540-73351-5_38
17. Quan, C., Ren, F.: Recognizing sentence emotions based on polynomial kernel method using Ren-CECps. In: Proceedings of the 5th International Conference on Natural Language Processing and Knowledge Engineering, NLPKE 2009, China, 2009, pp. 1–7 (2009). https://doi.org/10.1109/NLPKE.2009.5313834
18. Sadiku, J., Biba, M.: Automatic stemming of albanian through a rule-based approach. J. Int. Res. Publ. Lang. Individuals Soc. **6** (2012)
19. Shaila, S.G., Vadivel, A.: Cognitive based sentence level emotion estimation through emotional expressions. In: Selvaraj, H., Zydek, D., Chmaj, G. (eds.) Progress in Systems Engineering. AISC, vol. 366, pp. 707–713. Springer, Cham (2015). https://doi.org/10.1007/978-3-319-08422-0_100
20. Williams, G., Mahmoud, A.: Analyzing, classifying, and interpreting emotions in software users' tweets. In: 2nd IEEE/ACM International Workshop on Emotion Awareness in Software Engineering, Argentina, 2017, pp. 2–7 (2017). https://doi. org/10.1109/SEmotion.2017.1
21. Xu, J., Xu, R., Lu, Q., Wang, X.: Coarse-to-fine sentence-level emotion classification based on the intra-sentence features and sentential context. In: 21st ACM International Conference on Information and Knowledge Management, USA, pp. 2455–2458 (2012). https://doi.org/10.1145/2396761.2398665

A Novel Personalized Citation Recommendation Approach Based on GAN

Ye Zhang$^{(\boxtimes)}$, Libin Yang, Xiaoyan Cai, and Hang Dai

School of Automation, Northwestern Polytechnical University,
Xi'an 710072, China
yezhang@mail.nwpu.edu.cn

Abstract. With the explosive growth of scientific publications, researchers find it hard to search appropriate research papers. Citation recommendation can overcome this obstacle. In this paper, we propose a novel approach for citation recommendation by applying the generative adversarial networks. The generative adversarial model plays an adversarial game with two linked models: a generative model G that captures the data distribution, and a discriminative model D that estimates the probability which a sample came from the training data rather than G. The model first encodes the graph structure and the content information to obtain the content-based graph representation. Then, we encode the network structure and co-authorship to gain author-based graph representation. Finally, the concatenation of the two representations will be acted as the node feature vector, which is a more accurate network representation that integrates the author and content information. Based on the obtained node vectors, we propose a novel personalized citation recommendation approach called CGAN and its variation VCGAN. When evaluated on AAN dataset, we found that our proposed approaches outperform existing state-of-the-art approaches.

Keywords: Citation recommendation · Generative adversarial network
Latent representation · Deep learning

1 Introduction

Citation recommendation is the task of providing researchers a list of references satisfying their requirements. In simple terms, a good citation recommendation system can learn publication information and help to accelerate frequent tasks by intelligently identifying references. It is thus desirable to design a system that attempts to automatically generate high-quality candidate papers given a query document.

A variety of citation recommendation systems have been proposed in recent years. These research technologies include collaborative filtering [1], topic modeling [2, 3], machine learning [4] and deep learning [5]. Collaborative filtering mainly focuses on the rating matrices created from the researchers' readership or the adjacency matrix of the citation network. The probabilistic topic models have been widely used to recommend literatures due to its sound theoretical foundation and promising performance in this decade, such as LDA [2] and PLSA [3]. Duma et al. [4] explore new

© Springer Nature Switzerland AG 2018
M. Ceci et al. (Eds.): ISMIS 2018, LNAI 11177, pp. 268–278, 2018.
https://doi.org/10.1007/978-3-030-01851-1_26

recommendation algorithms using classification methods. Compared with other methods, deep learning algorithms perform more effectively in feature extraction and are scalable to large-scale data [5].

Graph embedding aims to output a vector representation for each node in the graph, such that nodes close to each other in network structure have similar vector representations in a low dimensional space. It is an effective scheme to represent graph data for further analysis. Many methods for network embedding have been proposed in recent years, such as DeepWalk [6], TADW [7] and TriDNR [8]. They aim to capture various connectivity information in a network to obtain representation results. These existing methods are effective in the network with different objectives, however, they suffer from lack of additional constraints to improve the robustness of the learned representations. Recently, many researchers have proposed to learn robust and reusable representations by training generative adversarial models [9, 10]. Moreover, some models have been designed to deal with graph data [11, 12].

In this paper, we propose a novel personalized citation recommendation approach called CGAN and its variation VCGAN based Generative Adversarial Network (GAN). Our approach leverages the principle of adversarial learning [9] to learn robust network representations. It first encodes the graph structure and the content information to obtain the content-based graph representation. Then, we encode the network structure and co-authorship to gain author-based graph representation. Finally, the concatenation of the two representations will be acted as the node feature vector. Based on the obtained nodes vectors, we propose a personalized citation recommendation approach. The experimental results demonstrate that our method outperforms other models in terms of both Recall and NDCG.

2 Related Work

2.1 Citation Recommendation

According to different scenarios of usages, citation recommendation can be divided into two main categories: global citation recommendation and local citation recommendation. This difference is the different queries according to which the citation should be made. The local citation recommendation aims to recommend a short list of papers according to the given context. The given context is the query and typically consists of one to three sentences. It usually ranks papers by measuring the relevance of paper content to the citation context [13]. On the contrary, the global citation recommendation suggests a list of papers to be the references for entire paper manuscript. The global citation recommendation approaches can produce effective query-oriented recommendation using rich information, including author, title, abstract, etc., thus can provide a comprehensive view of relevant references [14, 15]. We mainly focus on global citation recommendation in this paper.

2.2   Network Embedding Methods

Network embedding is an effective manner to represent the nodes in a network as low-dimensional vectors. Many network embedding methods have been proposed in recent years. These researches can be divided into three main categories: matrix factorization methods, probabilistic models and deep learning methods. The matrix factorization methods capture different kinds of high-order proximities by processing the adjacency matrix to obtain graph embedding [16, 17]. Probabilistic models attempt to learn graph embedding by employing manifold features of the graph [6, 18]. And researchers have proved that many probabilistic algorithms are equivalent to matrix factorization approaches [7].

These approaches above principally study the structure relationship or minimizing the reconstruction error. They have ignored the data distribution of the latent codes, and can easily result in poor representation in dealing with real-word sparse and noisy graph data. One common way to handle this problem is to introduce some regularization to the latent codes and enforce them to follow some prior data distribution. So generative adversarial based frameworks have been researched to learn robust latent representation. Dai et al. [11] exploit the strengths of generative adversarial networks in capturing latent features, and investigate its contribution in graph representation. And Pan et al. [12] propose an adversarial regularized graph auto-encoder for graph data.

3   Problem Definition

In this section, we introduce the definitions of problems and notations to be used in our citation recommendation system. Then, we will present details of our proposed citation recommendation approach based on generative adversarial network.

We formally define personalized citation recommendation task in this paper as follows: Given a bibliography dataset P and Q of the set of queries that contain titles and authors information. For every query, which consists of keywords, title or other information, we aim to recommend a high-quality list of references that have better correlation with the query. We first construct a bibliographic citation network, which is represented as $\mathcal{G} = (V, E, M, C)$ and $\mathcal{G}$ is a directed graph, where V is the paper nodes set, E is a set of edges representing the citation relation between the nodes. M is the adjacency matrix of graph $\mathcal{G}$ representing the network structure, C represents the node features information matrix, which $C_i \in C$ indicates the content features of the node v_i such as titles or authors information.

Given a network $\mathcal{G}$, our network embedding model aims to learn a low-dimensional vector $X_i \in R^d$ with the format as follows: f:$(M, C) \rightarrow X$, where X_i^T is the i-th row of the matrix $X \in R^{N \times d}$. N is the number of nodes and d is the dimension of embedding. X is the representation matrix that encodes the structure and features information of the citation network. We construct the generative adversarial network embedding model to obtain the graph embedding matrix that it is our hard core. Then, we recommend a small subset of target $p \in P$ for a query manuscript $q \in Q$ by ranking the papers via the similarity score list s(q, p). Figure 1 shows the overview of personalized citation

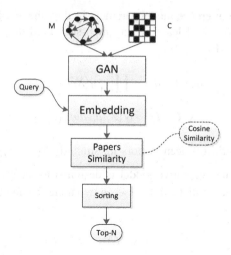

Fig. 1. Personalized citation recommendation model based on GAN

recommendation based on GAN. The details of every part will be presented in the following sections.

4   Proposed Method

4.1   Graph Convolutional Network

Graph convolutional network (GCN) [19] is an effective variant of convolutional network which operates directly on graph. We build a layer-wise convolution operation via a localized first-order approximation of spectral graph convolution. The form of this convolution propagation rule as follows:

$$X^{l+1} = \rho\left(\tilde{D}^{-\frac{1}{2}}\tilde{M}\tilde{D}^{-\frac{1}{2}}X^l W^l\right) \tag{1}$$

Here, X^l is the input of convolution layer and X^{l+1} is the output matrix in the l-th layer. The first layer input is $X^0 = C \in R^{N \times m}$ (N nodes and m features) in our models. $\tilde{M} = M + I$ represents the adjacency matrix of graph $\mathcal{G}$ with added self-connections and I is the identity matrix, $\tilde{D}$ is the diagonal matrix, $\tilde{D}_{ii} = \sum_j M_{ij}$, and W^l is a layer-specific weight matrix that we will train in our convolution neural network. ρ denotes the activation function. In this paper, we construct a two-layer GCN and we use ReLU and linear activations in the first and second layers respectively. Consequently, $X = X^{(2)}$ is the representation matrix of the nodes, which encodes both graph structure and nodes features.

Then, we obtain an inference model parameterized by the two-layer GCN, which is a framework for unsupervised learning on graph-structured data based on the variational auto-encoder (VAE) in [20]:

$$q(X|C,M) = \prod_{i=1}^{N} q(X_i|C,M) \qquad (2)$$

$$q(X_i|C,M) = N\left(X_i|\mu_i, diag\left(\sigma_i^2\right)\right) \qquad (3)$$

$\mu = X^{(2)}$ is the matrix of mean vectors μ_i, similarly $\log\sigma = \rho_{linear}(X^1, M|W^1)$.

Generative Model. The generative model is designed to reconstruct graph structure M. We define $p\left(\tilde{M}|X\right)$ to predict a link between two nodes by an inner product between latent variables:

$$p\left(\tilde{M}_{ij} = 1|X_i, X_j\right) = \sigma\left(X_i^T X_j\right) \qquad (4)$$

where σ is the sigmoid function. And we can derive the embedding matrix X and reconstruction adjacency matrix $\tilde{M}$ as follows:

$$\tilde{M} = \sigma\left(XX^T\right) \qquad (5)$$

Optimization. We optimize our network model as follows:

$$L_1 = E_{q(X|C,M)}[logp(M|X)] \qquad (6)$$

or by the variational lower bound:

$$L_2 = E_{q(X|C,M)}[logp(M|X)] - KL[q(X|C,M)||p(X)] \qquad (7)$$

Where $KL(q||p)$ is the Kullback-Leibler divergence between $q(\cdot)$ and $p(\cdot)$. And $p(X)$ is a Gaussian prior.

4.2   Adversarial Embedding Model

The generative adversarial model consists of a generator G and a discriminator D. The generator function is designed to capture the data distribution. In our framework, G is applied to transform the input network features to embedding vectors. The discriminator is trained to estimate the probability that a sample came from the training data rather than G. Specifically, the embedding vectors are acted as fake samples and the real data is selected from a prior distribution $p(X)$. In this process, we simultaneously train D to maximize the probability that assigns the correct label to both training examples and samples from G, and train G to minimize $log(1 - D(G(C,M)))$. This is a two-layer minimax game for the generator and discriminator playing against each other. The cost function $V(D, G)$ can be written as follows:

$$\min_{G} \max_{D} V(D, G) = E_{x \sim p_{data}(X)}[logD(X)] + E_{x \sim p_x(X)}[log(1 - D(G(C, M)))] \quad (8)$$

where $p_{data}(X)$ is a Gaussian prior. Algorithm 1 is the detailed description of models.

Algorithm 1 Citation network embedding models based on GAN

Input: Adjacency matrix M and paper-title matrix C_1 and paper-author matrix C_2;
The dimension of vector d;
The number of iterations K;
The number steps of iterating discriminator T.
Output: The embedding matrix X.
1. **for** iterator = 1, 2, 3, ..., K **do**
2.   Get the variable matrix X_i, $i = 1$ or 2 through Eq.(1)
3.   **for** k=1,2,3, ..., T **do**
4.   Sample s entities x^j, from latent matrix X_i;
5.   Sample s entities a^j, from the prior distribution P_x;
6.   Update the discriminator by ascending stochastic gradient Eq.(8)
      end for
7.   Update the graph convolutional network with its stochastic gradient Eq.(6) (called for CGAN) or Eq.(7) (called for VCGAN).
      end for
8. Get the representation matrix X, which is the concatenation of X_1 and X_2.
Return X

4.3   Citation Recommendation

After training the GAN model, we apply the embedding matrix to recommend citation papers for a query q. The cosine similarity is an effective and regular way to define the similarity between two feature vectors. Therefore, the recommending score $s(q, p_j)$ is computed by the cosine similarity of representation vector of q and a candidate paper p_j as follows:

$$s(q, p_j) = \frac{q \cdot p_j}{\|q\| \cdot \|p_j\|} \quad (9)$$

Here, $\|\cdot\|$ is the 2-norm. According to (9), we calculate $s(q, p_j)$ for all candidates and recommend papers with higher scores.

5   Experiments and Evaluation

In this section, we establish the experiments on the real-world publication dataset to evaluate the performance of our proposed methods.

5.1   Experiment Settings

Dataset. The AAN dataset [21] is a famous dataset that contains the complete information of papers in computational linguistic and natural language processing. We select a subset published from 1965 to 2012 that contains 13,394 papers as our experimental dataset and the references are also in the dataset for every paper. Then, there are 10715 papers published before 2012 to be used to train models, and the remaining 2679 papers published in 2012 are regarded as the testing data. Table 1 demonstrates the statistics of these datasets.

Table 1. Statistics of the dataset

	Papers	Authors	Citations
Train	10715	7540	63831
Test	2679	1716	16001

Evaluation Methods. To evaluate the recommendation accuracy and quality of our models, we use Recall and NDCG as evaluation metrics. Both metrics are commonly used to evaluate recommendation results.

- **Recall**, is the rate of the number of real cited papers in the Top-N recommendation list to the total number of cited articles. This metric is calculated as follows:

$$\text{Recall@N} = \frac{1}{Q} \sum_{j=1}^{Q} \left(\frac{R_q \cap T_q}{T_q} \right) \tag{10}$$

Here N is the length of the recommendation list and Q is the number of queries. R_q is the Top-N list based on a query paper q. T_q is the set of papers citing q.

- **NDCG**, normally, the highly relevant references should appear higher in the Top-N list. So, we use NDCG to measure the ranked recommendation list. It can be defined as:

$$\text{NDCG@N} = \frac{1}{Q} \sum_{j=1}^{Q} \left(\sum_{i=1}^{N} \frac{2^{r_i} - 1}{log_2(i+1)} / IDCG@N \right) \tag{11}$$

where r_i is the ratio of the i-th paper in the ranking list. The paper is relevant or not when $r_i = 1$ or $r_i = 0$. IDCG@N is the ideal ranking result.

Baselines. To validate the effectiveness of our models based on GAN, we compare them with other baseline approaches as follows:

- **DeepWalk:** It is a network representation method by using random walk and Skip-gram model which encodes relations into a continuous vector space.
- **TADW:** It is a state-of-the-art approach that incorporates text information and network to learn the network representation.
- **TriDNR:** It simultaneously considers paper network structure and paper vertex content to learn paper network representation.

Parameter Settings. We compare models based on their ability to recommend references in citation recommendation. For DeepWalk, TADW and TriDNR, the parameters are provided by the authors in their original papers. For instance, an embedding dimension of 128, 10 random walks of length 80 per node and window size of 10, trained for every epoch. For CGAN and VCGAN, we create a weight variable with Glorot and Bengio [22] and optimize the models using Adam [23] with a learning rate of 0.001. The dimension of the hidden layer is 32 and the latent variables is 16 in our experiments that we use the implementation from [20].

5.2 Results Analysis

We compare the proposed models and other baselines in terms of citation recommendation performance. The comparison results of different methods are summarized in Table 2.

It is obviously to see that CGAN and VCGAN outperform other baselines in terms of Recall and NDCG. The statistics demonstrate that generative adversarial graph network can effectively improve the robustness and discrimination of citation recommendation algorithms. Furthermore, the VCGAN model achieves higher recommendation performance than CGAN (on average of 18.42% with respect to Recall and 20.20% with respect to NDCG). DeepWalk, TADW and TriDNR algorithms perform fairly poor on citation recommendation network. This is mainly because DeepWalk learns network representation using network structure only and ignores the text information. TADW is a matrix factorization algorithm that factorizes a matrix M, which is a large sparse matrix. In practice, it is very computationally expensive so that TADW only factorizes an approximate M, which will affect its recommendation efficiency. Although TriDNR considers both network structure and vertex content, it ignores inter-relationship among vertices. So it performs worse than our proposed approaches.

Table 2. Performance comparison between different methods

Top-N	25		50		75		100	
Metrics	Recall	NDCG	Recall	NDCG	Recall	NDCG	Recall	NDCG
DeepWalk	0.27891	0.17256	0.34751	0.19428	0.39208	0.20682	0.42224	0.21509
TADW	0.26317	0.27469	0.34156	0.26371	0.38773	0.26402	0.42472	0.26607
TriDNR	0.28102	0.27351	0.36324	0.26901	0.40522	0.26237	0.43870	0.27119
CGAN	**0.28133**	**0.26094**	**0.38931**	**0.27023**	**0.41793**	**0.27274**	**0.44609**	**0.27491**
VCGAN	**0.42774**	**0.31312**	**0.45003**	**0.28062**	**0.44982**	**0.33013**	**0.44461**	**0.33086**

5.3   Case Study

To better understand the availability of our models, we give case study in this sub-section to compare the recommended results for a given query, which we use the title in our dataset as the query information. Because of the page limit, we only list top-5 retrieved papers obtained by VCGAN, CGAN and TriDNR approaches. Table 3 shows the top-5 recommended papers selected from the recommendation list.

As shown in Table 3, the results returned by the VCGAN approach have 4 records that match the ground truth citation list of the query, whereas the results returned by the CGAN and TriDNR approaches have 3 and 2 matching records, respectively. This observation demonstrates that VCGAN and CGAN obtain a better result in this case study since the manuscript author, title, as well as text information of training papers are utilized in our approaches. And the VCGAN model achieves higher performance because of the superior performance of the variational graph auto-encoders [20]. The KL-divergence term can then be interpreted as regularizing parameters, encouraging the approximate posterior to be close to the prior. It is a very effective technique to improve the accuracy of models.

Table 3. The top 5 recommended citations: the ground truths are ☑

Query: sentiment classification using automatically extracted subgraph features
VCGAN
(1) Recognizing contextual polarity in phrase-level sentiment analysis ☑
(2) Sentiment analysis of citations using sentence structure-based features ☑
(3) Sentiment classification and polarity shifting ☑
(4) Sentiment classification on customer feedback data: noisy data, large feature vectors and the role of linguistic analysis ☑
(5) Opinion and suggestion analysis for expert recommendations
CGAN
(1) Recognizing contextual polarity in phrase-level sentiment analysis ☑
(2) Sentiment classification on customer feedback data: noisy data, large feature vectors, and the role of linguistic analysis ☑
(3) Interactive annotation learning with indirect feature voting
(4) Seeing stars: exploiting class relationships for sentiment categorization with respect to rating scales ☑
(5) Going beyond traditional QA systems: challenges and keys in opinion question answering
TriDNR
(1) Learning subjective nouns using extraction pattern bootstrapping
(2) A sentimental education: sentiment analysis using subjectivity summarization based on minimum cuts ☑
(3) Identifying types of claims in online customer reviews
(4) Generalizing dependency features for opinion mining adding redundant features for CRFs-based sentence sentiment classification ☑
(5) User-directed sentiment analysis: visualizing the affective content of documents

6 Conclusion

Graph embedding is an effective approach to represent graph data into a low-dimensional space for further analysis. Generative adversarial network can estimate generative model via an adversarial process. In this paper, we propose a generative adversarial citation recommendation model combining network structure, content information and authorship. This framework can capture network structure and other abundant information to improve the robustness of recommendation. Experimental results demonstrate the effectiveness of our algorithms. In the future, we plan to explore the affection of the models parameters and the number of the convolution layers, and expand the construction to combine with venue information in the process of network representation.

References

1. McNee, S.M., Istvan, A., et al.: On the recommending of citations for research papers. In: Proceedings of ACM Conference on Computer Supported Cooperative Work, pp. 116–125 (2002)
2. Blei, D.M., Ng, A.Y., Jordan, M.I.: Latent Dirichlet allocation. J. Mach. Learn. Res. **3**, 993–1022 (2003)
3. Hofmann, T.: Unsupervised learning by probabilistic latent semantic analysis. Mach. Learn. **42**, 177–196 (2001)
4. Duma, D., Liakata, M., Clare, A., Ravenscroft, J., Klein, E.: Applying core scientific concepts to context-based citation recommendation. In: Proceedings of LREC (2016)
5. Ebesu, T., Fang, Y.: Neural citation network for context-aware citation recommendation. In: Proceedings of the 40th International ACM SIGIR Conference on Research and Development in Information Retrieval (2017)
6. Perozzi, B., Al-Rfou, R., Skiena, S.: DeepWalk: online learning of social representations. In: Proceedings of the 20th ACM SIGKDD International Conference on Knowledge Discovery and Data Mining, pp. 701–710. ACM (2014)
7. Yang, C., Liu, Z.Y., Zhao, D.L., Sun, M.S., Chang, E.Y.: Network representation learning with rich text information. In: International Joint Conference on Artificial Intelligence (2015)
8. Pan, S.R, Wu, J., Zhu, X.Q, Zhang, C.Q, Wang, Y.: Tri-party deep network representation. In: Proceedings of the 25th International Joint Conference on Artificial Intelligence, New York City, NY, USA, pp. 701–710 (2016)
9. Goodfellow, I.J., Pouget-Abadie, J., Mirza, M., et al.: Generative adversarial nets. In: NIPS, pp. 2672–2680 (2014)
10. Makhzani, A., Shlens, J., Jaitly, N., Goodfellow, I.: Adversarial autoencoders. In: ICLR Workshop (2016)
11. Dai, Q.Y., Li, Q., Tang, J., Wang, D.: Adversarial network embedding. arXiv preprint arXiv: 1711.07838 (2017)
12. Pan, S.R., Hu, R.Q., Long, G.D., Jiang, J., Yao, L., Zhang, C.Q.: Adversarially Regularized Graph Autoencoder. arXiv preprint arXiv:1802.04407v1 (2018)
13. Bethard, S., Jurafsky, D.: Who should I cite: learning literature search models from citation behavior. In: Proceedings of the 19th ACM Conference on Information and Knowledge Management (CIKM 2010), pp. 609–618 (2010)

14. Dai, T., Zhu, L., Cai, X.Y., Pan, S.R., Yuan, S.: Explore semantic topics and author communities for citation recommendation in bipartite bibliographic network. J. Ambient Intell. Hum. Comput. **9**, 957–975 (2017)
15. Shaparenko, B., Joachims, T.: Identifying the original contribution of a document via language modeling. In: Proceedings of the 32nd Annual International ACM SIGIR Conference on Research and Development in Information Retrieval, pp. 696–697 (2009)
16. Cao, S., Lu, W., Xu, Q.: Deep neural networks for learning graph representations. In: AAAI, pp. 1145–1152 (2016)
17. Ou, M., Cui, P., Pei, J., et al.: Asymmetric transitivity preserving graph embedding. In: KDD, pp. 1105–1114 (2016)
18. Grover, A., Leskovec, J.: node2vec: scalable feature learning for networks. In: KDD, pp. 855–864 (2016)
19. Kipf, T.N., Welling, M.: Semi-supervised classification with graph convolutional networks. arXiv preprint arXiv:1609.02907, (2016)
20. Kipf, T.N., Welling, M.: Variational graph auto-encoders. In: NIPS (2016)
21. Radev, D.R., Muthukrishnan, P., Qazvinian, V.: The ACL anthology network corpus. Lang. Resour. Eval. **47**(4), 919–944 (2013)
22. Glorot, X., Bengio, Y.: Understanding the difficulty of training deep feed forward neural networks. In: AISTATS, vol. 9, pp. 249–256 (2010)
23. King, D.P., Ba, J.L. Adam: a method for stochastic optimization. In: Proceedings of the International Conference on Learning Representations (ICLR) (2015)

Novelty Detection and Class Imbalance

Unsupervised LSTMs-based Learning for Anomaly Detection in Highway Traffic Data

Nicola Di Mauro[✉] and Stefano Ferilli

Dipartimento di Informatica, Università di Bari, Bari, Italy
nicola.dimauro@uniba.it

Abstract. Since road traffic is nowadays predominant, improving its safety, security and comfortability may have a significant positive impact on people's lives. This objective requires suitable studies of traffic behavior, to help stakeholders in obtaining non-trivial information, understanding the traffic models and plan suitable actions. While, on one hand, the pervasiveness of georeferencing and mobile technologies allows us to know the position of relevant objects and track their routes, on the other hand the huge amounts of data to be handled, and the intrinsic complexity of road traffic, make this study quite difficult. Deep Neural Networks (NNs) are powerful models that have achieved excellent performance on many tasks. In this paper we propose a sequence-to-sequence (Seq2Seq) autoencoder able to detect anomalous routes and consisting of an encoder Long Short Term Memory (LSTM) mapping the input route to a vector of a fixed length representation, and then a decoder LSTM to decode back the input route. It was applied to the TRAP2017 dataset freely available from the Italian National Police.

Keywords: Traffic understanding · Autoencoders
Recurrent Neural Networks

1 Introduction

As a consequence of a much easier and comfortable possibility of traveling nowadays, traffic on roads has become predominant, especially in some Countries, where road traveling is preferred to other options, such as railways or airways. People travel for work, for leisure, or for personal commitments. Actually, a large portion of our lives is spent on roads. As a consequence, roads have become an important component of our lives, and so improving road traffic may have a significant positive impact on our general lives. There are several perspectives on which traffic can be improved: safety (having to do with car accidents), security (related to road crimes), well-being (related to traffic flows, jams and facilities available along the road).

In order to set up appropriate improvements in these fields, a study of traffic behavior is needed. Unfortunately, studying traffic is quite difficult, both because

© Springer Nature Switzerland AG 2018
M. Ceci et al. (Eds.): ISMIS 2018, LNAI 11177, pp. 281–290, 2018.
https://doi.org/10.1007/978-3-030-01851-1_27

of the lack of publicly available data and because road traffic is much more complex than other means of transportation (e.g., trains or airplanes), due to several reasons:

- possible routes are fixed and quite simple in the other cases, while road vehicles can reach almost any place following any route they like;
- all details of routes, stops, timings, and coordination among vehicles are centrally planned and determined in the other cases, while in road vehicles they depend on a number of factors related to the end-users, and may even dynamically change during the journey;
- the number and variability in type of vehicles traveling on roads is much greater than for other means of transportation.

Hence, automatic techniques that may help stakeholders in carrying out such a study represent a precious resource to obtain the desired goals. For instance, one might want to predict traffic jams, or accidents, in order to place appropriate actions aimed at avoiding such events. Or, one might want to identify abnormal behavior of vehicles, that might be associated to suspect activities. In other cases, one might want to exploit the model to improve the patrolling plans and/or the environment in which traffic takes place, removing criticalities and optimizing the overall behavior.

These techniques may work at various levels: at a lower level, extracting non-trivial information that is relevant to stakeholders, by using data mining approaches, may allow them to react appropriately; at a higher level, obtaining human-readable models that allow stakeholders to have a wide and comprehensive picture of traffic behavior may allow them to identify criticalities and set up more general strategies for improving both the situation and their approach to it. Developing automatic decision support systems that use both the previous levels to carry out some kind of reasoning may help stakeholders in taking their decisions and setting up their plans.

Actual implementation of these techniques is possible today thanks to the pervasiveness of georeferencing and mobile technologies, allowing to both know the static position of relevant objects, and track the routes of moving entities. Not only technology today enables the collection of traffic data at a very fine-grained level. Many institutions and companies actually use this technology: e.g., on urban and extra-urban roads cameras are positioned that can monitor the road situation and car flow; on highways automatic systems can detect the presence of specific cars by reading their plate number—for billing or speed checking purposes; insurance companies provide their customers with GPS trackers; last but not least, everybody uses GPS-enabled mobile phones and navigators that may reveal their position and route.

Several stakeholders have also already started to collect the data sensed by these devices. Outstanding examples are insurance companies and, especially, National Traffic Polices. The Italian National Police, in particular, has recently made available for research purposes a dataset reporting vehicle positions on an Italian highway along one year, in occasion of the 1st Italian Conference on Traffic Police (TRAP2017) [8]. This was very important, because no other datasets with similar features—real data, big data, long detection timespan—are freely available.

In this paper, we propose a sequence-to-sequence (Seq2Seq) [14] deep learning approach based on autoencoder Long Short Term Memory (LSTM) [6] to detect anomalous routes—possibly associated to suspect or otherwise relevant behaviors. A qualitative and quantitative analysis of the experimental results of our approach, applied to the TRAP2017 dataset, shows that the system returns an acceptable number of relevant routes, and that the detected routes may indeed be worth attention by the traffic analysts.

This paper is organized as follows. After introducing some related works in Sect. 2, subsequent sections describe the reference dataset, the task and the proposed approach. Then, Sect. 5 discusses experiments and Sect. 6 concludes the paper.

2   Related Work

The research aimed at extracting information and models from mobility data is a recent and flourishing field in Artificial Intelligence [12]. Works are present focusing on people (persons, groups) or on territory (roads, cities, regions). They aim at providing support and/or recommendation to persons and groups, or helping institutional stakeholders (e.g., police or administration) in planning their activities. Since the approach proposed in this work will be applied to a dataset specifically concerned with highway traffic, and purposely collected to challenge researchers on extracting information that is useful to police activities [9], in the following we will focus on this particular setting and will quickly overview the work carried out so far on this specific topic and dataset, as presented at the First Italian Conference on Traffic Police in 2017.

In the perspective of 'traffic understanding', intended as the general task of obtaining a model of traffic on a given road or in a given region with specific purposes, [4] proposes a process mining approach to learn models of traffic flow that can be used for analysis, supervision, and prediction purposes. [13] suggests to prune the database in order to reduce its size while retaining interesting information, and to apply Formal Concept Analysis to obtain knowledge about the observed behavior patterns in criminal activities.

More oriented to traffic analysis, [1] proposes a scalable and fully automated procedure to effectively import and process traffic data in order to analyze them. After analyzing the TRAP2017 dataset, in [2] the same authors developed a clustering-based tool to model and classify routes and plates, and to identify itineraries possibly related to criminal intents. Their tool also allows some characterization of the clusters, useful for police officers to better understand them.

Other works focus on prediction. Although based on a different dataset, [10] presents a tensor completion based method for highway traffic prediction that can take into account multi-mode features (such as daily and weekly periodicity, spatial information, temporal variations, etc.). Trying to overcome the drawbacks of traditional models and algorithms that require a predefined and static length of past data, and do not take into account dynamic time lags and temporal autocorrelation, [5] explores the use of LSTMs obtaining higher accuracy and good generalization. [11] proposed a stacked autoencoder model for traffic flow prediction, which considers the spatial and temporal correlations inherently. Differently from [5,11], here we propose a Seq2Seq autoencoder based on LSTM in order to study anomalous routes.

3   Dataset Description and Task Definition

The TRAP2017 dataset includes a log of transits of vehicles on a limited portion of Italian highways. The data were collected in 365 days (year 2016) by the Italian National Police using automatic Number Plate Reading Systems placed on 27 'gates' spread along the road. Each of the 155,586,309 entries in the log reports: plate number (anonymized), gate id, lane id, timestamp, and plate nationality. An (anonimyzed) map of the considered portion of highways is shown in Fig. 1, reporting gates (along with their position expressed in kilometers), entrances/exits, tollbooths, and service areas.

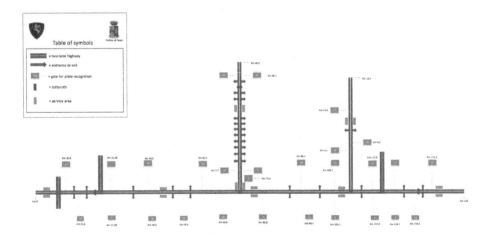

Fig. 1. Map of the highway considered by TRAP2017.

Among all possible tasks to be carried out on these data, we focused on spotting routes that can be considered anomalous for some reason, and thus may require further analysis. Since the dataset is organized by day of the year, we defined a route as the temporal sequence of gates passed by a given plate number in a day. As a consequence, we may have routes of different length. Since the dataset is not annotated, we decided to go for an unsupervised approach, and so from a technical viewpoint we set up an outlier detection task. This means that anomalous routes are not interpreted, and our definition for what is anomalous is simply something unexpected.

We aimed at proposing a general technique that may be applied to many situations. For this reason, we decided to use as few information as possible. Specifically, we decided to completely ignore the map, since it may not be available in some cases, and to only focus on the sequence of gates, ignoring all the other fields in the dataset—time, lane and nationality. In particular, we ignore time, which is expected to be a very relevant feature, both as regards how long it takes to go from a gate to the next one—too long or too short times may indicate anomalous behavior—and as regards the day, month or season—different days and periods of the year are typically associated to different traffic behaviors.

4   Proposed Approach

The Seq2Seq setting has received significant research attention. It first uses an encoder to encode a source sequence, and then applies a decoder to the encoded sequence in order to decode it into a target sequence. The goal is to estimate the conditional probability of generating the target sequence given the encoding of the source sequence.

4.1   Autoencoders

An autoencoder neural network, first introduced as a dimensionality reduction method, is an unsupervised learning algorithm that applies backpropagation, setting the target values $\mathbf{y}$ to be equal to the inputs $\mathbf{x}$. Internally, it has a hidden layer $\mathbf{h}$ that describes a code used to represent the input. In particular, it transforms an input vector $\mathbf{x}$ into a latent representation $\mathbf{h}$—the encoded representation—as follows:

$$\mathbf{h} = \sigma(\mathbf{W}\mathbf{x} + \mathbf{b}) \tag{1}$$

where $\mathbf{W}$ and $\mathbf{b}$ correspond to the weights and bias in the encoder, and σ is the sigmoid function. The latent representation $\mathbf{h}$ is then mapped back into the reconstructed feature space as follows:

$$\mathbf{y} = \sigma(\mathbf{W'}\mathbf{h} + \mathbf{b'}) \tag{2}$$

where $\mathbf{W'}$ and $\mathbf{b'}$ correspond to the weights and bias in the decoder. The autoencoder is trained by minimizing the reconstruction error $||\mathbf{y} - \mathbf{x}||$. Many autoencoders can be connected to build a stacked autoencoder, which can be used to learn multiple levels of non-linear features.

4.2   Recurrent Neural Network

Recurrrent Neural Networks (RNNs) are a generalization of feedforward NNs to sequences. Given an input sequence $\mathbf{x} = (\mathbf{x}_1, \ldots, \mathbf{x}_T)$, a standard RNN iteratively computes an output sequence $\mathbf{y} = (\mathbf{y}_1, \ldots, \mathbf{y}_T)$ using the following equations:

$$h_t = f(U x_t + W h_{t-1}) \tag{3}$$

$$y_t = V h_t, \tag{4}$$

where f is usually a nonlinearity such as tanh or Rectified Linear Unit (ReLU) activation function, being **h** the internal state of the RNN.

A RNNs Seq2Seq based approach consists in using a RNN to map an input sequence to a fixed-sized vector and then map such a vector to the output sequence with another RNN [3]. However, learning with this approach might be difficult due to the resulting long term dependencies, a problem solved by adopting Long Short-Term Memory (LSTM)—a special kind of RNN. LSTM overcome the vanishing gradients problem by replacing an ordinary neuron by a complex architecture called the LSTM unit or block.

4.3   The Model

In our proposed autoencoder LSTM model, the encoder is a LSTM NN that reads each symbol of an input sequence **x** sequentially. After reading the end of the sequence, the hidden state of the LSTM NN is a summary **c** of the whole input sequence **x**. The decoder is another LSTM NN whose initial hidden state is set to the representation **c** of **x**, and trained to generate the output sequence **y** [3,14]. Specifically, the input to the encoder is a sequence of integers (denoting the gates), each encoded as one-hot vectors with length 28 (the number of gates). Since here the aim is to correctly reconstruct the input, the decoder aims at reconstructing the same sequence of integers as the input. An anomaly is reported when the input sequence has not been correctly reconstructed by the autoencoder. In such a case, the system can also report an estimate of the difference between the actual sequence and the predicted one, which can be straightforwardly interpreted as the degree of anomaly of the actual sequence itself. In this way the autoencoder can be used to distinguish between normal and anomalous routes. The proposed model was implemented in Python using the Keras Deep Learning library[1]. The weights of the model, minimizing a categorical cross-entropy, were learned using the adam optimizer [7] for a number of epochs provided as input.

5   Experiments

We ran several experiments, with routes of different length $n \in [7, 13]$. Experiment associated to length n aimed at detecting anomalous routes of length n. This range was selected as representative of the more frequent route lengths, and allowed us to check the behavior of our approach on different inputs. The number of epochs for training was set to 5, as a trade-off between expected effectiveness and efficiency. After splitting the dataset into 70% training and 30% test sets, we ran our experiments from two different perspectives:

analytic aimed at extracting relevant routes in historical data;
predictive aimed at determining whether a new route, not included in historical data, is worth attention.

---

[1] https://keras.io/.

In the former perspective, autoencoding-based anomaly detection was applied to the training routes; in the latter perspective, it was applied to routes in the test set.

Table 1. Statistics on the dataset and experimental results.

Length		7	8	9	10	11	12	13
# Routes		2124515	1893074	733406	539718	308938	284695	260271
Training	# Routes	1487160	1325151	513384	377802	216256	199286	182189
	% Accuracy	96.52	97.51	98.73	96.67	92.61	93.99	94.84
	# Anomalous	51753	32996	6520	12581	15981	11977	9401
Test	# Routes	637355	567923	220022	161916	92682	85409	78082
	% Accuracy	96.48	97.49	98.57	96.49	92.37	93.65	94.35
	# Anomalous	22435	14255	3146	5683	7072	5423	4412

5.1 Quantitative Analysis of Results

Statistics on the dataset and quantitative experimental results, divided by route lengths, are reported in Table 1. Specifically, the number of routes (overall, used for training, and used for testing) is reported in rows labeled '# Routes'. For training and test outcomes, corresponding to the 'analytic' and 'predictive' settings respectively, both the percentage of correct reconstructions (% Accuracy) and the absolute number of anomalous routes proposed (# Anomalous, corresponding to the complement of Accuracy).

We first note that the percentages of accuracy on training and test sets are quite similar, indicating that we may expect to have the same performance on unseen routes as we have on known ones. This is important, because the 'predictive' setting is expected to be adopted continuously by the police officers, and so it must ensure high-quality results on which basing their decisions. Then, we note that the percentages of anomalous routes identified for the various lengths are very low, and that they are lower for lengths involving more routes. This is important in order to keep to a minimum the burden on police officers that are supposed to check these routes. The largest number of anomalous routes is about 70000, corresponding to length 7. This amounts to about 5850 a month, i.e., about 195 per day on average, which may be considered an acceptable rate.

5.2 Qualitative Analysis of Results

Let us now enter into more detail about the quality of our results, by showing some sample anomalous routes identified by the system. Specifically, in Table 2 we reported three representative anomalous routes for each possible length: the most anomalous one—having the highest reconstruction error estimate, the most frequent one—having the largest number of occurrences in the set of anomalous routes, and one which is apparently strange. Due to space considerations, we

Table 2. Three representative anomalous routes for each route length in [7,13].

Routes length	Anomalous routes
7	(15, 15, 3, 27, 5, 27, 1)
	(9, 27, 4, 16, 2, 20, 25)
	(9, 26, 10, 23, 2, 16, 4)
8	(27, 4, 16, 2, 20, 25, 1, 22)
	(27, 4, 16, 2, 20, 12, 1, 22)
	(21, 27, 4, 16, 2, 20, 25, 12)
9	(1, 1, 22, 25, 1, 22, 25, 12, 1)
	(4, 16, 2, 20, 25, 15, 23, 10, 26)
	(20, 20, 16, 16, 4, 4, 10, 18, 23)
10	(23, 2, 2, 18, 23, 2, 23, 2, 23, 2)
	(15, 23, 18, 10, 9, 21, 4, 16, 2, 20)
	(4, 16, 2, 20, 14, 5, 23, 18, 26, 9)
11	(12, 3, 5, 1, 5, 25, 12, 1, 25, 12, 1)
	(27, 26, 10, 23, 15, 5, 25, 20, 2, 16, 4)
	(22, 1, 12, 20, 2, 23, 15, 5, 8, 3, 13)
12	(6, 24, 4, 26, 6, 24, 4, 26, 4, 16, 17, 24)
	(1, 12, 25, 22, 1, 12, 25, 1, 12, 25, 1, 25)
	(5, 15, 23, 18, 10, 9, 21, 4, 16, 2, 20, 25)
13	(4, 26, 10, 4, 26, 4, 27, 9, 26, 4, 26, 4, 26)
	(10, 18, 23, 5, 8, 3, 13, 22, 1, 20, 2, 16, 4)
	(13, 3, 8, 5, 23, 18, 10, 16, 2, 25, 12, 1, 22)

cannot discuss all of them in depth. However, the reader may use the map in Fig. 1 to have an idea of the routes, and of how strange they may be considered. Consider, for instance, the following routes:

- route (4, 26, 10, 4, 26, 4, 27, 9, 26, 4, 26, 4, 26). It contains many loops between gates 4 and 26, with occasional passages from other gates. Since a service station is present exactly between those gates, there is a chance that this car is repeatedly stopping in this station, possibly to look for possible people to be robbed;
- another quite strange route is (20, 20, 16, 16, 4, 4, 10, 18, 23), where there are several pairs of occurrences of the same gate, and these gates are not close to each other. This raises the question about why a car should alternate in-highway and out-highway routes aimed at passing twice from the same place, and for very distant places. This is especially interesting since sometimes there is no apparent reason (e.g., a service station) for passing from those places;
- route (23, 2, 2, 18, 23, 2, 23, 2, 23, 2) basically consists of many passages from two gates—23, 2—without ever passing from intermediate gates that are present on the highway.

It is also interesting to note that some gates very frequently appear in the strange routes, while some other appear very seldom or do not appear at all. This is evident also in the tiny fraction of anomalous routes selected in Table 2. E.g., gates 2 and 20 are present in about 2/3 of the selected routes, and gate 2 in particular occurs 16 times in them. Also some combinations are more frequent and might deserve attention. E.g., pair (2,20) occurs in nearly half the selected routes; pair (26,4) occurs several times in two routes, in spite of these two gates being placed on opposite sides (and direction) on the highway, indicating that the vehicle performed several loops in a day on that portion of the road.

6   Conclusions and Future Work

Improving safety, security and comfort of road traffic may have a dramatic positive impact on people's lives, because most of our time is spent on roads. To effectively pursue this objective, stakeholders must rely on suitable studies of traffic behavior that provide them with non-trivial information, so that they may understand the traffic models and plan suitable actions. While, on one hand, the pervasiveness of georeferencing and mobile technologies allows us to know the position of relevant objects and track their routes, on the other hand the huge amounts of data to be handled, and the intrinsic complexity of road traffic compared to other means of transportation, make this study quite difficult. This calls for automatic techniques based on Artificial Intelligence approaches.

This paper proposed an approach to traffic mining aimed at identifying potentially anomalous behavior that is worth attention, and applied it to the TRAP2017 dataset, concerning a year's data about highway traffic, freely available from the Italian National Police. A quantitative analysis of the results showed that the number of anomalous routes identified is consistent with our expectations, and a qualitative analysis of the results revealed that the routes selected by our approach are indeed peculiar, and deserve further insight. Future work will proceed in several directions. A cooperation with police operators that may evaluate the outcomes and provide expert insight is planned. Also, a study of the relationships between the procedure's parameters and the quantity an quality of anomalous routes returned will be carried out. Finally, we plan to design further modules that, in cooperation with the proposed one, will provide police officers with further information on the available traffic data. If the good performance of the approach will be confirmed, we will build a system that will support police officers in their activities.

Acknowledgments. Work partially funded by the Italian PON 2007–2013 project PON02_00563_3489339 'Puglia@Service'.

References

1. Bernaschi, M., Celestini, A., Guarino, S., Lombardi, F., Mastrostefano, E.: Traffic data: exploratory data analysis with Apache Accumulo. In: Leuzzi, F., Ferilli, S. (eds.) TRAP 2017. AISC, vol. 728, pp. 129–143. Springer, Cham (2018). https://doi.org/10.1007/978-3-319-75608-0_10

2. Bernaschi, M., Celestini, A., Guarino, S., Lombardi, F., Mastrostefano, E.: Unsupervised classification of routes and plates from the TRAP 2017 dataset. In: Leuzzi and Ferilli [8], pp. 97–114

3. Cho, K., et al.: Learning phrase representations using RNN encoder-decoder for statistical machine translation. In: Proceedings of the 2014 Conference on Empirical Methods in Natural Language Processing (EMNLP), pp. 1724–1734 (2014)

4. Ferilli, S., Redavid, D.: A process mining approach to the identification of normal and suspect traffic behavior. In: Leuzzi and Ferilli [8], pp. 37–56

5. Fumarola, F., Lanotte, P.F.: Exploiting recurrent neural networks for gate traffic prediction. In: Leuzzi and Ferilli [8], pp. 145–153

6. Hochreiter, S., Schmidhuber, J.: Long short-term memory. Neural Comput. $9(8)$, 1735–1780 (1997). https://doi.org/10.1162/neco.1997.9.8.1735

7. Kingma, D.P., Ba, J.: Adam: a method for stochastic optimization. CoRR abs/1412.6980 (2014). http://arxiv.org/abs/1412.6980

8. Leuzzi, F., Ferilli, S. (eds.): 1st Italian Conference for the Traffic Police (TRAP-2017). Advances in Intelligent Systems and Computing. Springer, Heidelberg (2017). https://doi.org/10.1007/978-3-319-75608-0

9. Leuzzi, F., Del Signore, E., Ferranti, R.: Towards a pervasive and predictive traffic police. In: Leuzzi and Ferilli [8], pp. 19–36

10. Liao, J., Zhao, X., Tang, J., Zhang, C., He, M.: Efficient and accurate traffic flow prediction via fast dynamic tensor completion. In: Leuzzi and Ferilli [8], pp. 69–82

11. Lv, Y., Duan, Y., Kang, W., Li, Z., Wang, F.: Traffic flow prediction with big data: a deep learning approach. IEEE Trans. Intell. Transp. Syst. $16(2)$, 865–873 (2015). https://doi.org/10.1109/TITS.2014.2345663

12. Nanni, M.: Advancements in mobility data analysis. In: Leuzzi and Ferilli [8], pp. 11–16

13. Rodriguez-Jimenez, J.: Detecting criminal behaviour patterns in Spain and Italy using formal concept analysis. In: Leuzzi and Ferilli [8], pp. 57–68

14. Sutskever, I., Vinyals, O., Le, Q.V.: Sequence to sequence learning with neural networks. In: NIPS 2014, pp. 3104–3112 (2014)

SCUT-DS: Learning from Multi-class Imbalanced Canadian Weather Data

Olubukola M. Olaitan and Herna L. Viktor[✉]

School of Electrical Engineering and Computer Science, University of Ottawa,
Ottawa, ON, Canada
{oolai067,hviktor}@uottawa.ca

Abstract. Learning from multi-class imbalanced data streams with multiple minority classes, and varying degrees of skewed distributions, is an important problem in many real-world applications. However, to date, this aspect has received limited attention in the research community. Rather, the focus is on binary class problems or, alternatively, multi-class scenarios are decomposed into multiple binary sub-problems that are handled separately. Furthermore, the evolving nature of data streams make the task of correctly predicting minority instances challenging. In this paper, we introduce the SCUT-DS approach that combines multi-class synthetic oversampling and cluster-based under-sampling. SCUT-DS is a window-based method that balances the number of incoming instances of all classes directly, as the stream evolves. We present our experimental evaluation against a stream of Canadian weather data, with varying degree of skewed distributions and multiple classes. We demonstrate that our SCUT-DS algorithms consistently improve the recognition rates of the minority instances in this multi-class imbalanced setting. Our results are especially promising for difficult-to-learn minority classes, notably for predicting ice storms and glaze events.

Keywords: Class imbalance · Multi-class learning · Classification
Sampling · Online learning · Data streams · Adverse weather

1 Introduction

In numerous application domains such as weather monitoring, sensor networks, web logs and video surveillance, data are generated as streams. The distribution of instances of the classes in such streams may be significantly skewed and the number of class labels is often numerous. The task of learning in such multi-class imbalanced settings, where instances of some classes occur much more frequently than others, is arduous. In an evolving stream, this difficulty is further aggravated, due to the temporal and often interleaved rates of data arrival.

The inherent characteristics of data streams, namely, speed, volume, limited memory and time requirements, make it infeasible to store or to perform multi-scan analysis [2]. These factors contribute to making the prediction of minority

© Springer Nature Switzerland AG 2018
M. Ceci et al. (Eds.): ISMIS 2018, LNAI 11177, pp. 291–301, 2018.
https://doi.org/10.1007/978-3-030-01851-1_28

instances in skewed streams a difficult task. The above-mentioned learning difficulties increase when considering multi-class imbalanced data streams [13]. In most recent works regarding class imbalance, multi-class data sets are reduced into binary data sets either by combining one minority class with another class in the data set (OVO) or by combining one minority at a time with all the remaining classes in the data set (OVA) [13]. Such class decomposition makes it difficult to study the classes in their natural form, since the imbalanced ratio is increased, thus introducing potential bias [13].

In this paper, we present an approach that *directly* improve the recognition rates of the minority instances in multi-class imbalanced data stream, without class decomposition. Our SCUT-DS approach extends the SCUT algorithm [1], a multi-class imbalanced data sets classification approach for batch environments, using hybrid sampling techniques. Based on this observation, our SCUT-DS methodologies use a window-based sequencing evaluation approach, combining synthetic minority oversampling (SMOTE) and cluster-based under-sampling. This ensures that the models are built and updated using training sets that are representative of all sub-concepts and are balanced across all classes present. Our experimental evaluation against a Canadian weather case study confirms that our methodology have higher average recalls for the minority instances, most especially for the difficult-to-learn minority instances.

This paper is organized as follows. We discuss background concepts in Sect. 2, while the SCUT-DS methodologies are introduced in Sect. 3. In Sect. 4, our case study, experimental evaluation and results are presented. Section 5 concludes the paper and provides suggestions for future work.

2 Background

Related Work. There are few studies that directly address multi-class imbalanced data sets without decomposing the data sets into multiple binary-class data sets; more so, most of these studies are done in the batch environment. Related works include [5] where SMOTE was used to increase the minority instances. Also in SERA [3], MUSERA [4] and [6], previously seen minority instances were accumulated and used for augmenting the minority instances. The disadvantages of these methods are that the accumulation of all past minority instances might not be feasible, since a stream is assumed to be continuous. Secondly, selecting from these minority instance may overwhelm the system as their number increases. The computational time and resource requirements for storing, processing and selecting from all previously accumulated minority instances for the training will be expensive. Thirdly, the data sets used by these studies for experimenting are two-class in nature, while the data type was not disclosed in MuSERA. Another related work is WOS-ELM [10], which is a cost-sensitive method. It uses imbalance ratio to determine when to update the model. The use of imbalance ratio removed the need to wait for the window to be of a particular size before updating the model. It is a cost-sensitive approach; hence, its range of applicability may be reduced because domain knowledge is required to assign weights. Similarly, the algorithm was evaluated on a binary-class imbalanced data stream. The methods in [12] may be subjected to some extreme factors in highly imbalanced data set because the sampling techniques were not combined.

Factors that Make Learning Difficult. The poor predictive accuracy of existing algorithms on minority instances could be attributed to factors such as poor representation, overlapping classes and small disjuncts [8,9]. The sub-clusters in small disjuncts represent sub-concepts. These factors cause the training set to be lopsided because some classes are not well-represented. Poor representation or skewed distribution is known as 'between-class imbalance' while small disjuncts are referred to as 'within-class imbalance' [1].

Intuitively, data sets with between-class imbalance issue alone may not be that difficult to learn, because data regions do not overlap with the majority class such as in Fig. 1. Although instances with the red labels have poor representation, they are well separated from the other two classes and are likely to have high prediction accuracy using few training data, because there are no small disjuncts or overlap. However, if the instances of the minority class are dispersed and the class is besieged with both within-class and between-class imbalance as shown in Fig. 2, it may be difficult to obtain a highly accurate model. This is because the algorithms may not be presented with enough representative examples. Rather, instances may be categorized as belonging to the most-represented class in the region. For instance, instances of the blue class in Fig. 1, is likely to suffer from poor prediction accuracy, because it overlaps with a majority class. Also, when grouped into sub-clusters as presented in Fig. 2, the green class has more representatives in each of the sub-clusters. Based on these observations, we introduce the SCUT-DS approach for learning such minority concepts.

Fig. 1. Illustration of a well separated class with poor representation versus overlapping class with poor representation. (Color figure online)

Fig. 2. Illustration of class with poor representation and sub-concepts. (Color figure online)

3 SCUT-DS Methodologies

The two SCUT-DS variants, namely SCUT-DS+ and SCUT-DS++, extend the batch-setting SCUT algorithm [1], as discussed next.

3.1   SCUT Algorithm

The SCUT method systematically combines SMOTE-based oversampling of the minority classes, with cluster-based under-sampling of the majority classes, in order to address within-class and between-class imbalance in multi-class imbalanced static data without decomposing the data set [1]. Rather, SMOTE is applied on the minority classes, while cluster-based under-sampling is conducted on the majority classes after splitting the data sets using their classes. In the SCUT approach, the Expectation Maximization (EM) cluster analysis algorithm was used. EM is a model-based method that learns the soft clusters using a probabilistic mixture model where the probability to belonging to each cluster, or Gaussian distribution, is determined. The experimental evaluation show that the cluster-based under-sampling mechanism ensures that all data region are learned by selecting the training instances from all identified sub-concepts. Interested readers are referred to [1] for details of the over-sampling and under-sampling processes.

3.2   SCUT-DS Framework

The essence of the resampling technique in the SCUT-DS methodologies is to ensure that good representative and balanced training sets are presented to the classification algorithm. This prevents the construction of lopsided models. Similar to SCUT, the resampling technique addresses within and between class imbalances. The SCUT-DS methodologies are chunk-based and proceed as follows. Initially, when a particular window size is reached, the resampling algorithm is used to balance instances across classes in the chunk. The resampled chunk is used as the training set to build the model that will be used for predicting a new incoming chunk. The chunk that was used for testing will next be used to update the model, thus following an interleaved-test-then-train method [2].

Recall that the SCUT method employ the EM cluster analysis algorithm. EM is a soft cluster analysis algorithm, potentially leading to overlapping clusters. Also, the algorithm may result in numerous scans, while aiming to maximize the likelihood of instances belonging to a cluster. Multiple scans could potentially be computationally expensive in a streaming setting. In SCUT-DS, we therefore perform under-sampled using the k-means partitioning-based algorithm.

3.3   SCUT-DS+

SCUT-DS+ augments the minority instances by generating synthetic instances using SMOTE-based oversampling. Thus, for every chunk when a specified chunk size is reached, the resampling algorithm is applied. The sampling distribution, s, is calculated as the division of the chunk size, w, by the number of classes in the data set, n, i.e. $s = w/n$. The different samples of size s are combined before updating the model. SCUT-DS+ is illustrated in Fig. 3, where $n = 5$.

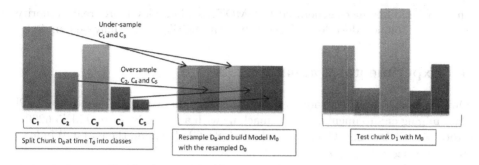

Fig. 3. SCUT-DS+ where the most recent chunk is used for sampling.

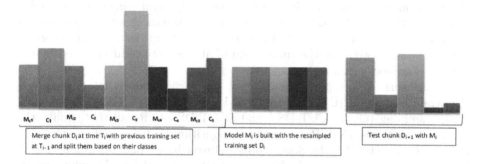

Fig. 4. SCUT-DS++ where a history of chunks are used for sampling.

3.4 SCUT-DS++ Framework

In SCUT-DS++, as shown in Fig. 4, the first number of chunks are accumulated to produce a larger training size. Thus, the differences between SCUT-DS+ and SCUT-DS++ are the accumulated chunks and how subsequent models are updated. To update the models in SCUT-DS++, the recent chunk is merged with the recent training instances and split according to their classes. The sampling distribution, s, is the number of instances in both the recent chunk (w) and the accumulated training data (t), divided by the number of classes in the chunk (n). The split and merged data sets are sampled based on the computed sampling distribution, s. Again, if the number of the merged instances is less than s, SMOTE-based oversampling is applied, while cluster-based under-sampling is applied for all majority classes. The reasoning behind this sampling approach is to ensure that there are sufficient training data and to prevent the situation whereby positive instances from some sub-concepts are not available in the training set, because they are not present in the recent chunk that are used to update the model. Note that the first variant, SCUT-DS+ is only reliant on the recently seen instances. SCUT-DS++ thus reduces the dependence on the synthetic instances generated by SMOTE and employ more 'real' minority instances. The pseudocode for SCUT-DS+ and SCUT-DS++ may be found at[1].

4   Experimental Evaluation

Our experiments were conducted by extending the WEKA [7] and MOA [2] data mining environments, using a machine with an Intel(R) Core(TM)i5-6200U processor, CPU @ 2.40 GHz processor with 8.0 GB RAM on the 64-bit Windows 10 Operating System (OS).

Our main goal in this paper is to improve the recognition rates of the minority instances in the Canadian weather case study. Recall is a single class measure, it gives the true picture of the ratio of the correct predictions per class by the classification algorithm [13]. In order to have an overall single metric and a measure that will be devoid of the effect of generalization due to the proportion of the majority classes, we selected the average recalls of the minority classes as the evaluation metric, i.e. the summation of all minority recalls divided by the number of minority classes.

In this work, we compared a number of approaches, namely interleaved-test-then-train evaluation (INTER), Random Under-sampling (RU), Cluster-based Under-sampling (CU), SCUT-DS+ and SCUT-DS++. Recall that, unlike the other four methods, INTER is a base-line algorithm that does not resample the training data. The CU algorithm applies cluster-based under-sampling on all the classes in order to resample the training data, while RU used random under-sampling. The sampling rates for both CU and RU were set to the size of the least represented class in the chunk.

The aims of our experimentation were four-fold. Our first aim was to explore a real-world Canadian weather data stream, as will be discussed below. Throughout this study, our goal was to assess the accuracy amongst five algorithms namely, SCUT-DS+, SCUT-DS++, INTER, CU and RU, in order to determine any benefits of using the SCUT-DS approaches. Also, we assessed the abilities of SCUT-DS+ versus SCUT-DS++ to generalize and to construct accurate models. Lastly, we compared the accuracy among six classification algorithms, namely Hoeffding Trees (HTs) [2], Naive Bayes (NB), and the OzaBag-HT, OzaBag-NB, OzaBoost-HT and OzaBoost-NB online ensembles [11], when utilizing the above-mentioned sampling/no-sampling methods.

4.1   Canadian Weather Data: Experimental Results

In this case study, we considered historical Canadian weather data that may be accessed through the Open Data Canada repository[2]. Here, the goal is to accurately predict the occurrence of adverse and dangerous weather conditions, in a timely manner. The subsets of data that were used for experimenting are

---

[1] https://github.com/BukkyOlaitan.
[2] http://climate.weather.gc.ca/historical_data/search_historic_data_e.html.

presented in Table 1. The reader will notice that the number of classes varies from three to five, and that the data are highly skewed, with data representing adverse weather conditions in the minority. In this experimental datasets, Fog, Ice and Snow are the minority classes while Cloudy and Clear are the majority classes. Ice storms (or so-called glaze events) result in at least 6.5 mm of ice on exposed surfaces. While these are infrequent occurrences, they often result in very dangerous driving conditions, power outages and flight cancellations. The same observation holds for adverse weather such as blizzards and heavy fog. To this end, our aim is to produce models to accurately predict these important minority classes accurately and in a timely manner. All parameters used were set through experimentation, with the chunk size w fixed at 1000, while the number of clusters was maintained at 5.

Table 1. The weather data characteristics.

Data sets	No of class	Distribution	Size
Cloudy-Clear-Ice	3	21146:27209:2581	50,936
Clear-Fog-Ice	3	27209:1557:2581	31,347
Cloudy-Snow-Ice	3	21146:4091:2581	27,818
Cloudy-Clear-Snow-Ice	4	21146:27209:4091:2581	55,027
Cloudy-Clear-Snow-Ice-Fog	5	21146:27209:4091:2581:1557	56,584

Table 2 shows the average recall rates obtained. As expected, the results clearly show the benefit of some form of sampling. That is, the base-line INTER method consistently lead to poorer minority class recalls. In summary, SCUT-DS+ and SCUT-DS++ yielded the most accurate results, when compared with CU and RU. (The only exception is for the Cloudy-Clear-Snow-Ice dataset, where CU on its own, together with OzaBag-NB and OzaBoost-HT, led to slightly higher results.)

These results may be attributed to the specific resampling techniques employed in SCUT-DS+ and SCUT-DS++. Specifically, the rebalancing of the training sets ensures that instances of the minority classes are adequately represented. Cluster-based under-sampling ensures that training instances were chosen from all identified sub-concepts.

The reader will notice that the accuracies of SCUT-DS+ and SCUT-DS++ are comparable. The improvement in SCUT-DS+ implies that some of the synthetic instances generated occupy newly discovered data space, which allowed for better generalization on the incoming stream. SCUT-DS++ relies on clustering and includes more real instances, making it highly dependent on the similarity between the accumulated past instances and those in recent chunk. From the perspective of the classification algorithms, it can be observed that the performance of the base learners and the ensembles are comparable, both for HTs and NBs classifiers. Additionally, there were improvement in the accuracies of ensembles algorithms compared to the HT and NB algorithms.

Table 2. Average minority recall rates of weather datasets.

Data set	Algorithm	HT	NB	OzaBag-HT	OzaBag-NB	OzaBoost-HT	OzaBoost-NB
		Average Recalls of Minority Classes in the Data set					
Cloudy-	INTER	0.2227	0.2565	0.2086	0.2619	0.2194	0.3060
Clear-	SCUT-DS+	**0.6599**	**0.6474**	**0.7090**	**0.6515**	**0.7065**	**0.6632**
Ice	SCUT-DS++	0.6490	0.5712	0.7052	0.5670	0.6803	0.5733
	RU	0.5275	0.5275	0.5100	0.5100	0.5212	0.5237
	CU	0.5983	0.5983	0.5874	0.5874	0.6074	0.6095
Clear-	INTER	0.5464	0.5711	0.5162	0.5706	0.5637	0.6039
Fog-	SCUT-DS+	**0.8261**	**0.7797**	**0.8593**	**0.7774**	0.7959	**0.7928**
Ice	SCUT-DS++	0.8114	0.7189	0.8236	0.7181	**0.8142**	0.7473
	RU	0.7577	0.7143	0.7634	0.7133	0.7737	0.7325
	CU	0.7984	0.7592	0.7955	0.7597	0.8014	0.7704
Cloudy-	INTER	0.5559	0.5668	0.5808	0.5653	0.5565	0.5964
Clear-	SCUT-DS+	0.7572	**0.7606**	0.7668	0.7610	0.7732	**0.7664**
Snow-	SCUT-DS++	**0.7844**	0.7320	**0.7898**	0.7325	0.7823	0.7408
Ice	RU	0.7437	0.7130	0.7356	0.7063	0.7666	0.7373
	CU	0.7746	0.7605	0.7756	**0.7617**	**0.7840**	0.7570
Cloudy-	INTER	0.6368	0.6225	0.6432	0.6232	0.6434	0.6506
Clear-	SCUT-DS+	0.8182	**0.7814**	0.8285	**0.7784**	0.8131	**0.7807**
Snow-	SCUT-DS++	**0.8495**	0.7523	**0.8517**	0.7518	**0.8307**	0.7759
Ice-	RU	0.7621	0.7219	0.7735	0.7211	0.7890	0.7413
Fog	CU	0.7863	0.7388	0.7949	0.7361	0.7760	0.7450
Cloudy-	INTER	0.7662	0.8307	0.7554	0.8312	0.7846	0.8226
Snow-	SCUT-DS+	0.8729	**0.8652**	**0.8772**	**0.8646**	**0.8848**	**0.8749**
Ice	SCUT-DS++	**0.8695**	0.8265	0.8641	0.8262	0.8651	0.8345
	RU	0.8491	0.8507	0.8497	0.8500	0.8724	0.8709
	CU	0.8536	0.8539	0.8489	0.8541	0.8687	0.8620

It may also be noticed that some minority labels are more difficult to predict than others, for example, the 'Ice' label. The percentage improvement in accuracy for SCUT-DS+ and SCUT-DS++, using INTER as the basis and compared to the other two techniques, are usually very high for data sets that contains such difficult-to-learn class. In the visualization graphs in Figs. 5, 6 and 7, the green represents the minority classes in each of these data sets. It can be noticed that of all the three minority instances in these graphs, the Ice class is the most dispersed. Although, all the three minority classes suffer poor representation, the Ice class as seen in Fig. 6 overlap indiscriminately with the majority classes. Their sub-clusters are not very good, when compared to the Fog and Snow classes in Figs. 5 and 7, respectively.

Fig. 5. Visualization of Cloudy-Clear-Fog Data Set. (Color figure online)

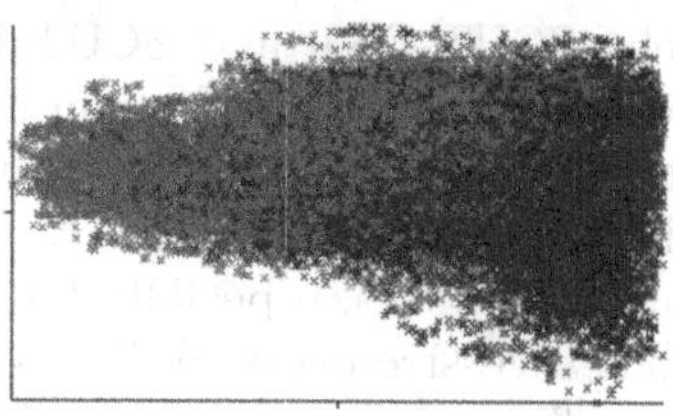

Fig. 6. Visualization of Cloudy-Clear-Ice Data Set. (Color figure online)

Fig. 7. Visualization of Cloudy-Clear-Snow Data Set. (Color figure online)

4.2 Discussion

The experimental results confirmed that the SCUT-DS+ and SCUT-DS++ approaches improved the average recognition rates of the minority instances. The most significant improvements were noticed in the individual recalls of the difficult-to-learn minority 'Ice' class, when compared to the individual recalls of the other two minority classes. The SCUT-DS methodologies address the factors that make imbalanced data classification difficult. The combination of under-sampling and oversampling addresses within-class and between-class imbalance and also prevented the excessive use of one sampling technique over the other, thus making the algorithms to be effective for highly imbalanced data sets and difficult-to-learn classes. The flexibility of the algorithms is also improved, because the average of the total instances in the classes is used as the resampling distribution rates. This removes the non-trivial task of determining the resam-pling size to use. Disjuncts, or within-class imbalances, are solved by cluster-based under-sampling and SMOTE-based oversampling. The problem of class overlap is partly addressed with cluster-based under-sampling, which reduces the number of instances lying at the class boundary. Cluster-based under-sampling aided in the selection of relevant minority instances instead of having to store all past seen minority examples, which could overburden the system.

5 Conclusion

This paper introduced the SCUT-DS methodology, a novel and flexible approach for effective learning in multi-class imbalanced data streams. The two variants, namely, SCUT-DS+ and SCUT-DS++, directly addressed multi-class imbalanced data stream classification using a hybrid sampling approach. The experimental evaluation against a stream of weather data confirmed that the SCUT-DS methodologies' algorithms consistently improved the recognition rates of the minority instances, especially for difficult-to-learn concepts. This was achieved without excessive oversampling, excessive under-sampling or the need to accumulate all previously seen minority instances. Our methods are sampling-based and thus their flexibility, and range of applicability, are extensive.

Learning of multi-class imbalanced data sets is an interesting problem which opens many opportunities for more research directions. Firstly, we plan to conduct a robust empirical experimental evaluation against multiple synthetic and real datasets. We will also consider highly skewed domains where the number of classes are high. Our framework may further be extended in a number of ways, notably in considering other cluster analysis methods and other evaluation metrics. It would further be worthwhile to consider the case when the majority in some chunks might become minority in other chunks, and vice versa. Finally, we plan to extend our case study by incorporating environmental and climate change indicators.

References

1. Agrawal, A., Viktor, H.L., Paquet, E.: SCUT: multi-class imbalanced data classification using SMOTE and cluster-based undersampling. In: 2015 7th International Joint Conference on Knowledge Discovery, Knowledge Engineering and Knowledge Management (IC3K), vol. 1, pp. 226–234. IEEE (2015)
2. Bifet, A., Holmes, G., Kirkby, R., Pfahringer, B.: MOA: Massive Online Analysis. J. Mach. Learn. Res. **11**, 1601–1604 (2010)
3. Chen, S., He, H.: SERA: selectively recursive approach towards nonstationary imbalanced stream data mining. In: International Joint Conference on Neural Networks, IJCNN 2009, pp. 522–529. IEEE (2009)
4. Chen, S., He, H., Li, K., Desai, S.: MUSERA: multiple selectively recursive approach towards imbalanced stream data mining. In: The 2010 International Joint Conference on Neural Networks (IJCNN), pp. 1–8. IEEE (2010)
5. Ditzler, G., Polikar, R., Chawla, N.: An incremental learning algorithm for nonstationary environments and class imbalance. In: 2010 20th International Conference on Pattern Recognition (ICPR), pp. 2997–3000. IEEE (2010)
6. Gao, J., Ding, B., Fan, W., Han, J., Philip, S.Y.: Classifying data streams with skewed class distributions and concept drifts. IEEE Internet Comput. **12**(6), 37–49 (2008)
7. Hall, M., Frank, E., Holmes, G., Pfahringer, B., Reutemann, P., Witten, I.H.: The WEKA data mining software: an update. ACM SIGKDD Explor. Newsl. **11**(1), 10–18 (2009)
8. He, H., Garcia, E.A.: Learning from imbalanced data. IEEE Trans. Knowl. Data Eng. **21**(9), 1263–1284 (2009)
9. Jo, T., Japkowicz, N.: Class imbalances versus small disjuncts. ACM SIGKDD Explor. Newsl. **6**(1), 40–49 (2004)
10. Mirza, B., Lin, Z., Toh, K.-A.: Weighted online sequential extreme learning machine for class imbalance learning. Neural Process. Lett. **38**(3), 465–486 (2013)

11. Oza, N.C.: Online bagging and boosting. In: 2005 IEEE International Conference on Systems, Man and Cybernetics, vol. 3, pp. 2340–2345. IEEE (2005)
12. Wang, S., Minku, L.L., Yao, X.: Dealing with multiple classes in online class imbalance learning. In: IJCAI, pp. 2118–2124 (2016)
13. Wang, S., Yao, X.: Multiclass imbalance problems: analysis and potential solutions. IEEE Trans. Syst. Man Cybern. Part B (Cybern.) **42**(4), 1119–1130 (2012)

An Efficient Algorithm for Network Vulnerability Analysis Under Malicious Attacks

Toni Mancini$^{(\boxtimes)}$, Federico Mari, Igor Melatti, Ivano Salvo, and Enrico Tronci

Computer Science Department, Sapienza University of Rome, Rome, Italy
tmancini@di.uniroma1.it

Abstract. Given a communication network, we address the problem of computing a lower bound to the transmission rate between two network nodes notwithstanding the presence of an intelligent malicious attacker with limited destructive power.

Formally, we are given a link capacitated network N with source node s and destination node t and a budget B for the attacker.

We want to compute the Guaranteed Maximum Flow from s to t when an attacker can remove at most B edges. This problem is known to be NP-hard for general networks.

For Internet-like networks we present an efficient ILP-based algorithm coupled with instance transformation techniques that allow us to solve the above problem for networks with more than 200 000 nodes and edges within a few minutes. To the best of our knowledge this is the first time that instances of this size for the above problem have been solved for Internet-like networks.

1 Introduction

Given a communication network, especially an Internet sub-network, we are often interested in analysing its vulnerability to intelligent malicious attacks, consisting in removal (destruction) of network connections.

More specifically, given a communication network, we are interested in computing a lower bound to the transmission rate between two network nodes (*e.g.*, between a server and a router), notwithstanding the presence of a malicious attacker. Such attacker has limited *destructive power* B, but is intelligent enough to compute the maximum damage it can be provoked with B.

If, as usual, we model a communication network as a graph with edge capacities, the above problem becomes that of computing the Guaranteed Maximum Flow (GMF) when a malicious attacker can remove edges from the graph.

Of course, if an attacker can remove *any* number of edges, nothing can be guaranteed, that is the GMF is just zero. However, fortunately, even malicious attackers have a limited, possibly large, budget (*e.g.*, destructive power). This has motivated research on the computation of the GMF with a limited budget for the attacker. We will refer to this problem as *Network Interdiction Problem (NIP)* [34,39].

© Springer Nature Switzerland AG 2018
M. Ceci et al. (Eds.): ISMIS 2018, LNAI 11177, pp. 302–312, 2018.
https://doi.org/10.1007/978-3-030-01851-1_29

1.1   Motivations

Many networking problems can be cast as NIP (for a complete survey, we refer the reader to [36]). For example, as for network security, the following problems have been cast as NIPs: DOS attacks [2], DOS-resistant authentication [3], and insider threat analysis [8]. Vulnerability analysis of infrastructural networks (such as electric power [9,24,26], water supply, critical infrastructure networks [32], etc.) can also be cast as a NIP [4,12,33]. Finally, NIP plays a role also in vulnerability-aware design of: communication networks [13], circuits [40], operating systems [37], and infrastructural networks [35].

1.2   Contributions

Unfortunately, NIP has been shown to be an NP-complete problem [34,39]. Nevertheless, many interesting NIPs can be solved casting NIP as an Integer Linear Programming (ILP) problem [39].

Here we focus on analysing vulnerability of meaningful Internet sub-networks, when some of their connections may be removed by an attacker. This entails solving NIPs for networks with hundreds of thousands of edges.

We propose an ILP-based algorithm to solve *very large* instances of the NIP which transforms the input instances in order to exploit the statistical structure of Internet-like networks.

Our experimental results (Sect. 4) show that, thanks to our NIP instance transformation techniques, our algorithm is able to analyse vulnerabilities of networks consisting of around 200 000 nodes in *a few minutes*. To the best of our knowledge, none of the previously proposed methods for NIP can handle Internet-like graphs of such size.

1.3   Paper Outline

The paper is organised as follows. In Sect. 2 we give some preliminaries and formally state our Network Interdiction Problem. In Sect. 3 we outline our NIP instance transformation techniques which are the key enabler for the efficient vulnerability analysis of very large Internet-like networks using an ILP-based approach, along the lines of [39]. In Sect. 4 we show the effectiveness of our instance transformation techniques for the vulnerability analysis of realistic Internet-like networks of very large size.

2   Preliminaries

In the following we denote with $\mathbb{R}^+$ and $\mathbb{N}^+$ the set of positive real and natural numbers, respectively.

2.1   Capacitated Networks

Here we recall the standard definitions of capacitated networks, flow, and maximum flow.

Definition 1 (Capacitated network). *A directed [undirected] capacitated network is a tuple $N = (V, E, s, t, c)$ where (V, E) is a directed [undirected] graph, $s, t \in V$ ($s \neq t$) are, resp., the* source *and* destination *nodes and $c : E \to \mathbb{R}^+$ defines capacities of the links E. The capacity of a link (u, v) will be denoted by $c(u, v)$ or equivalently by c_{uv}.*

In the following we will use the shorter term *network* to refer to a capacitated network as defined in Definition 1.

Definition 2 (Network flow). *A flow in a network $N = (V, E, s, t, c)$ is a function $f : E \to \mathbb{R}^+$ s.t. $\forall(i, j) \in E$ $f(i, j) \leq c_{ij}$ and, $\forall j \in (V \setminus \{s, t\})$, $\sum_{i \in V | (i,j) \in E} f(i, j) = \sum_{i \in V | (j,i) \in E} f(j, i)$.*

Definition 3 (Network flow value). *The value of a flow f in a network $N = (V, E, s, t, c)$ is $\mathrm{val}(f) = \sum_{v \in V | (s,v) \in E} f(s, v) = \sum_{v \in V | (v,t) \in E} f(v, t)$.*

Definition 4 (Network maximum flow). *The maximum flow of a network N is $\mathrm{MaxFlow}(N) = \max\{\mathrm{val}(f) | f \text{ is a flow in } N\}$.*

2.2   Network Interdiction Problem

Here we formally define the problem we are interested in.

Definition 5 (Network Interdiction Problem, NIP). *A Network Interdiction Problem, (NIP) is a triple (N, B) where $N = (V, E, s, t, c)$ is a network (directed or undirected) and B is a positive integer number (attacker budget).*
    An (N, B) attack is a map α from E to the set $\{0, 1\}$ s.t. $\sum_{(u,v) \in E} \alpha(u, v) \leq B$. We denote with $N|_\alpha$ the network (V, E, s, t, c') where $c'(u, v) = (1 - \alpha(u, v))c(u, v)$.
    A solution to the NIP (N, B) is an (N, B) attack α s.t. , for all (N, B) attacks β, $\mathrm{MaxFlow}(N|_\alpha) \leq \mathrm{MaxFlow}(N|_\beta)$.
    If α is a solution to the NIP (N, B), we call $\mathrm{MaxFlow}(N|_\alpha)$ the Guaranteed Maximum Flow (GMF) of (N, B) (notation: $\mathrm{GMF}(N, B)$).

In [39] it is proven that NIP is NP-hard. Various variants of the NIP have been proposed, envisioning, for example, the possibility for an attacker to reduce the capacity of edges (rather than destroying them completely), or different costs to destroy different edges (see, *e.g.*, [34,39]). In this paper, for space limitations and ease of presentation we focus on the core version of the problem. However, our approach can be generalised to most such variants.

2.3   An ILP Formulation Approach to NIP

Following the approach in [39], the NIP (N, B) for undirected networks $N = (V, E, s, t, c)$ can be formulated as the following Integer Linear Programming (ILP) problem.

$$\text{minimize} \sum_{(u,v)\in E} c_{uv} y_{uv}$$

$$\text{subject to:} \sum_{(u,v)\in E} z_{uv} \leq B$$

$$
\begin{aligned}
y_{uv} + z_{uv} &\geq d_u - d_v & \forall(u, v) \in E \quad (1) \\
y_{uv} + z_{uv} &\geq d_v - d_u & \forall(u, v) \in E \\
d_s = 0, d_t = 1, d_u &\in \{0, 1\} & \forall u \in V \setminus \{s, t\} \\
y_{uv}, z_{uv} &\in \{0, 1\} & \forall(u, v) \in E
\end{aligned}
$$

The main idea underneath such ILP problem is to compute a (s, t)-cut of N. Namely, the cut is identified by decision variables d_u (with value 1 if $u \in V$ is in the s partition and 0 otherwise) and y_{uv} (with value 1 if edge $(u, v) \in E$ crosses the cut and 0 otherwise).

Given this, the attack α, which is the solution to the instance of the NIP at hand, is defined as $\alpha(u, v) = z_{uv}$, for all $(u, v) \in E$, and the resulting GMF is MaxFlow($N|_\alpha$).

Note that, if the network N is directed, we just drop the set of constraints (1), as in directed networks flow goes only from s to t.

3   An Efficient NIP Algorithm on Internet-Like Networks

A direct use of state-of-the-art Integer Linear Programming (ILP) solvers (e.g., GLPK, www.gnu.org/software/glpk, or CPLEX, www.ilog.com/products/cplex) to perform network vulnerability analysis allows us to tackle input network graphs consisting of at most a few tens of thousands of nodes. For very-large networks, the main focus of this paper, this direct approach is unviable.

In this section we present proper *instance transformation* techniques aimed at solving the Network Interdiction Problem (NIP) on very large network instances.

We point out that our goal is *not* to select classes of networks for which NIP is polynomial, but rather to *effectively* solve the NIP for the class of networks of our interest, *i.e.*, Internet-like networks. In particular, Internet-like networks typically do *not* fall in the classes of networks for which NIP is known to be polynomial, see, *e.g.*, [34]. Hence, our approach is much in the spirit with which Mixed Integer Linear Programming (MILP) and ILP solvers (see, *e.g.*, [14, 15, 28]), model checkers (see, *e.g.*, [5, 17, 18, 20–23, 25, 27, 38]), controllers synthesizers (see, *e.g.*, [1, 6, 19, 29]), local search–based (see, *e.g.*, [10, 16, 31]), SAT and SMT solvers (see, *e.g.*, [30]) are built and exploited to solve large problem instances. Of course, being our problem NP-hard, our approach (if P $\neq$ NP) will still result in an exponential algorithm in the worst case. However, experimental results in Sect. 4 undoubtedly show that, by applying the instance transformation techniques outline next, *off-the-shelf* ILP solvers can be made able to perform vulnerability analysis of networks with 200 000 nodes in a just a few of minutes, with dramatic scalability improvements and huge speedups wrt. to a direct ILP encoding.

3.1 Structure of Internet-Like Networks

An *Internet-like* network is a network with a structure similar to the Internet graphs. In such networks, the distribution of node out-degrees (*i.e.*, the numbers of outgoing edges per node) is widely dispersed around the average value, in that most of the nodes have an out-degree much smaller than the average out-degree, and a few nodes (*hubs*) have an out-degree much larger than the average out-degree.

Our algorithm applies, in the given order, the following three network transformations to the input NIP instance. Such transformations are inspired to general networks processing techniques (see, *e.g.*, [7]), but have been carefully chosen and adapted to exploit the typical structure of Internet-like networks.

3.2 Phase 1: Connected Component Selection

Given a NIP instance (N, B) over network $N = (V, E, s, t, c)$, our first step transforms it in (N', B), where the new network $N' = (V', E', s, t, c')$ is the single connected component of N containing both the source node s and the destination node t, and $c'(u, v) = c(u, v)$ for $(u, v) \in E'$.

Of course, in case s and t belong to two different connected components of N, the problem answer can be immediately computed, as the network (GMF) is just zero.

3.3 Phase 2: Detour Elimination

Our second instance transformation step removes all *detour edges*, *i.e.*, edges which are *only* contained in cycles.

Formally, given our NIP instance (N', B) (as computed by the previous transformation step) with $N' = (V', E', s, t, c')$, we transform it in (N'', B), where network $N'' = (V'', E'', s, t, c'')$, with V'' induced by E'', $c''(u, v) = c'(u, v)$ for $(u, v) \in E''$ and:

$$E'' = \left\{ (u, v) \in E' \;\middle|\; \begin{array}{l} \text{exists a } (s, t)\text{-path in } N'' \text{ which} \\ \text{is not a cycle and contains edge } (u, v) \end{array} \right\}.$$

3.4   Phase 3: Chains Compaction

Our third and final transformation step focuses on maximal *chains*, *i.e.*, on the $\subseteq$-maximal non-singleton sets of nodes $\{v_1, \ldots, v_r\} \in V''$ ($r \geq 3$) of network N'' such that, for each $1 \leq i < r$, $(v_i, v_{i+1}) \in E''$ and, for each $1 < i < r$ the out-degree of node v_i in N'' is 1. Our algorithm transforms network N'' by compacting each $\subseteq$-maximal chain $C = \{v_1, \ldots, v_r\} \subseteq V''$ in a single edge $(v_{\tilde{\jmath}(C)}, v_{\tilde{\jmath}(C)+1})$, being $\tilde{\jmath}(C) \in \{1, \ldots, r-1\}$ s.t. $(v_{\tilde{\jmath}(C)}, v_{\tilde{\jmath}(C)+1})$ is the edge with the minimum capacity of all the edges in the removed chain and $\tilde{\jmath}(C)$ is the minimum index with such property (in case the minimum is attained in more than one edge of the chain).

More formally, the compacted network $N''' = (V''', E''', s, t, c''')$ is such that V''' is induced by E''', $c'''(u, v) = c''(u, v)$ for $(u, v) \in E'''$ and:

$$E''' = (E'' \setminus \{(v_i, v_{i+1}) \in E'' \mid \text{exists a } \subseteq\text{-maximal chain } C = \{v_1, \ldots, v_r\} \subseteq V'',$$
$$1 \leq i < r, i \neq \tilde{\jmath}(C)\})$$
$$\cup\{(v, v_{\tilde{\jmath}(C)}) \mid \text{exists a } \subseteq\text{-max. chain } C = \{v_1, \ldots, v_r\} \subseteq V'', (v, v_1) \in E''\}$$
$$\cup\{(v_{\tilde{\jmath}(C)}, v) \mid \text{exists a } \subseteq\text{-max. chain } C = \{v_1, \ldots, v_r\} \subseteq V'', (v_r, v) \in E''\}$$

Our problem instance (N'', B) (output of the previous phase) is transformed into (N''', B).

Applying the above sequence of transformations to the input instance (N, B) at hand results in a new problem instance (N''', B). The following result shows that from a (optimal) solution for the latter we can efficiently compute (in linear time in the network size) a (optimal) solution to the former (proof omitted for lack of space).

Theorem 1. *Let (N, B) be an instance of a NIP, with $N = (V, E, s, t, c)$. Moreover, let (N''', B) be the NIP instance computed by applying the three transformations above to (N, B), with $N''' = (V''', E''', s, t, c''')$.*

Let α be a (optimal) solution to (N''', B). Then, $\beta(u, v) = \alpha(u, v)$ if $(u, v) \in E'''$ else 0 is a (optimal) solution to (N, B).

Summing up, our instance transformation techniques allow us to solve a NIP instance (N, B) by reducing it to (N''', B), by solving it through the ILP of Sect. 2.2, and by efficiently computing back an optimal solution to the original problem from the optimal solution computed for the latter.

4   Experiments

In order to assess effectiveness of the Network Interdiction Problem (NIP) instance transformation techniques of Sect. 3, below we report the experimental results obtained on very large sub-networks of the Internet. Our goal is to assess the *marginal impact* of our instance transformation techniques on the performance of an off-the-shelf Integer Linear Programming (ILP) solver.

All our experiments have been run on a 3GHz Intel Xeon QuadCore Linux PC with 8GB of RAM. Our instance transformation algorithms have been implemented in C. Our ILP solver is the well-known GLPK (www.gnu.org/software/glpk).

4.1 Defining NIP Benchmark Instances

Our first step is to generate realistic Internet-like networks of various sizes to be used as instances for our experiments. We started from an Internet snapshot G downloaded from [11]. Although G is a directed graph, for the aims of our experiments, we read edges of G as undirected (*i.e.*, bidirectional) edges.

Starting from G, we generated 20 sub-graphs G_e (where $e \in [1, 20]$) by randomly removing from G 20 given numbers of edges. Figure 1 shows statistical properties of our generated instances: for each graph, the figure shows its number of nodes and edges. Note that, for each $i \in [1, 19]$, graph G_{i+1} has both a higher number of nodes and a higher number of edges than graph G_i.

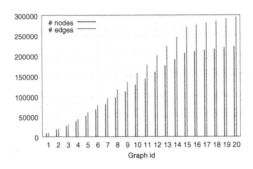

Fig. 1. Number of nodes and edges for our graphs $G_1, \ldots, G_{20}$.

The step above only defines nodes and edges for our graphs. In order to translate them in networks, we have to define, for each graph, the edges capacity function c and the source and destination nodes s, t. As for c, we define $c(u, v)$ to be proportional to the degree of u ($\deg(u)$) and v ($\deg(v)$). More specifically, for each graph, we set $c(u, v) = 1000 \frac{\deg(u) + \deg(v)}{\text{maxdeg}}$, where maxdeg is the maximum degree of a node in the graph. As for s and t, we randomly choose two distinct nodes among those having maximum degree in each graph.

As a result, we obtain 20 NIP instances. For each instance, value for B, defining the maximum number of links that the attacker can destroy (the attacker *budget*), has been fixed to the largest value for which the Guaranteed Maximum Flow (GMF) is greater than zero, thus placing us in the worst case (where the attacker has maximum freedom).

4.2 Experimental Results

We are now ready to assess the *marginal impact* of our instance transformation techniques on the performance of the GLPK ILP solver.

We solved each instance both by *directly* running the ILP problem defined in Sect. 2.2 (we call it the Direct NIP, dir-NIP, approach) and by first transforming the instance using our algorithm and then solving the ILP problem on the transformed instance (we call it the Transformed NIP, trans-NIP, approach).

For our comparisons we will focus on computation time, as memory requirements are always low (maximum 1 GB of RAM). Furthermore, as in our networks
the number of edges is nearly linear in the number of nodes (Fig. 1), we will measure computation times as a function of the number of network nodes only.

Figure 2a shows the *dramatic* improvements that are *consistently* obtained
by our approach (on *all* instances) based on applying the transformations of
Sect. 3 before running our off-the-shelf ILP solver (trans-NIP). In the figure, the
x axis shows the number of nodes of the networks $G_1, \ldots, G_{20}$ of Fig. 1, and the
y axis shows (in a logarithmic scale) the CPU computation time for trans-NIP
and dir-NIP on each network.

Note that trans-NIP succeeds in completing the analysis and computing the
GMF for *all* networks in our dataset in just *a few minutes*. On the contrary, dir-
NIP only succeeds in computing (in times which are up to *2 orders of magnitude
longer* than those of trans-NIP) the GMF for networks $G_1, \ldots, G_8$, which all
have less than 100 000 nodes. On larger networks, GLPK did not even terminate
after 24 h (our timeout).

Note that, while CPU times for dir-NIP are just ILP solving times, CPU times
for trans-NIP include both instance transformation and ILP solving times on the
transformed networks. To assess the impact of our instance transformations in
the overall trans-NIP computation time, Fig. 2b shows a breakdown (for each
instance) of the total trans-NIP solving time into instance transformation and
ILP solving time. The figure reveals that instance transformation indeed takes
negligible time wrt. ILP solving.

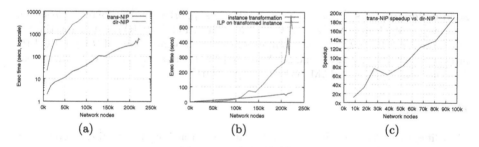

Fig. 2. (a) trans-NIP and dir-NIP computation times (seconds). (b) trans-NIP computation times breakdown. (c) trans-NIP speedup wrt. dir-NIP. All plots are drawn as
functions of the number of nodes of our benchmark networks.

Finally, Fig. 2c shows the average speedup of trans-NIP wrt. dir-NIP on each
network solved by both approaches (*i.e.*, $G_1, \ldots, G_8$). From the figure, we see
that the speedups achieved by trans-NIP grow almost linearly with the number
of nodes of the graph. Namely, the average speedup is at least $10\times$ (for network
G_1) and at most $190\times$ (for network G_8), with an average of approximately $90\times$.
This shows the feasibility and effectiveness of our approach on large and realistic
Internet-like networks.

5   Conclusions

In this paper we focused on the Network Interdiction Problem (NIP) and presented effective instance transformation techniques to compute the Guaranteed Maximum Flow (GMF) of very large meaningful Internet-like networks, when an intelligent malicious attacker with limited resources tries to destroy network edges. Namely, our methodology adds suitable network transformations explicitly tailored to Internet-like networks to standard Integer Linear Programming (ILP)-based algorithms. Our experimental results show that our instance transformations are able to enable off-the-shelf ILP solvers (GLPK) to successfully analyse vulnerabilities of Internet sub-networks consisting of 200 000 nodes in just a few minutes, whilst a direct ILP application does not scale beyond around 100 000 nodes. In all our test cases, our methodology enables much faster analysis. In particular, on small networks, speedup wrt. direct ILP solving is at least $10\times$, growing up to $190\times$ on the largest tested portions of Internet snapshots (consisting of around 200 000 nodes and 300 000 edges). On average, our algorithm is $90\times$ faster than direct ILP solving. This shows the effectiveness of our approach.

Our instance transformation techniques are *explicitly designed* to exploit the *statistical properties* of Internet-like networks and might not by so effective on networks with a very different statistical structure. Finally, although we addressed the core version of the problem, our techniques can be generalised to most of the existing problem extensions and variants, as they mostly focus on the network structure.

Acknowledgements. This work was partially supported by the following research projects/grants: Italian Ministry of University & Research (MIUR) grant "Dipartimenti di Eccellenza 2018–2022" (Dept. Computer Science, Sapienza Univ. of Rome); EC FP7 project SmartHG (Energy Demand Aware Open Services for Smart Grid Intelligent Automation, 317761); INdAM "GNCS Project 2018".

References

1. Alimguzhin, V., Mari, F., Melatti, I., Salvo, I., Tronci, E.: Linearizing discrete time hybrid systems. IEEE TAC **62**(10), 5357–5364 (2017)
2. Aura, T., Bishop, M ., Sniegowski, D.: Analyzing single-server network inhibition. In: Proceedings of CSFW 2000, p. 108. IEEE (2000)
3. Aura, T., Nikander, P., Leiwo, J.: DOS-resistant authentication with client puzzles. In: Christianson, B., Malcolm, J.A., Crispo, B., Roe, M. (eds.) Security Protocols 2000. LNCS, vol. 2133, pp. 170–177. Springer, Heidelberg (2001). https://doi.org/10.1007/3-540-44810-1_22
4. Church, R.L., Scaparra, M.P., Middleton, R.S.: Identifying critical infrastructure: the median and covering facility interdiction problems. Ann. Assoc. Am. Geogr. **94**(3), 491–502 (2004)
5. Clarke, E.M., Grumberg, O., Peled, D.A.: Model Checking. MIT Press (1999)
6. Della Penna, G., Intrigila, B., Magazzeni, D., Melatti, I., Tronci, E.: CGMurphi: automatic synthesis of numerical controllers for nonlinear hybrid systems. Eur. J. Control **19**(1), 14–36 (2013)

7. Evans, J.: Optimization Algorithms for Networks and Graphs. Routledge (2017)
8. Ha, D., Upadhyaya, S., Ngo, H., Pramanik, S., Chinchani, R., Mathew, S.: Insider threat analysis using information-centric modeling. In: Craiger, P., Shenoi, S. (eds.) DigitalForensics 2007. ITIFIP, vol. 242, pp. 55–73. Springer, New York (2007). https://doi.org/10.1007/978-0-387-73742-3_4
9. Hayes, B.P., Melatti, I., Mancini, T., Prodanovic, M., Tronci, E.: Residential demand management using individualised demand aware price policies. IEEE Trans. Smart Grid 8(3), 1284–1294 (2017)
10. Hoos, H.H., Stützle, T.: Stochastic Local Search: Foundations and Applications. Elsevier (2004)
11. The Internet Mapping Project: http://www.cheswick.com/ches/map/
12. Jeong, H.S., Qiao, J., Abraham, D.M., Lawley, M., Richard, J.-P., Yih, Y.: Minimizing the consequences of intentional attack on water infrastructure. Comp.-Aided Civil Infrastructure Eng. 21, 79–92 (2006)
13. Korkmaz, T., Krunz, M.: Multi-constrained optimal path selection. In: Proceedings of INFOCOM 2001, pp. 834–843 (2001)
14. Lin, Y., Austin, L.M., Burns, J.R.: An intelligent algorithm for mixed-integer programming models. Comp. Oper. Res. 19(6), 461–468 (1992)
15. Mancini, T.: Now or Never: negotiating efficiently with unknown or untrusted counterparts. Fundam. Inform. 149(1–2), 61–100 (2016)
16. Mancini, T., Flener, P., Pearson, J.: Combinatorial problem solving over relational databases: view synthesis through constraint-based local search. In: Proceedings of SAC 2012. ACM (2012)
17. Mancini, T., Mari, F., Massini, A., Melatti, I., Merli, F., Tronci, E.: System level formal verification via model checking driven simulation. In: Sharygina, N., Veith, H. (eds.) CAV 2013. LNCS, vol. 8044, pp. 296–312. Springer, Heidelberg (2013). https://doi.org/10.1007/978-3-642-39799-8_21
18. Mancini, T., et al.: Computing personalised treatments through in silico clinical trials. A case study on downregulation in assisted reproduction. In: Proceedings of RCRA 2018 (2018)
19. Mancini, T., Mari, F., Massini, A., Melatti, I., Salvo, I., Tronci, E.: On minimising the maximum expected verification time. IPL 122, 8–16 (2017)
20. Mancini, T., Mari, F., Massini, A., Melatti, I., Tronci, E.: Anytime system level verification via random exhaustive hardware in the loop simulation. In: Proceedings of DSD 2014. IEEE (2014)
21. Mancini, T., Mari, F., Massini, A., Melatti, I., Tronci, E.: System level formal verification via distributed multi-core hardware in the loop simulation. In: Proceedings of PDP 2014. IEEE (2014)
22. Mancini, T., Mari, F., Massini, A., Melatti, I., Tronci, E.: Anytime system level verification via parallel random exhaustive hardware in the loop simulation. Microprocess. Microsyst. 41, 12–28 (2016)
23. Mancini, T., Mari, F., Massini, A., Melatti, I., Tronci, E.: SyLVaaS: system level formal verification as a service. Fundam. Inform. 149(1–2), 101–132 (2016)
24. Mancini, T., et al.: Demand-aware price policy synthesis and verification services for smart grids. In: Proceedings of SmartGridComm 2014. IEEE (2014)
25. Mancini, T., et al.: Parallel statistical model checking for safety verification in smart grids. In: Proceedings of SmartGridComm 2018. IEEE (2018)
26. Mancini, T., et al.: User flexibility aware price policy synthesis for smart grids. In: Proceedings of DSD 2015. IEEE (2015)

27. Mancini, T., Tronci, E., Salvo, I., Mari, F., Massini, A., Melatti, I.: Computing biological model parameters by parallel statistical model checking. In: Ortuño, F., Rojas, I. (eds.) IWBBIO 2015. LNCS, vol. 9044, pp. 542–554. Springer, Cham (2015). https://doi.org/10.1007/978-3-319-16480-9_52

28. Mancini, T., et al.: Optimal fault-tolerant placement of relay nodes in a mission critical wireless network. In: Proceedings of RCRA 2018 (2018)

29. Mari, F., Melatti, I., Salvo, I., Tronci, E.: Model based synthesis of control software from system level formal specifications. ACM TOSEM **23**(1), 6 (2014)

30. Marques-Silva, J., Malik, S.: Propositional SAT solving. Handbook of Model Checking, pp. 247–275. Springer, Cham (2018). https://doi.org/10.1007/978-3-319-10575-8_9

31. Michel, L., Van Hentenryck, P.: Constraint-based local search. In: Martí, R., Panos, P., Resende, M. (eds.) Handbook of Heuristics, pp. 1–38. Springer, Cham (2017). https://doi.org/10.1007/978-3-319-07153-4_7-1

32. Murray, A.T., Grubesic, T.H.: Critical infrastructure protection: the vulnerability conundrum. Telemat. Inf. **29**(1), 56–65 (2012)

33. Phillips, C., Painton Swiler, L.: A graph-based system for network-vulnerability analysis. In: Proceedings of NSPW 1998, pp. 71–79. ACM (1998)

34. Phillips, C.A.: The network inhibition problem. In: Proceedings of STOC 1993, pp. 776–785. ACM (1993)

35. Shen, S.: Optimizing designs and operations of a single network or multiple interdependent infrastructures under stochastic arc disruption. Comput. Oper. Res. **40**(11), 2677–2688 (2013)

36. Smith, J.C., Prince, M., Geunes, J.: Modern network interdiction problems and algorithms. In: Pardalos, P.M., Du, D.-Z., Graham, R.L. (eds.) Handbook of Combinatorial Optimization, pp. 1949–1987. Springer, New York (2013). https://doi.org/10.1007/978-1-4419-7997-1_61

37. Tadayon, B., Smith, J.C.: Algorithms and complexity analysis for robust single-machine scheduling problems. J. Scheduling **18**(6), 575–592 (2015)

38. Tronci, E., et al.: Patient-specific models from inter-patient biological models and clinical records. In Proceedings of FMCAD 2014. IEEE (2014)

39. Wood, R.K.: Deterministic network interdiction. Math. Comp. Mod. **17**(2), 1–18 (1993)

40. Xiao, Y., Thulasiraman, K., Xue, G.: Constrained shortest link-disjoint paths selection: a network programming based approach. IEEE Trans. Circ. Sys. **53**(5), 1174–1187 (2006)

Social Data Analysis

An Instrumented Methodology to Analyze and Categorize Information Flows on Twitter Using NLP and Deep Learning: A Use Case on Air Quality

B. Juanals[1,2] and J. L. Minel[3(✉)]

[1] Centre Norbert Elias, Aix Marseille University - CNRS - EHESS, Marseille, France
[2] UMI IGlobes, CNRS - University of Arizona, Tucson, USA
[3] MoDyCo, University Paris Nanterre - CNRS, Nanterre, France
jean-luc.minel@u-paris10.fr

Abstract. This article focuses on the development of an instrumented methodology for modeling and analyzing the circulation message flows concerning air quality on the social network Twitter. This methodology aims at describing and representing, on the one hand, the modes of circulation and distribution of message flows on this social media and, on the other hand, the content exchanged between stakeholders. To achieve this, we developed Natural Language Processing (NLP) tools and a classifier based on Deep Learning approaches in order to categorize messages from scratch. The conceptual and instrumented methodology presented is part of a broader interdisciplinary methodology, based on quantitative and qualitative methods, for the study of communication in environmental health. A use case of air quality is presented.

Keywords: Air quality · Instrumented methodology
Circulation of information · Deep learning · Social network · Twitter

1 Introduction

This article proposes the development of an instrumented methodology for modeling and analyzing the flow of messages about air quality on the Twitter social network platform. In the areas of health and the environment, organizations' commitment to digital media and the social web is one of the new forms of mediation and media coverage developed for the public. Organizations have incorporated a social media activity into their communication policies; they follow the evolutions of the media practices of the public. However, the use of the socio-digital networks raises new methodological problems related to the description and analysis of new information-communication practices. They concern the specificities of the editorialization of information on these devices, the volume of messages and the interactions they allow.

© Springer Nature Switzerland AG 2018
M. Ceci et al. (Eds.): ISMIS 2018, LNAI 11177, pp. 315–322, 2018.
https://doi.org/10.1007/978-3-030-01851-1_30

The purpose of our work is to conceive, by relying jointly on methods anchored in social sciences and digital humanities, a representation of the modes of circulation and distribution of message flows on Twitter in relation with stakeholders and the content exchanged. Twitter was chosen because, besides being a widely used social media, it offers the possibility, through an API, of collecting messages from a set of user accounts from associated hashtags, which cannot be done on other social media platforms (Facebook, Instagram). The methodology is based on the contribution of work on the communication of organizations on health and environmental issues on the social web [17,19], social media platforms and media communication to the understanding of phenomena such as the mediation and circulation of information in these sociotechnical environments. The outline of this paper is the following. First, in Sect. 2, we present the literature review. In Sect. 3, we will present our methodology to collect and analyze flows of tweets. In Sect. 4, a use case on air quality is presented. Finally, we conclude in Sect. 5.

2   Literature Review

As mentioned in [5], tweet analysis has led to a large number of studies in many domains such as ideology prediction in Information Sciences [4], spam detection [20], dialog analysis in Linguistics [2], and natural disaster anticipation in Emergency [16], while work in Social Sciences and Digital Humanities has developed tweet classifications [18]. Recently, several studies on tweet classification have been carried out in NLP [10]. Basically, these analyses aim at categorizing open-domain tweets using a reasonable amount of manually classified data and either small sets of specific classes (e.g. positive versus negative classes in sentiment analysis) or larger sets of generic classes (e.g. News, Events and Memes classes in topic filtering). Associating NLP and machine learning techniques, [6] have classified institutional tweets in communication categories. Until recently, the most commonly used models were supervised learning, Support Vector Machine (SVM), Random Forest, Gradient Boosting Machine and Naive Bayes (NB) [13]. In supervised machine learning, features are extracted from tweets and metadata and then vectorized as training examples to build models. But lately, deep learning models applied to natural language processing tasks have achieved remarkable results [15]. Moreover, [8] reported on a series of experiments with convolutional neural networks and showed "that a simple CNN with little hyperparameter tuning and static vectors achieves excellent results on multiple benchmarks". Our model of deep learning classifier is largely inspired of this work.

3   An Instrumented Methodology

The proposed methodology aims to describe and analyze the informational and communicational dynamics at work on the Twitter platform. It explores the circulation patterns of message flows and exchanges, apprehended as a dynamic process, as well as the relationships which are established between different stakeholders. Our aim is to question the forms of engagement, participation and relationships between organizations and audiences by analyzing the flow of their messages. We consider that the field of environmental health and the socio-technical device Twitter contribute to configuring the relations and the interactions between the participants. In this perspective, hashtags on Twitter may be seen as meeting points of different categories of stakeholders interested in the same themes while being anchored in different spheres - the environment and health, politics, the media, industry, the economy, etc.

Our analysis focuses on Twitter messages (called tweets) sent by accounts of organizations and non-institutional stakeholders. Concerning messages, we will call a message sent by a twitter account an 'original tweet' and a message sent by an account different from the issuing account a 'retweet'. The current Twitter API gives access to the original tweet (and its sending account) of a retweet. The generic term tweet includes 'original tweet' and 'retweet'. Regarding stakeholder qualification, we distinguished accounts managed by institutions (called 'organizational account'), and accounts managed by individuals (called 'private account').

Our methodology is broken down into several steps. First, analyze the set of data collected by applying a t-SNE analysis to get a global picture of all the data. Second, train a classifier based on a convolutional neural networks (cf. below) to categorize the data and analyze the modes of involvement and interaction between organizational and private accounts. Third, carrying out a lexical analysis to identify the topics addressed. Finally, identifying networks and passing accounts. We implemented our whole methodology by building a workflow based on open access tools. We also developed some scripts python to manage the interoperability between all these different tools.

About the second step, the main drawback of shallow supervised machine learning approaches, is that they require a very time consuming step to identify linguistic or semiotic features and raise issues about the relevance of these features. Recently, new approaches based on Deep Learning techniques and especially on convolutional neural networks (convets), which no longer require researchers to identify features, have been proposed. A second advantage of convets is that they obtain better results in terms of accuracy than shallow machine learning systems [8,14]. It is for these reasons that we developed a classifier based on convets.

The architecture of our classifier is composed of several layers [7]. The first layer is a pre-trained word embedding as proposed by [15] with a kernel that matches the 5 words used as neighbors. A word embedding is a distributed representation where each word is mapped to a fixed-sized vector of continuous values. The benefit of this approach is that different words with a similar meaning will have a similar representation. A fixed-vector size of 100 was chosen. The following layers, a Conv1D with 200 filters and a MaxPooling1D are based on the works reported in [3,8]. The back-end of the model is a standard Multilayer Perceptron layer to interpret the convets features. The output layer uses a softmax activation function to output a probability for each of the three classes affected at the tweet processed. Finally, only the class with the highest probability is kept. The evaluation using the standard cross-validation 10-fold test [9] gave an accuracy of 0.97 in line with the state of the arts [8].

4   A Use Case on Air Quality

Environmental health is an emerging and hotly debated topic that covers several fields of study such as pollution in urban or rural environments and the consequences of these changes on health populations. The environmental factors analyzed fall into four broad thematic dimensions relating to polluted sites and soils, water quality and air quality and habitat. Among these factors, we focused on air quality, which was the subject of many alerts in major cities at the end of 2016 and which is becoming an international concern with the regular peaks of fine particles matter in urban areas.

The data acquisition stage consisted in harvesting tweets with the following hashtags: the hashtag #Air and one other hashtag among the following list: #pollution, #santé (health), #qualité(quality) or #environnement(environment). In this paper, we limit the analysis to French tweets, by using the "lang" features in Tweeter API, sent between the first of November 2017 and the 30th of July 2018. This period of time is considered as a proof of concept and we intent to use the classifier to process all the tweets that will be sent during the year 2018. The main figures are the following: 4 832 tweets of which 39% of original tweets and 61% of retweets sent by 2517 participants (405 organizational accounts, 2112 private accounts). More specifically: 41% of organizational accounts and only 30% of private accounts produced original tweets. Participation for private accounts was largely limited to the action of retweeting (75% of tweets) the messages sent by the institutional partners. In order to obtain global picture of all the accounts, we applied the t-SNE algorithm [12] based on 5 dimensions proposed by [6]: type of account, number of original tweets sent, relayed score, relaying score, mentioned score. The main interest of this algorithm is to take a set of points in a high-dimensional space and find a faithful representation of those points in a 2D plane (cf. Fig. 1). We tuned the algorithm features as recommended in [12] and finally perplexity = 30 and iteration = 500 were chosen. In other words, it offers a global vision of practices of production, diffusion and interaction (mention, retweet, quote).

Categorical analysis relying on automatic classification is a relevant processing method to characterize the semantics of the messages as it makes possible to analyze the modes of engagement of the stakeholders on Twitter. Automatic classification implies a prior human classification (supervised machine learning). Taking into account the size of the corpus of tweets, a human analysis would still have been possible but first, as mentioned in [11], the inter-coder minimum reliability is usually around 0.74, and secondly, we intend to process in real time all the tweets that will be sent during the year 2018. These two reasons argue in favour of developing an automatic classification.

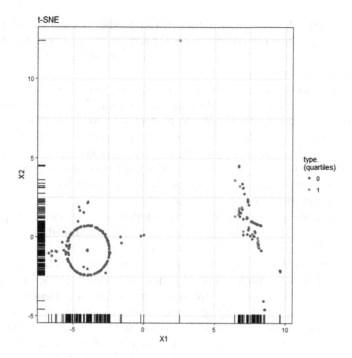

Fig. 1. t-SNE analysis

In order to build a classifier, a classification analysis of the contents of a sample of 350 randomly chosen original tweets was carried out in two stages [7]. First, these 350 tweets were annotated by hand by two experts using three classes: "informative","promoting", "humorous" (in the sense of emotional or expressive). In the second stage, a classifier, based on convets (see Sect. 3), was trained on the annotated samples. The training loss decreases with every epoch, and the training accuracy increases with every epoch; to prevent overfitting, the training was stopped after four epochs. The classifier was applied to the corpus of all tweets to categorize them. Table 1 show the findings. There is a main difference between institutional and individuals accounts concerning the "humorous" class. Private accounts sent twice as many "Humorous" tweets as institutional accounts. Similarly, institutional accounts sent much more "Promoting" tweets.

Table 1. Automatic classification of original tweets

Categorization	Institutional accounts	Private accounts
Informative	58%	53%
Promoting	32%	25%
Humorous	10%	22%

Observing the flow of information through the circulation of messages involves looking at the modes of stakeholder participation. They are materialized in information-communication practices. To answer this question, the analysis is based on the classification of the accounts.

In order to characterize Twitter accounts we used some of attributes proposed by [6]: "relayed", "relaying", "mentioned" and "passing". As pointed out by [6], the value of this index is not significant in itself; it simply provides a means of comparing accounts. We identified six passing accounts that had a significant score [7]. The analysis of these passing accounts makes it possible to identify some of their characteristics. These are all accounts of organizations with the exception of one influencer. It is remarkable these key influential accounts do not share their communities of accounts.

From the whole corpus, the data were partioned in restricted subcorpora built according to the criteria of the type of stakeholder (organizational or individual). These limited corpora enable future analyses (content analyses or discourse analyses) related to the status and role of the stakeholders to be carried out. In the space of this article, we focus on the analysis of several graphs. The aggregated communities are computed by applying the Louvain algorithm [1]; they highlights several points. From the whole graph (cf. on the left Fig. 2), if we select the organizational accounts (cf. on the middle Fig. 2), communities are linked by their retweet policy. One community is related with Anne Hidalgo (Mayor of Paris), a second one is related with Ambassad'AIr (a NGO), a third one is related with Air Paca (another NGO) and finally another community is related to the Prefect of the Occitanie Region (representing the French goverment). These communities, which share a few links between them, correspond to the territorial and administrative organization of the French regions whose communication activity is most apparent on twitter. If we select the private accounts from the whole graph (not shown), three accounts can be identified; they do not share any users. Private accounts do not communicate with each other, they do not mention or retweet each other (cf. on the far right Fig. 2); these three accounts are retweeted, quoted or mentioned. Their participation in the flow of information is limited to a subset of their followers.

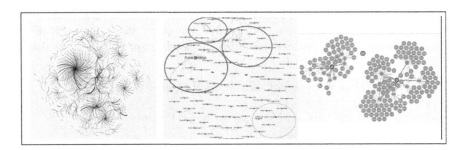

Fig. 2. Whole graph, organizational accounts graph and two communities

We computed the first ten significant terms and hashtags in the tweets to identify the main themes of the exchanges. The 10 most frequent terms in the text of the tweets (after deleting prepositions, conjunctions and hashtags) are the following. Two words "villes" (towns) and "intérieur" (inside) evoke places of observed pollution. Three words "défi" (challenge), "vigilance" (vigilance) "informer" (to inform), evoke the concerns about air quality and the concerns about air quality, monitoring pollution, and informing the public. Three words "préfectoral" (prefectural), "ballon" (balloon) and "dispositif" (device) evoke the administrative and technological tools used to monitor air quality. It is surprising that the widely discussed causes of air pollution related to "pesticides" (pesticides) and "particules" (particles matter) are hardly mentioned (they have a very low number of occurrences). The 10 most frequent hashtags in the text of the tweets (after deleting hashtags used to query the Twitter API) are the following. Five hashtags refer to geographical places. Three cities "Paris", "Rennes" and "Marseille" and two regions "Martinique" and "Haute Garonne" are the most frequently mentioned and show a higher communication activity related to political positions. Two hashtags refer to pollution events "PicPollution", (PeakPollution) and public or activist exhortations in favor of limiting activities that generate pollution "stopPollution". The hashtag "mobilité" (mobility) refers to the issue of the urban traffic.

5 Conclusions

We described an instrumented methodology for the analysis of the flow of messages on the Twitter platform. This methodology is based on a multi-dimensional approach in order to apprehend the complexity of the modes of circulation of messages. We developed a workflow to implement our methodology; an essential part of this workflow is a classifier based on convolutional neural networks which categorize the flow of tweets.

This research is part of a broader framework of ongoing work on the evolution of environmental communication in the public space. We are applying the same methodology on English tweets, during the same period of time, in order to carry out a comparative analysis between anglophone and francophone audience.

Acknowledgments. This study is partially funded by iGlobes (UMI 3157).

References

1. Blondel, V., Guillaume, J.L., Lambiotte, R., Lefebvre, E.: Fast unfolding of communities in large networks. J. Stat. Mech. **2008**(10), 1–12 (2008)
2. Boyd, D., Golder, S., Lotan, G.: Tweet, tweet, retweet conversational aspects of retweeting on twitter. In: 43rd Hawaii International Conference on System Sciences, HICSS, pp. 1–10 (2010)
3. Brownlee, J.: Deep Learning for Natural Language Processing. Machine Learning Mystery, Vermont, Australia (2017)

4. Djemili, S., Longhi, J., Marinica, C., Kotzinos, D., Sarfat, G.E.: What does twitter have to say about ideology? In: NLP 4 CMC: Natural Language Processing for Computer-Mediated Communication, pp. 16–25 (2014)

5. Foucault, N., Courtin, A.: Automatic classification of tweets for analyzing communication behavior of museums. In: LREC 2016, pp. 3006–3013 (2016)

6. Juanals, B., Minel, J.-L.: Analysing cultural events on twitter. In: Nguyen, N.T., Papadopoulos, G.A., Jędrzejowicz, P., Trawiński, B., Vossen, G. (eds.) ICCCI 2017. LNCS (LNAI), vol. 10449, pp. 376–385. Springer, Cham (2017). https://doi.org/10.1007/978-3-319-67077-5_36

7. Juanals, B., Minel, J.-L.: Categorizing air quality information flow on twitter using deep learning tools. In: Nguyen, N.T., Pimenidis, E., Khan, Z., Trawiński, B. (eds.) ICCCI 2018. LNCS (LNAI), vol. 11055, pp. 109–118. Springer, Cham (2018). https://doi.org/10.1007/978-3-319-98443-8_11

8. Kim, Y.: Convolutional neural networks for sentence classification. CoRR abs/1408.5882 (2014). http://arxiv.org/abs/1408.5882

9. Kohavi, R.: A study of cross-validation and bootstrap for accuracy estimation and model selection, pp. 1137–1143. Morgan Kaufmann (1995)

10. Kothari, A., Magdy, W., Darwish, K., Mourad, A., Taei, A.: Detecting comments on news articles in microblogs. In: Kiciman, E., et al. (eds.) 7th International Conference on Web and Social Media (ICWSM). The AAAI Press (2013)

11. Lachlan, K., Spence, P., Lin, X., Maria, D.G.: Screaming into the wind: examining the volume and content of tweets associated with hurricane sandy. Commun. Stud. **65**(5), 500–518 (2014)

12. Van der Maaten, L., Hinton, G.: Visualizing high-dimensional data using t-SNE. J. Mach. Learn. Res. **9**, 2579–2605 (2008)

13. Malandrakis, N., Falcone, M., Vaz, C., Bisogni, J.J., Potamianos, A., Narayanan, S.: Sail: sentiment analysis using semantic similarity and contrast features. In: 8th International Workshop SemEval, pp. 512–516 (2014)

14. Manning, C.D.: Computational linguistics and deep learning. Comput. Linguist. **41**(4), 701–707 (2015). https://doi.org/10.1162/COLI_a_00239

15. Mikolov, T., Chen, K., Corrado, G., Dean, J.: Efficient estimation of word representations in vector space. CoRR abs/1301.3781 (2013). http://arxiv.org/abs/1301.3781

16. Sakaki, T., Okazaki, M., Matsuo, Y.: Tweet analysis for real-time event detection and earthquake reporting system development. IEEE Trans. Knowl. Data Eng. **25**(4), 919–931 (2013)

17. Schmidt, C.: Trending now: using social media to predict and track disease outbreaks. Environ. Health Perspect. **120**(1), 30–33 (2012)

18. Shiri, A., Rathi, D.: Twitter content categorisation: A public library perspective. J. Inf. Knowl. Manage. **12**(4), 1350035-1–1350035-8 (2013)

19. Thackeray, R., Neiger, B., Smith, A., Van Wagenen, S.: Adoption and use of social media among public health departments. BMC Public Health **12**, 242 (2012)

20. Yamasaki, S.: A trust rating method for information providers over the social web service: a pragmatic protocol for trust among information explorers, and provider information. In: 11th Annual International Symposium on Applications and the Internet (SAINT 2011), pp. 578–582 (2011)

Market-Aware Proactive Skill Posting

Ashiqur R. KhudaBukhsh$^{(\boxtimes)}$, Jong Woo Hong, and Jaime G. Carbonell

Carnegie Mellon University, Pittsburgh, USA
{akhudabu,jongwooh,jgc}@cs.cmu.edu

Abstract. Referral networks consist of a network of experts, human or automated agent, with differential expertise across topics and can redirect tasks to appropriate colleagues based on their topic-conditioned skills. *Proactive skill posting* is a setting in referral networks, where agents are allowed a one-time local-network-advertisement of a subset of their skills. Heretofore, while advertising expertise, experts only considered their own skills and reported their strongest skills. However, in practice, tasks can have varying difficulty levels and reporting skills that are uncommon or rare may give an expert relative advantage over others, and the network as a whole better ability to solve problems. This work introduces *market-aware proactive skill posting* where experts report a subset of their skills that give them competitive advantages over their peers. Our proposed algorithm in this new setting, proactive-DIEL$_\Delta$, outperforms the previous state-of-the-art, proactive-DIEL$_t$ during the early learning phase, while retaining important properties such as tolerance to noisy self-skill estimates, and robustness to evolving networks and strategic lying.

Keywords: Active learning · Referral networks
Proactive skill posting

1 Introduction

A referral network [1] consists of experts, human or autonomous agents, where each expert (teacher, worker, agent) can redirect difficult tasks to appropriate expert colleagues. Such networks draw inspiration from real-world examples of networks of physicians or consultancy firms. The *learning-to-refer* challenge involves estimating topic-conditioned expertise of colleagues in a referral network in an active learning framework.

In this paper, we focus on *proactive skill posting* [2], a setting where experts perform a one-time (only at the beginning of simulation or when they join a network) local-network advertisement to network-connected colleagues of a subset of their skills, focusing on the expert top skills. In the real world, such skill advertisements are common as experts often tell their colleagues about tasks they are good at and often forge links with colleagues via social networks. However, such (potentially noisy) priors are private information, and experts may strategically lie or unknowingly overestimate or underestimate their own skills to

M. Ceci et al. (Eds.): ISMIS 2018, LNAI 11177, pp. 323–332, 2018.
https://doi.org/10.1007/978-3-030-01851-1_31

attract more referrals, i.e., profit from more business. Hence, a key component of algorithms designed for this setting is a mechanism to elicit truthful reporting of skills to make the system incentive compatible. However, when human experts select the set of topics to advertise, they not only consider their true absolute expertise at those particular topics, but also take their relative advantage over others into account. For instance, in a network of physicians, an accomplished brain surgeon would want to publish the fact that she is skilled at brain surgery, even though her success rate at diagnosing the common cold could be much higher. In a similar vein, in this work, we introduce the notion of *market-aware skill posting*, i.e. experts posting skills on topics they have relative advantage over others and propose proactive-DIEL$_\Delta$ for this setting.

Our key contributions are the following: First, we introduce *market-aware proactive skill posting* in referral networks, previously not addressed in the proactive skill posting literature. Second, we perform extensive empirical evaluations on existing data sets comparing against proactive-DIEL$_t$, the known state-of-the-art, and demonstrate that our newly-introduced algorithm, proactive-DIEL$_\Delta$, outperforms proactive-DIEL$_t$ in terms of early learning-phase advantage. We construct additional data sets with larger variance in task-difficulty and show that the performance gap between proactive-DIEL$_\Delta$ and proactive-DIEL$_t$ widens in the presence of tasks with varying difficulty levels. Finally, we show that like its predecessors, proactive-DIEL$_\Delta$ is robust to strategic lying, evolving networks, and noisy self-skill estimates.

Related Work: The referral framework draws inspiration from referral chaining, first proposed in [3] and subsequently extended in [4,5] (for further relevant literature, see, e.g., [1]). Recent research in referral networks has focused on three broad directions: identifying key algorithms and evaluating relative performance on uninformed prior settings [1], designing algorithms immune to strategic lying that incorporate partially available (potentially noisy) priors [2,6], and robustness to practical factors such as evolving networks, capacity constraints [1] and time-varying expertise [7].

Our work on *proactive skill posting* is related to the bandit literature with *side-information* [8,9] in the sense that algorithms in this setting do not start from scratch, but have leg-up based on task-relevant information. However, a key difference is that, instead of obtaining that side data from observed trials [9] or shape of the reward distribution [10], the *side-information* in our case is obtained in a decentralized manner through advertisement of skills by the experts themselves, who may in fact willfully misreport to attract more business. This ties our work broadly to the vast literature in adversarial machine learning [11] and truthful mechanism design [12–14]. For further relevant literature, see, e.g., [2,6].

2   Referral Networks

We summarize the basic notation, definitions, and assumptions and provide necessary background for *market-aware proactive skill posting*.

Referral Network: Represented by a graph (V, E) of size k in which each vertex v_i corresponds to an expert e_i $(1 \leq i \leq k)$ and each bidirectional edge $\langle v_i, v_j \rangle$ indicates a *referral link* which implies e_i and e_j can co-refer problem instances.

Subnetwork: of an expert e_i: The set of experts linked to an expert e_i by a referral link.

Referral Scenario: Set of m instances $(q_1, \ldots, q_m)$ belonging to n topics $(t_1, \ldots, t_n)$ addressed by the k experts $(e_1, \ldots, e_k)$ connected through a referral network (V, E).

Expertise: Expertise of an expert/question pair $\langle e_i, q_j \rangle$ is the probability with which e_i can solve q_j.

Referral Mechanism: Following previous proactive skill posting literature [2,6], for a per-instance query budget Q, we kept fixed to $Q = 2$ across all our current experiments. The referral mechanism consists of the following steps.

1. A user issues an *initial query* q_j to a randomly chosen *initial expert* e_i.
2. The initial expert e_i examines the instance and solves it if possible. This depends on the *expertise* of e_i wrt. q_j.
3. If not, a *referral query* is issued by e_i to a *referred expert* e_j within her subnetwork, with a query budget of $Q - 1$. *Learning-to-refer* involves improving the estimate of who is most likely to solve the problem.
4. If the referred expert succeeds, she sends the solution to the initial expert, who sends it to the user.

A detailed description of our assumptions can be found in [1,2]. Some of the important assumptions are: the network connectivity depends on (cosine) similarity between the topical expertise (guided by the observation that experts with similar expertise are more likely to know each other), and the distribution of topical-expertise across experts can be characterized by a mixture of Gaussian distributions. Note that, in our model, it is still possible that experts with very little overlap in skills are connected for reasons beyond similar expertise (e.g., same geolocation, common acquaintances, etc.), making them prime candidates for referrals. Also, an expert pair with substantial overlap in expertise areas may still have specific topics where one expert is stronger than the other, making referrals mutually beneficial. For topical-expertise distribution, a mixture of two Gaussians was used to represent the expertise of experts with specific training for the given topic (higher mean, lower variance), contrasted with the lower-level expertise (lower mean, higher variance) of the layman population.

We now present necessary background for *market-aware proactive skill posting* and describe what distinguishes it from traditional *proactive skill posting*.

Advertising Unit: A tuple $\langle e_i, e_j, t_k, \mu_{t_k} \rangle$, where e_i is the *target expert*, e_j is the *advertising expert*, t_k is the topic and μ_{t_k} is e_j's (advertised) topical expertise. Similar to rewards in our uninformative prior setting, an advertising unit is also locally visible, i.e., only the *target expert* gets to see the advertised prior for a given unit.

Advertising Budget: In practice, experts have limited time to socialize with different colleagues and get to know each other's experience. We incorporate this phenomenon through the notion of budget and assume each expert is allocated a budget of B advertising units, where B is twice the size of that expert's subnetwork. Effectively means that each expert reports her top two skills to everyone in her subnetwork.

Explicit and Implicit Bid: A topic that is advertised in an advertising protocol is an explicit bid. Similarly, a topic that is not advertised, for which an upper bound can be assumed, is an implicit bid. Top two skills are differently defined in traditional proactive skill posting and market-aware skill posting. We describe the difference in the advertising protocol next.

Advertising Protocol: A one-time advertisement that happens right at the beginning of the simulation or when an expert joins the network. The advertising expert e_j reports to each target expert e_i in her subnetwork the two tuples $\langle e_i, e_j, t_{best}, \mu_{t_{best}} \rangle$ and $\langle e_i, e_j, t_{secondBest}, \mu_{t_{secondBest}} \rangle$, i.e., the top two topics in terms of the advertising expert's topic means.

Now, we describe the primary distinction between traditional *proactive skill posting* and *market-aware proactive skill posting*. In traditional proactive skill posting, for a given expert, $\mu_{t_{best}}$ is simply her maximum topical expertise. In *market-aware proactive skill posting*, we propose that every expert has access to an estimate of $\overline{\mu_{t_k}}$ (average network skill on each topic t_k) and reports the skills with her largest relative advantage μ_Δ (where for a given expert/topic pair $\langle e_j, t_k \rangle$, $\mu_{\Delta_{e_j,t_k}} = \mu_{e_j,t_k} - \overline{\mu_{t_k}}$).

Next, we illustrate the difference described above with the following example. Consider a referral network of N experts and across five different topics, $t_1, \ldots, t_5$, the average network expertise are respectively, 0.1, 0.3, 0.8, 0.9, 0.4. Now, consider an expert e whose expertise on the aforementioned five topics are 0.4, 0.3, 0.7, 0.65, 0.5, respectively. In traditional proactive skill posting, for every colleague e_i of e, e will have the following two advertising units: $\langle e_i, e, t_3, 0.7 \rangle$ and $\langle e_i, e, t_4, 0.65 \rangle$, reporting her skills on t_3 and t_4, the two topics she has highest expertise in an absolute scale. However, notice that e is unlikely to have any substantial relative advantage over other in those two topics as her expertise on those two topics is sufficiently lower than the network expertise. Also, t_1 is the hardest topic where average expertise of the network is 0.1 and e is relatively stronger in t_1 with $\mu_{\Delta_{e,t_1}} = 0.3$. Hence, in a market-aware skill posting setting, e will report her skills in t_1 and t_5, the two topics where she has relative advantage with the two advertisement units $\langle e_i, e, t_1, 0.4 \rangle$ and $\langle e_i, e, t_5, 0.5 \rangle$.

3   Distributed Referral Learning

Considering a single expert and a given topic, learning-to-refer is an action selection problem, and each expert maintains an action selection thread for each topic in parallel. If we think in the context of multi-armed bandit (MAB), an action or arm corresponds to a referral choice, i.e., picking an appropriate expert from

---

Algorithm 1. DIEL(e, T)

Initialization: $\forall i, n_i \leftarrow 2$, $\mathbf{r}_{i,n_i} \leftarrow (0, 1)$
Loop: Select expert e_i who maximizes

$$score(e_i) = m(e_i) + \frac{s(e_i)}{\sqrt{n_i}}$$

Observe reward r
Update $\mathbf{r}_{i,n_i}$ with r, $n_i \leftarrow n_i + 1$

---

the subnetwork. In order to describe an action selection thread, we first name the topic T and expert e. Let $q_1, \ldots, q_N$ be the first N *referral queries* belonging to topic T issued by expert e to any of her K colleagues in her subnetwork denoted by $e_1, \ldots, e_K$. For each colleague e_i, e maintains a reward vector $\mathbf{r}_{i,n_i}$ where $\mathbf{r}_{i,n_i} = (r_{i,1}, \ldots, r_{i,n_i})$, i.e., the sequence of rewards observed from expert e_i on issued n_i referred queries. Understandably, $N = \sum_{i=1}^{K} n_i$. Let $m(e_i)$ and $s(e_i)$ denote the sample mean and sample standard deviation of $\mathbf{r}_{i,n_i}$.

DIEL: First proposed in [15], Interval Estimation Learning (IEL) has been extensively used in stochastic optimization and action selection problems. Action selection using Distributed Interval Estimation Learning (DIEL) works in the following way [2]. As described in Algorithm 1, at each step, DIEL [2] selects the expert e_i with highest $m(e_i) + \frac{s(e_i)}{\sqrt{n_i}}$. The intuition is that high mean selects for best performance, and high variance selects for unexplored expert capability on topic, thus optimizing for amortized performance, as variance decreases over time, and best mean is selected reliably among the top candidates. DIEL operates in an uninformed prior setting, and every action is initialized with two rewards of 0 and 1, allowing us to initialize the mean and variance and making all experts equally likely to get picked in the beginning.

We now describe proactive-DIEL$_\Delta$, our proposed new algorithm and proactive-DIEL$_t$, our baseline.

3.1 Initialization

proactive-DIEL$_t$ initialization : Rather than DIEL sets $reward(e_i, t_k, e_j)$ for each i, j and k with a pair $(0, 1)$ in order to initialize mean and variance, proactive-DIEL$_t$ initializes $reward(e_i, t_k, e_j)$ for each advertisement unit $\langle e_i, e_j, t_k, \mu_{t_k} \rangle$ with two rewards of μ_{t_k} (explicit bid).

To initialize topics for which no advertisement units are available (implicit bid), we initialize the rewards as if the expert's skill was the same as on her second best topic, that is, with two rewards of $\mu_{t_{secondBest}}$, effectively being an upper bound on the actual value.

proactive-DIEL$_\Delta$ Initialization : A similar prior-bounding technique can be used in proactive-DIEL$_\Delta$ with the following modification. Recall that, each expert has knowledge about $\overline{\mu_{t_k}}$, $\forall k$ – the average network expertise across all topics. Let t_{best}^Δ and $t_{secondBest}^\Delta$ be the two explicit bids for an expert e with $\mu_{t_{secondBest}^\Delta}$ –

$\overline{\mu}_{t^{\Delta}_{secondBest}} \leq \mu_{t^{\Delta}_{best}} - \overline{\mu}_{t^{\Delta}_{best}}$, and $t^{\Delta}_{implicit}$ be any implicit bid. The following inequality holds,

$$\mu_{t^{\Delta}_{implicit}} - \overline{\mu}_{t^{\Delta}_{implicit}} \leq \mu_{t^{\Delta}_{secondBest}} - \overline{\mu}_{t^{\Delta}_{secondBest}} \tag{1}$$

since the relative advantage of any implicit bid must be less than or equal to the relative advantage of $t^{\Delta}_{secondBest}$. Rearranging Eq. 1, we get

$$\mu_{t^{\Delta}_{implicit}} \leq \mu_{t^{\Delta}_{secondBest}} + \overline{\mu}_{t^{\Delta}_{implicit}} - \overline{\mu}_{t^{\Delta}_{secondBest}} \tag{2}$$

All terms of the right-hand side of the Eq. 2 are known, and every implicit bid is initialized with two rewards of $\mu_{t^{\Delta}_{secondBest}} + \overline{\mu}_{t^{\Delta}_{implicit}} - \overline{\mu}_{t^{\Delta}_{secondBest}}$.

3.2   Penalty on Distrust

proactive-DIEL$_{\Delta}$ follows the same penalty mechanism, *Penalty on Distrust*, as proactive-DIEL$_t$[1]. In this approach, in addition to assigning a binary reward depending on the task outcome, we assign a penalty. For a given instance, if the reward is r and the penalty p, the effective reward is $r - p$. The assigned penalty incorporates a factor we may call *distrust*, as it estimates a likelihood the expert is lying, given our current observations. Further details can be found in [6].

4   Experimental Setup

Baselines: We used DIEL (the parameter-free version first presented in [2]), the known best-performing algorithm in uninformed prior setting, and proactive-DIEL$_t$, the best-performing algorithm in proactive skill setting as our baselines.

Data Set: Our test set for performance evaluation is the same data set, $\mathcal{D}$, used in [1,2][2], which consists of 1000 *referral scenarios*. Each *referral scenario* consists of a network of 100 experts and 10 topics.

In addition to $\mathcal{D}$, we constructed two data sets inducing larger variance in task-difficulty. Recall that, for topical-expertise distribution, we consider a mixture of two Gaussians (with parameters $\lambda = \{w_i^t, \mu_i^t, \sigma_i^t\}$ $i = 1, 2.$). One of them ($\mathcal{N}(\mu_2^t, \sigma_2^t)$) has a greater mean ($\mu_2^t > \mu_1^t$), smaller variance ($\sigma_2^t < \sigma_1^t$) and lower mixture weight ($w_2^t \ll w_1^t$). For a given topic t_i, we modify the topical expertise for all experts in the following way:

$\mu_{t_i, e_j} = d_{factor}\, \mu_{t_i, e_j} \forall j$, where $d_{factor} \sim U[C, 1]$. The multiplicative factor ensures that the initial property of being sampled from a mixture of Gaussians holds. Different values for the parameter C allows us to vary the difficulty level of a task. We generated two additional data sets using $C = 0.25$ (denoted as $\mathcal{D}_{0.25}$) and $C = 0.5$ (denoted as $\mathcal{D}_{0.5}$).

---

[1] The subscript t stands for trust.

[2] The data set can be downloaded from https://www.cs.cmu.edu/~akhudabu/referral-networks.html.

Performance Measure: Following previous proactive skill posting litera-ture [2,6], we use two performance measures – overall task accuracy of our multi-expert system and *ICFactor*, an empirical measure for evaluating Bayesian-Nash incentive compatibility (a weaker form of incentive compatibility where being truthful is weakly better than lying). If a network receives n tasks of which m tasks are solved (either by the *initial expert* or the *referred expert*), the overall task accuracy is $\frac{m}{n}$. As an empirical measure for evaluating Bayesian-Nash incen-tive compatibility, we use *ICFactor* (described in [2]); an *ICFactor* value greater than 1 implies truthfulness in expectation, i.e., truthful reporting fetched more referrals than strategic lying. For evolving networks and noisy skill estimates, we used the same experimental setting as described in [6].

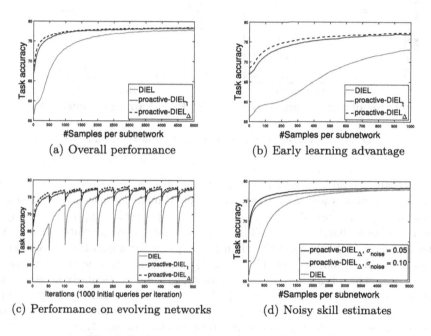

(a) Overall performance

(b) Early learning advantage

(c) Performance on evolving networks

(d) Noisy skill estimates

Fig. 1. Performance comparison on data set $\mathcal{D}$.

5   Results

Figure 1 and Table 1 summarize main experimental results of our proposed new algorithm, proactive-DIEL$_\Delta$. The results highlight the following. First, Fig. 1(a) demonstrates that we can use similar prior-bounding technique to initialize proactive-DIEL$_\Delta$ and a reward adjustment mechanism described in [6] and obtain improved cold-start performance than proactive-DIEL$_t$, the known state-of-the-art. Particularly, as shown in Fig. 1(b), the advantage during the early phase of learning is superior which widens as the tasks get harder (see, Fig. 2). A paired t-test reveals that during the early learning phase (1000 samples or less per sub-network), proactive-DIEL$_\Delta$ is better than both the baselines with p-value less

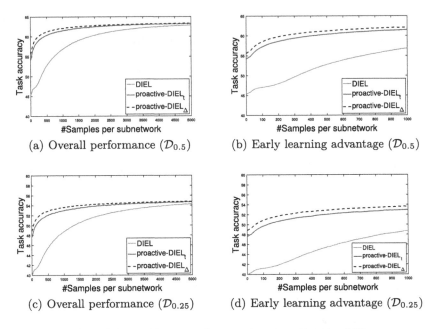

Fig. 2. Performance comparison on data sets $\mathcal{D}_{0.25}$ and $\mathcal{D}_{0.50}$.

than 0.0001. Second, in a real-world setting, it is easy to imagine situations where new entrants will join a network while old members leave. In a situation, where at regular interval a sizable chunk of the network composition changes, Fig. 1(c) shows that the early-phase learning advantage translates into better adaptation to evolving networks. Third, our proposed algorithm is tolerant to noisy self-skill estimates (shown in Fig. 1(d)), thus absolutely accurate estimates of own skill, which is an impractical assumption, is not particularly necessary for the algorithm to succeed. Table 1 lists *ICFactor* for all possible strategy combinations of reporting an expert's top two skills. We found that, across all three data sets, proactive-DIEL$_\Delta$ exhibited robustness to strategic lying.

We conclude our results section with a summary of our main results and an outlook to future research directions. Our results demonstrate: (1) Market-aware skill posting gives larger early-learning phase advantage than previously known state-of-the-art, (2) our proposed algorithm, proactive-DIEL$_\Delta$ is robust to noise in self-skill estimates and strategic lying, and (3) the performance gap between proactive-DIEL$_\Delta$ and proactive-DIEL$_t$ widens when we use larger variance in task difficulty (4) the early learning-phase advantage is particularly useful in evolving networks. Future research directions may include: (1) extending market-aware skill posting to other MAB algorithms (e.g., ϵ-Greedy, Q-Learning) (2) in presence of a larger amount of noise, or when $\overline{\mu_{t_k}}$ is not known, regularizing the advertised priors relative to the subnetwork and (3) designing proactive skill posting algorithms robust to time-varying expertise.

Table 1. Empirical analysis of Bayesian-Nash incentive-compatibility on three data sets, $\mathcal{D}$, $\mathcal{D}_{0.5}$ and $\mathcal{D}_{0.25}$. Each row represents a specific combination of strategies an expert can use to report her best and second-best skills. All values indicate that being truthful is no worse than lying.

$\mu_{t^{\Delta}_{best}}$	$\mu_{t^{\Delta}_{secondBest}}$	$\mathcal{D}$	$\mathcal{D}_{0.5}$	$\mathcal{D}_{0.25}$
Truthful	Overbid	1.0174	1.0006	1.0026
Overbid	Truthful	1.1892	1.4393	1.7843
Overbid	Overbid	1.2159	1.7993	1.7620
Truthful	Underbid	1.0976	1.0617	1.0860
Underbid	Truthful	1.1301	1.1586	1.1333
Underbid	Underbid	1.3672	1.1900	1.1588
Underbid	Overbid	1.1213	1.0982	1.1404
Overbid	Underbid	1.2835	1.4303	1.4616

References

1. KhudaBukhsh, A.R., Carbonell, J.G., Jansen, P.J.: Robust learning in expert networks: a comparative analysis. J. Intell. Inf. Syst. **51**(2), 207–234 (2018)
2. KhudaBukhsh, A.R., Carbonell, J.G., Jansen, P.J.: Proactive skill posting in referral networks. In: Kang, B.H., Bai, Q. (eds.) AI 2016. LNCS (LNAI), vol. 9992, pp. 585–596. Springer, Cham (2016). https://doi.org/10.1007/978-3-319-50127-7_52
3. Kautz, H., Selman, B., Milewski, A.: Agent amplified communication, pp. 3–9 (1996)
4. Yolum, P., Singh, M.P.: Dynamic communities in referral networks. Web Intell. Agent Syst. **1**(2), 105–116 (2003)
5. Yu, B.: Emergence and evolution of agent-based referral networks. Ph.D. thesis, North Carolina State University (2002)
6. KhudaBukhsh, A.R., Carbonell, J.G., Jansen, P.J.: Incentive compatible proactive skill posting in referral networks. In: European Conference on Multi-Agent Systems. Springer (2017)
7. KhudaBukhsh, A.R., Carbonell, J.G.: Expertise drift in referral networks. In: Proceedings of the 17th International Conference on Autonomous Agents and MultiAgent Systems, pp. 425–433. International Foundation for Autonomous Agents and Multiagent Systems (2018)
8. Langford, J., Strehl, A., Wortman, J.: Exploration scavenging. In: Proceedings of the 25th International Conference on Machine learning, pp. 528–535. ACM (2008)
9. Shivaswamy, P., Joachims, T.: Multi-armed bandit problems with history. In: Artificial Intelligence and Statistics, pp. 1046–1054 (2012)
10. Bouneffouf, D., Feraud, R.: Multi-armed bandit problem with known trend. Neurocomputing **205**, 16–21 (2016)
11. Huang, L., Joseph, A.D., Nelson, B., Rubinstein, B.I., Tygar, J.: Adversarial machine learning. In: Proceedings of the 4th ACM Workshop on Security and Artificial Intelligence, pp. 43–58. ACM (2011)
12. Babaioff, M., Sharma, Y., Slivkins, A.: Characterizing truthful multi-armed bandit mechanisms. SIAM J. Comput. **43**(1), 194–230 (2014)

13. Tran-Thanh, L., Stein, S., Rogers, A., Jennings, N.R.: Efficient crowdsourcing of unknown experts using multi-armed bandits. In: European Conference on Artificial Intelligence, pp. 768–773 (2012)

14. Biswas, A., Jain, S., Mandal, D., Narahari, Y.: A truthful budget feasible multi-armed bandit mechanism for crowdsourcing time critical tasks. In: Proceedings of the 2015 International Conference on Autonomous Agents and Multiagent Systems, pp. 1101–1109. International Foundation for Autonomous Agents and Multiagent Systems (2015)

15. Kaelbling, L.P.: Learning in Embedded Systems. MIT Press, Cambridge (1993)

Evidential Multi-relational Link Prediction Based on Social Content

Sabrine Mallek[1,2(✉)], Imen Boukhris[1], Zied Elouedi[1], and Eric Lefevre[2]

[1] LARODEC, Institut Supérieur de Gestion de Tunis, Université de Tunis,
Tunis, Tunisia
sabrinemallek@yahoo.fr, imen.boukhris@hotmail.com, zied.elouedi@gmx.fr
[2] Univ. Artois, EA 3926, Laboratoire de Génie Informatique et d'Automatique de
l'Artois (LGI2A), 62400 Béthune, France
eric.lefevre@univ-artois.fr

Abstract. A novel framework to address the link prediction problem in multiplex social networks is introduced. In this framework, uncertainty found in social data due to noise, missing information and observation errors is handled by the belief function theory. Despite the numerous published studies on link prediction, few research are concerned with social data imperfections issues which cause distortions in social networks structures and probably inaccurate results. In addition, most works focus on similarity scores based on network topology whereas social networks include rich content which may add semantic to the analysis and enhance results. To this end, we develop a link prediction method that combine network topology and social content to predict new links existence along with their types in multiplex social networks. Structural and social neighbors information are gathered and pooled using belief function theory combination rules. It is subsequently revised according to global information about the multiplex. Experiments performed on real world social data show that our approach works well and enhances the prediction accuracy.

Keywords: Link prediction · Multiplex social networks
Belief function theory · Uncertainty · Social content

1 Introduction

Recent years have witnessed the huge growth of the World Wide Web, enabling access to rich social content. Personalized recommendations [7] are effective tools to alleviate the massive data. Recommendation in social networks [5] comes to a vogue in fields such as network science and computer science (i.e., data mining and machine learning). In particular, potential connections recommendation is generally performed by evaluating similarities between social entities. Such task is handled by link prediction. Classical works use network topology to perform predictions, discarding valuable social content. Various social networks include information about the entities. For example, personal information such as age,

© Springer Nature Switzerland AG 2018
M. Ceci et al. (Eds.): ISMIS 2018, LNAI 11177, pp. 333–343, 2018.
https://doi.org/10.1007/978-3-030-01851-1_32

gender and address, behavioral information like smoking and sporting activities and beliefs and opinions such as political preferences and favorite sports team. Obviously, considering such patterns adds semantic to prediction and allows personalized recommendations in multiplex social networks where nodes connect via multiple relationships types simultaneously.

Social networks structure highly depends on the accurate nature of data. Yet, such data are usually missing and incorporate erroneous labels due to noise and observation errors [2,14]. This induces uncertainties regarding the network structure. As pointed out in [29], one can encounter two kinds of uncertainty in social networks (i) whether a specific node is distinct or not (ii) or whether a link actually exists between two nodes. Furthermore, multiplex social networks are more affected with such uncertainties. According to [27], inaccuracies concerning multiplex social networks components are very common due to unreliable experimental settings and technical issues. To avoid data imperfections, one may discard some of the data risking the loss of useful information or use all uncertain data [14]. In both cases, this will inadequately affect the analysis.

To cope with uncertainty problems, we embrace the belief function theory [8,26]. Such theory enables us not only to present and manage data imperfections but it also allows tools to update, pool evidence and make decisions. We develop a fruitful link prediction framework that takes both topological features and social content in multiplex social network into account. Furthermore, our framework considers the worthwhile information given by relationships mixture and use it properly to study social linkage.

The rest of the paper is organized as follows. Section 2 gives some related works on link prediction. In Sect. 3, belief function basic concepts are re-called. Section 4 details our proposals for evidential link prediction in multiplex social networks. Section 5 reports the experimental results and Sect. 6 concludes the paper.

2   Link Prediction Related Work

Let $G(V,E)$ be a social network graph where V is the set of nodes and E is the set of edges. Link prediction is the problem of determining the potential existence of a link between unconnected nodes (u,v) given an observed state of the network. Many link prediction methods have been proposed that can be classified into three families: structural similarity based methods, supervised methods and latent models based methods.

Structural Similarity Approaches. Generally, similarity based methods use two types of topological information: local information and global information. Local methods explore node neighborhoods where popular measures include Common Neighbors [23], Adamic/Adar [1], the Jaccard Coefficient [11], Preferential Attachment [23] and Resource Allocation [37]. For instance, the common neighbors method, denoted by CN, computes the number of common neighbors between (u,v). By contrast, global methods use structural topology of the global network. Well known scores include the shortest path, Katz score [13] and Sim-Rank [12]. Even though global methods give generally better results than local based methods [15], they have limited use. Actually, computing global score require information of the whole network which can be time consuming. Furthermore, global topological information is not always accessible. Above all, the additional complexity, by comparison with local methods, does not certainly pay off since the latter methods are very able to give great performance results. Meanwhile, other works use alternative information for link prediction. The authors in [34] combine network proximity to node attributes to predict links in weighted social networks by promoting associations between nodes with similar features. Yet, the approach only applies to weighted networks. In [10], link prediction is performed using authors features to predict co-authorships. However, some attributes are not convenient to assess similarities between authors.

Supervised Learning Based Approaches. Link prediction can be reformulated into a binary classification problem where the objective is to predict the classes of query links. For the most part, topological scores are employed as predicator attributes and the existence of links represent class labels [4,10]. Some other works use other features, for example the authors in [3] used network motifs as features for supervised learning link prediction. In [25], community information, connectivity, interaction and trust information are were used as features for supervised link prediction. In [24], co-participation in events over time are predicted using both network and entity features under supervised learning. Although, link prediction under supvised learning benefits from all the advantages of classification, some issues need to be handled such as the convenient classification model, the appropriate combination of features and class imbalance.

Latent Models Based Approaches. Such methods assume a particular organizing principles of the network structure, with some rules and parameters captured by maximizing the likelihood of the observed structure. Predicted links are determined on the basis of their probabilities computed according to these rules and parameters. Many methods combine matrix factorization and latent features to predict new links [21]. Others use Markov Random Field [30,35], Bayesian nonparametric models [31] or stochastic relational models [33]. Yet, this family of methods suffer from high computational costs due to matrix decomposition, factor matrices and latent features learning which limits their application to large networks.

Challenges. Obviously, using social content enhances link prediction as it adds semantic to the task. In online social networks, users information is generally undisclosed for legal, ethical or practical concerns. Furthermore, information about the social entities cannot be considered fully accurate i.e., when collecting social data from surveys, the asked people can lie to some questions, when gathering data from online social networks, users can put wrong and misleading information or even invent fake profiles. Link prediction algorithms are sensitive to such noisy and missing data especially in complex networks such as social networks [36] and biological networks [6]. To handle such ambiguity, we take uncertainty into account in link prediction by embracing the belief function theory which provides effective tools to manage imperfect knowledge. We draw on local methods and consider social content of node attributes to enhance predictions and add semantic.

Actually, we tackled the link problem under the belief function theory framework in previous works. In [18,19], two link prediction approaches are proposed based on network topology solely in uniplex and multiplex social networks. In [20], a link predicator is presented which handles uncertainty in social data and uses node attributes. Yet, the latter method operates only on uniplex social networks. It can obviously be applied to multiplex networks by considering links types' separately however relevant information concerning the multiplex will be neglected. This motivated us to develop a new link prediction framework for multiplex social networks that considers uncertainty and social content and takes into account worthwhile information of the multiplex structure.

3   Belief Function Theory Basic Concepts

The belief function theory [8,26] has well-defined mathematical concepts managing uncertain knowledge. A problem is represented by a finite set of exhaustive and mutually exclusive events called the frame of discernment denoted by Θ. Evidence in the belief function theory is quantified through a mass function $m : 2^{\Theta} \rightarrow [0,1]$, called basic belief assignment (bba), which satisfies:

$$\sum_{A \subseteq \Theta} m(A) = 1. \tag{1}$$

Evidence induced from two reliable and distinct sources of information is pooled using the conjunctive rule of combination [28] denoted by $\bigcirc$. It is applied as follows:

$$m_1 \bigcirc m_2(A) = \sum_{B,C \subseteq \Theta : B \cap C = A} m_1(B) \cdot m_2(C). \tag{2}$$

Sometimes one needs to combine two bba's m_1 and m_2 defined on two disjoint frames Θ and Ω. To do this, the vacuous extension is applied. First, the bba's are extended to the product space $\Theta \times \Omega = \{(\theta_i, \omega_k), \forall i \in \{1, \ldots, |\Theta|\}, \forall k \in \{1, \ldots, |\Omega|\}\}$. Then, the vacuous extension, denoted by $\uparrow$, is applied as follows:

$$m^{\Theta \uparrow \Theta \times \Omega}(C) = \begin{cases} m^{\Theta}(A) & \text{si } C = A \times \Omega, \\ & A \subseteq \Theta, C \subseteq \Theta \times \Omega \\ 0, & \text{otherwise} \end{cases} \tag{3}$$

When the source of information is not completely trustworthy, a discounting operation [26] can be applied to m using a discounting rate $\alpha \in [0,1]$. The discounting operator is defined as following where $^{\alpha}m$ is the discounted bba:

$$\begin{cases} {}^{\alpha}m(A) = (1 - \alpha) \cdot m(A), \forall A \subset \Theta \\ {}^{\alpha}m(\Theta) = \alpha + (1 - \alpha) \cdot m(\Theta). \end{cases} \quad (4)$$

On the other hand, in some cases, a bba m may be revised by reinforcing the evidence committed to an element A of the frame. The belief function theory offers a mechanism of reinforcement [22] which operates according to a reinforcement rate $\beta \in [0, 1]$. It is defined by:

$$\begin{cases} {}^{\beta}m(A) = (1 - \beta) \cdot m(A) + \beta \\ {}^{\beta}m(B) = (1 - \beta) \cdot m(B), \forall B \subseteq \Theta \text{ and } B \neq A. \end{cases} \quad (5)$$

To specify the relation between two different frames of discernment Θ and Ω, the multi-valued mapping mechanism [8] is applied. It is function, denoted by τ, which assigns the subsets $B \subseteq \Omega$ that may correspond to a subset $A \subseteq \Theta$ as follows:

$$m_{\tau}(A) = \sum_{\tau(B)=A} m(B). \quad (6)$$

Decision in the belief function framework can be made according to the pignistic probability denoted by $BetP$. It is computed on the basis of the bba m as follows [28]:

$$BetP(A) = \sum_{B \subseteq \Theta} \frac{|A \cap B|}{|B|} \frac{m(B)}{(1 - m(\emptyset))}, \forall A \in \Theta. \quad (7)$$

4 Evidential Link Prediction in Multiplex Social Networks

We introduce our proposals for evidential multi-relational link prediction using social contents. The novel method draws on local information approaches from literature. We consider a multiplex social network graph with uncertainties encapsulated at the links structure presented in Fig. 1. A multiplex social network is given by a graph $G(V, E_1, \ldots, E_n)$ where $V = \{v_1, \ldots, v_{|V|}\}$ is the set of nodes, and $E_1, \ldots, E_n$ are the sets of links where n is the number of connection types. Each link $uv \in E_i$ has assigned a bba defined on the frame of discernment $\Theta_i^{uv} = \{E_{uv}, \neg E_{uv}\}$ denoted by m_i^{uv}. The elements E_{uv} and $\neg E_{uv}$ mean respectively that uv exists or is absent. The mass function m_i^{uv} represents the degree of evidence about the existence of a link of type i between (u, v).

In the following, we present our framework to predict the existence of a link uv of type i in the detailed five steps below.

4.1 Step 1: Evaluating Similarity

Our proposals are based on local information methods. To evaluate the likelihood of existence of uv_i, we use social content of node attributes and structural

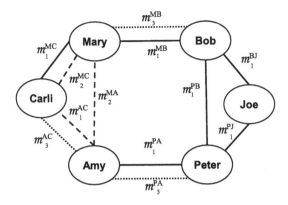

Fig. 1. Evidential multiplex social network

information. The sets of neighbors $\tau(u)$ and $\tau(v)$ linked to u and v according to at least a link of type i are extracted. Then, we match the features of u to those of the nodes in $\tau(v)$, and those of v to $\tau(u)$. Consequently, both semantic and structural information are used by the predicator. Accordingly, the similarity between $node_1$ and $node_2$ is assessed as follows:

$$S_{node_1,node_2} = \frac{\#matched\ attributes}{\#total\ attributes} \tag{8}$$

According to assessed similarities, we detect the most similar node $u' \in \tau(u)$ and $v' \in \tau(v)$ to respectively v and u. This is motivated by the idea that social entities with similar connections are likely to connect. For instance, when the characteristics or behavior of a person is very similar to another one's friend i.e., same hometown, same college and same work, then they are more likely to be friends.

4.2   Step 2: Reliability Evaluation

When the neighbors are not common neighbors or not completely similar, they are not considered fully trustworthy. Thus, their evidence is revised using a discounting operation. The bba's $m_i^{uu'}$ and $m_i^{vv'}$ are discounted respectively using $\alpha' = 1 - S_{uv'}$ and $\alpha'' = 1 - S_{vu'}$. For instance, for $m_i^{uu'}$, discounting is computed as follows:

$$\begin{cases} \alpha'm_i^{uu'}(\{E_{uu'}\}) = (1-\alpha') \cdot m_i^{uu'}(\{E_{uu'}\}) \\ \alpha'm_i^{uu'}(\{\neg E_{uu'}\}) = (1-\alpha') \cdot m_i^{uu'}(\{\neg E_{uu'}\}) \\ \alpha'm_i^{uu'}(\Theta_i^{uu'}) = \alpha' + (1-\alpha') \cdot m_i^{uu'}(\Theta_i^{uu'}) \end{cases} \tag{9}$$

4.3   Step 3: Information Pooling

In order to pool evidence given by similar neighbors, we need to unify our referential, for that a vacuous extension (Eq. 3) is performed on the product space $\mathcal{PS} = \Theta^{uu'} \times \Theta^{vv'}$. The obtained masses are fused using the conjunctive rule (Eq. 2). We get the overall bba $m_{\bigcirc}^{\mathcal{PS}}$ as follows:

$$m_{\bigcirc}^{\mathcal{PS}} = m^{uu_i' \uparrow \mathcal{PS}} \bigcirc m^{vv_i' \uparrow \mathcal{PS}} \tag{10}$$

Next, the obtained bba's are transfered to the frame Θ_i^{uv} by applying a multi-valued mapping mechanism (Eq. 6) according to the procedure presented in [19].

4.4   Step 4: Evidence Update

The next step is to take into account the multiplex structure. When u and v already have common neighbors, we evaluate the rate of common neighbors connected by the links of type i denotes by CN^{uv_i} according to the total number of common neighbors denotes by CN_{tot}^{uv}, we get: $\delta^{uv_i} = \frac{|CN^{uv_i}|}{|CN_{tot}^{uv}|}$. Then, m^{uv_i} is discounted using $\beta = 1 - \delta^{uv_i}$. We obtain $^\beta m^{uv_i}$.

On the other hand, when u and v already connect via m link(s) ($m \in [1, \gamma-1]$) (γ is the number of link types), the rate S_{*i}^{m+1} of simultaneous links of more than two types in the multiplex is computed, where $* = \{1, \ldots, m\}$ are the types of the shared links. If $S_{*j}^{m+1} \neq 0$, the bba is reinforced on the element "exists" using $\phi = \frac{S_{*j}^m}{S_G^m}$ as a reinforcement rate. Finally, we obtain the mass $^\gamma m_i^{uv}$.

4.5   Step 5: Link Selection

Decision about the link existence is made according to the score $BetP^{uv_i}(E)$. Local information approaches from literature rank similarity scores according to a decreasing order and predict links with highest score values. In the same way, link selection is made by ranking query links according to pignistic probabilities on the element "exists" where the top k ones are predicted.

4.6   Complexity

Our link prediction framework is based on local information approaches as we only consider neighboring nodes for prediction. However, unlike these methods, our framework handles uncertainty found in social data. Link prediction methods based on local information are known to have the lowest computational complexity among existing methods from literature [32] especially compared to global methods. Actually, local similarity scores are easy and simple to compute. As our method builds upon local methods, the theoretical complexity would be approximately the same. For the most part, it is based on the common neighbors method. The theoretical complexity of the latter is $\mathcal{O}(N.k^2)$ [9], where N denotes the number of nodes and k is the average degree of all nodes in the network.

The additional computational costs for manipulating and pooling evidence are minor since we consider frames of discernment with only two elements. Indeed, the multiplex structure adds a layer of complexity as we count simultaneous links which costs $\mathcal{O}(N)$. Thus, the overall theoretical complexity is $\mathcal{O}(N^2.k^2)$.

5   Experimental Study

To evaluate the performance of our novel framework, we perform tests on the relationships network collected from the social evolution dataset [16]. It contains 21 K links connecting 84 persons according to 5 link types. Data include information regarding the social entities collected through sociometric surveys such as political opinions, diet attitudes, favorite music genres or smoking behavior.

Data are pre-processed to construct the evidential multiplex social network by simulating bba's according to the procedure presented in [17]. The technique is based on graph sampling and simulation methods popular in literature. We conduct a comparative study of our novel method, denoted by mLPNA, with three other link prediction in evidential social networks. The first method, denoted uLPLI, is an evidential link prediction approach based on local information for uniplex social networks [18]. It is inspired from the Common Neighbors method and uses solely structural properties. uLPLI is applied to each layer of the multiplex apart then overall performance is averaged over layers. The second method, denoted mLPLI, is the multiplex version of the uLPLI approach [19]. The third baseline method is the uniplex version of our proposed approach [20], denoted by uLPNA. All methods apply to evidential networks. What we are mainly concerned about here, is to evaluate the contribution of considering social content of node attributes and handling multiplex structure in the prediction task. Precision and recall results are reported in Fig. 2.

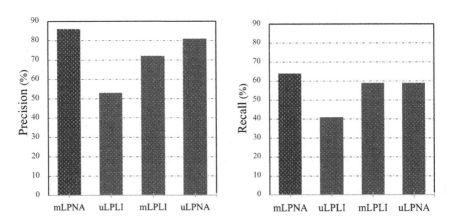

Fig. 2. Precision and recall results measured on the Relationships network

As illustrated in Fig. 2, we notice that mLPNA reaches the highest precision and recall results i.e., 86% precision and 64% recall. On the other hand, it outperforms its uniplex version i.e., 59% for uLPNA. mLPLI and uLPLI have lowest results which proves that considering node attributes information enhances performances. The same observation is made for recall results. That is, performance results in terms of both precision and recall are highest when social content and multiplex structure are considered in prediction.

The results are consistent with our assumptions. Social content combined to network topology add a semantic meaning to the link predicator. That is to say, the novel framework is more effective and accurate for predicting new links between social entities across layers. It improves the prediction quality. Furthermore, it is more suitable to multiplex social networks as it tackles social data uncertainty issues and handles such complex structure. The empirical results demonstrate that our approach has better predictive power than other methods that do not take social content and/or multiplex structure information into account. In addition to social content, our work opens up another dimension for addressing the link prediction problem.

6 Conclusion

In this paper, we have developed a framework to address the link prediction problem in multiplex social networks by combining topological properties and social content. Social data imperfections are handled thanks to the belief function theory. Similarities are assessed by matching nodes features where common neighbors are endorsed as trustworthy sources of information. Overall evidence is revised according to the global multiplex structure and potential links are predicted through a fusion procedures. The results of our framework confirm that semantic information based on nodes attributes promotes link prediction results and permits to understand relevant clues and better discern social entities linkage.

A straightforward direction for future work would be to study the case where there are missing attribute values which may be due to hidden profile information or to noise generated from anonymization. Therefore, we get uncertain node attributes and links structures in the network.

References

1. Adamic, L., Adar, E.: Friends and neighbors on the web. Soc. Netw. **25**(3), 211–230 (2003)
2. Adar, E., Ré, C.: Managing uncertainty in social networks. Data Eng. Bull. **30**(2), 23–31 (2007)
3. Aghabozorgi, F., Khayyambashi, M.R.: A new similarity measure for link prediction based on local structures in social networks. Phys. A Stat. Mech. Appl. **501**, 12–23 (2018)
4. Ahmed, N.M., Chen, L.: An efficient algorithm for link prediction in temporal uncertain social networks. Inf. Sci. **331**, 120–136 (2016)

5. Bao, J., Zheng, Y., Wilkie, D., Mokbel, M.: Recommendations in location-based social networks: a survey. GeoInformatica **19**(3), 525–565 (2015)
6. Ceci, M., Pio, G., Kuzmanovski, V.: Deroski, S.: Semi-supervised multi-view learning for gene network reconstruction. Plos One **10**(12), 1–27 (2015)
7. Chen, L., Zhang, Z., Liu, J., Gao, J., Zhou, T.: A novel similarity index for better personalized recommendation. CoRR abs/1510.02348 (2016)
8. Dempster, A.P.: Upper and lower probabilities induced by a multivalued mapping. Ann. Math. Stat. **38**, 325–339 (1967)
9. Gao, F., Musial, K., Cooper, C., Tsoka, S.: Link prediction methods and their accuracy for different social networks and network metrics. Sci. Program. **2015**, 1–13 (2015)
10. Hasan, M.A., Chaoji, V., Salem, S., Zaki, M.J.: Link prediction using supervised learning. In: Proceedings of the 6th Workshop on Link Analysis, Counter terrorism and Security, pp. 1–10 (2006)
11. Jaccard, P.: Étude comparative de la distribution florale dans une portion des Alpes et du Jura. Bulletin de la Société Vaudoise des Sciences Naturelles **37**, 547–579 (1901)
12. Jeh, G., Widom, J.: Simrank: A measure of structural-context similarity. In: Proceedings of the Eighth ACM SIGKDD International Conference on Knowledge Discovery and Data Mining, KDD 2002, pp. 538–543 (2002)
13. Katz, L.: A new status index derived from sociometric analysis. Psychometrika **18**(1), 39–43 (1953)
14. Kossinets, G.: Effects of missing data in social networks. Soc. Netw. **28**, 247–268 (2003)
15. Liben-Nowell, D., Kleinberg, J.: The link prediction problem for social networks. In: Proceedings of the Twelfth Annual ACM International Conference on Information and Knowledge Management, pp. 556–559 (2003)
16. Madan, A., Cebrian, M., Moturu, S., Farrahi, K., Pentland, A.: Sensing the health state of a community. Pervasive Comput. **11**(4), 36–45 (2012)
17. Mallek, S., Boukhris, I., Elouedi, Z., Lefevre, E.: Evidential link prediction based on group information. In: Prasath, R., Vuppala, A.K., Kathirvalavakumar, T. (eds.) MIKE 2015. LNCS (LNAI), vol. 9468, pp. 482–492. Springer, Cham (2015). https://doi.org/10.1007/978-3-319-26832-3_45
18. Mallek, S., Boukhris, I., Elouedi, Z., Lefevre, E.: The link prediction problem under a belief function framework. In: Proceedings of the IEEE 27th International Conference on the Tools with Artificial Intelligence, pp. 1013–1020 (2015)
19. Mallek, S., Boukhris, I., Elouedi, Z., Lefevre, E.: An evidential method for multi-relational link prediction in uncertain social networks. In: Proceedings of the 5th International Symposium on Integrated Uncertainty in Knowledge Modelling and Decision Making, pp. 280–292 (2016)
20. Mallek, S., Boukhris, I., Elouedi, Z., Lefevre, E.: Evidential link prediction in uncertain social networks based on node attributes. In: Benferhat, S., Tabia, K., Ali, M. (eds.) IEA/AIE 2017. LNCS (LNAI), vol. 10350, pp. 595–601. Springer, Cham (2017). https://doi.org/10.1007/978-3-319-60042-0_65
21. Menon, A.K., Elkan, C.: Link prediction via matrix factorization. In: Proceedings of the 2011 European Conference on Machine Learning and Knowledge Discovery in Databases, pp. 437–452 (2011)
22. Mercier, D., Denœux, T., Masson, M.H.: Belief function correction mechanisms. In: Bouchon-Meunier, B., Magdalena, L., Ojeda-Aciego, M., Verdegay, J.L., Yager, R.R. (eds.) Foundations of Reasoning under Uncertainty. Studies in Fuzziness and Soft Computing, vol. 249, pp. 203–222 (2010)

23. Newman, M.E.J.: Clustering and preferential attachment in growing networks. Phys. Rev. E **64**, 025102 (2001)
24. O'Madadhain, J., Hutchins, J., Smyth, P.: Prediction and ranking algorithms for event-based network data. SIGKDD Explor. Newsl. **7**(2), 23–30 (2005)
25. de Sa, H.R., Prudêncio, R.B.C.: Supervised link prediction in weighted networks. In: The 2011 International Joint Conference on Neural Networks, pp. 2281–2288 (2011)
26. Shafer, G.R.: A Mathematical Theory of Evidence. Princeton University Press, Princeton (1976)
27. Sharma, R., Magnani, M., Montesi, D.: Missing data in multiplex networks: a preliminary study. In: 2014 Tenth International Conference on Signal-Image Technology and Internet-Based Systems (SITIS), pp. 401–407 (2014)
28. Smets, P.: Application of the transferable belief model to diagnostic problems. Int. J. Intell. Syst. **13**(2–3), 127–157 (1998)
29. Svenson, P.: Social network analysis of uncertain networks. In: Proceedings of the 2nd Skövde Workshop on Information Fusion Topics (2008)
30. Wang, C., Satuluri, V., Parthasarathy, S.: Local probabilistic models for link prediction. In: Proceedings of the 2007 Seventh IEEE International Conference on Data Mining, ICDM 2007, pp. 322–331 (2007)
31. Williamson, S.A.: Nonparametric network models for link prediction. J. Mach. Learn. Res. **17**(202), 1–21 (2016)
32. Yang, J., Yang, L., Zhang, P.: A new link prediction algorithm based on local links. In: Proceedings of the 2015 International Workshops on Web-Age Information Management, pp. 16–28 (2015)
33. Yu, K., Chu, W.: Gaussian Process Models for Link Analysis and Transfer Learning, pp. 1657–1664. Curran Associates Inc. (2008)
34. Yu, Z., Kening, G., Feng, L., Ge, Y.: A new method for link prediction using various features in social networks. In: Proceedings of the 11th Web Information System and Application Conference, WISA 2014, pp. 144–147 (2014)
35. Zalaghi, Z.: Link prediction in social networks using Markov random field. Ciłncia e Natura **37**, 125–132 (2015)
36. Zhang, P., Wang, X., Wang, F., Zeng, A., Xiao, J.: Measuring the robustness of link prediction algorithms under noisy environment. Sci. Rep. **6**, 18881 (2016)
37. Zhou, T., Lü, L., Zhang, Y.C.: Predicting missing links via local information. Eur. Phys. J. B-Condens. Matter Complex Syst. **71**(4), 623–630 (2009)

Spatio-temporal Analysis

Predicting Temporal Activation Patterns via Recurrent Neural Networks

Giuseppe Manco, Giuseppe Pirrò[(✉)], and Ettore Ritacco

ICAR - CNR, via Pietro Bucci 8/9C, 87036 Arcavacata di Rende (CS), Italy
{giuseppe.manco,giuseppe.pirro,ettore.ritacco}@icar.cnr.it

Abstract. We tackle the problem of predict whether a target user (or group of users) will be active within an event stream before a time horizon. Our solution, called PATH, leverages recurrent neural networks to learn an embedding of the past events. The embedding allows to capture influence and susceptibility between users and places closer (the representation of) users that frequently get active in different event streams within a small time interval. We conduct an experimental evaluation on real world data and compare our approach with related work.

1 Introduction

There is an increasing amount of streaming data in the form of sequences of events characterized by the time in which they occur and their mark. This general model has instantiations in many contexts, from sequences of tweets characterized by a (re)tweet time and identity of the (re)tweeter and/or the topic of the tweet, to sequences of locations characterized by the time and location of each check-in. We focus on influence-based activation networks, that is, event sequences where the occurrence of an event can boost or prevent the occurrence of another event. Understanding the structural properties of these networks can provide insights on the complex patterns that govern the underlying evolution process and help to forecast future events.

The problem of inferring the topical, temporal and network properties characterizing an observed set of events is complicated by the fact that, typically, the factors governing the influence of activations and their dependency from times are hidden. Indeed, we only observe activation times (e.g. retweet time) and related marks, while, activations can depend on several factors including the stimulus provided by the ego-network of a user or his attention/propensity towards specific themes.

The goal of this paper is to introduce PATH (Predict User Activation from a Horizon), which focuses on scenarios where there is the need to *predict whether a target user (or group of users) will be active before a time horizon T_h*. PATH can be used, for instance, in market campaigns where target users are the potential influencers that if active, before T_h, can contribute to further spread an advertisement and trigger the activation of influencees that can be made aware of a certain product/service. PATH learns an embedding of the past event

M. Ceci et al. (Eds.): ISMIS 2018, LNAI 11177, pp. 347–356, 2018.
https://doi.org/10.1007/978-3-030-01851-1_33

history via Recurrent Neural Networks that also cater for the diffusion memory. The embedding allows to capture influence and susceptibility between users and places closer (the representation of) users that frequently get active in different streams within a small time interval.

1.1 Related Work

A number of proposals have addressed the problem of modeling streams of events via neural networks: (i) approaches like DeepCas [7] and DeepHawkes [2] tackle the problem of predicting the length that a cascade will reach within a timeframe or its incremental popularity; (ii) approaches like Du et al. [3] and Neural Hawkes Process (NHP) [8] model and predict time event markers and time; (iii) Survival Factorization (SF) [1] leverages influence and susceptibility for time and event predictions. PATH adopts a different departure point from these approaches: it focuses on predicting the activation of (groups of) users before a time horizon instead of the exact activation time.

Differently from (i) PATH considers time and uses an embedding to capture both influence and susceptibility between users and predict future activations. Moreover, (i) focuses on the prediction of cumulative values only (e.g., cascade size). Differently from (ii), we do not assume that time and event are independent and capture their interdependencies via the embedding and cascade history. Besides, (ii) focuses on predicting event types only (e.g., popular users), which is not enough in the scenarios targeted by PATH (e.g., targeted market campaigns) where one is interested in predicting the behavior of specific users instead of their types. A for (iii), it fails in capturing the cumulative effect of history while PATH captures by using an embedding.

The contributions of the paper are as follows: **(i)** PATH, a classification-based approach based on recurrent neural networks allowing to model the likelihood of observing an event as a combined result of the influence of other events; **(ii)** an experimental evaluation and a comparison with related work.

The remainder of the paper is organized as follows. We introduce the problem in Sect. 2. We present PATH in Sect. 3. We compare our approach with related research in Sect. 4. We conclude and sketch future work in Sect. 5.

2 Problem Definition

We focus on network of individuals who react to solicitations along a timeline. An activation network can be viewed as an instance of a marked point process (e.g., information cascades) on the timeline, defined as a set $\mathcal{H} = \{\mathcal{H}^c\}_{1 \leq c \leq m}$. Here, $\mathcal{H}^c = \{(t_i, u_i)\}_{1 \leq i \leq m_c}$ represents a cascade where $u_i \in \mathcal{U}^c$ is a user in a set of all users that get active in the cascade $\mathcal{H}^c$ at time $t_i \in \mathbf{t}^c$ where $\mathbf{t}^c$ is the projection over the timestamps in $\mathcal{H}^c$. We denote by $\mathcal{U}$ the set of users in all cascades.

Given a cascade $\mathcal{H}^c$, we denote by $\mathcal{H}^c_{<t}$ (resp. $\mathcal{H}^c_{\leq t}$) the set of events $e_i \in \mathcal{H}^c$ such that $t_i < t$ (resp., $t_i \leq t$). The terms $t^c_{<t}$ and $\mathcal{U}^c_{<t}$ can be defined accordingly.

2.1 Modeling Diffusion

We start from the observation that what is likely to happen in the future (viz. which user will be active and when) depends on what happened in the past (viz. the chain of previously active users). One important point to take into account is the susceptibility of users, that is, the extent to which they are influenced by specific previously activated users. Our model should be flexible enough to reflect both *exciting* and *inhibitory* effects. While the former boosts the likelihood of observing u active in c, the latter actually could prevent it to do so.

Given a cascade $\mathcal{H}^c$, a timestamp $t \geq 0$ and a user $u \notin \mathcal{U}_{<t}^c$, the goal is to obtain an estimate of the density function $f(t, u|\mathcal{H}_{<t}^c)$, which can be used to model the following evolution scenario: *given a cascade c and time horizon T_h^c; how likely is it that u will become active in c within T_h^c?*

The challenge, at this point, is how to concretely formulate the density f. We can decouple its specification as follows:

$$f(t, u|\mathcal{H}_{<t}^c) = g(t|u, \mathcal{H}_{<t}^c) \cdot h(u|\mathcal{H}_{<t}^c), \tag{2.1}$$

where the first component represents the likelihood that u becomes active within t, given $\mathcal{H}_{<t}^c$, and the second component represents the likelihood that u activates (independent of the time) as a reaction to the current history.

The modeling of $\mathcal{H}_{<t}^c$ can include different pieces of information, among which, the sequence of user activations, their activation times, the relative activation speed, and possibly the topic of the cascade. Nevertheless, our assumption is that $\mathcal{H}_{<t}^c$ can also encode latent information including susceptibility and influence between users that can be derived, for instance, from neighborhood information in a networks (e.g., follower/followee relations in Twitter) or user behaviors (e.g., users that retweet after a certain set of other influential users (re)tweet). This is exactly what we want to unveil in our modeling.

2.2 Embedding History

We want to learn and embedding of users in a latent K-dimensional space such that users in the same cascade are closer in the embedding, and users within different cascades are distant.

The embedding is based on the idea of a virtual graph $G = (\mathcal{V}, \mathcal{E})$ where $(u, v) \in \mathcal{E}$ if and only if there is a cascade c such that $v \prec_c u$ (viz. u precedes v in the cascade c). Thus, the influence exerted by v on u represents the probability that, any random walk on G starting from v, eventually touches u. By assuming that $\alpha_{u,v}$ represents such a probability, we factorize it as $\alpha_{u,v} = \mathbf{s}_u \cdot \mathbf{a}_v$, where $\mathbf{S} = [\mathbf{s}_1, \ldots, \mathbf{s}_N], \mathbf{A} = [\mathbf{a}_1, \ldots, \mathbf{a}_N] \in \mathbb{R}^{N \times K}$ are the susceptibility and influence matrices, respectively. Matrices are computed by relying on the standard network architecture borrowed from the word2vec paradigm [9]:

$$\mathbf{a}_k = \mathbf{W}_e \mathbf{u}_k \qquad \mathbf{s}_k = \mathbf{V}_e \mathbf{u}_k$$

Here, $\mathbf{u}, \mathbf{v}$ represents the one-hot encodings of u and v and $k \in [1, \ldots, K]$. The matrices $\mathbf{W}_e, \mathbf{V}_e$ represent the embeddings, obtained by minimizing an adapted form of contrastive loss [4] that penalizes the distance of users within the same cascades and the closeness of users in different cascades.

2.3 Capturing the Diffusion Memory

To encode temporal relationship within $\mathcal{H}_{<t}^c$ we use recurrent neural networks (RNNs). An RNN is a recursive structure that, at the current step, gets as input the previous network state (the outputs form the hidden units) along with the current input to compute a new state. The following picture provides an overview of a simple RNN cast to our context.

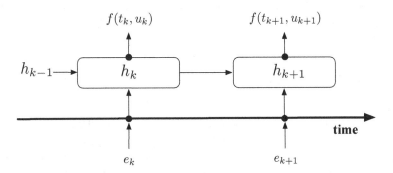

At each step k, we feed into the network an event $(t_k, u_k) \in \mathcal{H}^c$ that encodes the current user (u_k) and its activation time (t_k). The learned hidden state (h_k) represents the non-linear dependency between these components and past events, which can be used to model $f(t_k, u_k | \mathcal{H}_{<t_k}^c)$. In the following, we adopt the LSTM instantiation of the RNN framework [5]. The idea of an LSTM unit is to reliably transmitting important information many time steps into the future. At every time step, the unit modifies the internal status by deciding which part to keep or replace with new information coming from the current input. In the following we shall use the shortcut $\mathbf{h}_k = \text{LSTM}(\mathbf{z}_k, \mathbf{h}_{k-1})$ to denote a basic functional architecture that elaborates an input $\mathbf{z}_k$ and outputs the updated state.

3   PATH: Predicting User Activation from a Horizon

We now introduce PATH (Predicting User Activation from a Horizon), which focuses on simplifying $f(t, u | \mathcal{H}_{<t}^c)$ as the binary response function $\mathbb{I}(t \leq T_h | u, \mathcal{H}_{<t}^c)$ that denotes whether u becomes active in c within T_h. We focus on events $(t_k, u_k) \in \mathcal{H}^c$ and consider as an additional feature the time delay $\delta_k = t_k - t_{k-1}$ relative to the previous activation within the cascade. This allows us to capture the property that cascades may have intrinsically different diffusion speeds causing some of them to concentrate users' activations in a short timeframe while others in a more extended interval. In what follows we consider a cascade $\mathcal{H}^c$ enhanced with information about time delays, that is,

$\mathcal{H}^c = \{(t_k, u_k, \delta_k)_{1 \leq k \leq m_c}$. Given a partially observed cascade $\mathcal{H}^c_{\leq t_l}$ (with $t_l < T^c_h$ representing the timespan of the observation window), our objective is to predict, for a given entity $u \notin \mathcal{U}^c_{t_l}$, whether $u \in \mathcal{U}^c_{T^c_h}$. In order to uncover all the characteristics of the activations within cascades, we consider a model built on all possible prefixes of the available cascades.

In other words, for a given cascade c of length m_c, we feed into the network cascade prefixes starting from 2 elements until reaching $m_c - 1$ elements. Notice that, we do not consider the 1 element prefix, which we assume becomes "spontaneously" active. Moreover, we also add negative examples as follows: for each $u \notin \mathcal{U}^c$, we associate the cascades $\mathcal{H}^c_{\leq t_j} \cup \{(t_j, u, \delta_j)\}$ (with $1 \leq j \leq m_c - 1$) and $\mathcal{H}^c \cup \{(T^c_h, u, (T^c_h - t_{m_c}))\}$ with negative labels. Again, the intuition is that, since u is not active no partial cascade provides the sufficient intensity to activate u within the given time horizon.

Adding negative examples in the data preparation represent an effective data augmentation process, which enlarges the training data by inferring new inputs in the training set. This is crucial to let the approach better fine tune separation between active and inactive users, as well as better characterize the true activation time of active users. Let $\mathcal{T}_C$ denote the set of all pairs $\langle \mathcal{H}^c_{\leq t_i}, y_i \rangle$ that can be built as described above. Our idea is to exploit the embedding and LSTM tools described in the previous section to solve the supervised problem at hand.

Figure 1 illustrates the basic architecture of the model where arrows represent inputs and boxes the elements of the architecture. The input layer (bottom part) takes as input the user for which we want to predict the activation (u_n) along with information about already active users represented by triples of the form of (u_k, t_k, δ_k) and the network history (h_k) (recurrent layer in the figure).

Given a pair $\langle \mathcal{H}^c_{\leq t_i}, y_i \rangle \in \mathcal{T}_C$ with $|\mathcal{H}^c_{\leq t_i}| = n$ and by considering $(t_k, u_k, \delta_k) \in \mathcal{H}^c_{\leq t_i}$ (with $1 \leq k \leq n$), the architecture of the network can be captured by the following equations:

$$\mathbf{a}_k = \mathbf{W}_e \mathbf{u}_k \tag{3.1}$$

$$\mathbf{h}_k = \text{LSTM}\left([\mathbf{a}_k, t_k, \delta_k], \mathbf{h}_{k-1}\right) \tag{3.2}$$

$$\hat{y}_i = \sigma\left(\mathbf{W}_o \mathbf{h}_n\right) \tag{3.3}$$

$$\tilde{y}_i = \exp\left\{-\left\|\mathbf{a}_n - \sum_{k=1}^{n-1} \mathbf{a}_k\right\|^2\right\} \tag{3.4}$$

In the output layer, $\hat{y}_i$ represents the probability that y_i is positive, as provided by the network: that is, it encodes the probability that u_n becomes active within t_n; $\tilde{y}_i$ encodes the affinity between u_n and all users preceding it within $\mathcal{H}^c_{\leq t_i}$. The distance $\|\mathbf{a}_n - \sum_{k=1}^{n-1} \mathbf{a}_k\|$ plays a crucial role here: since the target user is on the tail of the cascade, the embedding should emphasize the similarities with the predecessors that trigger an activation, and by the converse minimize the similarities with those ones which do not trigger it.

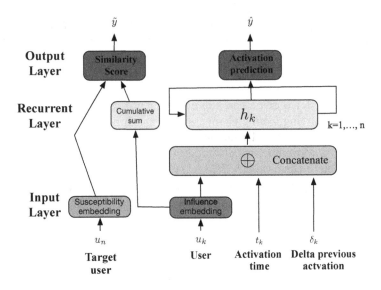

Fig. 1. Overview of the architecture of PATH.

The loss is a combination of cross-entropy and the embedding loss previously described:

$$\mathcal{L} = \sum_{\substack{\langle \mathcal{H}, y \rangle \in \mathcal{T}_C \\ |\mathcal{H}| = n}} \{ y \left(\gamma \log(\hat{y}) + \beta \log \tilde{y} \right) + (1 - y) \left(\gamma \log(1 - \hat{y}) + \beta \log(1 - \tilde{y}) \right) \} \qquad (3.5)$$

where γ and β are weights balancing cross-entropy and embedding.

4   Experiments

We validate our approach by analysing the algorithm on real-life datasets. In particular, we analyse the capability of the algorithm at predicting the activation time of users within an information cascade. The implementation we use in the experiments can be found at https://github.com/gmanco/PATH.

4.1   Datasets

We evaluated PATH by exploiting two real-world datasets containing propagation cascades crawled from the timelines of Twitter[1] and Flixster.[2] In particular, Twitter includes ∼32K nodes with ∼9K cascades while Flixster includes ∼2K nodes with ∼5K cascades. The propagation mechanism on Twitter is expressed by retweeting, in other words a chain of repetitions and transmissions of a tweet from a set of users to their neighbors in a recursive process.

---

[1] http://www.twitter.com/.
[2] http://www.flixster.com/.

Each activation corresponds to a retweet. An activation in `Flixster` happens when a user rates a movie, while a cascade is composed by all the activations related to the same movie.

The two datasets differ essentially for the following characteristics: `Twitter` includes a larger number of users and shorter delays than `Flixster`. In addition, retweets intuitively highlight two relevant aspects, namely the importance of the topic and the single influence of the individual from which the retweet is performed. By contrast, movie ratings are more likely to exhibit a cumulative effect: popular movies are more likely to be considered than unpopular ones.

4.2   Evaluation Methodology

We evaluate `PATH` against two baseline models, both relying on Survival Analysis [6]. The first instantiation implements a Cox proportional hazard model (*CoxPh* in the following). We implement the model using the `lifelines`[3] package and extract, for each event $(t_k, u_k, \delta_k) \in \mathcal{H}^c$, the following features: (1) size of the prefix; (2) last activation time; (3) average delay for each active user so far; (4) number of neighbors in the history, and (5) coverage percentage of them within the history; (6) the activation time of the most recent neighbor, if any; (7) correlation between the activation of the current user an its neighbors within the history, computed in previous cascades. This model represents an intuitive baseline where features are manually engineered and include a mix of external information (coming from the underlying network neighborhood) and information derived from the cascade itself.

The second instantiation is given by the *Survival Factorization* (*SF* in the following) framework described in [1]. The comparison is important since *SF* relies on the same guiding ideas of `PATH` (influence and susceptibility) with the substantial difference that there is no cumulative effect of $\mathcal{H}^c_{<t_k}$, but instead an influential user has to be detected for each activation.

To evaluate the approaches, we proceed as follows: given training and test sets $\mathcal{C}_{train}$ and $\mathcal{C}_{test}$, we train the model on $\mathcal{C}_{train}$ and measure the accuracy of the predictions on $\mathcal{C}_{test}$. The two sets are obtained by randomly splitting the original dataset by ensuring that there is no overlap among the cascades of the two sets, but there is no entity in the test that has not been observed in the training. We used 70% of data for training and 30% for testing.

For the evaluation, we chronologically split each cascade $c \in \mathcal{C}_{test}$ into c_1 and c_2 such that, for each $u \in c_1$ and $v \in c_2$, we have that $u \prec_c v$. Next, we pick a random subsample $c_3 \subseteq \mathcal{U} - \mathcal{U}^c$. Then, given a target horizon T_h^c, we measure *TP*, *FP*, *TN* and *FN* by feeding the models on c_1 and then predicting the activation within T_h^c for each element in $c_2 \cup c_3$.

The choice of T_h^c can follow different strategies; Fixed horizon (Fixed horizon (**FH**): setting T_c^h as the maximum observed activation time $T_h^{test} = \max\{t | t \in \mathbf{t}^c, c \in \mathcal{C}_{test}\}$; Variable horizon (**VH**): varying T_h^c from the smallest to the largest activation time and computing the activation probabilities associated to each

---

[3] See http://lifelines.readthedocs.io for details.

possible value; Actual Time (**AT**): a particular case of the VH strategy, where $T_h^c \triangleq T_h^{u,c}$ is relative to the true activation time in c of each user $u \in c_2 \cup c_3$.

In the experiments, we plot the ROC and the F-Measure curves relative to the above alternatives and report the AUC and F values. Notice that, for *PATH*, the encoding of sequences as described in Sect. 3 already presumes that users are evaluated on intermediate timestamps prior to their actual activation. Thus, VH and AT roughly coincide in this case. Since both *CoxPh* and *SF* are capable of inferring, for each $(u, \mathcal{H})$ pair, the probability $S_u(t|\mathcal{H}^c)$, the comparison with PATH is accomplished by computing $1 - S_u(\hat{t}|\mathcal{H}^c)$ where $\hat{t}$ is the above described horizon timestamp.

The parameter space for PATH was explored by grid-search, measuring the loss on a on a separate portion of the training set by 5-fold cross-validation. In the following we report 5 different instantiations, which differ from the number of cells in the LSTM (32/64), the dimensionality of the embedding (32/64) and the batch size in the training (128/256/512). Concerning *SF*, the number of factors was set to 16 for both datasets.

4.3   Evaluation Results

Figures 2 and 3 report the ROC curves for the experiments. We can observe that PATH consistently outperforms the baselines and in particular exhibits a very good accuracy on all configurations. This is especially true on Flixster, where by the converse *SF* does not seem capable of correctly correlating previous activations times. The cumulative influence effect is evident here, as a natural consequence of the underlying domain where cascading effects are more likely as a consequence of a "word of mouth" process. On Twitter, where the activation is more likely due to the influence of a single user (as testified by the good performance of *SF*), PATH still achieves the best scores, thus proving the capability of the recurrent layer to adapt the influence to a single user.

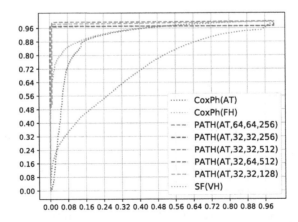

Fig. 2. ROC curves for PATH, *CoxPh* and *SF* on Flixster.

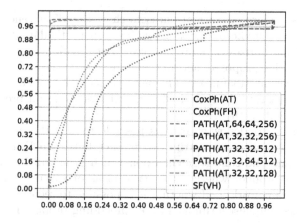

Fig. 3. ROC curves for PATH, *CoxPh* and *SF* on Twitter.

Figure 4(a) and (b) display the F-measure curves for varying values of the threshold on the probabilities. Here, we can observe that, contrary to the baselines, higher thresholds do not cause a significant drop of the recall. The only exception is *CoxPh* (FH), which seems more stable on Flixster. This is a clear sign that the probabilities associated with active and inactive users in PATH differ substantially, and in particular active events are associated with significantly higher probabilities than inactive events.

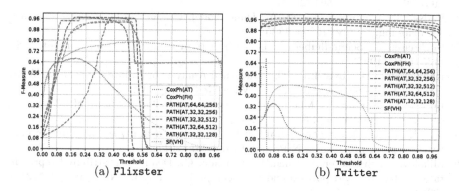

Fig. 4. F-Measure curves for PATH, *CoxPh* and *SF* on both datasets.

5   Concluding Remarks and Future Work

We focused on the problem of predicting user activations in a given time horizon and show that the embedding of the user activation history, where users that become active on the same cascades are placed close, can be effectively learned via recurrent neural networks.

Experiments performed on real datasets show the effectiveness of the approach in accurately predicting next activations. In particular, it emerged that focusing on a time horizon is more effective than predicting actual activation times, and that the proposed network architecture successfully captures the resulting classification capabilities. The network is capable of summarizing both cumulative and separate influential effects, and moreover can predict based on the properties based on the history.

It is natural to wonder whether it is possible to cast the intuitions behind our approach in a generative setting, to predict both which user is likely to become active, and the time segment upon which s/he will become active. It is also natural to wonder whether the basic ideas of survival analysis can be cast within our framework, and whether it is possible to recover the distinction between influence and susceptibility in the embedding proposed in the model.

References

1. Barbieri, N., Manco, G., Ritacco, E.: Survival factorization on diffusion networks. In: Ceci, M., Hollmén, J., Todorovski, L., Vens, C., Džeroski, S. (eds.) ECML PKDD 2017. LNCS (LNAI), vol. 10534, pp. 684–700. Springer, Cham (2017). https://doi.org/10.1007/978-3-319-71249-9_41
2. Cao, Q., et al.: DeepHawkes: bridging the gap between prediction and understanding of information cascades. In: ACM CIKM, pp. 1149–1158 (2017)
3. Du, N., Dai, H., Trivedi, R., Upadhyay, U., Gomez-Rodriguez, M., Song, L.: Recurrent marked temporal point processes: embedding event history to vector. In: ACM SIGKDD, pp. 1555–1564 (2016)
4. Hadsell, R., Chopra, S., LeCun, Y.: Dimensionality reduction by learning an invariant mapping. In: CVPR, pp. 1735–1742 (2006)
5. Hochreiter, S., Schmidhuber, J.: Long short-term memory. Neural Comput. 9(8), 1735–1780 (1997)
6. Kalbfleisch, J.D., Ross, L.P.: The Statistical Analysis of Failure Time Data. Wiley Series in Probability and Statistics. Wiley, New York (2002)
7. Li, C., Ma, J., Guo, X., Mei, Q.: DeepCas: an end-to-end predictor of information cascades. In: WWW, pp. 577–586 (2017)
8. Mei, H., Eisner, J.M.: The Neural Hawkes Process: a neurally self-modulating multivariate point process. In: NIPS, pp. 6757–6767 (2017)
9. Mikolov, T., Sutskever, I., Chen, K., Corrado, G., Dean, J.: Distributed representations of words and phrases and their compositionality. In: NIPS, pp. 3111–3119 (2013)

Handling Multi-scale Data via Multi-target Learning for Wind Speed Forecasting

Annalisa Appice[1,2](✉), Antonietta Lanza[1], and Donato Malerba[1,2]

[1] Dipartimento di Informatica, Università degli Studi di Bari Aldo Moro,
via Orabona, 4, 70126 Bari, Italy
{annalisa.appice,antonietta.lanza,donato.malerba}@uniba.it
[2] Consorzio Interuniversitario Nazionale per l'Informatica - CINI, Bari, Italy

Abstract. Wind speed forecasting is particularly important for wind farms due to cost-related issues, dispatch planning, and energy markets operations. This paper presents a multi-target learning method, in order to model historical wind speed data and yield accurate forecasts of the wind speed on the day-ahead (24 h) horizon. The proposed method is based on the analysis of historical data, which are represented at multiple scales in both space and time. Handling multi-scale data allows us to leverage the knowledge hidden in both the spatial and temporal variability of the shared information, in order to identify spatio-temporal aided patterns that contribute to yield accurate wind speed forecasts. The viability of the presented method is evaluated by considering benchmark data. Specifically, the empirical study shows that learning multi-scale historical data allows us to determine accurate wind speed forecasts.

1 Introduction

Wind power has increased in recent years to meet the growing energy demand. The uncertainty caused by the discontinuous nature of wind energy affects the power grid. Hence, forecasting wind behavior (e.g., wind speed) is important for energy managers and electricity traders. These predictions are employed for optimal operation policies and operative costs, load balancing, site and capacity planning and unit commitment for electricity market [17]. However, the variable nature of wind speed [10] provides a challenge for the actual integration of a wind speed forecasting service to electricity grid. As the wind variability occurs in time, as well as in space scales, the profiles of the available power of wind sources depend on the geographic location, the season (or time of the year), the time of the day and other physical parameters.

In this paper, the wind speed forecasting task is addressed by considering wind speed data measured every 10 min along day-ahead time horizons. A time series approach is formulated by resorting to machine learning, in order to learn a forecasting model from the historical data only. A particular contribution is the consideration of the multi-scale representation of the historical data, in order

© Springer Nature Switzerland AG 2018
M. Ceci et al. (Eds.): ISMIS 2018, LNAI 11177, pp. 357–366, 2018.
https://doi.org/10.1007/978-3-030-01851-1_34

to handle the wind speed variability at both the space and time resolutions. The implications of learning multi-scale data on the accuracy of the forecasting operation are explored in a day-ahead forecasting task, that is here formulated as a multi-target learning problem [3]. In particular, a multi-target model is considered, in order to learn a single model that predicts multiple output variables at the same time – one variable for each time point over the 24-ahead horizon. The decision of learning multi-target models is supported by various studies which have repeatedly proved that these models are typically easier to interpret, perform better, and overfit less than single-target predictions [3,5,20].

Neighboring and windowing mechanisms are adopted, in order to represent the historical data at the various scales in both the space and the time, respectively. These mechanisms are combined with the standard deviation operator that is used to quantify the spatial and/or temporal variability of the data. This multi-scale representation of the wind speed variability contributes to define new input variables, as well as new output variables. In this way, a multi-target model can be learned by accounting for the variability of measurements at different sites and times. The viability of the proposed method is assessed in the multi-target framework by comparing the accuracy of traditional multi-target models to the accuracy of multi-scale multi-target models in a benchmark scenario. The sensitivity of the accuracy of the models learned is evaluated along the size of the scale. Finally, the accuracy gained in by taking into account appropriate patterns of the data variability is explored.

The paper is organized as follows. In the next Section, the state-of-the-art of the time series analysis is explored for the wind speed forecasting. In Sect. 3, the basics of this study are introduced. In Sect. 4, the multi-scale multi-target learning phase is described. In Sect. 5 the benchmark dataset considered for the empirical evaluation is presented and the relevant results are illustrated. Finally, Sect. 6 draws some conclusions and outlines some future work.

2   Related Work

In the literature, different forecasting horizons have been investigated: long-term (from one day to one week ahead), medium-term (from 6 h to one day ahead), short-term (from 30 min to 6 h ahead) and very short-term (few seconds to 30 min ahead) [19]. On the other hand, various approaches have been developed for wind speed forecasting in renewable energy systems. In particular, three main predictive categories are described in the literature [2]: the physical [9], the time series [1,4,7,12,13] and the hybrid [6] approaches. The physical approach describes a physical relationship between wind speed, atmospheric conditions, local topography and the output from the wind power turbine. The time series approach consists of time series forecasts, which are based on the historical data (the wind speed collected at a specific site), while neglect commonly the meteorological data. Finally, the hybrid approach applies a combination of physical and time series models.

By focusing the attention on the time series approach (which is the most popular in practice and also the subject of this paper), a wide plethora of time-series methods employ a general class of statistical models, that is, the Auto-Regressive Moving Average (ARMA) or Auto-Regressive Integrated Moving Average (ARIMA), in order to estimate future observations of a wind farm through a linear combination of the past data. The literature [11,18] has shown that these auto-regressive models are very well suited to capture short range correlations. Hence, they have been used extensively in a variety of (very) short-term forecasting applications (less than 6 h). Recent studies have also proved that auto-regressive models can be profitably extended, in order to account for spatial characteristics of time series data and gain in accuracy [14–16]. In alternative, the time series approach also involves the use of machine learning techniques, which are commonly well suited to produce accurate prediction in medium-term and long-term forecasting applications. Examples of artificial intelligence wind speed forecasting methods apply Support Vector Machine [17], K-Nearest Neighborhood [4] and Cluster analysis [13].

By investing in machine learning, this paper explores the use of the multi-target learning in a medium-term wind speed forecasting application (24 h ahead). We note that the benefits of multi-target learning in time series forecasting have been recently assessed in [5] considering the problem of deriving 24 h ahead solar radiation forecasts. Differently from this seminal study, that has modeled the spatial "correlation" of the solar radiation, in order to define new "input" variables only, we leverage here the power of a model of both the spatial and temporal "variability" of the wind speed, in order to define new "input" and "output" variables, which contribute to yield accurate 24 h ahead forecasts of the wind speed.

3   Basics

Premises. Without loss of generality, the applicative scenario considered in this work is described by the following four premises. First, the spatial location of a wind farm is modeled by means of 2-D point coordinates[1] (e.g. latitude and longitude). Second, the spatial locations of the wind farms are known, distinct and invariant. Third, wind farms transmit measurements of the wind speed and they are synchronized in the transmission time. Finally, transmission time points are equally spaced in time.

Learning Task. Based upon these premises, the task we intend to perform is to forecast wind speed at each farm of the grid. The forecasting model is that learned from the input historical data of the wind speed, as they are collected from a grid of wind farms, every 10 min, over $m + 1$ consecutive days. As additional input information, the wind speed variability at the spatial, temporal and spatio-temporal scales are considered. The output of the learning phase is a multi-target predictive model that allows us to yield fine-grained forecasts for the next day (24 h) at 10 min intervals, based on the input historical wind speed data as they are measured at 10 min over the past m days.

---

[1] Multi-dimensional representations of geographic space can be equally dealt.

Input and Output Variables. Formally, let k_i be the i-th farm, (X_i, Y_i) are the geographic coordinates of k_i. Let us consider the historical wind speed data, measured from k_i, over days $1, \ldots, m, m+1$. They are transformed into a training example, that is represented by vectors $\mathbf{x_i}$, $\mathbf{x_i^S}$, $\mathbf{x_i^T}$ and $\mathbf{x_i^{ST}}$, which cover the role of independent input variables, and vectors $\mathbf{y_i}$, $\mathbf{y_i^S}$, $\mathbf{y_i^T}$ and $\mathbf{y_i^{ST}}$, which cover the role of dependent output (or target) variables, respectively. We note that the input variables are calculated over days $1, \ldots, m$, while the output variables are calculated over day $m+1$. These input and output variable vectors are formally described in the following.

Vector $\mathbf{x_i}$ is defined as follows:

$$\mathbf{x_i} = (x_{i_1}, \ldots, x_{i_{144}}, x_{i_{145}}, \ldots, x_{i_{288}}, \ldots, x_{i_{144m}}), \tag{1}$$

where x_{i_t} denotes the wind speed measured from k_i at time t with $t = 1, \ldots, 144m$ (i.e. every day is divided into 144, ten minutes spaced, time points so that t denotes the time point that occurs every $10\,\mathrm{min}$ at days $1, \ldots, m$). Similarly, vector $\mathbf{y_i}$ is defined as follows:

$$\mathbf{y_i} = (y_{i_{144m+1}}, \ldots, y_{i_{144(m+1)}}), \tag{2}$$

where y_{i_t} represents the wind speed measured from k_i at time t with $t = 144m + 1, \ldots, 144(m+1)$ (every $10\,\mathrm{min}$ at day $m+1$).

By applying the standard deviation operator in combination with the neighboring and/or windowing mechanisms, new data vectors can be defined. They represent the variability of the wind speed data considered at spatial, temporal and spatio-temporal scales. In particular, the spatial scale is defined by the neighboring mechanism, the temporal scale is defined by the windowing mechanism, while the spatio-temporal scale is define by combining the neighboring and windowing mechanisms.

Given radius R, applying the neighboring mechanism to k_i, a circular neighborhood of k_i is constructed. This is a set of wind farms k_j so that $d(k_i, k_j) \leq R$ where $d(\cdot, \cdot)$ denotes the geographic distance. Considering the spatial scale defined by this neighboring mechanism, we define vectors $\mathbf{x_i^S}$ and $\mathbf{y_i^S}$, which represent farm k_i at the space scale with radius R over days $1 \ldots, m$ and day $m+1$, respectively. Specifically,

$$\begin{aligned} \mathbf{x_i^S} &= (x_{i_1}^S, \ldots, x_{i_{144}}^S, x_{i_{145}}^S, \ldots, x_{i_{288}}^S, \ldots, x_{i_{144m}}^S), \\ \mathbf{y_i^S} &= (y_{i_{144m+1}}^S, \ldots, y_{i_{144(m+1)}}^S) \end{aligned} \tag{3}$$

where:

$$\begin{aligned} x_{i_t}^S &= stdev(\{x_{j_t} | d(k_i, k_j) \leq R\})\ with\ t = 1 \ldots, 144m, \\ y_{i_t}^S &= stdev(\{y_{j_t} | d(k_i, k_j) \leq R\})\ with\ t = 144m + 1 \ldots, 144(m+1). \end{aligned} \tag{4}$$

Given length L so that L is a factor of 144, the windowing mechanism transforms the sequence of consecutive time points $t_1, \ldots, t_{144(m+1)}$ into the sequence of $\frac{144}{L}(m+1)$ consecutive time windows so that:

$$windowing[1\ldots144(m+1)] =$$

$$= \underbrace{W_1[1 \to L], W_2[L+1 \to 2L], \ldots, W_{\frac{144}{L}}[(\frac{144}{L} - 1)L + 1 \to 144]}_{day\ 1}, \underbrace{\ldots}_{day\ 2}, \ldots, \underbrace{\ldots}_{day\ m},$$

$$\underbrace{W_{\frac{144}{L}m+1}[t_{\frac{144}{L}m} + 1 \to t_{\frac{144}{L}m+L}], \ldots, W_{\frac{144}{L}(m+1)}[t_{\frac{144}{L}m} + (\frac{144}{L} - 1)L + 1 \to t_{\frac{144}{L}(m+1)}]}_{day\ m+1},$$

$$(5)$$

where each each window covers L consecutive time points. Considering the temporal scale defined by this windowing mechanism, we can define vectors $\mathbf{x_i^T}$ and $\mathbf{y_i^T}$, which represent farm k_i at the time scale with length L over days $1\ldots, m$ and day $m+1$, respectively. Specifically,

$$\mathbf{x_i^T} = (x_{i_1}^T, \ldots, x_{i_{\frac{144}{L}}}^T, x_{i_{\frac{144}{L}+1}}^T, \ldots, x_{i_{\frac{144}{L}2}}^T \cdots x_{i_{\frac{144}{L}(m-1)+1}}^T, \ldots, x_{i_{\frac{144}{L}m}}^T),$$

$$\mathbf{y_i^T} = (y_{i_{\frac{144}{L}m+1}}^T, \ldots, y_{i_{\frac{144}{L}(m+1)}}^T) \tag{6}$$

where:

$$x_{i_t}^T = stdev(\{x_{i_r}|r \in W_t\})\ with\ t = 1, \ldots, \frac{144}{L}m,$$

$$y_{i_t}^T = stdev(\{y_{i_r}|r \in W_t\})\ with\ t = \frac{144}{L}m+1, \ldots, \frac{144}{L}(m+1). \tag{7}$$

Finally, given radius R and length L, we define vectors $\mathbf{x_i^{ST}}$ and $\mathbf{y_i^{ST}}$, which represent farm k_i at the space scale with radius R and the time scale with length L over days $1, \ldots, m$ and day $m+1$, respectively. Specifically,

$$\mathbf{x_i^{ST}} = (x_{i_1}^{ST}, \ldots, x_{i_{\frac{144}{L}}}^{ST}, x_{i_{\frac{144}{L}+1}}^{ST}, \ldots, x_{i_{\frac{144}{L}2}}^{ST} \cdots x_{i_{\frac{144}{L}(m-1)+1}}^{ST}, \ldots, x_{i_{\frac{144}{L}m}}^{ST})$$

$$\mathbf{y_i^{ST}} = (y_{i_{\frac{144}{L}m+1}}^{ST}, \ldots, y_{i_{\frac{144}{L}(m+1)}}^{ST}) \tag{8}$$

where:

$$x_{i_t}^{ST} = stdev(\{x_{j_r}|d(k_i, k_j) \le R\ and\ r \in W_t\})\ with\ t = 1, \ldots, \frac{L}{144}m,$$

$$y_{i_t}^{ST} = stdev(\{y_{j_r}|d(k_i, k_j) \le R\ and\ r \in W_t\})\ with\ t = \frac{144}{L}m+1, \ldots, \frac{144}{L}(m+1).$$

$$(9)$$

4 Multi-scale Multi-target Learning

Let us consider a wind farm grid $\mathcal{K}$, which is composed of N wind farms $k_1, k_2, \ldots, k_N$, and a historical dataset $\mathcal{D}$, which comprises wind speed measurements collected from $\mathcal{K}$ over $m+1$ days. Adopting the notation introduced in Sect. 3, $\mathcal{D}$ is spanned over an independent input space $\mathbf{X} \times \mathbf{X^S} \times \mathbf{X^T} \times \mathbf{X^{TS}}$ and a dependent output space $\mathbf{Y} \times \mathbf{Y^S} \times \mathbf{Y^T} \times \mathbf{Y^{TS}}$. The multi-target predictive model f^{+ST} can be learned from $\mathcal{D}$ so that:

$$f^{+ST} : \mathbf{X} \times \mathbf{X^S} \times \mathbf{X^T} \times \mathbf{X^{TS}} \rightarrow \mathbf{Y} \times \mathbf{Y^S} \times \mathbf{Y^T} \times \mathbf{Y^{TS}}, \qquad (10)$$

This predictive model is a "multi-scale" upgrade of the traditional multi-target predictive model [3,20].[2] We note that the output space of $f^{+ST}(\cdot)$ yields 24 h forecasts of the fine-grained wind speed ($\mathbf{Y}$), as well as 24 h ahead forecasts of the winds speed variability at the space and time scales considered ($\mathbf{Y^S}$, $\mathbf{Y^T}$ and $\mathbf{Y^{TS}}$).[3] However, this study aims at yielding accurate fine-grained forecasts of wind speed; hence the empirical study will explore the accuracy of model $f^{+ST}(\cdot)$ along $\mathbf{Y}$ only.

In this study, predictive model $f^{+ST}(\cdot)$ is learned as a tree, i.e. a hierarchy of clusters (Predictive Clustering Trees (PCTs)): the top node corresponds to one cluster containing all the data, which is recursively partitioned into smaller clusters while moving down the tree. CLUS, including PCTs for multi-target regression [8], is available at clus.sourceforge.net.

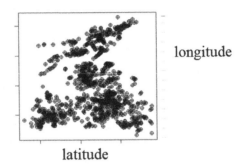

longitude

latitude

Fig. 1. Wind farm at Easter US in 2004.

5   Experimental Study

The experiments are carried out using real world data publicly provided by the DOE/NREL/ALLIANCE3 (http://www.nrel.gov/). The data consist of wind speed measurements from 1326 different locations at 80 m of height in the Eastern region of the US (see Fig. 1). The data were collected in 10 min intervals during the year of 2004. This wind farm grid was able to produce 580 GW, and each farm produces between 100 MW and 600 MW. For the evaluation of the

---

[2] The traditional multi-target predictive model $f \colon \mathbf{X} \rightarrow \mathbf{Y}$ can be learned by neglecting information on the data variability.

[3] Information on the wind speed variability is included in the learning setting as a constraint to improve the predictive ability of the forecasting model learned.

results, we consider the root mean squared error (RMSE), computed over the grid at each time point, as an indicator of the predictive performance. We derive twelve (training and testing) datasets, which are constructed as follows: for every month, days 1–11 define a training dataset that is processed to learn the forecasting model (with $m = 10$), while days 15–25 defines the testing set used to evaluate the performance of the forecasting model learned on the corresponding training dataset. The 24 h ahead forecasting errors, averaged on the twelve datasets, are analyzed. For this empirical evaluation, the multi-scale information is modeled at the spatial scale with radius $R = 10$ km or $R = 50$ km, as well as at the temporal scale with length $L = 1$ h (6 consecutive time points) or $L = 3$ h (18 consecutive time points). The performance of the standard deviation operator is compared to that of the sum and mean operators. Finally, the forecasting performance of the multi-scale multi-target predictive model $(f^{+ST}(\cdot))$ is compared to the performance of the baseline multi-target model $(f(\cdot))$.

Evaluating Scale Size. We start by analyzing the performance of the multi-scale multi-target predictive models learned along the size of the space (R) and time (L) scales. We run the proposed algorithm with the standard deviation operator computed over neighborhoods with radius $R = 10$ km or $R = 50$ km and windows with length $L = 1$ h or $L = 3$ h. Average RMSE results are plotted in Fig. 2. They show that the accuracy of the forecasting model is more sensitive to the time scale than to the spatial scale. In any case, the decrease in the forecasting accuracy performance, due to a large scale in the temporal resolution, is observed starting from forecasts produced 12 h far from the current time point. Therefore, selecting the appropriate scale size is a crucial issue to yield accurate long-term forecasts. Based on these preliminary results, we select $R = 10$ km and $T = 1$ h for the remaining of this study.

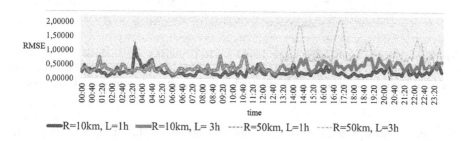

Fig. 2. Multi-scale multi-target predictive models ($R = 10, 50$ km, $L = 1, 3$ h): testing RMSE (averaged on twelve testing datasets - axis Y) plotted with respect to 10 min-spaced time points (over 24 h ahead horizon - axis X).

Evaluating Multi-scale Operator. We proceed by exploring the performance of the multi-scale multi-target predictive models learned along the selection of the multi-sclare operator used to express the wind speed variability. Considering $R = 10$ km and $T = 1$ h, we compare the forecasting accuracy achieved when the data variability model is computed through the standard deviation operator to the accuracy achieved when the variability model is computed through the sum or mean operators. Average RMSE results are plotted in Fig. 3. These results empirically support the effectiveness of our choice of resorting to the standard deviation as the most appropriate second order statistic to model the wind speed variability in both space and time. It actually contributes to gain in forecasting accuracy in this application.

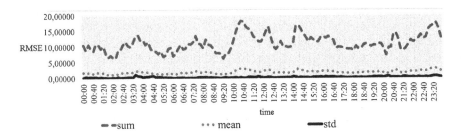

Fig. 3. Multi-scale multi-target predictive models (multi-scale operator analysis - sum, mean and standard deviation): testing RMSE (averaged on twelve testing datasets - axis Y) plotted with respect to 10 min-spaced time points (over 24 h ahead horizon - axis X).

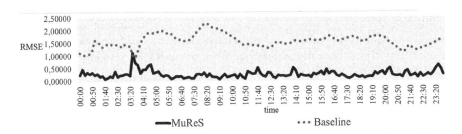

Fig. 4. Multi-scale multi-target predictive models (multi-scale analysis - multi-scale multi-target model (MuReS) vs multi-target model (Baseline): testing RMSE (averaged on twelve testing datasets - axis Y) plotted with respect to 10 min-spaced time points (over 24 h ahead horizon - axis X).

Evaluating Multi-scale Learning Schema. We complete this study by comparing the performance of the multi-scale multi-target predictive models learned with $R = 10$ km, $L = 1$ h and the standard deviation as multi-scale operator to the performance of the baseline multi-target predictive models learned neglecting data variability at space and time scales. Average RMSE results are plotted in Fig. 4. These results show empirically the viability of the main idea inspiring this study: the accuracy of multi-target learning in the wind speed forecasting can be greatly improved by augmenting both the input and output spaces of the learning problem with multi-scale information modeling the data variability of the wind speed observed at both space and time scales.

6   Conclusion

This paper studies the problem of the medium-term (24 h ahead) wind speed forecasting by considering different dimensions of analysis: data variability, spatial and temporal resolution, multi-target learning, aiming to investigate the relevant implications for the dealing with the multi-scale representation of the data variability for the problem at hand. Results in a benchmark dataset clearly show that accounting for the data variability at space and time scales allows us to learn prediction models, which are much more accurate than traditional models that neglect the multi-scale information. Moreover, experimental results confirm that the standard deviation is an appropriate multi-scale operator of the data variability to be taken into account in this specific application. Finally, defining the size of the scale (particularly in the time scale) can be crucial issue to guarantee accurate long-term forecasts. As future work, we intend to explore the implications of these models of the data variability in statistical time series models (e.g Arima or Var). We also plan to investigate more sophisticated incremental learning methods that are able to update the forecasting model as new historical data are collected, in order to fit the learned models to drifting data. Finally, we plan to investigate deep learning framework for wind speed variability feature learning.

Acknowledgments. Authors thank Enrico Laboragine for his support in developing the algorithm presented and running the experiments. This work is carried out in partial fulfillment of the research objectives of ATENEO Project 2014 on "Mining of network data" and ATENEO Project 2015 on"Models and Methods to Mine Complex and Large Data" funded by the University of Bari Aldo Moro.

References

1. Almeida, V., Gama, J.: Collaborative wind power forecast. In: Bouchachia, A. (ed.) ICAIS 2014. LNCS (LNAI), vol. 8779, pp. 162–171. Springer, Cham (2014). https://doi.org/10.1007/978-3-319-11298-5_17
2. Ambach, D., Vetter,P.: Wind speed and power forecasting - a review and incorporating asymmetric loss. In: 2016 Second International Symposium on Stochastic Models in Reliability Engineering, Life Science and Operations Management (SMRLO), pp. 115–123 (2016)
3. Appice, A., Džeroski, S.: Stepwise induction of multi-target model trees. In: Kok, J.N., Koronacki, J., Mantaras, R.L., Matwin, S., Mladenič, D., Skowron, A. (eds.) ECML 2007. LNCS (LNAI), vol. 4701, pp. 502–509. Springer, Heidelberg (2007). https://doi.org/10.1007/978-3-540-74958-5_46
4. Appice, A., Pravilovic, S., Lanza, A., Malerba, D.: Very short-term wind speed forecasting using spatio-temporal lazy learning. In: Japkowicz, N., Matwin, S. (eds.) DS 2015. LNCS (LNAI), vol. 9356, pp. 9–16. Springer, Cham (2015). https://doi.org/10.1007/978-3-319-24282-8_2
5. Ceci, M., Corizzo, R., Fumarola, F., Malerba, D., Rashkovska, A.: Predictive modeling of PV energy production: how to set up the learning task for a better prediction? IEEE Trans. Ind. Inf. **13**(3), 956–966 (2017)

6. Hui, Z., Bin, L., Zhuo-qun, Z.: Short-term wind speed forecasting simulation research based on ARIMA-LSSVM combination method. In: ICMREE 2011, vol. 1, pp. 583–586 (2011)
7. Kavasseri, R.G., Seetharaman, K.: Day-ahead wind speed forecasting using f-ARIMA models. Renewable Energy **34**(5), 1388–1393 (2009)
8. Kocev, D., Vens, C., Struyf, J., Dzeroski, S.: Tree ensembles for predicting structured outputs. Pattern Recognit. **46**(3), 817–833 (2013)
9. Lange, M., Focken, U.: New developments in wind energy forecasting. In: 2008 IEEE Power and Energy Society General Meeting - Conversion and Delivery of Electrical Energy in the 21st Century, pp. 1–8 (2008)
10. Negnevitsky, M., Mandal, P., Srivastava, A.K.: An overview of forecasting problems and techniques in power systems. In: 2009 IEEE Power Energy Society General Meeting, pp. 1–4 (2009)
11. Palomares-Salas, J., De la Rosa, J., Ramiro, J., Melgar, J., Aguera, A., Moreno, A.: Arima vs. neural networks for wind speed forecasting. In: IEEE International Conference on Computational Intelligence for Measurement Systems and Applications, CIMSA 2009, pp. 129–133 (2009)
12. Pravilovic, S., Appice, A., Lanza, A., Malerba, D.: Mining cluster-based models of time series for wind power prediction. In: Greco, S., Picariello, A. (eds.) 22nd Italian Symposium on Advanced Database Systems, SEBD 2014, pp. 9–20 (2014)
13. Pravilovic, S., Appice, A., Lanza, A., Malerba, D.: Wind power forecasting using time series cluster analysis. In: Džeroski, S., Panov, P., Kocev, D., Todorovski, L. (eds.) DS 2014. LNCS (LNAI), vol. 8777, pp. 276–287. Springer, Cham (2014). https://doi.org/10.1007/978-3-319-11812-3_24
14. Pravilovic, S., Appice, A., Malerba, D.: An Intelligent technique for forecasting spatially correlated time series. In: Baldoni, M., Baroglio, C., Boella, G., Micalizio, R. (eds.) AI*IA 2013. LNCS (LNAI), vol. 8249, pp. 457–468. Springer, Cham (2013). https://doi.org/10.1007/978-3-319-03524-6_39
15. Pravilovic, S., Appice, A., Malerba, D.: Integrating cluster analysis to the ARIMA model for forecasting geosensor data. In: Andreasen, T., Christiansen, H., Cubero, J.-C., Raś, Z.W. (eds.) ISMIS 2014. LNCS (LNAI), vol. 8502, pp. 234–243. Springer, Cham (2014). https://doi.org/10.1007/978-3-319-08326-1_24
16. Pravilovic, S., Bilancia, M., Appice, A., Malerba, D.: Using multiple time series analysis for geosensor data forecasting. Inf. Sci. **380**, 31–52 (2017)
17. Santamara-Bonfil, G., Reyes-Ballesteros, A., Gershenson, C.: Wind speed forecasting for wind farms: a method based on support vector regression. Renewable Energy **85**, 790–809 (2016)
18. Shi, J., Qu, X., Zeng, S.: Short-term wind power generation forecasting: direct versus indirect arima-based approaches. Int. J. Green Energy **8**(1), 100–112 (2011)
19. Soman, S.S., Zareipour, H., Malik, O., Mandal, P.: A review of wind power and wind speed forecasting methods with different time horizons. In: North American Power Symposium 2010, pp. 1–8 (2010)
20. Stojanova, D., Ceci, M., Appice, A., Malerba, D., Dzeroski, S.: Dealing with spatial autocorrelation when learning predictive clustering trees. Ecol. Inf. **13**, 22–39 (2013)

Temporal Reasoning with Layered Preferences

Luca Anselma[1(✉)], Alessandro Mazzei[1], Luca Piovesan[2],
and Paolo Terenziani[2]

[1] Dipartimento di Informatica, Università di Torino,
Corso Svizzera 185, 10149 Turin, Italy
{anselma,mazzei}@di.unito.it
[2] DISIT, Università del Piemonte Orientale "A. Avogadro", Alessandria, Italy
{luca.piovesan,paolo.terenziani}@uniupo.it

Abstract. Temporal representation and temporal reasoning is a central in Artificial Intelligence. The literature is moving to the treatment of "non-crisp" temporal constraints, in which also preferences or probabilities are considered. However, most approaches only support numeric preferences, while, in many domain applications, users naturally operate on "*layered*" scales of values (e.g., Low, Medium, High), which are domain- and task-dependent. For many tasks, including decision support, the evaluation of the *minimal network* of the constraints (i.e., the tightest constraints) is of primary importance. We propose the first approach in the literature coping with *layered preferences on quantitative temporal constraints*. We extend the widely used *simple temporal problem* (STP) framework to consider layered user-defined preferences, proposing (i) a formal representation of quantitative constraints with layered preferences, and (ii) a temporal reasoning algorithm, based on the general algorithm *Compute-Summaries*, for the propagation of such temporal constraints. We also prove that our temporal reasoning algorithm evaluates the minimal network.

Keywords: Temporal constraints with preferences · Temporal reasoning
Temporal constraint propagation

1 Introduction

Representing and reasoning about time is fundamental in many "intelligent" tasks and activities, such as planning, scheduling, human–machine interaction, natural language understanding, diagnosis, robotics and data management. Since the Eighties, many different approaches to *quantitative* (i.e., considering the metric of time; e.g., *action B must be started at least* 1 h *after the end of action A*) and\or *qualitative* (i.e., considering only the relative position of actions\events; e.g., *A before B*) temporal constraints have been developed (see, e.g., the surveys in [1]). A milestone approach, regarding quantitative constraints, is STP [2], in which constraints of the form $P_1[c,d]$ P_2 model the minimal (c) and maximal (d) distance between a pair of time points (notably, the above constraints are equivalent to *Bounds on Differences* (BoD) constraints of the form $c \leq P_2 - P_1 \leq d$, and [c,d] is usually called **admissibility interval**). Different AI approaches have been provided which, given a set of STP constraints, propose **reasoning** algorithms that propagate such constraints, to check

© Springer Nature Switzerland AG 2018
M. Ceci et al. (Eds.): ISMIS 2018, LNAI 11177, pp. 367–376, 2018.
https://doi.org/10.1007/978-3-030-01851-1_35

their *consistency*, and\or to find a *solution* (i.e., an instantiation of all the variables such that all constraints are satisfied), or to make explicit the *minimal network* of constraints (i.e., a set of constraints which has exactly the same solutions of the original one, and in which the minimum and maximum implied distances between each pair of variables are made explicit – see [2]).

Example 1. Let t_1, t_2 and t_3 be time points, and let KB be the following set of STP constraints: KB = $\{t_1 [10, 15] t_2, t_2[20,30]t_3, t_1[25,40]t_3\}$. KB is consistent, and $\{t_1 = 0, t_2 = 10, t_3 = 30\}$ is a *solution* (also termed *scenario*) of KB. The *tightest* constraints implied by KB (the so called *minimal network* [2]) are KB$' = \{t_1 [10, 15] t_2, t_2[20,30]t_3, t_1[30,40]t_3\}$ (in particular, the minimum distance between t_1 and t_3 is 30). ■

While in several tasks the goal is to find a scenario, in others, such as in decision support or mixed-initiative approaches, the minimal network of the constraints must be determined, to provide users with a compact representation of all the possible solutions (since the choice of a specific solution has to be left to the users).

So far, we have only considered "crisp" constraints, in the sense that they represent a set of "equally possible" distances between variables. All of these approaches rely on the framework of classical Constraint Satisfaction Problem (CSP), inheriting from it a number of fundamental limitations, mainly related to a lack of flexibility and a limited representation of uncertainty [3]. A paradigmatic example is the execution of clinical treatments with temporal constraints. Usually, expressing "crisp" temporal constraints is not possible/useful in the medical context: constraints are expressed in the form of recommendations, to be followed as much as possible.

Example 2. As a running example, we consider a case of a comorbid patient suffering from both urinary tract infection and gastroesophageal reflux. Nalidixic acid (NA) is an antibiotic used for the treatment of urinary tract infections. As maintenance therapy, it should be administered two times a day, with an interval of 12 h between each pair of consecutive administrations. However, such a recommendation with a "high" preference is not strict since the interval can also be of 11 or 13 h with a "medium" preference, or of 10, 14 or 15 h with a "low" preference. On the other hand, calcium carbonate (CC) is used to treat gastroesophageal reflux and it should be taken with "high" preference after lunch and dinner, but it can also be taken with a "medium" preference during meals. Since the concurrent administration of the two drugs can lead to a decrease of the effect of NA, to avoid interactions the first administration of NA (henceforth NA1), in the morning, must be administered at least 1 h before the first administration of CC (CC1). Analogously, the delay between the second administration of CC (CC2) and the second administration of NA (NA2), in the evening, should be at least 3 h with "low" preference or 4 h with "high" preference. Besides the above constraints, our approach can also consider constraints that may arise from patient's lifestyle preferences. In particular, in this example we suppose that the patient may want to take the first NA dose in the morning at 5am with "low" preference, or after 6am with "high" preference. Moreover, we hypothesize that the patient has lunch time preferences (which are 12pm or 1pm with "high" preference, 11am or 2pm with "medium" preference, 3pm with "low" preference), and dinner time preferences (which are 6pm, 7pm or 8pm with "high" preference, 9pm with "medium" preference, 5pm or 10pm with "low" preference). ■

To deal with issues related to preferences in constraints, a huge stream of research has extended the CSP formalism in a fuzzy direction, by replacing classical "crisp" constraints with soft "not-crisp" constraints modeled by fuzzy relations.

Concerning **qualitative temporal constraints**, in their seminal work Badaloni and Giacomin [4] have defined a new formalism in which the "crisp" qualitative temporal constraints in Allen's Interval Algebra are associated with a degree of *plausibility*, and have proposed temporal reasoning algorithms to propagate such constraints. In [5] Dubois et al. have proposed the calculus of fuzzy Allen relations (including the composition table) and the patterns for propagating uncertainty about (fuzzy) Allen relations in a possibilistic way. Ryabov et al. [6] attach a *probability* to each of Allen's basic interval relations. A similar probabilistic approach has been proposed more recently by Mouhoub and Liu [7], as an adaptation of the general probabilistic CSP framework. Finally, in [8] Dubois et al. have shown how possibilistic temporal uncertainty can be handled in the setting of point algebra. *"Non-crisp" quantitative temporal constraints* have been considered by Khatib et al. [9], that extended the STP and the TCSP framework [2] to consider temporal *preferences*. An analogous approach has been recently proposed in [10]. However, such approaches only consider numeric preferences while, in many application domains, experts express preferences in term of a "layered" scale of "qualitative" preferences (e.g., <Low, Medium, High>). The approach in this paper overcomes such a limitation, being parametric with respect to the scale (see Sect. 4 for detailed comparisons with the literature).

2 Representing STP Constraints with Layered Preferences

In current approaches in the literature augmenting STP with preferences, given a constraint $P_1[c,d]P_2$, a numeric value of preference is associated with each possible distance in the admissibility interval [c,d]. However, in many areas and applications, preferences distribute in a "regular" way over the intervals, forming a sort of *"pyramid"* of **nested** admissibility intervals, in which the top interval has the highest preference and the bottom the lowest one. Example 2 is just one paradigmatic example, but this is the case in all situations in which preferences are "centered" on a given set of temporal values, and decrease while getting far from this center. In this paper, we focus on such nested distributions of preferences (that we will call *"pyramid"* for short), showing that considering such distributions provide several advantages with respect to the association of preferences with each possible distance in the admissibility intervals. First, we introduce the notion of "layered" scale of preferences.

Definition 1. Scale of Qualitative Preferences (SQP). An SQP (or "scale", for short) S_r of **cardinality** r is composed by an enumerative set $\{p_1, \ldots, p_r\}$ of r labels (r > 0), and a strict and total ordering relation < over the set. For simplicity, we denote an SQP by an ordered list $\langle p_1, \ldots, p_r \rangle$, such that $\forall i, 1 \leq i < r, p_i < p_{i+1}$. ∎

Terminology. Given an SQP S_r of **cardinality** r, we indicate by $S_r(i)$ the ith value in the scale $S_r(1 \leq i \leq r)$ and we denote by $\mathbb{S}$ the domain of SQPs. ∎

For example, an SQP coping with Example 2 is S_3^{ex}: <low, medium, high>.

We can now formally define the notion of preference function, focusing specifically on "pyramid" preferences. Intuitively speaking, a pyramid preference function over an admissibility interval D is a function such that the preference values (expressed as a SQP) (non-strictly) increase until the maximum preference value is reached at a certain value $v \in D$, and then (non-strictly) decrease. Formally:

Definition 2. Pyramid preference function (PPF). A pyramid preference function $Prefs_{S_r,D}$ with scale $S_r \in \mathbb{S}$ over an admissibility interval D is a total function $Prefs_{S_r,D}$: $D \rightarrow S_r$ such that:

$$\exists v \in D, \forall v_1, v_2 \in D,$$
$$((v_1 \leq v \wedge v_2 \leq v \wedge v_1 \leq v_2) \Rightarrow Prefs_{S_r,D}(v_1) \leq Prefs_{S_r,D}(v_2))$$
$$\wedge ((v \leq v_1 \wedge v \leq v_2 \wedge v_1 \leq v_2) \Rightarrow Prefs_{S_r,D}(v_1) \geq Prefs_{S_r,D}(v_2))$$

If p_s is the maximum value of a PPF (over a scale $S_r \in \mathbb{S}$), we say the PPF has height s. If we term $\mathbb{PP}$ the domain of PPFs, STP constraints with pyramid preferences can be abstractly defined as follows.

Definition 3. STP constraint with Pyramid preferences (PyP_STP). Given a scale $S_r \in \mathbb{S}$, a PyP_STP is a constraint $\langle x, y, [c, d], Prefs_{S,[c,d]} \rangle$ where $Prefs_{S,[c,d]}$ is a PPF in $\mathbb{PP}$. ∎

Terminology. We say that the *height* of PyP_STP is the *height* of its preference function.

We can exploit the fact that preferences form a pyramid of height s of nested *admissibility intervals* by proposing a compact and "user-friendly" representation of PyP_STP constraints (of height s).

Definition 4. Compact representation of STP constraints with Pyramid preferences. Given a scale $S_r \in \mathbb{S}$ of cardinality r, a PyP_STP of height s (s ≤ r) can be compactly represented by a constraint $\langle x, y, \langle\langle [c_1, d_1], p_1 \rangle, \dots, \langle [c_s, d_s], p_s \rangle\rangle \rangle$ where $[c_i, d_i]$, $i = 1, \dots, s$ are admissibility intervals such that $c_i \leq c_{i+1} \wedge d_i \geq d_{i+1}$, $1 \leq i < s$ and $p_1, \dots, p_s \in S_r$ denote the s lowest qualitative values in the scale S_r (i.e., $p_i = S_r(i)$, $1 \leq i \leq s$). ∎

For instance, the constraint between two consecutive administrations of NA (NA1 and NA2) in Example 2 can be represented through the "compact" PyP_STP constraint $\langle NA1, NA2, \langle\langle [10, 15], low \rangle, \langle [11, 13], medium \rangle, \langle [12, 12], high \rangle\rangle \rangle$.

Let $\mathbb{P} \cap \mathbb{P}\_STP$ denote the domain of the "compact" PyP_STPs as defined above.

The **semantics** of a "compact" PyP_STP of the form $\langle x, y, \langle\langle [c_1, d_1], p_1 \rangle, \dots, \langle [c_s, d_s], p_s \rangle\rangle \rangle$ over a scale $S_r \in \mathbb{S}$ of cardinality r (s ≤ r) is that it compactly represents a constraint $\langle x, y, [c_1, d_1], Prefs_{S,[c,d]} \rangle$ (see Definition 3) where $Prefs_{S,[c,d]}$ is such that:

(i) $\forall v \in [c_s, d_s] Prefs_{S,D}(v) = S_r(s)$
(ii) $\forall i \, 1 \leq i < s \, \forall v \in ([c_i, d_i] - [c_{i+1}, d_{i+1}]) Prefs_{S,D}(v) = S_r(i)$

The intuitive meaning of a "compact" PyP_STP $\langle x, y, \langle\langle [c_1, d_1], p_1 \rangle, \dots, \langle [c_s, d_s], p_s \rangle\rangle \rangle$ over a scale S_r (s ≤ r) is that the *difference* $y - x$ between y and x is in

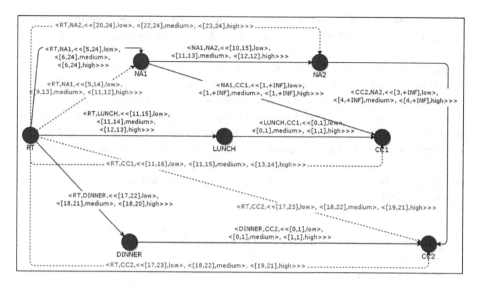

Fig. 1. Graphical representation of the constraints in Example 2. The solid edges represent the input constraints. The dashed edges represent part of the output of temporal reasoning.

$[c_s, d_s]$ with preference $S_r(s)$, or in $[c_{s-1}, d_{s-1}] - [c_s, d_s]$ with preference $S_r(s - 1)$, or .... or in $[c_1, d_1] - [c_2, d_2]$ with preference $S_r(1)$.

For example, the "compact" PyP_STP $\langle NA1, NA2, \langle\langle[10, 15], low\rangle, \langle[11, 13], medium\rangle, \langle[12, 12], high\rangle\rangle\rangle$ represents the fact that the difference between NA2 and NA1 has preference "high" if it is exactly 12, "medium" if it is 11 or 13, and "low" if it is 10, or between 14 and 15.

As in the general case, a graph representation can be provided for a set of "compact" PyP_STP constraints (as defined in Definition 4).

Definition 5. Graph of PyP_STP (PyP_STP_G). Given an SQP S_r of cardinality r, a set B of "compact" PyP_STPs $\langle x, y, \langle\langle[c_1, d_1], p_1\rangle, \ldots, \langle[c_s, d_s], p_s\rangle\rangle\rangle$ can be represented by an oriented graph $G = \langle V, E\rangle$ with a labelling function λ, where V (the set of nodes) represents the set of variables in B, E $\subseteq$ V $\times$ V, and $\lambda : E \rightarrow \mathbb{P} \frown \text{P\_STP}$. ∎

In Fig. 1, we graphically represent the PyP_STP_G arising from Example 2, at a granularity of hours. We represent the constraints in Example 2 with solid edges labelled with blue lettering. Notice that, in the figure, the time points RT, LUNCH and DINNER represent respectively the reference time (which, in our example, is 12am of the current day), and lunch and dinner times.

3   Temporal Reasoning

Instead of inventing a new algorithm and then proving that it computes the *tightest* constraints, we adopt a different strategy. We exploit the general algorithm *Compute-Summaries*(λ, V, E, $\oplus$, $\odot$, 0, 1) in [11], which is shown in Fig. 2 below. *Compute-Summaries* is indeed a highly parametric *all-to-all shortest paths* algorithm to solve different problems concerning oriented paths in a graph. Such an algorithm takes as input a graph *(V, E)*, a labelling function $\lambda: E \rightarrow L$ operating on the edges of the graph, an "*extension*" operator $\odot$, a "*resume*" operator $\oplus$, the *identity* for $\odot$ (indicated by 1), and the *identity* for $\oplus$ (indicated by 0). It is a dynamic algorithm, which, in case $\odot$, $\oplus$, 1 and 0 are defined in such a way that $\langle L, \oplus, \odot, 0, 1 \rangle$ is a *closed semiring* (see Property 1), evaluates the *all-to-all shortest paths* (i.e., the *minimal network*) [11]. Thus, our idea is to define the *extension* $\odot^P$ and *resume* $\oplus^P$ operators (and their *identities*) operating on "compact" PyP_STP constraints (i.e., on both *admissibility intervals* and *preferences*) in such a way that the structure $\langle L, \oplus^P, \odot^P, 0, 1 \rangle$ is a *closed semiring*. Then, we adopt a specific instance of *Compute-Summaries*, by instantiating its parameters with our operators. In such a way, we achieve an algorithm that computes the *minimal network*, as desired.

Compute-Summaries (λ,V,E,$\oplus$,$\odot$,0,1) algorithm

1. n $\leftarrow$ |V|
2. **for** i$\leftarrow$1 **to** n **do**
3.   **for** j$\leftarrow$1 **to** n **do**
4.   **if** j=i **then** $L_{ij}^0 \leftarrow 1 \oplus \lambda(i,j)$
5.   **else** $L_{ij}^0 \leftarrow \lambda(i,j)$
6. **for** k$\leftarrow$1 **to** n **do**
7.   **for** i$\leftarrow$1 **to** n **do**
8.   **for** j$\leftarrow$1 **to** n **do**
9. $L_{ij}^k \leftarrow L_{ij}^{k-1} \oplus (L_{ik}^{k-1} \odot L_{kj}^{k-1})$

Fig. 2. Compute-Summaries algorithm

We adopt the notation in [11]: (i) 1,2,...,n indicate the vertices in V, where n = |V|; (ii) we extend the notion of labelling function to paths: the label of a path p = $<v_1, v_2, ..., v_t>$ is $\lambda(p) = \lambda(v_1, v_2) \odot \lambda(v_2, v_3) \odot ... \odot \lambda(v_{t-1}, v_t)$; (iii) L_{ij} denotes the application of the resume operator $\oplus$ to all the paths from i to j in the graph (V,E) (i.e., $L_{ij} = \oplus_{p=i \rightarrow j} \lambda(p)$); (iv) L_{ij}^k denotes the application of the resume operator $\oplus$ to all the paths from i to j in the graph (V,E) traversing the nodes 1,...,k only (i.e., $L_{ij}^k = \oplus_{p \in Q} \lambda(p)$, where Q is the set of all paths in (V,E) connecting i to j and traversing only the nodes in {1,...,k}). We assume that $\lambda(i,j) = 0$ if $(i,j) \notin E$.

Whenever the application of $\oplus$ and $\odot$ operators in line 9 obtains an inconsistent constraint (see below), the algorithm stops and signals the inconsistency.

Complexity. The complexity of Compute-Summaries is $\Theta(n^3 \cdot (T_\oplus + T_\odot))$, where $T_\oplus$ and $T_\odot$ denote the time required to evaluate $\oplus$ and $\odot$ respectively [11].

We can still adopt the above algorithm to perform correct and complete propagation, but we have to extend it with a proper definition of the $\oplus$ and $\odot$ operators (that we term $\oplus^P$ and $\odot^P$), to deal with "compact" PyP_STP constraints. As regards the operations on the admissibility intervals, we adopt the "standard" ones adopted in the STP framework [2]: the resume of two admissibility intervals $[c_1,d_1]$ and $[c_2,d_2]$ is their intersection $[\max(c_1,c_2), \min(d_1,d_2)]$, while their extension is the sum of their starting and ending points $[c_1 + c_2, d_1 + d_2]$. However, in the case of "compact" PyP_STP constraints, such operations have to be iterated on the admissibility intervals at each level of the pyramids. As regards preferences, they can be easily evaluated by stating that, at each layer i, the preference $p_i = S(i)$ is retained.

In the constraint propagation algorithm, in case a degenerate interval is computed at a level i (i.e., the intersection is empty), all admissibility intervals higher than or equal to i can be dropped from the set of "compact" PyP_STPs since this corresponds to an inconsistency at level i. If the inconsistency is at the first level of the PyP_STPs, then the entire set of PyP_STPs is inconsistent.

Our formal definition of the resume operator $\oplus^P$ is reported below. At each level i of the PyP_STPs C' and C", the intersection between the admissibility intervals at level i in C' and C" is computed, and its result is paired with the preference associated with that level, until the highest level common to both constraints is reached or an inconsistency at a given level is detected.

Definition. Resume ($\oplus^P$). Given a scale S_r, and given two "compact" PyP_STPs
$C' = \langle x, y, \langle\langle [c'_1, d'_1], p_1\rangle, \ldots, \langle [c'_k, d'_k], p_k\rangle\rangle\rangle$ and $C'' = \langle x, y, \langle\langle [c''_1, d''_1], p_1\rangle, \ldots,$
$\langle [c''_l, d''_l], p_l\rangle\rangle\rangle$ (where $1 \le k \le r$ and $1 \le l \le r$), their resume is obtained as follows:

$$C' \oplus^P C'' = \langle x, y, \langle\langle [\max(c'_i, c''_i), \min(d'_i, d''_i)], p_i\rangle, i = 1, \ldots, \min(k, l)\rangle\rangle$$

For instance, the resume operation applied to the two following "compact" PyP_STP constraints between the points RT and NA2 produces a new restricted PyP_STP constraint as result

$\langle RT, NA2, \langle\langle [0, 24], low\rangle, \langle [0, 24], medium\rangle, \langle [0, 24], high\rangle\rangle\rangle \oplus^P$
$\langle RT, NA2, \langle\langle [15, 39], low\rangle, \langle [17, 37], medium\rangle, \langle [18, 36], high\rangle\rangle\rangle$
$= \langle RT, NA2, \langle\langle [15, 24], low\rangle, \langle [17, 24], medium\rangle, \langle [18, 24], high\rangle\rangle\rangle$

In the extension operator $\odot^P$, at each level i of the pyramids, the new admissibility interval is computed by summing pairwise the starting and ending points of the input admissibility intervals at level i. The resulting admissibility interval is paired with the preference associated with that level, until the highest level, i.e., the highest level common to both constraints, is reached.

Definition. Extension ($\odot^P$). Given a scale S_r, and given two PyP_STPs $C' = \langle x, y, \langle\langle [c'_1, d'_1], p_1\rangle, \ldots, \langle [c'_k, d'_k], p_k\rangle\rangle\rangle$ and $C'' = \langle y, z, \langle\langle [c''_1, d''_1], p_1\rangle, \ldots, \langle [c''_l, d''_l], p_l\rangle\rangle\rangle$ (where $1 \le k \le r$ and $1 \le l \le r$), their extension is obtained as follows:

$$C' \odot^P C'' = \langle x, z, \langle\langle [c'_i + c''_i, d'_i + d''_i], p_i\rangle, i = 1, \ldots, \min(k, l)\rangle\rangle$$

For instance, the resume operation applied to the two following "compact" PyP_STP constraints, concerning RT and NA1 and concerning NA1 and NA2 respectively, produces as a result a new PyP_STP constraint concerning RT and NA2

$\langle RT, NA1, \langle\langle[5,24], low\rangle, \langle[6,24], medium\rangle, \langle[6,24], high\rangle\rangle\rangle \odot^P$
$\langle NA1, NA2, \langle\langle[10,15], low\rangle, \langle[11,13], medium\rangle, \langle[12,12], high\rangle\rangle\rangle$
$= \langle RT, NA2, \langle\langle[15,39], low\rangle, \langle[17,37], medium\rangle, \langle[18,36], high\rangle\rangle\rangle$

Complexity of the $\oplus^P$ and $\odot^P$ operators. Since the intersection and the sum of two intervals can be computed in constant time and they are computed for each preference value, both the $\oplus^P$ and $\odot^P$ operators can operate in time $\Theta(r)$, where r is the number of preference values in S_r, thus $T_\oplus = T_\odot = \Theta(r)$.

Thus, the complexity of the Compute-Summaries algorithm is $\Theta(n^3 \cdot r)$.

Identities for $\oplus^P$ and $\odot^P$. Given a Scale S_r with cardinality r, the identity for $\odot^P$ is $\bot = \langle\langle[0,0],p_1\rangle, \ldots, \langle[0,0],p_r\rangle\rangle$ since, given any PyP_STP $C = \langle\langle[c_1,d_1],p_1\rangle, \ldots, \langle[c_r,d_r],p_r\rangle\rangle$, $C \odot^P \langle\langle[0,0],p_1\rangle, \ldots, \langle[0,0],p_r\rangle\rangle = \langle\langle[0,0],p_1\rangle, \ldots, \langle[0,0],p_r\rangle\rangle \odot^P C = C$.

The identity for ^{P} is $\top = \langle\langle[-\infty,+\infty],p_1\rangle, \ldots, \langle[-\infty,+\infty],p_r\rangle\rangle$ since, given any PyP_STP $C = \langle\langle[c_1,d_1],p_1\rangle, \ldots, \langle[c_r,d_r],p_r\rangle\rangle$, $C \oplus^P \langle\langle[-\infty,+\infty],p_1\rangle, \ldots, \langle[-\infty,+\infty],p_r\rangle\rangle = \langle\langle[-\infty,+\infty],p_1\rangle, \ldots, \langle[-\infty,+\infty],p_r\rangle\rangle \oplus^P C = C$. $\langle\langle[-\infty,+\infty],p_1\rangle, \ldots, \langle[-\infty,+\infty],p_r\rangle\rangle$ is also the annihilator for $\odot^P$. In fact, given any PyP_STP $C = \langle\langle[c_1,d_1],p_1\rangle, \ldots, \langle[c_r,d_r],p_r\rangle\rangle$, $C \odot^P \langle\langle[-\infty,+\infty],p_1\rangle, \ldots, \langle[-\infty,+\infty],p_r\rangle\rangle = \langle\langle[-\infty,+\infty],p_1\rangle, \ldots, \langle[-\infty,+\infty],p_r\rangle\rangle \odot^P C = \langle\langle[-\infty,+\infty],p_1\rangle, \ldots, \langle[-\infty,+\infty],p_r\rangle\rangle$.

Intuitively, the identity for $\oplus^P$ represents the non-existing constraint, while the identity for $\odot^P$ corresponds to the distance between a point and itself.

Property 1. $\langle \mathbb{P} \frown \mathbb{P}\_STP, \oplus^P, \odot^P, \bot, \top \rangle$ is a *closed semiring*.

Proof (sketch). By definition, both $\oplus^P$ and $\odot^P$ are **closed** over $\mathbb{P} \frown \mathbb{P}\_STP$. $\oplus^P$ is **associative**, since it operates considering each preference level, and, at each level, it computes the maximum and the minimum of the endpoints of the admissibility intervals, and the maximum and the minimum operators are associative. For the same reasons, $\oplus^P$ is also **commutative** and **idempotent**. Thus, since $\bot$ is the identity for $\oplus^P$, $\langle \mathbb{P} \frown \mathbb{P}\_STP, \oplus^P, \bot \rangle$ is a **commutative and idempotent monoid**. $\odot^P$ is **associative**, since at each preference level it performs the pairwise sum of the endpoints of the admissibility intervals, and the sum is associative. Thus, since $\top$ is the identity for $\odot^P$, $\langle \mathbb{P} \frown \mathbb{P}\_STP, \odot^P, \top \rangle$ is a **monoid**. Moreover, $\top$ is an annihilator for $\odot^P$ and $\odot^P$ **distribute** over finite and countably infinite $\oplus^P$: this is due to the fact that pairwise sum distributes over the minimum and the maximum operators. Thus, $\langle \mathbb{P} \frown \mathbb{P}\_STP, \oplus^P, \odot^P, \bot, \top \rangle$ is a *closed semiring*. ∎

As an example of application of our version of the Compute-Summaries algorithm, in Fig. 1 we report with dashed edges labelled with red lettering some of the "compact" PyP_STP constraints obtained from the application of the algorithm to Example 2 (note that, due to graphic reasons, we cannot represent all the PyP_STP constraints of the minimal network). For example, the Compute-Summaries algorithm adopting our $\oplus^P$ and $\odot^P$ operators has restricted the original constraint between RT and NA1 to the constraint $\langle RT, NA1, \langle\langle[6,14], low\rangle, \langle[9,13], medium\rangle, [11,12], high\rangle\rangle\rangle$, which limits the first administration of NA between 6am and 2pm with a "low" preference, between 9am and 1pm with a "medium" preference, and between 11am and 12pm with an "high" preference; moreover, a new constraint $\langle RT, NA2, \langle\langle [20, 24],$ $low\rangle, \langle [22, 24], medium\rangle, [23, 24], high\rangle\rangle\rangle$ has been derived between RT and NA2, which limits the second administration of NA between 8pm and 12am with a "low" preference, between 10pm and 12am with a "medium" preference, and between 11pm and 12am with a "high" preference.

We have implemented our approach in Java (JDK 8), and run experimental tests on an Intel Core i7-6700HQ CPU, considering different numbers of nodes (from 200 to 1000), and scales of different cardinalities (4, 8, 12, 16, and 20). Table 1 shows the computation times. Results show that the execution time, starting from 600 nodes, is cubic in the number of the nodes and grows linearly in the cardinality of the scales.

Table 1. Experimental results (computation times in ms).

# levels	# nodes				
	200	400	600	800	1000
4	91	508	864	2102	3965
8	47	524	1602	3869	7976
12	82	717	2402	5722	11133
16	113	985	3300	7468	15511
20	127	1192	3849	9132	18262

4 Comparisons and Conclusions

Until now, only few AI approaches have considered quantitative temporal constraints with preferences. Among them, the approach in this paper is the first one that has focused on user-defined layered preferences and the evaluation of the tightest constraints. The approach in the literature which is the closest one to the one in this paper is the approach by Terenziani et al. [10]. In [10], the authors have extended STP by associating a quantitative (numeric) preference with each possible distance between time points. As in this paper, they have exploited *closed semirings* to grant that the tightest constraints are evaluated by the reasoning algorithm. The approach in this paper generalizes and extends the approach in [10] to two main respects: (i) it generalizes it to operate also on continuous domains for the variables (while [10] only copes with discrete domains), (ii) it supports user-defined layered preferences (while only numeric preferences were considered in [10]).

Notably, the treatment of layered preferences provides several practical and theoretical advantages. First of all, it facilitates users, that are not required to assign a specific numeric value of preference to each distance. Second, it reduces the space complexity of the representation of constraints (in [10], each possible value in each constraint distance must be explicitly stored, together with its preference). Third, it reduces the temporal complexity of the algorithm computing the tightest constraints (which in [10] is $\Theta\left(n^3 \cdot |D|^2\right)$, where D is the domain of the distance values).

Although our approach is domain- and task-independent, we aim at applying it mainly within our GLARE project [12], to support physicians in the treatment of comorbid patients [13], by integrating medical preferences in the reasoning process. Moreover, we also wish to investigate its integration in temporal relational databases for dealing with temporal indeterminacy [14–18].

References

1. Vila, L.: A survey on temporal reasoning in artificial intelligence. AI Commun. **7**, 4–28 (1994)
2. Dechter, R., Meiri, I., Pearl, J.: Temporal constraint networks. Artif. Intell. **49**, 61–95 (1991)
3. Dubois, D., Fargier, H., Prade, H.: Possibility theory in constraint satisfaction problems: handling priority, preference and uncertainty. Appl. Intell. **6**, 287–309 (1996)
4. Badaloni, S., Giacomin, M.: The algebra IAfuz: a framework for qualitative fuzzy temporal reasoning. Artif. Intell. **170**, 872–908 (2006)
5. Dubois, D., HadjAli, A., Prade, H.: Fuzziness and uncertainty in temporal reasoning. J. UCS **9**, 1168 (2003)
6. Ryabov, V., Trudel, A.: Probabilistic temporal interval networks. In: Proceedings of the 11th International Symposium on TIME 2004, pp. 64–67. IEEE (2004)
7. Mouhoub, M., Liu, J.: Managing uncertain temporal relations using a probabilistic interval algebra. In: 2008 IEEE International Conference on Systems, Man and Cybernetics, SMC 2008, pp. 3399–3404. IEEE (2008)
8. Dubois, D., HadjAli, A., Prade, H.: A possibility theory-based approach to the handling of uncertain relations between temporal points. Int. J. Intell. Syst. **22**, 157–179 (2007)
9. Khatib, L., Morris, P., Morris, R., Rossi, F.: Temporal constraint reasoning with preferences. In: Proceedings of the IJCAI, pp. 322–327. Morgan Kaufmann (2001)
10. Terenziani, P., Andolina, A., Piovesan, L.: Managing temporal constraints with preferences: representation, reasoning, and querying. IEEE Trans. Knowl. Data Eng. **29**, 2067–2071 (2017)
11. Cormen, T.H., Leiserson, C.E., Rivest, R.L.: Introduction to Algorithms. The MIT Press and McGraw-Hill Book Company, New York (1989)
12. Bottrighi, A., Terenziani, P.: META-GLARE: a meta-system for defining your own computer interpretable guideline system—architecture and acquisition. Artif. Intell. Med. **72**, 22–41 (2016)
13. Anselma, L., Piovesan, L., Terenziani, P.: Temporal detection and analysis of guideline interactions. Artif. Intell. Med. **76**, 40–62 (2017)
14. Anselma, L., Bottrighi, A., Montani, S., Terenziani, P.: Extending BCDM to cope with proposals and evaluations of updates. IEEE Trans. Knowl. Data Eng. **25**, 556–570 (2013)
15. Anselma, L., Stantic, B., Terenziani, P., Sattar, A.: Querying now-relative data. J. Intell. Inf. Syst. **41**, 285–311 (2013)
16. Anselma, L., Bottrighi, A., Montani, S., Terenziani, P.: Managing proposals and evaluations of updates to medical knowledge: theory and applications. J. Biomed. Inf. **46**, 363–376 (2013)
17. Anselma, L., Piovesan, L., Terenziani, P.: A 1NF temporal relational model and algebra coping with valid-time temporal indeterminacy. J. Intell. Inf. Syst. **47**, 345–374 (2016)
18. Anselma, L., Piovesan, L., Sattar, A., Stantic, B., Terenziani, P.: A comprehensive approach to "Now" in temporal relational databases: semantics and representation. IEEE Trans. Knowl. Data Eng. **28**, 2538–2551 (2016)

Granular and Soft Clustering

An Adaptive Three-Way Clustering Algorithm for Mixed-Type Data

Jing Xiong and Hong Yu[✉]

Chongqing Key Laboratory of Computational Intelligence, Chongqing University
of Posts and Telecommunications, Chongqing 400065, China
yuhong@cqupt.edu.cn

Abstract. The three-way clustering is different from the traditional two-way clustering. Instead of using two regions to represent a cluster by a single set, a cluster is represented by a pair of sets, and there are three regions such as the core region, fringe region and trivial region. The three-way representation intuitively shows that which objects are fringe to the cluster and it is proposed for dealing with uncertain clustering. However, the three-way clustering algorithm usually needs an appropriate evaluation function and corresponding thresholds. It is not scientific and efficient method for setting the thresholds in advance. Meanwhile, there is a large amount of mixed-type data in real life. Therefore, this paper proposes an adaptive three-way clustering algorithm for mixed-type data, which adjusts the three-way thresholds during the clustering process based on the idea of universal gravitation by excavating more detailed ascription relation between objects and clusters. The experimental results show that the proposed algorithm has good performance in indices such as the accuracy, F-measure, RI and NMI.

Keywords: Three-way clustering · Adaptive · Mixed-type data

1 Introduction

Clustering, used mostly as an unsupervised learning method, is a common technique for data mining. It is easy to discover that there are usually three relationships between an object and a cluster, namely, belong-to definitely, uncertain and not belong-to definitely. An object might belong to one or more clusters, and there are no clear boundaries between clusters in this situation. In fact, there are some approaches such as fuzzy clustering, interval clustering and rough clustering [2,3,8], are researched to deal with this kind of uncertain relationship. Yao [9] introduced the Bayes risk decision making into rough sets and proposed the decision-theoretic rough set model, then proposed the concept of three-way decisions. The theory of three-way decisions extends binary-decisions in order to overcome some drawbacks of binary-decisions.

Inspired by the theory three-way decisions, we have proposed a framework of three-way cluster analysis [11] and applied it to dealing with some problems such as overlapping incremental clustering, community detection and high-dimensional data clustering [10]. Instead of using two regions to represent a

© Springer Nature Switzerland AG 2018
M. Ceci et al. (Eds.): ISMIS 2018, LNAI 11177, pp. 379–388, 2018.
https://doi.org/10.1007/978-3-030-01851-1_36

cluster by a single set, a cluster is represented by a pair of sets, and there are three regions such as the core region, fringe region and trivial region. Objects in the core region are typical elements of the cluster and objects in the fringe region might belong to the cluster, and the objects in the trivial region certainly does not belong to the cluster. The three-way representation intuitively shows which objects are fringe to the cluster [2,3,8].

However, the three-way clustering algorithm usually needs an appropriate evaluation function and corresponding thresholds. Thresholds settings usually are based on expert experience, or iterated by giving a certain range and step size. When faced with large-scale data sets, it is difficult for experts to give appropriate thresholds. The method of iteratively obtaining thresholds can cause huge resource consumption. Meanwhile, large quantity of mixed-type data, containing categorical, ordinal and numerical attributes have commonly existed in real world. In order to overcome the above problems, we propose an adaptive three-way clustering algorithm for mixed-type data, abbreviated as ATWC-MD in this paper.

The basic idea of the ATWC-MD algorithm is a two stages processing. The first stage is to obtain an initial clustering result which contains the two-way clustering result and undecided objects through an appropriate two-way clustering algorithm using the measurement of distance for the mixed-type data. In our previous work [12], we construct the weighted tree according to the semantics of attributes, the number of attribute values and the occurrence frequency of attribute values. The second stage is to obtain the three-way clustering result through the proposed adaptive three-way clustering algorithm (ATWC). The basic idea of the ATWC is to automatically assign the uncertain objects or noise objects into the corresponding core regions, fringe regions or trivial regions of their neighbor clusters, by adjusting the three-way thresholds during the clustering process based on the idea of universal gravitation.

2 Preliminary

2.1 The Three-Way Clustering

Let a universe be $\mathbf{U} = \{\mathbf{x_1}, \cdots, \mathbf{x_n}, \cdots, \mathbf{x_N}\}$ with N objects. Each object has D-dimensional attributes, represented as $\mathbf{x_i} = \{x_i^1, \cdots, x_i^d, \cdots, x_i^D\}$. The clustering result could be represented as $\mathbf{C} = \{C_1, \cdots, C_k, \cdots, C_K\}$ with K clusters.

In contrast to the general crisp representation of a cluster, we represent a three-way cluster C_k as a pair of sets:

$$C_k = (Co\,(C_k), Fr\,(C_k)), \qquad (1)$$

Here, $Co\,(C_k) \subseteq \mathbf{U}$ and $Fr\,(C_k) \subseteq \mathbf{U}$. let $Tr\,(C_k) = \mathbf{U} - Co\,(C_k) - Fr\,(C_k)$.

If $\mathbf{x} \in Co\,(C_k)$, the object $\mathbf{x}$ belongs to the cluster C_k definitely; if $\mathbf{x} \in Fr\,(C_k)$, the object $\mathbf{x}$ might belong to C_k; if $\mathbf{x} \in Tr\,(C_k)$, the object $\mathbf{x}$ does not belong to C_k definitely. If $Fr(C_k) = \emptyset$, the representation of C_k in Eq. (1) turns into $C_k = Co(C_k)$ which is a single set and $Tr(C_k) = \mathbf{U} - Co(C_k)$. And this is a representation of two-way clustering. In other words, the classic two-way representation is a special case of three-way representation. In another way, for $1 \leq k \leq K$, we can define a cluster scheme by the following properties:

$$(i) \ \text{for} \ \forall k, \ Co(C_k) \neq \emptyset;$$
$$(ii) \ \bigcup_{k=1}^{K}(Co(C_k) \cup Fr(C_k)) = U. \tag{2}$$

Property (i) implies that a cluster cannot be empty. This makes sure that a cluster is physically meaningful. Property (ii) states that any object of U must definitely belong to or might belong to a cluster, which ensures that every object is properly clustered.

With respect to the family of clusters, $\mathbf{C}$, we have the following family of clusters formulated by three-way representation as:

$$\mathbf{C} = \{(Co\,(C_1)\,, Fr\,(C_1))\,, \cdots , (Co\,(C_k)\,, Fr\,(C_k))\,, \cdots , (Co\,(C_K)\,, Fr\,(C_K))\}. \tag{3}$$

In the approaches based on the theory of three-way decisions, an evaluation function $v\,(\mathbf{x}, C_k)$ is usually designed and where two thresholds α and β are set in advance; then, the three-way decision rules can be constructed as Eq. (4).

$$\begin{aligned} &if \ v\,(\mathbf{x}, C_k) \geq \alpha, &decide \ \mathbf{x} \ \text{to} \ Co\,(C_k); \\ &if \ \beta \leq v\,(\mathbf{x}, C_k) < \alpha, &decide \ \mathbf{x} \ \text{to} \ Fr\,(C_k); \\ &if \ v\,(\mathbf{x}, C_k) < \beta, &decide \ \mathbf{x} \ \text{to} \ Tr\,(C_k). \end{aligned} \tag{4}$$

In fact, the evaluation function $v(\mathbf{x}, C_k)$ can be a risk decision function, a similarity function, a distance function and so on. In other words, the evaluation function will be specified accordingly when an algorithm is devised. Obviously, the thresholds α and β have a direct influence on the finial clustering result.

2.2   The Measurement of Distance for Mixed-Type Data

For any object $\mathbf{x_n}$ in a universe $\mathbf{U}$, where $\{x_n^1, \cdots, x_n^P\}$, $\{x_n^{P+1}, \cdots, x_n^{P+Q}\}$, $\{x_n^{P+Q+1}, \cdots, x_n^{P+Q+T}\}$ represent the categorical attribute values, ordinal attribute values, numerical attribute values of $\mathbf{x_n}$ respectively. P, Q and T represent the number of different attribute types. Where $P + Q + T = D$.

The distance between the categorical attribute values is represented by the weighted tree structure. The leaf nodes are composed of the attribute values. The distance between any two categorical values is the distance between the two leaf nodes in the weighted tree [12]. Let $w(node_i^h, child_j(node_i^h))$ be the weight between $node_i^h$ and its j-th child node $child_j(node_i^h)$. It is calculated as follows:

$$w(node_i^h, child_j(node_i^h)) = \dfrac{f\,(node_i^h) \cdot \dfrac{fr(child_j(node_i^h))}{fr(node_i^h)}}{h^\eta}. \tag{5}$$

where h is the depth of the node, $\eta \in N+$. $f\,(node_i^h)$ represents the weight function between $node_i^h$ and $child_j(node_i^h)$. The function compute the weight coefficient in view of influence by the number of children on the node. $fr(node_i^h)$ represents $node_i^h$ occurrence frequency in the data set. When it is a leaf node, $fr(node_i^h)$ is the frequency of the value. When it is an internal node, $fr(node_i^h)$ is the sum of the occurrence frequency of its all children.

The distance between any categorical attribute values $Cdist(Cat_u^d, Cat_v^d)$ is equal to the sum of the weights values. Cat_u^d and Cat_v^d represent the leaf node in the weighted tree.

The measurement for ordinal attribute values is similar to categorical attributes, which is also based on a weighted tree structure. Let Ord^d be the d-th ordinal attribute, its attribute values have meaning of order or rank. Let $w\left(Ord^d\right)$ be the weight function between two adjacent nodes in the tree. the $w\left(Ord^d\right)$ could be defined as:

$$
w\left(Ord^d\right) = \begin{cases} 1, & y \leq \vartheta_1 \\ 1 - \zeta_1\left(y - \vartheta_1\right), & \vartheta_1 < y \leq \vartheta_2 \\ 1 - \zeta_1\left(\vartheta_2 - \vartheta_1\right) - \zeta_2\left(y - \vartheta_2\right), & y > \vartheta_2 \end{cases} \tag{6}
$$

where $\zeta_1, \zeta_2 \in (0,1)$, $\vartheta_1, \vartheta_2 \in N+$.

The distance $Odist(Ord_u^d, Ord_v^d)$ between Ord_u^d and Ord_v^d is equal to the sum of the weights values.

We generally use Euclidean distance to calculate the distance between numerical attributes after Z-score normalization.

Let $Dist\left(\mathbf{x}_i, \mathbf{x}_j\right)$ be the distance between objects $\mathbf{x_i}$ and $\mathbf{x_j}$. The measurement of distance as follows:

$$
Dist\left(\mathbf{x}_i, \mathbf{x}_j\right) = \sqrt{\sum_{p=1}^{P} Cdist^2\left(x_i^p, x_j^p\right)} + \sqrt{\sum_{q=1}^{Q} Odist^2\left(x_i^{P+q}, x_j^{P+q}\right)}
$$
$$
+ \sqrt{\sum_{t=1}^{T} Ndist^2\left(x_i^{P+Q+t}, x_j^{P+Q+t}\right)}. \tag{7}
$$

3   The Proposed Method

3.1   Framework of ATWC-MD Algorithm

The proposed ATWC-MD Algorithm concludes two stages, the process is shown in Fig. 1. In the first stage, for all data objects in the universe $\mathbf{U}$, the initial clustering result which contains the two-way clustering result and undecided objects are obtained through one appropriate two-way clustering algorithm using the measurement of distance for the mixed-type data. In the initial clustering result, the objects with a clear and definite attribution in the two-way clustering result, namely the fringe region of a cluster is empty. We denote the two-way clustering result as $\mathbf{TwoC} = \{TwoC_1, \cdots, TwoC_k, \cdots, TwoC_K\}$. For the objects without a clear and definite attribution, namely, they are not assigned to any clusters in the two-way clustering algorithm, we denote them as $\mathbf{U} - \mathbf{TwoC}$.

In the following experiments, we choose the novel density peaks clustering (DPC) algorithm [5] or DPC-KNN algorithm [1] accordingly as the two-way clustering algorithm. Of course, the distance between the mixed-type data objects are computed by the proposed equation 7.

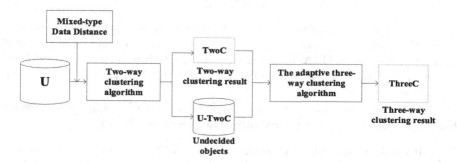

Fig. 1. The framework of the proposed method

In the second stage, we obtain the three-way clustering result through the adaptive three-way clustering algorithm (ATWC) based on the two-way clustering result and the undecided objects. We denote the three-way clustering result as **ThreeC** = $\{ThreeC_1, \cdots, ThreeC_k, \cdots, ThreeC_K\}$. The core region of $ThreeC_k$ is initialized as the corresponding $TwoC_k$. Then, the ATWC Algorithm utilizes the gravitation formula as the evaluation function, and assigns the objects in the $\mathbf{U} - \mathbf{TwoC}$ to the core regions, fringe regions or trivial regions of corresponding neighboring clusters according to the size of gravity. During clustering process, the three-way thresholds α and β are updated adaptively for every undecided object. The following subsection will describe the ATWC algorithm.

3.2 The ATWC Algorithm

First, in order to get the local mass of each object in $\mathbf{U}$, we should calculate the distance between each two objects $\mathbf{x_i}$ and $\mathbf{x_j}$ according to Eq. 7.

Definition 1: Local Mass Based on Cutoff Distance. The local mass based on cutoff distance of the object $\mathbf{x_i}$, M_i, is the sum of mass of object whose distance with $\mathbf{x_i}$ is less than the cutoff distance, and it is formalized as Eq. (8).

$$M_i = \sum_{\mathbf{x_j} \in \mathbf{U}} m_j \chi \left(dist\left(\mathbf{x_i}, \mathbf{x_j}\right) - dist_{cutoff}\right), \tag{8}$$

where m_j is the initial quality of the object $\mathbf{x_j}$, if there is no clear definition in the actual problem, we set $m_j = 1$, and the $dist_{cutoff}$ is a parameter which means the cutoff distance. If $x \le 0$, $\chi(x) = 1$; if $x > 0$, $\chi(x) = 0$.

Definition 2: Local Mass Based on KNN. The local mass based on KNN of $\mathbf{x_i}$ is computed by Eq. (9).

$$M_i = exp\left(-\frac{1}{k}\sum_{\mathbf{x_j}\in KNN(\mathbf{x_i})} m_j dist\left(\mathbf{x_i}, \mathbf{x_j}\right)\right), \tag{9}$$

where the k is a parameter which means the number of nearest neighbors, according to the reference [1], $k = p \times N$, where p is a parameter and N is the total number of objects in $\mathbf{U}$. $NN_k(\mathbf{x_i})$ represents the k-th nearest neighbors of $\mathbf{x_i}$, we set $KNN(\mathbf{x_i}) = \{\mathbf{x_j} \in \mathbf{U}|dist(\mathbf{x_i}, \mathbf{x_j}) \leq dist(\mathbf{x_i}, NN_k(\mathbf{x_i}))\}$, it represents the k-nearest neighbors of $\mathbf{x_i}$.

We need to note that if there is a predefined value of $dist_{cutoff}$, we choose Eq. (8) to compute the local mass; otherwise, we choose Eq. (9).

Definition 3: Local Mass-Contours. Let $TwoC_k^{center}$ be the cluster center of any one cluster $TwoC_k$. The farthest distance between $TwoC_k^{center}$ and an object in $TwoC_k$ is denoted as $\max_{\mathbf{x_i}\in TwoC_k}(dist(\mathbf{x_i}, TwoC_k^{center}))$.

Let L_k be the number of the local mass-contour of $TwoC_k$. Then we determine L_k using the following Eq. (10).

$$L_k = \left\lceil \frac{\max\limits_{\mathbf{x_i}\in TwoC_k}(dist(\mathbf{x_i}, TwoC_k^{center}))}{\tau} \right\rceil + 1, \tag{10}$$

where $\tau = dist_{cutoff}$ if the local mass is computed by Eq. (8), or when the local mass is calculated by Eq. (9) $\tau = \frac{1}{N}\sum_{i=1}^{N} dist(\mathbf{x_i}, NN_k(\mathbf{x_i}))$. Meanwhile, $L_k \in N^*$ and $L_k \geq 2$. The second item, "+1", is to guarantee there are at least two local mass-contours in a cluster.

Let $\min\limits_{\mathbf{x_i}\in TwoC_k} M_i$ and $\max\limits_{\mathbf{x_i}\in TwoC_k} M_i$ represent the minimum and the maximum values of local mass in $TwoC_k$ respectively. Let MC_k^l be the value of the l-th local mass-contour of $TwoC_k$. So MC_k^l could be computed as Eq. (11).

$$MC_k^l = \max\limits_{\mathbf{x_i}\in TwoC_k} M_i - l \times \frac{\max\limits_{\mathbf{x_i}\in TwoC_k} M_i - \min\limits_{\mathbf{x_i}\in TwoC_k} M_i}{L_k}, \tag{11}$$

where $l \in N^*$ and $l \leq L_k$. In particular, $\max\limits_{\mathbf{x_i}\in TwoC_k} M_i$ usually is the local mass of $TwoC_k^{center}$, since the cluster center usually has the biggest local mass in a cluster.

The local mass-contours of a cluster are defined as some closed and disjoint curves in the cluster, according to the local mass distribution in its two-way clustering result. A local mass-contour satisfies the following properties: (1) the local masses of objects on the same local mass-contour are equal; (2) the local masses of objects inside the local mass-contour are greater than or equal to those of objects on the local mass-contour; (3) the difference between the local masses of two adjacent local mass-contours is equal.

The basic idea of the proposed method is to assign the undecided objects to the nearest clusters, which are called as "neighbor clusters". Let $\mathbf{x}'$ be an undecided object, namely $\mathbf{x}' \in \mathbf{U} - \mathbf{TwoC}$. Let $\mathbf{Nei}\,(\mathbf{x}')$ be the set of neighbor cluster of $\mathbf{x}'$. The k-th neighbor cluster of $\mathbf{x}'$ is denoted as $Nei\,(\mathbf{x}')_k$, where $k \geq 0$.

Definition 4: The Set of Neighbor Cluster Nei$\,(\mathbf{x}')$ are calculated as follows.

$$\mathbf{Nei}\,(\mathbf{x}') = \{TwoC_k | dist\,(\mathbf{x}', \mathbf{x_j}) \leq R \wedge \mathbf{x_j} \in TwoC_k\}, \qquad (12)$$

where $R = \max\limits_{k \in K}\left(\max\limits_{\mathbf{x_i} \in TwoC_k}(dist\,(\mathbf{x_i}, TwoC_k^{center}))\right)$, K represents the truth number of clusters.

Assuming $Nei\,(\mathbf{x}')_k$ be the k-th neighbor cluster of $\mathbf{x}'$, and it has L_k local mass-contours totally. Let $Nearest_{L_k}$ represent the nearest object to $\mathbf{x}'$ in $Nei\,(\mathbf{x}')_k$. Let $Nearest_{L_k-1}$ represent the nearest object to $Nearest_{L_k}$ in the $(L_k - 1)$-th local mass-contour. Accordingly, let $Nearest_{l-1}$ represent the nearest object to $Nearest_l$ in the $(l - 1)$-th local mass-contour. Meanwhile, in order to guarantee the consistency of representation, let $Nearest_0$ represent the object with the maximal local mass in $Nei\,(\mathbf{x}')_k$.

As we have mentioned before that it is necessary to design an appropriate evaluation function and determine the corresponding thresholds α and β when carrying out the three-way clustering algorithm based on the two-way clustering result. In this paper, we use the formula of universal gravitation as the evaluation function which is defined as follows.

Definition 5: Universal Gravitation. Let M_i, M_j be the local mass of $\mathbf{x_i}, \mathbf{x_j}$ respectively. $F\,(\mathbf{x_i}, \mathbf{x_j})$ represents the universal gravitation between M_i and M_j. So $F\,(\mathbf{x_i}, \mathbf{x_j})$ is defined as Eq. (13).

$$F\,(\mathbf{x_i}, \mathbf{x_j}) = \frac{G M_i M_j}{dist^2\,(\mathbf{x_i}, \mathbf{x_j})}, \qquad (13)$$

where G is the constant of universal gravitation.

Then, the thresholds α and β are determined by Eq. (14).

$$\begin{aligned}\alpha &= \max\limits_{l \in \{1,\ldots,L_k\}} F\,(Nearest_{l-1}, Nearest_l), \\ \beta &= \min\limits_{l \in \{1,\ldots,L_k\}} F\,(Nearest_{l-1}, Nearest_l).\end{aligned} \qquad (14)$$

We choose the maximum value of gravitation between objects on two adjacent local mass-contours as α. If the gravitation between the undecided object $\mathbf{x}'$ and the nearest object $Nearest_{L_k}$ in $Nei\,(\mathbf{x}')_k$ is larger than α, $\mathbf{x}'$ is divided into core region of neighbor cluster $Nei\,(\mathbf{x}')_k$. In the same way, we choose the minimum value of gravitation between objects on two adjacent local mass-contours as β. If the gravitation between the undecided object $\mathbf{x}'$ and the nearest object $Nearest_{L_k}$ in $Nei\,(\mathbf{x}')_k$ is smaller than α and larger than β, then $\mathbf{x}'$ is divided into fringe region of neighbor cluster $Nei\left(\mathbf{x}'\right)_k$. The three-way decision rules are defined as follows.

$$
\begin{aligned}
&if\ F\left(Nearest_{L_k}, \mathbf{x}'\right) \geq \alpha, && decide\ \mathbf{x}'\ to\ Co\left(Nei\left(\mathbf{x}'\right)_k\right); \\
&if\ \beta \leq F\left(Nearest_{L_k}, \mathbf{x}'\right) < \alpha, && decide\ \mathbf{x}'\ to\ Fr\left(Nei\left(\mathbf{x}'\right)_k\right); \quad (15) \\
&if\ F\left(Nearest_{L_k}, \mathbf{x}'\right) < \beta, && decide\ \mathbf{x}'\ to\ Tr\left(Nei\left(\mathbf{x}'\right)_k\right).
\end{aligned}
$$

Thus, we can determine the thresholds α and β adaptively according to Eq. (14) and obtain the three-way result according to the three-way decision rules in Eq. (15), the detail processing of ATWC is described in Algorithm 1.

Algorithm 1. The ATWC Algorithm

Input: **TwoC** and **U − TwoC**;
Output: The three-way clustering result **C** =
$\qquad \{(Co\left(C_1\right), Fr\left(C_1\right)), \cdots, (Co\left(C_k\right), Fr\left(C_k\right)), \cdots, (Co\left(C_K\right), Fr\left(C_K\right))\}$.
Calculate the local mass of the objects in **U** ;
Initialize the core region of three-way clusters
for *every* $TwoC_k$ *in* **TwoC do**
$\quad \llcorner\ Co\left(ThreeC_k\right) \leftarrow TwoC_k;$
Assigns the objects in the **U − TwoC** to the core regions, fringe regions or trivial regions of corresponding neighboring clusters;
for *every* $\mathbf{x}'$ *in* **U − TwoC do**
$\quad$ calculate the **Nei**$\left(\mathbf{x}'\right)$ of $\mathbf{x}'$ according to Eq. (12);
$\quad$ **if Nei**$\left(\mathbf{x}'\right) \neq \varnothing$ **then**
$\qquad$ **for** *every* $Nei\left(\mathbf{x}'\right)_k$ *in* **Nei**$\left(\mathbf{x}'\right)$ **do**
$\qquad\quad$ calculate L_k according to Eq. (10);
$\qquad\quad$ **for** *every local mass-contour of* $Nei\left(\mathbf{x}'\right)_k$ **do**
$\qquad\qquad \llcorner$ calculate the value of local mass-contour according to Eq. (11);
$\qquad\quad$ **for** *every local mass-contour of* $Nei\left(\mathbf{x}'\right)_k$ **do**
$\qquad\qquad \lvert$ calculate the nearest object $Nearest_l$ in the l-th local
$\qquad\qquad \llcorner$ mass-contour from outside to inside;
$\qquad\quad$ assign $\mathbf{x}'$ into the core region, fringe region or trivial region of
$\qquad\quad \llcorner$ $Nei\left(\mathbf{x}'\right)_k$ according to the three-way clustering rules as Eq. (15).

---

4   Experimental Results

This section describes experiments on 4 UCI data sets [6] with the proposed method and some other algorithms in order to compare their performance in indices such as Accuracy, F-measure, RI [4] and NMI [7]. Table 1 describes the detail information of the tested data sets.

The DPC algorithm [5] and the DPC-KNN algorithm [1] are used here as the two-way clustering compared algorithms. The proposed ATWC-MD methods based on DPC algorithm, DPC-KNN algorithm are called as DPC+ATWC-MD,

Table 1. The UCI data sets

Data sets	Size	Attribute number			Clusters
		Cat	Ord	Num	
Iris	150	0	0	4	3
Patient	90	7	0	1	3
Teaching assistant	151	4	0	1	3
Credit	690	9	0	6	2

Table 2. Experimental results on the UCI data sets

Dataset	Indicator	DPC+ATWC-MD	DPC-KNN+ATWC-MD	DPC-KNN	DPC
Iris	Parameter	$dist_{cutoff} = 0.45$	$p = 0.06$	$p = 0.04$	$dist_{cutoff} = 0.475$
	Accuracy	**0.813**	**0.887**	0.6731	0.72
	F-measure	**0.823**	**0.811**	0.741	0.805
	RI	0.682	**0.7121**	0.501	0.536
	NMI	0.650	**0.7161**	0.6349	0.707
Patient	Parameter	$dist_{cutoff} = 4$	$p = 0.94$	$p = 0.94$	$dist_{cutoff} = 0.4$
	Accuracy	**0.711**	**0.778**	0.067	0.711
	F-measure	0.591	**0.625**	0.125	0.591
	RI	0.52	**0.546**	0.003	0.522
	NMI	0.019	0.129	**0.141**	0.008
Teacher assistant	Parameter	$dist_{cutoff} = 0.95$	$p = 0.61$	$p = 0.19$	$dist_{cutoff} = 0.75$
	Accuracy	**0.848**	0.795	0.391	0.655
	F-measure	**0.864**	**0.879**	0.319	0.752
	RI	**0.718**	0.630	0.316	0.455
	NMI	0.689	**0.817**	0.482	0.65
Credit	Parameter	$dist_{cutoff} = 2.1$	$p = 0.012$	$p = 0.015$	$dist_{cutoff} = 1.2$
	Accuracy	0.469	**0.556**	0.351	0.301
	F-measure	**0.614**	0.42	0.368	0.382
	RI	0.219	**0.506**	0.144	0.059
	NMI	**0.405**	0.089	0.054	0.234

DPC-KNN+ATWC-MD, respectively. In order to guarantee the fairness and objectivity of the experiments, we choose the optimal results after running each algorithm under different parameters for each data set. The Table 2 records the detailed experimental results. The line of "Parameter" record the parameters used in the corresponding algorithms. The optimal results are in bold.

From the above experimental results, it is obvious that the proposed methods, DPC+ATWC-MD and DPC-KNN+ATWC-MD, perform better in most cases than the compared algorithms in all indices, which shows that the proposed method is effective and feasible.

5   Conclusions

The proposed method can deal with the thresholds problem in the three-way clustering, it can automatically assign the undecided objects into the corresponding core regions, fringe regions or trivial regions of their neighbor clusters. Meanwhile, the proposed method can preserve the basic shape information of the two-way clusters, and deal with the fringe objects effectively. However, when the algorithm deals with mixed-type data, the algorithm still has many difficulties in measuring the distance of the categorical values, it needs to artificially set the internal nodes when semantic information between attributes are not enough. In other words, when the relationship between the values of the categorical attribute is not clear in advance, it is difficult to construct a weighted tree structure. That is a further work in future.

Acknowledgments. This work was supported in part by the National Natural Science Foundation of China under Grant Nos. 61751312, 61533020 and 61379114.

References

1. Du, M., Ding, S., Jia, H.: Study on density peaks clustering based on k-nearest neighbors and principal component analysis. Knowl. Based Syst. **99**, 135–145 (2016)
2. Lingras, P., West, C.: Interval set clustering of web users with rough k-means. J. Intell. Inf. Syst. **23**(1), 5–16 (2004)
3. Lingras, P., Peters, G.: Rough clustering. Wiley Interdiscip. Rev. Data Min. Knowl. Discov. **1**(1), 64–72 (2011)
4. Rand, W.M.: Objective criteria for the evaluation of clustering methods. Publ. Am. Stat. Assoc. **66**(336), 846–850 (1971)
5. Rodriguez, A., Laio, A.: Machine learning. Clustering by fast search and find of density peaks. Science **344**(6191), 1492–1496 (2014)
6. UCI Machine Learning Repository. http://archive.ics.uci.edu/ml
7. Wang, Y., Chen, L., Mei, J.P.: Incremental fuzzy clustering with multiple medoids for large data. IEEE Trans. Fuzzy Syst. **22**(6), 1557–1568 (2014)
8. Wang, P.X., Yao, Y.Y.: CE3: a three-way clustering method based on mathematical morphology. Knowl. Based Syst. **155**, 54–65 (2018)
9. Yao, Y.Y., Deng, X.F.: Three-way decisions with probabilistic rough sets. Inf. Sci. **180**(3), 341–353 (2010)
10. Yu, H., Wang, X.C., Wang, G.Y., Zeng, X.H.: An active three-way clustering method via low-rank matrices for multi-view data. Inf. Sci. **000**, 1–17 (2018)
11. Yu, H.: A framework of three-way cluster analysis. In: Polkowski, L., et al. (eds.) IJCRS 2017. LNCS (LNAI), vol. 10314, pp. 300–312. Springer, Cham (2017). https://doi.org/10.1007/978-3-319-60840-2_22
12. Yu, H., Chang, Z.H., Zhou, B.: A novel three-way clustering algorithm for mixed-type data. In: IEEE International Conference on Big Knowledge, pp. 119–126. IEEE (2017)

Three-Way Spectral Clustering

Hong Shi[1], Qiang Liu[1], and Pingxin Wang[1,2(✉)]

[1] School of Computer Science, Jiangsu University of Science and Technology,
Zhenjiang 212003, China
`wangpingxin@just.edu.cn`
[2] College of Mathematics and Information Science, Hebei Normal University,
Shijiazhuang 050024, China

Abstract. In recent years, three-way clustering has shown promising performance in many different fields. In this paper, we present a new three-way spectral clustering by combining three-way decision and spectral clustering. In the proposed algorithm, we revise the process of spectral clustering and obtain an upper bound of each cluster. Perturbation analysis is applied to separate the core region from upper bound and the differences between upper bound and core region are regarded as the fringe region of specific cluster. The results on UCI data sets show that such strategy is effective in reducing the value of DBI and increasing the values of ACC and AS.

Keywords: Spectral clustering · Three-way decision
Three-way clustering · Three-way spectral clustering

1 Introduction

Clustering plays a key role in identifying the internal structure of data. The purpose of clustering is divide a group of unmarked samples into different clusters in light of similarity, such that the clusters have high intra-class similarity and low inter-class similarity. Cluster analysis is an unsupervised learning method, which has been positively applied in image processing [1], web search [2], security assurance [3] and bioinformatics [4].

Roughly speaking, hierarchical clustering and partition clustering are the two most commonly used clustering approaches [5]. We focus on the latter in this paper. As we know, the k-means [6] is the most frequently used partition clustering method. It has converged when the centroids no longer change. But the original k-means algorithm is easy to converge to a local optimal solution and sensitive to the initial data. With the purpose of clustering on arbitrary shape sample space and converge to global optimal solution, the spectral clustering [7, 8] method was proposed. The idea of that is to view objects as vertices and similarities between objects as weighted edges. It turns a clustering problem into a graph segmentation problem that makes the weights of the edges connecting different clusters as low as possible and the weights of the edges within a cluster as high as possible.

© Springer Nature Switzerland AG 2018
M. Ceci et al. (Eds.): ISMIS 2018, LNAI 11177, pp. 389–398, 2018.
https://doi.org/10.1007/978-3-030-01851-1_37

The k-means and spectral clustering method are hard clustering [5], which assign each object to exactly one cluster. Hard clustering methods do not address satisfactorily the relationships between an element and a cluster. It is obviously that there may be three types of relationships between an element and a cluster, namely, belong-to fully, belong-to partially (i.e., also not belong-to partially), and not belong-to fully. To represent the three types of relationship, it is necessary to introduce a notion of three-way clustering [9–11], as an alternative to conventional hard clustering. In hard clustering, cluster is represented by a set that divides the space into two regions. Objects belong to the cluster if they are in the set, otherwise they do not belong to the cluster. In three-way clustering, a cluster is represented by pair of sets called the core region and fringe region, respectively. The core region, the fringe region, the complement of the core region and the fringe region give rise to a trisection of the space. A trisection captures the three types of relationships between a cluster and an object. With this understanding of clusters, Yu et al. [9–11] initiated studies on three-way clustering by drawing ideas from a theory of three-way decision [13–15].

In this paper, we propose the three-way spectral clustering by combining three-way decision and spectral clustering. Different from the classical spectral clustering, the result of three-way spectral clustering uses core region and fringe region to represent a cluster. The rest of this paper is organized as follows. Section 2 reviews some basic concepts of spectral clustering, three-way decision and three-way clustering. Section 3 presents the process of three-way spectral clustering (TWSC for short). Section 4 reviews several evaluation functions. Experimental results are reported in Sect. 5.

2   Preliminaries

In this section, we briefly introduce the background of the proposed method, which consists of spectral clustering, three-way decision and three-way clustering.

2.1   Spectral Clustering

Spectral clustering was first suggested by Donath and Hoffman [16] in 1973 to construct graph partitions based on eigenvectors of the adjacency matrix. A comprehensive review of spectral clustering can be found in [17]. We introduce some related concepts of spectral clustering and one popular spectral clustering algorithm: NJW algorithm, which was presented by Ng, Jordan and Weiss [18].

Let $G = (V, E)$ be an undirected weighted graph with vertex set $V = \{v_1, \cdots, v_n\}$. Each vertex v_i in this graph represents a data point and nonnegative weight w_{ij} represents the similarity of two vertices v_i and v_j. The weighted adjacency matrix of the graph is the matrix $W = (w_{ij})(i, j = 1, \cdots, n)$. As G is undirected grape, it means W is symmetric, i.e., $w_{ij} = w_{ji}$. The basic idea of spectral clustering is to find a partition of the graph such that the edges between different groups have a low similarity and the edges within a group have a high similarity.

Most of the spectral clustering algorithms can be divided to three steps. The first one is to construct the similarity matrix representing the data sets. The second one is to compute eigenvalues and eigenvectors of the Laplacian matrix and map each point to a lower-dimensional representation based on one or more eigenvectors. The third step is to assign points to two or more classes, based on the new representation. Taking NJW algorithm for example, we give the procedure of spectral clustering in Algorithm 1.

Algorithm 1. NJW algorithm

Input: $V=\{v_1, v_2, \cdots, v_n\} \in R^l$, the number of clusters k and parameter σ.
Output: Clusters $C=\{C_1, C_2, \cdots, C_k\}$.

1. Form the similarity matrix W defined by $w_{ij} = \exp(-\frac{d^2(v_i,v_j)}{2\sigma^2}), i \neq j$, and $w_{ii} = 0$.
2. Define D to be the diagonal matrix whose (i,i)-element is the sum of W's i-th row, and construct the matrix $L = D^{-\frac{1}{2}} W D^{\frac{1}{2}}$.
3. Find the largest k eigenvectors of L (chosen to be orthogonal to each other in the case of repeated eigenvalues), and form the matrix X by stacking the eigenvectors in columns $X = (x_1, x_2, \cdots, x_k) \in R^{n \times k}$.
4. Form the matrix Y from X by normalizing each of X's rows to have unit length, $Y_{ij} = \frac{X_{ij}}{(\sum_j X_{ij}^2)^{\frac{1}{2}}}$.
5. Treat each row of Y as a point in R^k and classify them into k classes via k-means algorithm.
6. Assign the original points v_i to cluster j if and only if row i of the matrix Y was assigned to cluster j.

2.2 Three-Way Decision

The theory of three-way decision was first proposed by Yao [13–15] on the basis of the commonly binary-decision model through adding a third option and used to interpret rough set. Due to a positive or negative decision may not be made if there is insufficient evidence. In this case, people can choose an alternative decision that is neither yes nor no, which also be called a deferment decision that requires further judgments. Three-way decision divides the universe into the positive, negative and boundary regions, denoting the regions of acceptance, rejection and non-commitment for ternary classifications, which has been successfully applied in different fields

Three-way decision investigates scientifically a common problem-solving and information-processing practice. Unlike a certain decision making, three-way decision newly increase a deferment strategy. Many soft computing models for leaning uncertain concepts, such as interval sets, rough sets, fuzzy sets and shadowed sets, have the tri-partitioning properties and can be reinvestigated within the framework of three-way decisions [19]. Recently, a trisecting-and-acting model is propose by Yao [20], which can be depicted by Fig. 1. According to the model, there are two basic tasks in three-way decision. The first one is how to divide a universal into three parts and the second one is to develop different strategy for different part.

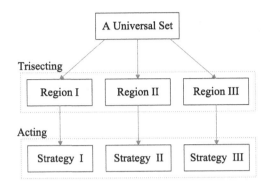

Fig. 1. Trisecting-and-acting model

2.3   Three-Way Clustering

By combining three-way decision with cluster analysis, Yu et al. [9–11] presented a framework of three-way clustering which represents a cluster by a pair of sets called core region and fringe region. Wang and Yao [12] proposed a framework of a contraction-and-expansion based three-way clustering called CE3 based on the ideas of erosion and dilation from mathematical morphology. Here we summarize the basic concepts of three-way clustering. Assume $C = \{C_1, \cdots, C_k\}$ is a family clusters of universe $V = \{v_1, \cdots, v_n\}$. In contrast to the general crisp representation of a cluster, three-way clustering represents a three-way cluster C_i as a pair of sets:

$$C_i = (\mathrm{Co}(C_i), \mathrm{Fr}(C_i)),$$

where $\mathrm{Co}(C_i) \subset V$ and $\mathrm{Fr}(C_i) \subset V$. In this paper, we adopt the following properties on $\mathrm{Co}(C_i)$ and $\mathrm{Fr}(C_i)$:

(I) $\mathrm{Co}(C_i) \neq \phi, (i = 1, \cdots, k)$,

(II) $\cup_{i=1}^k (\mathrm{Co}(C_i) \cup \mathrm{Fr}(C_i)) = V$,

(III) $\mathrm{Co}(C_i) \cap \mathrm{Co}(C_j) = \phi \ (i \neq j)$.

Property (I) demands that each cluster cannot be empty. Property (II) states that it is possible that an element $v \in V$ belongs to more than one cluster. Property (III) requires that the core regions of clusters are pairwise disjoint. Base on the above discussion, we have the following family of clusters to represent the result of three-way clustering.

$$\mathbb{C} = \{(\mathrm{Co}(C_1), \mathrm{Fr}(C_1)), (\mathrm{Co}(C_2), \mathrm{Fr}(C_2)), \cdots, (\mathrm{Co}(C_k), \mathrm{Fr}(C_k))\}.$$

Note that the properties of $\mathbb{C}$ can be obtained from Properties (I)–(III).

3   Three-Way Spectral Clustering

Traditional spectral clustering assumes that a cluster must be represented by a set with crisp boundary. However, assigning uncertain elements into a cluster will increase decision risk. Three-way clustering puts the identified elements into the core region and the uncertain elements into the fringe region to reduce decision risk. In this section, we present the three-way spectral clustering algorithm by combining spectral clustering and three-way decision.

Suppose $V = \{v_1, v_2, \cdots, v_n\}$ is a given data set and k is the given number of the clusters. The main steps of three-way spectral clustering are as follows. Firstly, getting the corresponding similarity matrix W, diagonal matrix D, Laplace matrix L based on NJW algorithm [18]. Secondly, computing the largest k eigenvectors of L and form the matrix X by stacking the eigenvectors in columns $X = [x_1, x_2, \cdots, x_k] \in R^{n \times k}$. Thirdly, normalizing X achieve matrix Y, which each row $y_i(i = 1, 2, \cdots, n)$ is considered as a sample point. All the above steps are the same with spectral clustering. In order to get three-way clustering result, we use the following strategy on $Y = \{y_1, y_2, \cdots, y_n\}$ to obtain the core region and the fringe region. For each object y and randomly selected k centroids $z_1, \cdots, z_k$, let $d(y, z_j)$ be the distance between itself and the centroid z_j. Suppose $d(y, z_i) = \min_{1 \le j \le k} d(y, z_j)$ and $T = \{j : d(y, z_j) - d(y, z_i) \le \varepsilon_1$ and $i \ne j\}$, where ε_1 is a given parameter. Then,

1 If $T \ne \phi$, then $y \in C_i^u$ and $y \in C_j^u$.
2 If $T = \phi$, then $y \in C_i^u$.

The modified centroid calculations for above procedure are given by:

$$z_i = \frac{\sum_{y_k \in C_i^u} y_k}{|C_i^u|} \tag{1}$$

where $i = 1, \cdots, k$, $|C_i^u|$ is the number of objects in upper bound of cluster C_i. Repeat this process until modified centroids in the current iteration are identical to those that have been generated in the previous one, namely, the prototypes are stabilized. From the above procedure, we get a family of overlapping clusters, which are regard as the unions of core regions and fringe regions. We use center perturbation distance by adding weights of elements to separate the core region from C_i^u.

For a given upper bound $C_i^u(i = 1, \cdots, k)$, we classify the elements of C_i^u into two types.

$$\text{Type I} = \{y \in C_i^u \mid \exists j = 1, \cdots, k, j \ne i, y \in C_j^u\},$$
$$\text{Type II} = \{y \in C_i^u \mid \forall j = 1, \cdots, k, j \ne i, y \notin C_j^u\}.$$

Suppose the centroid of C_i^u is z_i and m_i is the number of elements in C_i^u. We adapt different strategies for different types. For objects in type I, we assign them into fringe region of C_i because they belong to two clusters at least. For

---

Algorithm 2. Three-way spectral clustering algorithm

Input: $V = \{v_1, v_2, \cdots, v_n\}$, the number of clusters k and parameter $\varepsilon_1, \varepsilon_2, \sigma$.
Output: Clusters $\mathbb{C} = \{(\mathrm{Co}(C_1), \mathrm{Fr}(C_1)), (\mathrm{Co}(C_2), \mathrm{Fr}(C_2)), \cdots, (\mathrm{Co}(C_k), \mathrm{Fr}(C_k))\}$.

1   Execute step 1-4 of Algorithm 1 and obtain matrix Y. View all rows of Y as $\{y_1, y_2, \cdots, y_n\}$.
2.   Randomly select k cluster centroid $z_1, z_2, \cdots, z_k$ in $\{y_1, y_2, \cdots, y_n\}$;
3.   **for** $i \leftarrow 1$ to n **do**
4.   **repeat**
5.   for an object y_i, determine its closet centroid y_h: $d(y_i, z_h) = min_{1 \leq c \leq k} d(y_i, z_c)$;
6.   determine set $T = \{j : d(y_i, z_j) - d(y_i, z_h) \leq \varepsilon_1 \wedge j \neq h\}$;
7.   **if** $T = \emptyset$ **then**
8.   assign y_i to C_h^u ;
9.   **else**
11.   assign y_i to C_h^u and C_j^u;
12.   **end if**
13.   calculate the new centroid for each cluster using Eq. (1);
14.   **until** modified centroids in the current iteration are identical to those of the previous;
15. **end for**
16. **for** $i \leftarrow 1$ to k **do**
17.   for each $y \in C_i^u$, determine set $G = \{j : j \neq i \wedge x \in C_j^u\}$;
18.   **if** $G \neq \emptyset$ **then**
19.   assign y to the fringe region of C_i, i.e. $x \in Fr(C_i)$;
20.   **else**
21.   add m_i times of y into C_i^u and denote the new cluster by $C_i^{u^*}$, where m_i is the number of elements in C_i^u ;
22.   calculate the new centroid z_i^* of $C_i^{u^*}$ using Eq. (1) and the difference $|z_i - z_i^*|$;
23.   **if** $|z_i - z_i^*| \leq \varepsilon_2$ **then**
24.   assign y to the core region of C_i, i.e. $y \in \mathrm{Co}(C_i)$;
25.   **else**
26.   assign y to the fringe of C_i, i.e. $y \in \mathrm{Fr}(C_i)$;
27.   **end if**
28.   **end if**
29. **end for**
30. Assign the original points v_i to the core region or the fringe region of $j-$cluster if and only if y_i is assigned to the core region or the fringe region of $j-$cluster.
31. **return** $\mathbb{C} = \{(\mathrm{Co}(C_1), \mathrm{Fr}(C_1)), (\mathrm{Co}(C_2), \mathrm{Fr}(C_2)), \cdots, (\mathrm{Co}(C_k), \mathrm{Fr}(C_k))\}$.

---

objects $y \in C_i^u$ in type II, we add m_i times y into C_i^u and denote the new cluster by C_i^{u*}. Calculate the new centroid z_i^* of C_i^{u*} by (1) and the differences between z_i^* and z_i. For a given parameter ε_2, if $|z_i^* - z_i| \leq \varepsilon_2$, y is assigned to core region of C_i, otherwise, y is assigned to fringe region of C_i. Assign the original points v_i to the core region or the fringe region of $j-$cluster if and only if row y_i is assigned to the core region or the fringe region of $j-$cluster. The process of three-way spectral clustering algorithm can be designed as Algorithm 2.

4   Evaluation Index of Algorithm

4.1   Accuracy

The Accuracy is a common external index for evaluating clustering results. Where the higher the accuracy rate, the better the performance of clustering.

Definition 1. *Accuracy (ACC hereafter).*

$$ACC = \frac{1}{N} \sum_{i=1}^{k} \theta_i$$

where N is the number of objects within each cluster, θ_i is the number of samples that are correctly divided into the i-th cluster and k represent clustering number.

4.2   Davies-Bouldin Index

Davies-Bouldin Index was proposed by Davide L. Davies and Donald W. Bouldin [21], also named DBI. Where the smaller the DBI rate, the better the performance of clustering.

Definition 2. *Davies-Bouldin Index (DBI hereafter).*

$$DBI = \frac{1}{k} \sum_{i=1}^{k} max_{i \neq j} \left(\frac{\overline{c_i} + \overline{c_j}}{||w_i - w_j||_2} \right)$$

where w_i and w_j are represent the clustering centers of the cluster C_i and C_j, $\overline{c_i}$ is the average distance from all the samples in the i-th cluster to w_i, $||w_i - w_j||_2$ denote the euclidean distance between w_i and w_j, k is the clustering number.

4.3   Average Silhouette Coefficient

Silhouette coefficient was developed by Peter J. Rousseeuw [22] in 1986. The silhouette Coefficient value range from -1 to 1, where the higher value indicates that the better the performance of clustering.

Definition 3. *Silhouette Coefficient of single sample d_i.*

$$S_i = \frac{b_i - a_i}{max(a_i, b_i)}$$

where a_i is the average distance from d_i to all other samples in its cluster, interpret a_i as intra-class similarity. $b_i = min\{D(d_i - c_j)\}$ is the smallest average distance from d_i to all samples in c_j, interpret b_i as inter-class similarity.

Definition 4. *Average Silhouette Coefficient (AS hereafter).*

$$AS = \frac{1}{N} \sum_{i=1}^{N} S_i$$

where N is the total number of samples. The average silhouette coefficient is represented by the mean value of the silhouette coefficient of all samples.

5   Experimental Illustration

This paper applies k-means algorithm, spectral clustering algorithm and three-way spectral clustering algorithm to verify the clustering performance by clustering five groups of UCI [23] dataset as shown in Table 1.

Through many times experiments, we can gain the value of parameters ε_1 and ε_2 ($\varepsilon_1 = 0.005, \varepsilon_2 = 0.26$), the average values and the best values, in which the average value is used to compare the overall performance of the algorithm and the best value is used to compare the best performance of the algorithm. The experimental results are shown in Table 2. Where KM, SC and TWSC represent k-means, spectral clustering, three-way spectral clustering respectively.

Table 1. Datasets used in experiments

Datasets	Numbers	Features	Categories
Iris	150	4	3
Wdbc	569	30	2
Wine	178	13	3
Hill	1212	100	2
Bank	1372	4	2

Table 2. Experimental results on UCI datasets

Datasets	Algorithm	Average value			Best value		
		DBI	AS	ACC	DBI	AS	ACC
Iris	KM	0.7800	0.6835	0.8502	0.7610	0.6959	**0.8867**
	SC	0.7691	0.6897	0.8659	0.7228	0.6900	0.8733
	TWSC	**0.7177**	**0.7308**	**0.8830**	**0.6944**	**0.7311**	0.8865
Wdbc	KM	1.1363	0.5765	0.9279	1.1363	0.5765	0.9279
	SC	1.1754	0.5406	0.9262	1.1754	0.5406	0.9262
	TWSC	**1.0881**	**0.5851**	**0.9474**	**1.0881**	**0.5851**	**0.9474**
Wine	KM	1.3128	0.4748	0.9435	**1.0729**	0.4764	0.9663
	SC	1.3069	0.4748	0.9663	1.3069	0.4748	0.9663
	TWSC	**1.1911**	**0.5370**	**0.9813**	1.1911	**0.5370**	**0.9812**
Hill	KM	**0.4048**	**0.9376**	0.5091	**0.4048**	**0.9376**	0.5091
	SC	0.4992	0.8926	0.5132	0.4992	0.8926	0.5132
	TWSC	0.4612	0.9124	**0.5144**	0.4605	0.9128	**0.5147**
Bank	KM	1.1913	0.5002	0.5747	1.1911	0.5004	0.5758
	SC	1.2002	0.4908	0.6144	1.2002	0.4908	0.6144
	TWSC	**1.1256**	**0.5305**	**0.6211**	**1.1256**	**0.5305**	**0.6211**

By observing the experimental results in Table 2, we have marked the best solution in bold. Firstly, the average value and the best value of TWSC in the index of DBI, AS and ACC are clearly superior to the KM and SC algorithm. But due to the instability of the KM algorithm, it is normal that several values surpass the TWSC clustering result in the best value. In addition, we can find that the performance of TWSC in the Hill dataset is not better than the KM and SC results in the index of DBI and AS, although the performance of TWSC in the index of ACC is a little more than the KM and SC results. In a word, the clustering performance of TWSC on most datasets is obviously better than the KM and SC. Meanwhile, we also find the shortcomings of this algorithm, which need further research and improvement.

6   Conclusion

With the view of improving the clustering performance, we developed a new clustering algorithm named three-way spectral clustering that combining three-way decision and spectral clustering algorithm. Moreover, the experimental results demonstrate that the new algorithm can significantly improve the structure of classification results by comparing with the traditional spectral clustering algorithm. But this algorithm needs to be further modification and improvement, which makes the clustering performance better. This will also be our next research direction.

References

1. Elalami, M.E.: Supporting image retrieval framework with rule base system. Knowl. Based Syst. **24**, 331–340 (2011)
2. Martin-Guerrero, J.D., Palomares, A., Balaguer-Ballester, E.: Studying the feasibility of a recommender in a citizen Webportal based on user modeling and clustering algorithms. Expert. Syst. Appl. **30**, 299–312 (2006)
3. Kalyani, S., Swarup, K.S.: Particle swarm optimization based k-means clustering approach for security assessment in power systems. Expert. Syst. Appl. **38**, 10839–10846 (2011)
4. Sebiskveradze, D., Vrabie, V., Gobinet, C., Durlach, A., Bernard, P.: Automation of an algorithm based on fuzzy clustering for analyzing tumoral heterogeneity in human skin carcinoma tissue sections. Lab. Investig. J. Tech. Methods Pathol. **91**, 799–811 (2011)
5. Sun, J.G., Liu, J., Zhao, L.Y.: Clustering algorithms research. J. Softw. **19**(1), 48–61 (2008)
6. Macqueen, J.B.: Some methods for classification and analysis of multivariate observations. In: Proceedings of 5th Berkeley Symposium on Mathematical Statistics and Probability conference, pp. 281–297 (1966)
7. Luxburg, U.V.: A tutorial on spectral clustering. Stat. Comput. **17**, 395–416 (2007)
8. Cai, X.Y., Dai, G.Z., Yang, L.B.: Survey on spectral clustering algorithms. Comput. Sci. **35**(7), 14–18 (2008)
9. Yu, H., Zhang, C., Wang, G.Y.: A tree-based incremental overlapping clustering method using the three-way decision theory. Knowl. Based Syst. **91**, 189–203 (2016)

10. Yu, H., Jiao, P., Yao, Y.Y., Wang, G.Y.: Detecting and refining overlapping regions in complex networks with three-way decisions. Inf. Sci. **373**, 21–41 (2016)

11. Yu, H., Wang, X.C., Wang, G.Y.: An active three-way clustering method via low-rank matrices for multi-view data. Inf. Sci. (2018). https://doi.org/10.1016/j.ins.2018.03.009

12. Wang, P.X., Yao, Y.Y.: CE3: a three-way clustering method based on mathematical morphology. Knowl. Based Syst. **155**, 54–65 (2018)

13. Gao, C., Yao, Y.Y.: Actionable strategies in three-way decisions. Knowl. Based Syst. **133**, 183–199 (2017)

14. Yao, Y.Y.: The superiority of three-way decisions in probabilistic rough set models. Inf. Sci. **181**, 1080–1096 (2011)

15. Yao, Y.: An outline of a theory of three-way decisions. In: Yao, J. (ed.) RSCTC 2012. LNCS (LNAI), vol. 7413, pp. 1–17. Springer, Heidelberg (2012). https://doi.org/10.1007/978-3-642-32115-3_1

16. Donath, W.E., Hoffman, A.J.: Lower bounds for the partitioning of graphs. IBM J. Res. Dev. **17**, 420–425 (1973)

17. Shi, J.B., Malik, J.: Normalized cuts and image segmentation. IEEE Trans. Pattern Anal. Mach. Intell. **22**, 888–905 (2000)

18. Ng, A.Y., Jordan, M.I., Weiss, Y.: On spectral clustering: analysis and an algorithm. In: International Conference on Neural Information Processing Systems: Natural and Synthetic, pp. 849–856. MIT Press (2001)

19. Yao, Y.: Three-way decision: an interpretation of rules in rough set theory. In: Wen, P., Li, Y., Polkowski, L., Yao, Y., Tsumoto, S., Wang, G. (eds.) RSKT 2009. LNCS (LNAI), vol. 5589, pp. 642–649. Springer, Heidelberg (2009). https://doi.org/10.1007/978-3-642-02962-2_81

20. Yao, Y.Y.: Three-way decisions and cognitive computing. Cogn. Comput. **8**, 543–554 (2016)

21. Maulik, U., Bandyopadhyay, S.: Performance evaluation of some clustering algorithms and validity indices. IEEE Trans. Pattern Anal. Mach. Intell. **24**, 1650–1654 (2002)

22. Rousseeuw, P.J.: Silhouettes: a graphical aid to the interpretation and validation of cluster analysis. J. Comput. Appl. Math. **20**, 53–65 (1987)

23. Asuncion, A.: UCI machine learning repository (2013). www.ics.uci.edu/~mlearn/MLRepository.html

The Granular Structures in Formal Concept Analysis

Ruisi Ren and Ling Wei$^{(\boxtimes)}$

School of Mathematics, Northwest University,
Xi'an 710069, People's Republic of China
ruisiren_rose@163.com, wl@nwu.edu.cn

Abstract. Granular analysis in formal concept analysis is a newly proposed interesting topic. In this paper, we study the granular structures in formal concept analysis. According to the principle of multiview, we define the granules from viewpoints of objects and attributes, respectively. On the basis of the semantic meaning and the function in lattice construction, ten different kinds of granules are given. These ten kinds of granules construct the multilevel granular structures. Then, we define the similarity between each pair of granules in the same level. On the basis of the similarity measurement, we show how to transform granules from one level to another. Finally, an example is presented to illustrate the results we obtain.

Keywords: Formal concept analysis · Granular structure
Similarity measurement · Granule transformation

1 Introduction

Formal concept analysis (FCA) is an efficient tool for data analysis [1]. The two basic notions in FCA are formal context and formal concept [2]. Formal context is the data base of FCA and the knowledge obtained from these data is reflected by a set of formal concepts which form a lattice called concept lattice.

The idea of granular computing (GrC) was obtained from the human-inspired and nature-inspired ways to abstracting and processing information [3]. The term "granular computing" was introduced by Zadeh in 1997 [4] and became a new study field [3]. Pedrycz [5] stated that "granular computing is geared toward representing and processing basic chunks of information-information granules." The granules and granular structures are two basic notions in GrC. A granule is a block consisting of a group of entities. Different granules have different granulation which can reflect the level of abstraction of granules. Yao [6] focused on high-level conceptual understanding of GrC by drawing results across multiple disciplines. The principle of multilevel emphasizes on understanding and representation at multiple levels of granules. The principle of multiview stresses understanding and representation from multiple angles.

© Springer Nature Switzerland AG 2018
M. Ceci et al. (Eds.): ISMIS 2018, LNAI 11177, pp. 399–408, 2018.
https://doi.org/10.1007/978-3-030-01851-1_38

Recently, GrC in FCA has become an interesting topic [7–11]. Zhi and Li [12] defined three different kinds of granules in FCA and discussed the methods for the description of these granules. Li and Wu [13] studied many topics of GrC in FCA. Based on the principle of multiview, Qi et al. [14] presented granules in FCA from viewpoints of objects and attributes, respectively. From each viewpoint, Qi et al. [14] gave five different types of granules, which are classified granules, pictorial granules, elementary granules, essential granules and atomic granules with different semantics and structures, and discussed the relationships among these different types of granules.

Granular structure gives a hierarchical structure of different granules. In this paper, we plan to study the granular structure of granules according to the degree of abstraction. After giving the granular structure, we show how to transform granules from one level to another. Specifically, we define the similarity measurement between two granules in the same level. On the basis of the similarity value, we show how to cluster a group of pictorial granules into an elementary granules, give the method to compressing the elementary granules into essential granules, and present the way to eliminating non-atomic granules to obtain the atomic granules from the essential granules.

The organization of this paper is as follows. Section 2 reviews some basic notions in FCA. Section 3 introduces ten types of granules in FCA. Section 4 defines the similarity measurement between different granules and shows the process of transformation among different levels of granules. An example is given in Sect. 5 to illustrate the results we obtain. Finally, the paper is concluded with a summary in Sect. 6.

2   Preliminaries

In formal concept analysis [1,2], the data set is represented by a triple named formal context.

Definition 1 [2]. *A formal context (G, M, I) consists of two sets G and M and a relation I between G and M. The elements of G are called the objects and the elements of M are called the attributes of the context. In order to express that an object g is in a relation I with an attribute m, we write gIm or $(g, m) \in I$ and read it as "the object g has the attribute m".*

Based on formal context (G, M, I), a pair of operators $*$ on $X \subseteq G$ and $A \subseteq M$ are defined as follows [2]:

$$X^* = \{m \in M \mid \forall g \in X, (g, m) \in I\}, \quad A^* = \{g \in G \mid \forall m \in A, (g, m) \in I\}.$$

For the convenience, $\forall g \in G, \forall m \in M$, $\{g\}^*$ and $\{m\}^*$ are denoted as g^* and m^*, respectively.

A formal context (G, M, I) is called clarified, if for any objects $g, h \in G$ from $g^* = h^*$ it always follows that $g = h$ and, correspondingly, $m^* = n^*$ implies $m = n$ for all $m, n \in M$. For the simplicity of description, we claim the contexts in this paper are clarified.

Based on the operators $*$, the notion of a formal concept is proposed as follows.

Definition 2 [2]. *A formal concept of the context (G, M, I) is a pair (X, A) with $X^* = A$ and $X = A^*$ $(X \subseteq G, A \subseteq M)$. We call X the extent and A the intent of the formal concept (X, A).*

The family of all formal concepts of context (G, M, I) forms a complete lattice, called concept lattice and denoted by $L(G, M, I)$. For any $(X_1, A_1), (X_2, A_2) \in L(G, M, I)$, the partial order is defined as:

$$(X_1, A_1) \leqslant (X_2, A_2) \Leftrightarrow X_1 \subseteq X_2 \ (\Leftrightarrow A_2 \subseteq A_1).$$

The infimum and the supremum are defined by:

$$(X_1, A_1) \wedge (X_2, A_2) = (X_1 \cap X_2, (A_1 \cup A_2)^{**}),$$
$$(X_1, A_1) \vee (X_2, A_2) = ((X_1 \cup X_2)^{**}, A_1 \cap A_2).$$

Then, we introduce some basic notions in lattice theory.

Definition 3 [15]. *Let L be a lattice. An element $x \in L$ is join-irreducible if*
1. $x \neq 0$ (in case L has a zero);
2. $x = a \vee b$ implies $x = a$ or $x = b$ for all $a, b \in L$.

Definition 4 [15]. *Let P be an ordered set and let $Q \subseteq P$. Then Q is called join-dense in P if for every element $a \in P$ there is a subset A of Q such that $a = \bigvee_P A$.*

The meet-irreducible element and meet-dense can be defined dually.

For any object $g \in G$, the pair (g^{**}, g^*) is a formal concept, called an object concept. Similarly, for any attribute $m \in M$, the pair (m^*, m^{**}) is also a formal concept, called an attribute concept. The object concepts and attribute concepts are very important in constructing a common formal concept, since every concept (X, A) can be constructed by a set of object concepts or a set of attribute concepts as follows:

$$\bigvee_{g \in X} (g^{**}, g^*) = (X, A) = \bigwedge_{m \in A} (m^*, m^{**}) \tag{1}$$

3   Multi-granules in Formal Concept Analysis

Based on the principle of multiview, two different views of granules are proposed [14]. One is from the view of objects and the other is from the view of attributes.

From the viewpoint of objects, different granules in FCA are defined to reflect different knowledge. If we regard a formal context as an information system, the whole set of objects will be classified into different equivalence classes. Hence, the equivalence class of objects can be regarded as one kind of granule in FCA. This kind of granule is called the classified granule and denoted as

$$[g]_M = \{g_i \in G \mid g^* = g_i^*\} \qquad (2)$$

The set of all object equivalence classes is denoted as $\mathcal{OBE}(G, M, I)$.

If the context is clarified, then an equivalence class is a single point set with one object. However, its related attribute information can not be reflected in this equivalence class. Thus, for every equivalence class of objects $\{g\}$ ($g \in G$), we add the related attribute information into it, resulting in an object induced pair (g, g^*) ($g \in G$). The object induced pair can be regarded as another kind of granule which can reflect the original binary relation between the object g and the attributes. The set of all object induced pairs of context (G, M, I) is denoted as $\mathcal{OBI}(G, M, I) = \{(g, g^*) \mid g \in G\}$. Since $\mathcal{OBI}(G, M, I) = \{(g, g^*) \mid g \in G\}$ can be directly shown as an object pictorial diagram, the object induced pair is called the pictorial granule.

In formal concept analysis, all knowledge is reflected as a concept lattice. In concept lattice, there are some important and special concepts, which reflect different kinds of knowledge. The atom concept is an upper neighbour of the bottom and shows the basic information about the lattice; the join-irreducible concept is indispensable to construct the lattice; the object concept is a kind of important basic concept to form every common concept and the construction method is shown in Eq. (1). Thus, these three different kinds of concepts can be regarded as three different types of granules. They are called atomic granule, essential granule and elementary granule, respectively, and the sets of these three granules are denoted as $\mathcal{A}(G, M, I)$, $\mathcal{J}(G, M, I)$ and $\mathcal{OB}(G, M, I)$, respectively.

Similar to above discussion, five types of granules from the viewpoint of attributes are proposed. The first one is the attribute equivalence class. That is $\{m\}$ ($m \in M$), called the classified granule. After adding the object information, we can get the attribute induced pair (m^*, m). That is the second kind of granule, called pictorial granule. The third kind of granule is the coatom concept which is a lower neighbors of the unit element and is called atomic granule. The forth kind of granule is the meet-irreducible concept and is called the essential granule. The last type of granule is the attribute concept (m^*, m^{**}) and is called the elementary granule. The sets of these five kinds of granules are denoted as $\mathcal{ATE}(G, M, I)$, $\mathcal{ATI}(G, M, I)$, $\mathcal{CA}(G, M, I)$, $\mathcal{M}(G, M, I)$ and $\mathcal{AT}(G, M, I)$, respectively.

Based on the sources of producing granules, the relationships among ten kinds of granules from the viewpoints of both objects and attributes are shown as Fig. 1.

4   Transformations Among Multi-granules

In Sect. 3, ten different kinds of granules are defined and each type is in a granular level. In this section, we discuss how to transform granules from one level to another. Since in classified contexts, the equivalence class of objects is a single

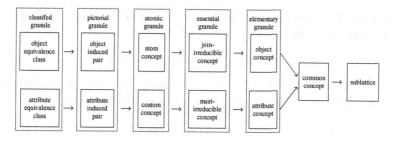

Fig. 1. Granule chain

point set with one object, the object induced pair just adds a set of attributes into the equivalence class of objects. Thus, in this paper, we regard the classifed granules and pictorial granules equivalently and just talk about the pictorial granules.

4.1   Similarity Measurement Between Two Granules

Based on above discussion, we know that different kinds of granules are in different granular levels and in one fixed granular level, there are different granules belonging to the same type. For each granule, there are some granules more similar to it than other granules. In this section, we present the similarity measurement between two granules in the same level. Since the granules defined from the view of objects (attributes) are described by a set of attributes (objects), the more similar the attribute (object) description sets are, the more similar the granules are.

Definition 5. Let (X_1, A_1) and (X_2, A_2) be two granules in the same granular level with $X_1, X_2 \subseteq G$ and $A_1, A_2 \subseteq M$. If (X_1, A_1) and (X_2, A_2) are granules from the view of objects, then the similarity of granule (X_1, A_1) to granule (X_2, A_2) is defined as:

$$Sim_G((X_1, A_1), (X_2, A_2)) = \frac{|A_1 \cap A_2|}{|A_2|}. \tag{3}$$

If (X_1, A_1) and (X_2, A_2) are two granules from the view of attributes, then the similarity of granule (X_1, A_1) to granule (X_2, A_2) is defined as:

$$Sim_M((X_1, A_1), (X_2, A_2)) = \frac{|X_1 \cap X_2|}{|X_2|}. \tag{4}$$

Remark: The similarity measurements in Definition 5 are not symmetric. That is, $Sim_G((X_1, A_1), (X_2, A_2)) = Sim_G((X_2, A_2), (X_1, A_1))$ and $Sim_M((X_1, A_1), (X_2, A_2)) = Sim_M((X_2, A_2), (X_1, A_1))$ may not hold.

According to the value of similarity measurement, all granules in one fixed level can form a poset. Specifically, if (X_1, A_1) and (X_2, A_2) are two granules from the view of objects and they are in the same granular level, then the partial order between these two granules is defined as:

$$(X_1, A_1) \leq_G (X_2, A_2) \Leftrightarrow A_2 \subseteq A_1 \Leftrightarrow Sim_G((X_1, A_1), (X_2, A_2)) = 1.$$

Similarly, if (X_1, A_1) and (X_2, A_2) are two granules from the view of attributes, then the partial order between these two granules is defined as:

$$(X_1, A_1) \leq_M (X_2, A_2) \Leftrightarrow X_1 \subseteq X_2 \Leftrightarrow Sim_M((X_2, A_2), (X_1, A_1)) = 1.$$

That is, based on the partial order, granules in one level also have a structure, which can be reflected by a Hasse diagram.

4.2 Transformations Among Multi-granules

In this section, we discuss how to transform granules from one level to another. Now, we talk about the granules from the view of objects. First, we present how to cluster a group of pictorial granules and transform them into an elementary granule. Let (x, x^*) be a pictorial granule. A denotation E_x is given as $E_x = \{x_i \mid Sim_G((x_i, x_i^*), (x, x^*)) = 1\}$ which is a set of objects of all pictorial granules whose similarity value to granule (x, x^*) is 1.

Theorem 1. *Let (x, x^*) be a pictorial granule. We have (E_x, x^*) is an object concept induced by x. That is (E_x, x^*) is an elementary granule.*

The next theorem discusses how to compress a group of elementary granules into a group of essential granules. Let (x^{**}, x^*) be an elementary granule. A denotation $S_{(x^{**}, x^*)}$ is given as $S_{(x^{**}, x^*)} = \{(x_i^{**}, x_i^*) \mid Sim_G((x_i^{**}, x_i^*), (x^{**}, x^*)) = 1$ and $(x_i^{**}, x_i^*) \neq (x^{**}, x^*)\}$. That is, $S_{(x^{**}, x^*)}$ is a set of elementary granules whose similarity values to granule (x^{**}, x^*) are 1 and which are different from the granule (x^{**}, x^*).

Theorem 2. *Let (x^{**}, x^*) be an elementary granule. If $\mid S_{(x^{**}, x^*)} \mid \geq 2$ holds and there exist $(x_1^{**}, x_1^*), (x_2^{**}, x_2^*) \in S_{(x^{**}, x^*)}$ satisfying $Sim_G((x_1^{**}, x_1^*), (x_2^{**}, x_2^*)) \neq 1$ and $Sim_G((x_2^{**}, x_2^*), (x_1^{**}, x_1^*)) \neq 1$, then (x^{**}, x^*) is not an essential granule.*

Finally, we presented how to compress essential granules into atomic granules. Let (X, A) be an essential granule. A denotation $P_{(X,A)}$ is given as $P_{(X,A)} = \{(X_i, A_i) \in \mathcal{J}(G, M, I) \mid Sim_G((X_i, A_i), (X, A)) = 1$ and $(X_i, A_i) \neq (X, A)\}$. $P_{(X,A)}$ is a set of essential granules whose similarity values to granule (X, A) are 1 and which are different from the granule (X, A).

Theorem 3. *Let (X, A) be an essential granule. If $P_{(X,A)} \neq \emptyset$, then (X, A) is an atomic granule.*

Analogous to the transformations among multi-granules from the view of objects, we will present the transformations among multi-granules from the view of attributes next. The granules in the following part are all defined from the view of attributes.

Let (m^*, m) be a pictorial granule. A denotation I_m is given as $I_m = \{m_j \mid Sim_M((m_j^*, m_j), (m^*, m)) = 1\}$.

Theorem 4. *Let (m^*, m) be a pictorial granule. We have (m^*, I_m) is an attribute concept induced by m. That is, (m^*, I_m) is an elementary granule.*

Next, we discuss how to transform elementary granules into essential granules. Let (m^*, m^{**}) be an elementary granule. A denotation $S_{(m^*, m^{**})}$ is given as $S_{(m^*, m^{**})} = \{(m_j^*, m_j^{**}) \mid Sim_M((m_j^*, m_j^{**}), (m^*, m^{**})) = 1 \text{ and } (m_j^*, m_j^{**}) \neq (m^*, m^{**})\}$.

Theorem 5. *Let (m^*, m^{**}) be an elementary granule. If $\mid S_{(m^*, m^{**})} \mid \geq 2$ holds and there exist $(m_1^*, m_1^{**}), (m_2^*, m_2^{**}) \in S_{(m^*, m^{**})}$ satisfying $Sim_M((m_1^*, m_1^{**}), (m_2^*, m_2^{**})) \neq 1$ and $Sim_M((m_2^*, m_2^{**}), (m_1^*, m_1^{**})) \neq 1$, then (m^*, m^{**}) is not an essential granule.*

Finally, we present how to transform essential granules into atomic granules. Let (X, A) be an essential granule. A denotation $Q_{(X,A)}$ is given as $Q_{(X,A)} = \{(X_j, A_j) \in \mathcal{M}(G, M, I) \mid Sim_M((X_j, A_j), (X, A)) = 1 \text{ and } (X_j, A_j) \neq (X, A)\}$.

Theorem 6. *Let (X, A) be an essential granule. If $Q_{(X,A)} \neq \emptyset$, then (X, A) is an atomic granule.*

4.3   The Granular Structures in Formal Concept Analysis

The granular structure gives the hierarchical structure of different types of granules according to the degree of abstraction. Based on the discussion in Sect. 4.2, we can give the granular structures in FCA as follows.

From the view of objects, we can see that the pictorial granules are object induced pairs which contain the detailed information of each object. Hence they are the most concrete granules and in the lowest level of granular structure. By clustering the most similar pictorial granules whose values of similarity measurements are 1, we can obtain the elementary granules. The elementary granules are the object concepts which contain the attribute information of a group of similar objects rather than the attribute information of each object. That is, the elementary granules are more abstract than the pictorial granules. Thus, the elementary granules are in the second lowest level of granular structure. After compression, we can pick the essential granules from the elementary granules. The essential granules are join-irreducible concepts which contain not only the attribute information of a group of similar object but also the information of lattice construction. Thus, essential granules are more abstract and in the third lowest level of granular structure. Finally, through compression, we can get the atomic granules which are the most abstract ones and in the highest level of granular structure. Similarly, the granular structure of different granules from the view of attributes can also be given. The granules in the lowest level contain the most concrete knowledge and the granules in the highest level contain the most abstract knowledge. Moreover, the upper level granules can be obtained from the lower level granules. These two granular structures in FCA are summarized and directly shown in Fig. 2.

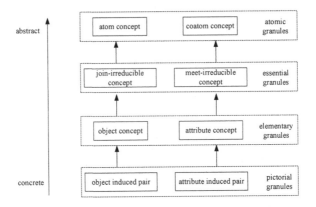

Fig. 2. Granular structures in formal concept analysis

5   An Example

In this section, we use an example to illustrate the results we got in previous sections. Table 1 shows a formal context (G, M, I). The object set is $G = \{1, 2, 3, 4, 5, 6, 7, 8\}$ and the attribute set is $M = \{a, b, c, d, e, f, g, h, i\}$. The concept lattice $L(G, M, I)$ is shown in Fig. 3. Consider the transformations of four different kinds of granules from the view of objects. There are 8 pictorial granules: $(1, abg)$, $(2, abgh)$, $(3, abcgh)$, $(4, acghi)$, $(5, abdf)$, $(6, abcdf)$, $(7, acde)$, $(8, acd)$. The values of similarity measurements between each pair of pictorial granules are shown in Table 2. That is, the value of $Sim_G((x_i, x_i^*), (x_j, x_j^*))$ is at the cell with i th row and j th column of Table 2. According to Theorem 1 and Table 2, we can get 8 elementary granules: $(123, abg)$, $(23, abgh)$, $(3, abcgh)$, $(4, acghi)$, $(56, abdf)$, $(6, abcdf)$, $(7, acde)$, $(678, acd)$. Then, we calculate the similarity between each pair of elementary granules and present these values in Table 3. According to Theorem 2 and Table 3, we get 7 essential granules: $(123, abg)$, $(23, abgh)$, $(3, abcgh)$, $(4, acghi)$, $(56, abdf)$, $(6, abcdf)$, $(7, acde)$. Finally, according to Theorem 3, we obtain 4 atomic granules: $(3, abcgh)$, $(4, acghi)$, $(6, abcdf)$, $(7, acde)$.

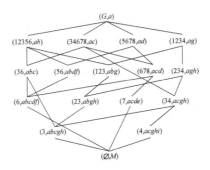

Fig. 3. Concept lattice $L(G, M, I)$

Table 1. A formal context (G, M, I)

	a	b	c	d	e	f	g	h	i
1	×	×					×		
2	×	×					×	×	
3	×	×	×				×	×	
4	×		×				×	×	×
5	×	×		×		×			
6	×	×	×	×		×			
7	×		×	×	×				
8	×		×	×					

Table 2. Similarity measurement between pictorial granules

	$(1,abg)$	$(2,abgh)$	$(3,abcgh)$	$(4,acghi)$	$(5,abdf)$	$(6,abcdf)$	$(7,acde)$	$(8,acd)$
$(1,abg)$	1	$\frac{3}{4}$	$\frac{3}{5}$	$\frac{2}{5}$	$\frac{2}{4}$	$\frac{2}{5}$	$\frac{1}{4}$	$\frac{1}{3}$
$(2,abgh)$	1	1	$\frac{4}{5}$	$\frac{3}{5}$	$\frac{2}{4}$	$\frac{2}{5}$	$\frac{1}{4}$	$\frac{1}{3}$
$(3,abcgh)$	1	1	1	$\frac{4}{5}$	$\frac{2}{4}$	$\frac{3}{5}$	$\frac{2}{4}$	$\frac{2}{3}$
$(4,acghi)$	$\frac{2}{3}$	$\frac{3}{4}$	$\frac{4}{5}$	1	$\frac{1}{4}$	$\frac{2}{5}$	$\frac{2}{4}$	$\frac{2}{3}$
$(5,abdf)$	$\frac{2}{3}$	$\frac{2}{4}$	$\frac{2}{5}$	$\frac{1}{5}$	1	$\frac{4}{5}$	$\frac{2}{4}$	$\frac{2}{3}$
$(6,abcdf)$	$\frac{2}{3}$	$\frac{2}{4}$	$\frac{3}{5}$	$\frac{2}{5}$	1	1	$\frac{3}{4}$	1
$(7,acde)$	$\frac{1}{4}$	$\frac{1}{4}$	$\frac{2}{5}$	$\frac{2}{5}$	$\frac{2}{4}$	$\frac{3}{5}$	1	1
$(8,acd)$	$\frac{1}{3}$	$\frac{1}{4}$	$\frac{2}{5}$	$\frac{2}{5}$	$\frac{2}{4}$	$\frac{3}{5}$	$\frac{3}{4}$	1

Table 3. Similarity measurement between elementary granules

	$(123,abg)$	$(23,abgh)$	$(3,abcgh)$	$(4,acghi)$	$(56,abdf)$	$(6,abcdf)$	$(7,acde)$	$(678,acd)$
$(123,abg)$	1	$\frac{3}{4}$	$\frac{3}{5}$	$\frac{2}{5}$	$\frac{2}{4}$	$\frac{2}{5}$	$\frac{1}{4}$	$\frac{1}{3}$
$(23,abgh)$	1	1	$\frac{4}{5}$	$\frac{3}{5}$	$\frac{2}{4}$	$\frac{2}{5}$	$\frac{1}{4}$	$\frac{1}{3}$
$(3,abcgh)$	1	1	1	$\frac{4}{5}$	$\frac{2}{4}$	$\frac{3}{5}$	$\frac{2}{4}$	$\frac{2}{3}$
$(4,acghi)$	$\frac{2}{3}$	$\frac{3}{4}$	$\frac{4}{5}$	1	$\frac{1}{4}$	$\frac{2}{5}$	$\frac{2}{4}$	$\frac{2}{3}$
$(56,abdf)$	$\frac{2}{3}$	$\frac{2}{4}$	$\frac{2}{5}$	$\frac{1}{5}$	1	$\frac{4}{5}$	$\frac{2}{4}$	$\frac{2}{3}$
$(6,abcdf)$	$\frac{2}{3}$	$\frac{2}{4}$	$\frac{3}{5}$	$\frac{2}{5}$	1	1	$\frac{3}{4}$	1
$(7,acde)$	$\frac{1}{4}$	$\frac{1}{4}$	$\frac{2}{5}$	$\frac{2}{5}$	$\frac{2}{4}$	$\frac{3}{5}$	1	1
$(678,acd)$	$\frac{1}{3}$	$\frac{1}{4}$	$\frac{2}{5}$	$\frac{2}{5}$	$\frac{2}{4}$	$\frac{3}{5}$	$\frac{3}{4}$	1

6   Conclusion

Granular structure is one of the most important notions in GrC. In this paper, considering the obtaining process and the abstraction level of granules, the granular structures in FCA are given. And the transformation methods of granules from one level to another are discussed. If we consider the semantics both "commonly possessing" and "commonly not possessing", the results of this paper can be extended in 3-way formal concept analysis.

Acknowledgements. The authors gratefully acknowledge the support of the Natural Science Foundation of China (No. 61772021 and No. 11371014).

References

1. Wille, R.: Restructuring lattice theory: an approach based on hierarchies of concepts, vol. 83, pp. 445–470. Reidel, Dordrecht-Boston (1982)
2. Ganter, B., Wille, R.: Formal Concept Analysis. Springer, Heidelberg (1999). https://doi.org/10.1007/978-3-642-59830-2
3. Yao, Y.Y.: The rise of granular computing. J. Chongqing Univ. Posts Telecommun. (Nat. Sci. Edn.) **20**, 299–308 (2008)
4. Zadeh, L.A.: Toward a theory of fuzzy information granulation and its centrality in human reasoning and fuzzy logic. Sets Syst. **90**, 111–127 (1997)
5. Pedrycz, W.: Granular computing: an introduction. In: Proceedings of the 9th Joint IFSA World Congress and 20th NAFIPS International Conference, pp. 1349–1354 (2001)
6. Yao, Y.Y.: A triarchic theory of granular computing. Granul. Comput. **1**(2), 145–157 (2016)
7. Wu, W.Z., Leung, Y., Mi, J.S.: Granular computing and knowledge reduction in formal contexts. IEEE Trans. Knowl. Data Eng. **21**(10), 1461–1474 (2009)
8. Shao, M.W., Leung, Y.: Relations between granular reduct and dominance reduct in formal contexts. Knowl.-Based Syst. **65**, 1–11 (2014)
9. Xu, W.H., Li, W.T.: Granular computing approach to two way learning based on formal concept analysis in fuzzy datasets. IEEE Trans. Cybern. **46**(2), 366–379 (2016)
10. Li, J.H., Mei, C.L., Xu, W.H., et al.: Concept learning via granular computing: a cognitive viewpoint. Inf. Sci. **298**, 447–467 (2015)
11. Li, J.H., Huang, C.C., Qi, J.J., et al.: Three-way cognitive concept learning via multigranularity. Inf. Sci. **378**, 244–263 (2017)
12. Zhi, H.L., Li, J.H.: Granular description based on formal concept analysis. Knowl.-Based Syst. **104**, 62–73 (2016)
13. Li, J.H., Wu, W.Z.: Granular computing approach for formal concept analysis and its research outlooks. J. Shandong Univ. (Nat. Sci.) **52**(7), 1–12 (2017)
14. Qi, J.J., Wei, L., Wan, Q.: Multi-level granularity in formal concept analysis. Granul. Comput. (2018). https://doi.org/10.1007/s41066-018-0112-7
15. Davey, B.A., Priestley, H.A.: Introduction of Lattices and Order, 2nd edn. Cambridge University Press, Cambridge (2002)

Fuzzy RST and RST Rules Can Predict Effects of Different Therapies in Parkinson's Disease Patients

Andrzej W. Przybyszewski[1,2(✉)]

[1] Polish-Japanese Institute of Information Technology, 02-008 Warsaw, Poland
przy@pjwstk.edu.pl
[2] Department of Neurology, UMass Medical School, Worcester, USA

Abstract. Neurodegenerative disorders (ND) such as Parkinson's disease (PD) are increasing in frequency with ageing, but we still do not have cure for ND.

In the present study, we have analyzed results of: neurological, psychological and eye movement (saccadic) tests in order to discover patterns (KDD) and to predict disease progression with fuzzy rough set (FRST) and rough set (RST) theories. It is a longitudinal study in which we have repeated our measurements every six months and estimated disease progression in three different groups of patients: BMT-group: medication only; DBS-group medication and deep brain stimulation (DBS); and POP–group same as DBS but with several years longer period of DBS. With help of above KDD methods, we have predicted UPDRS (Unified Parkinson's Disease Rating Scale) values in the following two visits on the basis of the first visit with the accuracy of 0.7 for both BMT visits; 0.56 for DBS, and 0.7-0.8 for POP visits. We could also predict UPDRS of DBS patients by rules obtained from BMT-group with accuracy of 0.6, 0.8, and 0.7 for three following DBS visits. Using FRTS we have predicted UPDRS of DBSW3 from DBSW2 with accuracy of 0.5. We could not predict by RST disease progression of POP patients from other groups but with FRST we could predict POPW1 on the basis of DBSW1 results (with accuracy of 0.33). In summary: long-term DBS (POP-group) in contrast to other-groups has changed brain mechanisms and only FRST found similarities between POP and other-groups in disease progressions.

Keywords: Neurodegenerative disease · Rough set · Decision rules
Granularity

1 Introduction

Parkinson (PD) is the second after Alzheimer most popular neurodegenerative diseases (ND) is caused by death of cells primary in the substantia nigra (SN) and lack of the dopamine. Therefore, the first help is to increase the level of the dopamine by its precursor L-Dopa. With disease development, there are adaptive mechanisms that inactivate high level of the dopamine and patients have to increase dosages of their medication till getting so-called ON-OFF effect. This effect can be cured by DBS. About 30 years ago, Alain Benabid in Grenoble (France) introduced the deep brain

© Springer Nature Switzerland AG 2018
M. Ceci et al. (Eds.): ISMIS 2018, LNAI 11177, pp. 409–416, 2018.
https://doi.org/10.1007/978-3-030-01851-1_39

stimulation (DBS) by targeting the subthalamic nucleus (STN). It was claimed as "the most important discovery since L-Dopa". But PD progression is also related with neurodegeneration in other structures like prefrontal cortex and related cognitive problems and limbic system leading to depressions and also related to lack of the dopamine as reward transmitter. However, the main symptoms in PD are movement disorders measured mainly by the UPDRS.

As disease starts about 20 years before first symptoms each patient has different plastic brain mechanism, and rate of the disease progression, which gives a problem to find an optimal therapy that depends on the doctor's experiences and precision of his/her tests. As results of different doctor's tests are partly subjective, we propose to use objective, doctor independent test of reflexive eye movement measurements that can be performed automatic without doctor's time.

In this study, we have developed methods of rules discovery (KDD) and symptoms prediction on the basis of the object classifications found in the visual system of primates [1]. There are many similarities of the unknown complex object classification to recognition of the disease complexity. Visual brain plasticity (learning mechanisms) mechanisms are universal in the whole brain. In general, classification processes are related to two different logic systems as we have demonstrated in the visual brain [1]. The brain solves problems of imprecisions and contradictions by using approach similar to rough set theory [2]. It means that descending and ascending pathways, by using different logics, interact in order to minimize the boundary region between known and actual object's properties. It is similar to testing hypothesis by neurologist: what are the most important symptoms of the actual patient.

This study is expansion of our previous work [3] by using additional to RST, fuzzy RST that are more universal in finding similarities. In the perspective, our methods should lead to the remote diagnosis and treatments (telemedicine).

2   Methods

We have analyzed tests from PD patients from three groups: (a) BMT – group: 23 patients that have therapy limited to medication; (b) DBS – group: 24 patients on medications and with the Deep Brain Stimulation (DBS) of subthalamic nucleus. For this group surgery was performed during the study; (c) POP-group: 15 patients with surgery performed earlier before our study. Patients from all groups were tested in the following sessions: MedOn/Off sessions: ses.1/3: without/after medication. In addition, patients from DBS and POP groups were tested in StimOn/Off sessions, StimOn: ses.2/4 without/with med. All tests: neurological, neuropsychological, and eye movement tests were performed in dept. of Neurology, Brodno Hospital, Warsaw Medical University. In this study, we have measured parameters of the fast, reflexive saccadic eye movements (EM). Detailed methodology was described earlier [3, 4]. In short, each patient has to follow horizontally moving randomly $10°$ to the right or $10°$ to the left dot. We have estimated the following saccadic parameters: delay – it is a time difference between the beginning of the stimulus and eye movements; relative amplitude: amplitude of the saccade related to the amplitude of the light spot movements; max saccade velocity; duration of the saccade as a time between the end and the beginning of each saccade. Eye movements were recorded by the head-mounted saccadometer (Ober Consulting, Poland).

2.1 Theoretical Basis

Our rough set (RS) data mining analysis is based on the Pawlak's concept of RS theory (RST) (Zdzislaw Pawlak [5]). An information system [5] a pair $S = (U, A)$, where U, A are nonempty finite sets called the universe of objects U and the set of attributes A. If $a \in A$ and $u \in U$, the value $a(u)$ is a unique element of V (where V is a value set).

We define as in [5] the *indiscernibility relation* of any subset B of A or $IND(B)$ as: $(x, y) \in IND(B)$ or $xI(B)y$ iff $a(x) = a(y)$ for every $a \in B$ where the value of $a(x) \in V$. It is an equivalence relation $[u]_B$ that we understand as a *B-elementary granule*. The family of $[u]_B$ gives the partition U/B containing u will be denoted by $B(u)$. The set $B \subset A$ of information system S is a reduct $IND(B) = IND(A)$ and no proper subset of B has this property [6]. In most cases, we are only interested in such reducts that are leading to expected rules (classifications). On the basis of the reduct we have generated rules using four different ML methods (RSES 2.2): exhaustive algorithm, genetic algorithm [7], covering algorithm, or LEM2 algorithm [8].

A *lower approximation* of set $X \subseteq U$ in relation to an attribute B is defined as $\underline{B}X = \{u \in U : [u]_B \subseteq X\}$. The *upper approximation* of X is defined as $\overline{B}X = \{u \in U : [u]_B \cap X \neq \phi\}$. The difference of $\overline{B}X$ and $\underline{B}X$ is the boundary region of X that we denote as $BN_B (X)$. If $BN_B (X)$ is empty then set than X is *exact* with respect to B; otherwise if $BN_B (X)$ is not empty and X is not *rough* with respect to B.

A decision table (training sample in ML) for S is the triplet: $S = (U, C, D)$ where: C, D are condition and decision attributes [9]. Each row of the information table gives a particular rule that connects condition and decision attributes for a single measurements of a particular patient.

However, FRTS (fuzzy rough set theory) replaced defined above 'crisp' dependences by a tolerance or similarity relations $R_a(x, y)$ as a value between two observations x and y. As summarized in [10] there are several tolerance relationships as normalized difference (so-called 'Eq. 1') or Gaussian or exponential differences [10]. There are formulas related to normalized differences between pairs of attributes. The most common are *Lukasiewicz* and *t.cos t-norms* [10]. As decision attributes are nominative we used crisp relations between them.

We define B-*lower* and B-*upper* approximations for each observation x in FRST as following: B-*lower* approximation as: $(R_B \downarrow X)(x) = \inf_{y \in U} I(R_B(x, y), X(y))$, where I is an *implicator* [10]. The B-lower approximation for the observation x is then the set of observations which are the most similar to observation x and it can predict the decision attribute with the highest confidence, based on conditional attributes B.

The B-upper approximation is defined by $(R_B \uparrow X)(x) = \sup_{y \in U} \tau(R_B(x, y), X(y))$, where τ is the t-norm. The B-upper approximation is a set of observations for which the prediction of decision attribute has the smallest confidence [10].

Also rules in FRST have different construction than in RST. They are based on the tolerance classes and appropriate decision concepts. The *fuzzy rule* is a triple *(B, C, D)*, where B is a set of conditional attributes that appear in the rule, C stands for fuzzy tolerance class of object and D stands for decision class of object. There are important differences that will be demonstrated in our results.

We have used the RSES 2.2 (Rough System Exploration Program) [9] with implementation of RST rules to process our data and "RoughSets" version 1.3-0 implemented in R [11]. As we have demonstrated earlier that the RST method is superior to other classical methods [3].

3 Results

As described in the Methods section, total 62 PD patients were divided into three groups: BMT-group (only medication), DBS-group (medication and STN stimulation, surgery during the study) and POP-group (medication and STN stimulation, long after the surgery). In BMT-group: 23 patients, the mean age was 57.8 ± 13 (SD) years; disease duration was 7.1 ± 3.5 years, UPDRS was 48.3 ± 17.9.

In POP-group: 15 patients, the mean age was 63.1 ± 18.2 (SD) years and disease duration was 13.5 ± 3.6 years (stat. diff. from BMT $p < 0.025$, and from DBS-group: $p < 0.015$), UPDRS was 59.2 ± 24.5 (stat. diff. than BMT-group: $p < 0.0001$).

In DBS-group: 24 patients, the mean age of 53.7 ± 9.3 years, disease duration was 10.25 ± 3.9 years; UPDRS was 62.1 ± 16.1 (stat. diff. than BMT-group: $p < 0.0001$).

These statistical data are related to the data obtained during the first session for each group: BMT W1 (visit one), DBS W1 (visit one) and POP W1 (visit one). It is clear that in visit one W1 UPDRS in different groups are different.

In DBSW1 visit before the surgery, there were only two sessions (MedOFF, MedON). In ses.1 mean UPDRS was 62.2 ± 16.1, in ses.3 was 29.9 ± 13.3 strongly ($p < 0.0001$) different from ses.1 (effect of medication). UPRDS of DBSW2 after the surgery in ses.1 is larger than before the surgery 65.3 ± 17.6 but there are not stat. sig. diff. UPDRS ses.1 of DBSW3 is 68.7 ± 17.7 and stat. diff. ($p < 0.03$) than in W2.

In POP-group UPDRS values are similar. There is an increase of the UPDRS ses.1 from W1: 63.1 ± 18.2 to W2: 68.9 ± 20.3 to W3: $74,2 \pm 18.4$ but there were smaller differences for ses.4 (both med. and stim. on) W1: 21 ± 11.3 to W2: 23.3 ± 9.5 to W3: $23,8 \pm 10.7$. It seems that groups DBS and POP are similar.

In BMT group UPDRS ses.1 W1: 48.3 ± 17.9; W2: 57.3 ± 16.8 ($p < 0.0005$ diff than W1); W3: 62.2 ± 18.2 ($p < 0.05$ diff. than W2). In ses.3 UPDRS was W1: 23.6 ± 10.3; W2: 27.8 ± 10.8; W3: 25 ± 11.6 (no stat. diff. between visits for ses.3).

3.1   FRST and RST Rules and ML Results for BMT Group

There were patient only on medication with two different sessions (MedOFF and MedON) measured every six months.

Using RST, after the discretization, we have divided UPDRS into 4 ranges: "(-Inf, 24.0)", "(24.0, 36.0)", "(36.0, 45.0)", "(45.0, Inf)", with help of the feature selection RSES software and by using machine learning and RST algorithms [7] we have obtained rules from BMTW1 and use them to predict UPDRS for BMTW2 and BMTW3. As in [3] we have used: parameters related to psychological testing (PDQ39

– quality of life, Epworth – quality of sleep), and parameters of saccades where in this only latency was significant (SccLat), decision attribute: UPDRS [3].

As each row gives a particular rule and by using RSES we have obtained more general rules like:

$$(S\# = 3)\&\left(PDQ39 = {}''(-Inf, 50.5)''\right) = > \left(UPDRS = {}''(-Inf, 24.0)''\right) \quad (1)$$

It states that if the session for ses.3 and PDQ39 is smaller than 50.5 then UPDRS will be smaller than 24.0.

In FRST also decision attribute was UPDRS and "in agreement with discretization in RST we got four values: (-Inf, 24.0) - > "1", "(24.0, 36.0)" - > "2", "(36.0, 45.0)" - > "3", (45.0, Inf)) - > "4". An example of analog FSRT rule:

$$(S\# = 3)\&\left(PDQ39 = {}''(40)''\right)\&\left(RSAmp = {}''(9.5)''\right)\&\left(RSLat = {}''(281)''\right)\&(RSDur = {}''(45)'' \\ = > (UPDRS = 1) \quad (2)$$

It states that if the session for ses.3 and PDQ39 is near 40 and RSAmp near 9.5 and RSLat is near 281 and RSDur is near 45 then UPDRS is 1. Where RS are parameters of the reflexive saccades: RSAmp is amplitude, RSLat is saccade latency, and RSDur is saccade duration.

We have used these rules to predict UPDRS in BMTW2 and W3 using 6-fold cross validation we have obtained for both visits global accuracy 0.7, and the global coverage 1.0. For FRST we have obtained accuracy 0.63.

3.2   FRST and RST Rules and ML Approach for DBS Group

We have predicted UPDRS of DBSW3 by rules from DBSW2 (both groups have 4 sessions), and we have obtained the global accuracy 0.56 and global coverage 1.

In the next step, we have applied the same BMTW1 rules to the DBS group. It was successful for DBSW1 pre-operative patients as they were also in two sessions with a high dosage of medication. We have obtained the global accuracy 0.64 with the global coverage 0.5.

As we have noticed that there are large difference between UPDRS in ses.1 W1 therefore prediction of UPDRS from BMT group for DBSW2 and W3 groups were not possible as there are different numbers of sessions. Therefore, we have divided DBSW2, W3 patients into two subgroups: one without stimulation (StimOff) and another one with StimOn. We could not predict UPDRS in DBS groups without stimulation (StimOff) only with stimulation StimOn, and all our predictions for DBS and POP groups are only for StimOn.

UPDRS of DBSW2 were predicted from BMTW1 rules with global accuracy 0.85, but with coverage was 0.3 and some classes were not at all predicted (for UPDRS larger than 63). We have obtained similar results for UPDRS of DBSW3 from BMTW1 rules; the global accuracy was 0.74 but the global coverage 0.56.

3.3   RST and FRST Rules and ML Approach for POP Group

On the basis of rules obtained from POPW1 we have predicted UPDRS in POPW2 with the accuracy 0.67; UPDRS in POPW3 from POPW1 with accuracy 0.8 and global coverage 0.97 (for details see [3]).

In contrast to the DBS group we were not successful in using rules from the BMT patients to predict UPDRS of the POP group patients. In order to find possible reasons we have compared rules from the BMTW1 patients with rules of POPW3 group as UPDRS values for both groups were similar (see Methods).

BMTW1 rules:

$$\left(PDQ39 = {}^{''}(Inf, 50.5)^{''}\right) \& \left(RSLat = {}^{''}(264.0, Inf)^{''}\right) \& \left(dur = {}^{''}(Inf, 5.65)^{''}\right) = >$$
$$\left(UPDRS = \{{}^{''}(-Inf, 33.5)^{''}[4], {}^{''}(33.5, 43.0)^{''}[1]\}\right)5$$

$$(3)$$

POPW3 rules:

$$(PDQ39 = {}^{''}(-Inf, 48.0)^{''}) \& \left(RSLat = {}^{''}(301.0, Inf)^{''}\right) \& \left(RSPeak = {}^{''}(403.5, 522.0)^{''}\right) = >$$
$$\left(UPDRS = {}^{''}(66.0, Inf)^{''}[1]\right) 1$$

$$(4)$$

Above rules (3) and (4) are contradictory as the same values of conditional attributes, limited by other attributes, give opposite results of the UPDRS.

We have performed computations for comparison of DBSW1 and POPW1 groups using FRTS. As decision attributes are nominal, we have changes range of UPDRS to small, medium, large and very large that was implemented as 1, 2, 3, 4 (Table 1).

We have obtained our predictions using the hybrid-fuzzy rules with aggregation by the *t.norm Lukasiewicz, tolerance Eq.* 3 (modified Gaussian from [10]), and as *relation Gaussian kernel* (with 0.2 as a parameter). As example of fuzzy rules are:

Table 1. Confusion matrix for UPDRS of POPW1-subgroup (StimOn) by rules obtained from DBSW1-subgroup (StimOn) with FRST.

Actual	Predicted				
	"1"	"2"	"3"	"4"	ACC
"1"	5.0	0.0	0.0	4.0	0.556
"2"	3.0	0.0	0.0	6.0	0.0
"3"	3.0	0.0	0.0	4.0	0.0
"4"	1.0	0.0	0.0	4.0	0.8
TPR	0.42	0.0	0.0	0.29	

TPR: True positive rates for decision classes; ACC: Accuracy for decision classes: the global coverage was 1 and the global "accuracy was 0.33, the coverage for all decision classes was 1. Where: (-Inf, 17.0) - > "1", "(17.0, 27.0)" - > "2", "(27.0, 36.0)" - > "3", (36.0, Inf)) - > "4".

$$(Epworth\ 'is\ around'\ 11)\&(RSDur\ 'is\ around'\ 66) = > (UPDRS = 3) \quad (5)$$

$$(RSLat'is\ around'\ 424)\&(RSDur\ 'is\ around'\ 48) = > (UPDRS = 2) \quad (6)$$

$$(RSLat\ 'is\ around'\ 286)\ v\ (RSLat\ 'is\ around'\ 292)\ v\ (RSLat\ 'is\ around'\ 251) \\ = > (UPDRS = 1) \quad (7)$$

Above fuzzy rules (5–7) are different that RST crisp rules (3), (4) but they have 'fuzzy' properties that seems to be more general and this way might cover some contradictions between rough set rules.

4 Discussion

There is always actual discussion how can we achieve that all medical procedures would be optimal for an individual patient? Thanks to technology is significant progress in medical science and neurology with new procedures improving PD patient's treatments. But generally, the long lasting neurodegeneration processes with the compensatory; specific for each person plastic changes it is extremely difficult question. We propose to solve this problem by using similar mechanisms that were found in the visual system to categorize complex objects. We have used rough and fuzzy rough set theories to fit our granules in optimal way to the complexity of the disease symptoms. We were successful in prediction of the symptoms development in time (longitudinal study) for different therapies such as medication and DBS. However, we had problems to predict symptoms of patients with the long lasting brain stimulation (POP-group). We were partly successful by using more general FRST rules, but might try to increase number of attributes in order to find what are mechanisms of the long-lasting brain stimulation. We expect that in the near future more systems will be replacing, at least in part, medical doctors as they are more objective, automatic, more precise and they do not take doctor's time. Such system, when intelligent, could also help in doctor's decision and help in treatment optimization.

In summary, we have tested three groups of patients: BMT-group (medication only); DBS-group (medication and short brain stimulation), and POP-group (medication with long lasting DBS) in the longitudinal study. Mean UPDRS without medication and stimulation were:

BMT: W1-48; W2-57; W3–62; with the disease duration of 7 years

DBS: W1-62; W2-65; W3–69; 10 years

POP: W1-63; W2-69; W3–74; 13.5 years

Mean UPDRS increase for half of the year for BMT: 7, for DBS: 3.5, for POP: 5.5. From above, it is evident that deep brain stimulation is slowing down the progression of the disease as the disease duration for POP group is almost twice as for BMT group. These is also our major problem with prediction of UPDRS from BMT group to DBS and POP more advanced groups with significant difference in UPDRS. As electric brain stimulation significantly lowered UPDRS, we may think that DBS resets brain and symptoms in time into the beginning of the disease. It is why it was possible to predict symptoms development in more advanced disease stages from earlier one. It might be the direction in finding the optimal treatment.

References

1. Przybyszewski, A.W.: Logical rules of visual brain: from anatomy through neurophysiology to cognition. Cogn. Syst. Res. **11**, 53–66 (2010)
2. Przybyszewski, A.W.: The Neurophysiological Bases of Cognitive Computation Using Rough Set Theory. In: Peters, James F., Skowron, A., Rybiński, H. (eds.) Transactions on Rough Sets IX. LNCS, vol. 5390, pp. 287–317. Springer, Heidelberg (2008). https://doi.org/10.1007/978-3-540-89876-4_16
3. Przybyszewski, A.W., Szlufik, S., Habela, P., Koziorowski, D.M.: Rules determine therapy-dependent relationship in symptoms development of Parkinson's Disease Patients. In: Nguyen, N.T., Hoang, D.H., Hong, T.-P., Pham, H., Trawiński, B. (eds.) ACIIDS 2018. LNCS (LNAI), vol. 10752, pp. 436–445. Springer, Cham (2018). https://doi.org/10.1007/978-3-319-75420-8_42
4. Przybyszewski, A.W., Kon, M., Szlufik, S., Szymanski, A., Koziorowski, D.M.: Multimodal learning and intelligent prediction of symptom development in individual Parkinson's Patients. Sensors, **16**(9), 1498 (2016). https://doi.org/10.3390/s16091498
5. Pawlak, Z.: Rough Sets: Theoretical Aspects of Reasoning About Data. Kluwer, Dordrecht (1991)
6. Bazan, J., Nguyen, H.S., Trung, T., Nguyen, Skowron A., Stepaniuk, J.: Desion rules synthesis for object classification. In: Orłowska, E. (ed.) Incomplete Information: Rough Set Analysis, pp. 23–57. Physica – Verlag, Heidelberg (1998). https://doi.org/10.1007/978-3-7908-1888-8_2
7. Bazan, J., Nguyen, H.S., Nguyen, S.H., Synak, P., Wróblewski, J.: Rough set algorithms in classification problem. In: Polkowski, L., Tsumoto, S., Lin, T. (eds.) Rough Set Methods and Applications, pp. 49–88. Physica-Verlag, Heidelberg New York (2000). https://doi.org/10.1007/978-3-7908-1840-6_3
8. Grzymała-Busse, J.: A new version of the rule induction system LERS. Fundamenta Informaticae **31**(1), 27–39 (1997)
9. Bazan, J.G., Szczuka, M.: The rough set exploration system. In: Peters, J.F., Skowron, A. (eds.) Transactions on Rough Sets III. LNCS, vol. 3400, pp. 37–56. Springer, Heidelberg (2005). https://doi.org/10.1007/11427834_2
10. Riza, L.S., et al.: Implementing algorithms of rough set theory and fuzzy rough set theory in the R package "RoughSets". Inf. Sci. **287**, 68–69 (2014)
11. Bazan, J.G., Szczuka, M.: RSES and RSESlib - a collection of tools for rough set computations. In: Ziarko, W., Yao, Y. (eds.) RSCTC 2000. LNCS (LNAI), vol. 2005, pp. 106–113. Springer, Heidelberg (2001). https://doi.org/10.1007/3-540-45554-X_12

From Knowledge Discovery to Customer Attrition

Katarzyna Tarnowska[1(✉)] and Zbigniew Ras[2,3]

[1] Department of Computer Science, San Jose State University,
1 Washington Sq, San Jose, CA 95192, USA
katarzyna.tarnowska@sjsu.edu
[2] Department of Computer Science, University of North Carolina at Charlotte,
9201 University City Blvd., Charlotte, NC 28223, USA
ras@uncc.edu
[3] Polish-Japanese Academy of Information Technology, 02-008 Warsaw, Poland
ras@pjwstk.edu.pl

Abstract. This article presents a novel approach to handle customer attrition problem with knowledge discovery methods. The data mining is performed on the customer feedback data, which was labelled by means of temporal transactional invoice data in terms of customer activity. The problem was raised within industry-academia collaboration project at University of North Carolina at Charlotte by one of the companies from the heavy equipment repair industry. They expressed interest in gaining better insight into this problem, already having their own active CRM program implemented. The goal and motivation within this topic is to determine whether there are markers in the sales trends that might suggest a customer is getting ready to defect. Observing the behavior of customers who left a company, one might be able to identify customers who may leave as well.

Keywords: Knowledge discovery · Data mining · Action rules
Customer attrition

1 Introduction

Customer attrition is an important problem in many industries: banking, mobile service providers, insurance companies, financial service companies etc. It was discovered [3] that on average, most US corporations lose half of their customers every five years. Another fact is that the longer a customer stays with the organization, the more profitable the customer becomes. The cost of attracting new customers is five to ten times more than retaining existing ones. Also, about 14% to 17% of the accounts are closed for reasons that can be controlled like price or service. Reducing the outflow of the customers by 5% can double a typical company's profit [3].

The goal is to retain as many customers as possible. A loyal customer might be worth much more than a new customer. Many companies drive customers away with poor customer retention initiatives (or a lack of a customer retention strategy).

© Springer Nature Switzerland AG 2018
M. Ceci et al. (Eds.): ISMIS 2018, LNAI 11177, pp. 417–425, 2018.
https://doi.org/10.1007/978-3-030-01851-1_40

2   Problem Statement

Customer attrition problem is a new subtopic that was raised by our industrial partner company within a broader frames of collaboration which include building a user-friendly data-driven recommender system for improving Net Promoter Score (customer loyalty) of a company [6]. The considered client company has implemented an active customer attrition program and already had some results in this area. Therefore, the results within this work would complement or allow to compare the results of the research done by the company independently.

According to their implemented CRM program, first it is important to know if the customer base is growing or shrinking. The problem known as churn or customer attrition is defined as the annual turnover of the market base. Secondly, it is important to know which customers a company has lost recently. Targeted approaches to fight customer churn rely on identifying customers who are likely to churn, and then provide them with an incentive to stay.

The other questions that the initiative helps in answering are: How effective are we at recovering lost customers? Can we predict which customers will leave us? To answer this question, the company collects data to determine the number of customers in each activity status. However, the prediction capabilities of the current system are very limited. They are based on simple analysis of NLS (Net Loyalty Status) of dormant or inactive customers. Therefore, predictive capabilities are limited here as it is a simple numeric analysis based on historical data. The sample data analysis findings might include: "Even if a customer scored 9 or 10 on survey, 26% stopped doing business for a year", "37% left when they scored between 1–8". The analysis also tries to predict what impact dormant customers have on the company's business (i.e. "10–11% of Service Revenue goes Dormant each year"), taking into account the amount of business customers do. Also, the idea is to target the most lucrative customers.

The remaining questions that need to be answered using the collected data are:

– What marketing efforts should we implement to attempt recovery?
– Which marketing is having the most impact?
– Add credit line and machine ownership info to data to identify low revenue/high opportunity customers.
– How well are we cross selling?
– Breakdown by Healthy customers (Revenue > 1000USD).

3   Assumptions

The survey data on the considered company collected by the consulting company was made available for the years 2010–2017 (total 26,454 records). Additionally, sales (transactional) data on company's customers was made available directly from that company from the years 2010–2017.

The customer feedback data reflects the structure of the surveys with each question being an attribute in the dataset called "benchmark". There are usually many questions developed for surveys but only a subset of them is asked to the end customers because of the telephone survey's time constraints. The dataset also contains details about the company, service type, as well as details about the surveyed customer: contact, location, etc.

The transactional (invoice) data contains the unique customer number and a number of transactions (invoices) in a given year. This dataset was exported from the company's sales system.

The survey (customer) data should be labelled first, in terms of customer activity, as the current labels are only related to the *Promoter Status* related to the metric called Net Promoter Score. A new decision attribute, denoting whether customer left or not will be added. This can be a binary attribute (*Leaving-yes/Leaving-no*) or a categorical attribute *CustomerStatus* - with values: *Active, Leaving, Lost*.

Each customer in the dataset is identifiable with a unique *Customer Number* (while there can be several contact names associated with each customer number). For each customer we can prepare temporal data for all their yearly surveys. Based on temporal records in the dataset available for that customer in the subsequent years, an algorithm can be applied for automatic labelling. The algorithm labels a customer temporal record based on the presence of that customer in transactional data in the years following the survey. An additional column *Customer Status* is added as a decision attribute.

4   Strategy and Overall Approach

The idea is to use survey data additionally to the sales data. This approach is novel in a way, that for the purpose of determining profile of churning customers mostly demographics or transactional data was used for mining. It is known that most customers stop doing business with a company because of bad customer service or their needs were ignored. However, customer retention efforts have also been costing organizations large amounts of resources. The customer feedback data should help in identifying and addressing the particular issues behind customer attrition. The survey data should be mined for markers of customers that are on the verge of leaving. Historical and temporal data on customer surveys should be used for that purpose. Correctly predicting that a customer is going to churn and then successfully convincing him to stay can substantially increase the revenue of a company, even if a churn prediction model produces a certain number of false positives.

The novelty of this approach lies in using transactional data to apply automatic labeling of customers in survey data as either *Active, Leaving* or *Lost*. Secondly, such automatically labelled data would be mined for actionable knowledge on how to change customer behavior from leaving to non-leaving.

4.1   Automatic Data Labelling

The approach lies in applying automatic labeling of customers in a survey dataset in terms of retention based on transactional data. Only a chosen subset of transactions are subject to telephone survey. So the records in the survey data are not good indicators whether a customer left or not, because the lack of records for the customer in the following years might simply result from that customer not being surveyed.

Additional data on transactions for each customer per year was made available. The algorithm for automatic labeling of the survey data (implemented in PL/SQL), labelled a customer as defected when: the customer appeared in the given year in the survey data, but has not reappeared at some point in the subsequent years in transactional data. As the transactional data up to 2017 was made available, survey data from years 2010–2016 could be labelled.

Also, the labeling algorithm for labeling customers into three categories was proposed and implemented, according to the business rules proposed by the domain experts (business consultants). "Active" customers are these actively making transactions each subsequent year after they were surveyed. "Leaving" customers are these who might be at risk of losing them, as the transactional data showed that they stopped doing business for one year (no transactional records in a year following the survey). Finally, "Lost" customers are these who stopped doing business continuously for two or more years after the survey. This labelling scheme is more similar to the one used by the company for its Customer Attrition Program. Also, other reasons for a customer not making transactions were taken into account: customers might have gone out of business, they might have merged with another company. These cases are indicated in the dataset as the accounts that have been deleted from the CRM system. Labelling algorithm takes into account these records and labels them as NULL (unknown) whenever an account has been deleted in the years following the survey for reasons outside of customer loyalty area.

4.2   Pattern Mining

KDD (Knowledge Discovery in Databases) is defined as the "nontrivial process of identifying valid, novel, potentially useful and ultimately understandable patterns in data". The goal is to describe each customer surveyed as potentially churner or potentially non-churner. Mining patterns in the responses data would suggest likelihood to defect. The example pattern characterizing leaving customer profile is listed below.

Listing 1.1. A sample pattern for a leaving customer profile

```
IF  Benchmark1 = 3  AND Benchmark2 = 7
       THEN "Leaving−yes"
```

Action rules suggest a way how to change the values of flexible attributes to get a desired state. Action rule concept was firstly proposed by Ras and Wieczorkowska in [4], and since then they have been successfully applied in many domain areas including business [4], medical diagnosis and treatment [8], music automatic indexing and retrieval [5]. The purpose is to analyze data to improve understanding of it and seek specific actions (recommendations) to enhance the decision-making process.

Decision attribute is a distinguished attribute [4], while the rest of the attributes are partitioned into stable and flexible attributes. In nomenclature, action rule is defined as a term: $[(\omega) \wedge (\alpha \rightarrow \beta) \rightarrow (\Phi \rightarrow \Psi)]$, where ω denotes conjunction of fixed stable attributes, $(\alpha \rightarrow \beta)$ are proposed changes in values of flexible attributes, and $(\Phi \rightarrow \Psi)$ is a desired change of decision attribute (action effect). In the considered problem area an example suggestion to change would be the actions to decrease customer attrition. So, in the considered domain, decision attribute is *CustomerStatus* (with values *Active, Leaving, Lost*). The flexible attributes would be benchmark attributes that denote area of customer service that should be potentially ameliorated to increase customer retention. The stable attribute could be a survey type, since we want to discover knowledge for a particular business area (Service, Parts, Rentals, etc.). The sample actionable pattern is listed below:

Listing 1.2. A sample pattern mined for actionable knowledge

```
IF  Benchmark1 (3->6) AND Benchmark2 (7->9)
      THEN Lost => Active
```

4.3   Clustering

In order to prevent customers from leaving the company, we should find all the customers similar in their profile to these who already left. The similarity can be defined as distance showing difference in answering the benchmark questions. The shorter the distance the more similar customers are. After identifying the customers whose profile matches these who already left the preventive measures should be taken by a company to decrease the customer's churn.

4.4   Sequence Mining

The available data - transactional and survey data is timestamped (date is available as an *Interview Date*.) Therefore, it is possible to detect trends in customer behavior and changes in responses over time. Sequence mining can be investigated for the prediction of customer events.

Sequence mining was originally introduced for market basket analysis where temporal relations between retail transactions are mined. The goal here is to use sequence mining for classification. If a certain sequence of responses was identified, leading to a customer defected with a high confidence, the same sequence of responses should be used for classifying customers displaying the same sequence.

4.5   Meta Action Mining and Triggering

Meta-actions are understood as higher-level actions. While an action rule is understood as a set of atomic actions that need to be made for achieving the expected result, meta-actions are the actions that need to be executed in order to trigger corresponding atomic actions. The strategy of automatic meta action mining from the text was proposed in [1] and triggering mechanism for the problem of improving Net Promoter Score was proposed in [2].

The approach for mining the action rules, meta actions and triggering in the customer attrition domain given the dataset is proposed as following. Action rules showing the change in customer attrition should be mined separately for different categories of customers: Promoters, Detractors and Passives. The next goal is to find groups of meta actions $M1$ triggering a rule $R1$ that guarantee the defection of the customers $X1$. It is assumed that if $M1$ does not happen, customers $X1$ will not defect. Analogously, find a group of meta-actions $M2$ triggering rule $R2$ that guarantee the defection of customers $X2$. As a result, for the customer group *(X1 AND X2)* the most probable set of actions causing them not to leave would be *(M1 OR M2)*.

The next steps within this research is to implement meta action triggering mechanism, that is, finding a minimum set of meta-actions guaranteeing that customers will not leave. The meta actions would be mined from the text in a similar way as it was for the "Promoter Status" version of the problem area [1,6]. The results should be compared with the results of Customer Attrition Program implemented by the company.

5   Evaluation

The evaluation included initial data analysis, implementing and validating automatic labelling algorithm, feature selection, classification and action rule mining for the churn and non-churn customers. Own PL/SQL scripts, WEKA and LISP-Miner were used in these experiments.

5.1   Data Labelling

About 23,000 customer surveys were available asked to the customers of the considered company. 8,593 distinct customers (as identifiable by *Customer Number*) were surveyed in total. Applying the labelling algorithm described in the previous sections resulted in the following distribution of the decision attribute ("Customer Status"):

- 76% of customer surveys were labelled as "Active";
- About 10% customers left: 4% left for one year and then returned and 6% did not return at all;
- 15% of records were labelled as UNKNOWN (either 2017 survey data or customer account deleted).

5.2   Attribute Selection

Some initial experiments with data included attribute selection. The goal here was to determine which survey benchmark questions are the most important (the most predictive) in terms of classifying a "defected" customer. Two methods using WEKA package were tested: Best First and Information Gain. Both methods indicated "Overall Satisfaction","NPS", "Likelihood to Repurchase" and general benchmarks (for the "All" category) as the most predictive in terms of attrition ("Customer Status").

5.3 Classification Model

Other experiments in WEKA included building a classification model. The chosen rule-based classifier - JRIP produced rules classifying customers status in terms of attrition. A sample classification rule is listed below.

Listing 1.3. A sample rule classifying a customer "Lost".

```
(Benchmark: Service − Final Invoice Matched Expectations >= 1)
and (Likelihood to Repurchase <= 6)
and (Likelihood to Repurchase <= 4)
and (Benchmark: All − Dealer Communication <= 4)
and (Benchmark: All − Ease of Contact >= 6)
 => CustomerStatus=Lost (16.0/6.0)
```

5.4 Action Rule Mining

The actionable patterns were defined and mined in LISP-Miner [7]. "Survey Type" was selected as a stable attribute, and all the"Benchmark" attributes were chosen as flexible attributes. The goal of the defined pattern was to find reasons behind the customer defection (changing the status from Active to Lost or from Active to Leaving). The sample patterns with the highest confidence are listed below.

Listing 1.4. Sample rules showing reasons behind customer defection.

```
SurveyType(Field) :    (BenchmarkAllEaseofContact(10)
−> BenchmarkAllEaseofContact(8)) =>
(CustomerStatus(Active)  −>  CustomerStatus(Leaving))
Conf = 0.82

SurveyType(Field) :    (BenchmarkAllDealerCommunication(10)
−> BenchmarkAllDealerCommunication(8)) =>
(CustomerStatus(Active)  −>  CustomerStatus(Leaving))
Conf = 0.82

SurveyType(Field) :
(BenchmarkServiceRepairCompletedCorrectly(10)
−> BenchmarkServiceRepairCompletedCorrectly(9)) =>
(CustomerStatus(Active)  −>  CustomerStatus(Leaving))
Conf = 0.81
```

The found action rules associate deteriorating of the certain "Benchmark" questions, such as "Ease of Contact", "Dealer Communication", "Overall Satisfaction", "Service Completed Timely", "Service Completed Correctly" as reasons beyond customers stopped doing the transactions (leaving). The extracted rules show minimal changes (in scoring the benchmarks) guaranteeing the defection of the customers with a certain probability (confidence). The minimal support for the rules was set to 20 (customers).

6   Conlusions

In this paper methods of Knowledge Discovery were presented to tackle the problem of Customer Attrition. A real-world data sets on customer feedback and customer transactions from the collaborating company were used for the experiments. The research resulted from the real business need and need to augment the already implemented active Customer Attrition Program of the company. The goal was to propose new more advanced analytical methods beyond simple numerical analysis. In particular actionable knowledge was discovered in the experiments showing the reasons or changes in customer feedback behavior leading to the customer defection. Also, the methodology was proposed for the further work involving text mining, using meta action concepts and implemeting the results into a knowledge-based recommender system that would recognize customers close to defect and propose actionable solutions to keep these customers.

References

1. Kuang, J., Raś, Z.W., Daniel, A.: Personalized meta-action mining for NPS improvement. In: Esposito, F., Pivert, O., Hacid, M.-S., Raś, Z.W., Ferilli, S. (eds.) ISMIS 2015. LNCS (LNAI), vol. 9384, pp. 79–87. Springer, Cham (2015). https://doi.org/10.1007/978-3-319-25252-0_9
2. Kuang, J., Raś, Z.W.: In search for best meta-actions to boost businesses revenue. Flexible Query Answering Systems 2015. AISC, vol. 400, pp. 431–443. Springer, Cham (2016). https://doi.org/10.1007/978-3-319-26154-6_33
3. Rombel, A.: CRM shifts to data mining to keep customers. Glob. Financ. **15**, 97–98 (2004)
4. Ras, Z.W., Wieczorkowska, A.: Action-rules: how to increase profit of a company. In: Zighed, D.A., Komorowski, J., Żytkow, J. (eds.) PKDD 2000. LNCS (LNAI), vol. 1910, pp. 587–592. Springer, Heidelberg (2000). https://doi.org/10.1007/3-540-45372-5_70
5. Ras, Z.W., Dardzinska, A.: From data to classification rules and actions. Int. J. Intell. Syst. **26**(6), 572–590 (2011)
6. Ras, Z.W., Tarnowska, K.A., Kuang, J., Daniel, L., Fowler, D.: User friendly NPS-based recommender system for driving business revenue. In: Polkowski, L., Yao, Y., Artiemjew, P., Ciucci, D., Liu, D., Ślęzak, D., Zielosko, B. (eds.) IJCRS 2017. LNCS (LNAI), vol. 10313, pp. 34–48. Springer, Cham (2017). https://doi.org/10.1007/978-3-319-60837-2_4

7. Simunek, M.: Academic KDD project LISp-miner. In: Abraham, A., Franke, K., Köppen, M. (eds.) Intelligent Systems Design and Applications. Advances in Soft Computing, vol. 23, pp. 263–272. Springer, Heidelberg (2003)

8. Tarnowska, K.A., Ras, Z.W., Jastreboff, P.J.: Mining for actionable knowledge in tinnitus datasets. In: Skowron, A., Yao, Y., Ślęzak, D., Polkowski, L. (eds.) Thriving Rough Sets, Chap. 18. SCI, vol. 708, pp. 367–396. Springer, Cham (2017). https://doi.org/10.1007/978-3-319-54966-8_18

Initial Analysis of Multivariate Factors for Prediction of Shark Presence and Attacks on the Coast of North Carolina

Sonal Kaulkar[1], Lavanya Vinodh[1], and Pamela Thompson[1,2](✉) (iD)

[1] Department of Computer Science, University of North Carolina at Charlotte, Charlotte, NC, USA
plthomps@uncc.edu
[2] Catawba College, Salisbury NC, USA

Abstract. Classification, association rules and clustering are used in the study to improve understanding of the presence of sharks in near shore waters during tourist seasons in middle Atlantic coastal waters, specifically North Carolina. The Global Shark Attack File combined with data on environmental, biotic and meteorological factors is prepared for analysis using the CRISP-DM process. In future work, combined inputs including a standardized hashtag for twitter mining, real time weather and water information, and data on crab and turtle presence will provide real-time input to an app or a dashboard providing early warning of shark presence.

Keywords: Shark attack · Clustering · Early warning shark attack app
Knowledge discovery process · Association rules · Multivariate analysis
Classification · Balancing data

1 Introduction

Shark attack incidents are one of the most well-known animal related phenomenon with wide spread documentation on the news and social media. Exploratory data analysis of the Global Shark Attack File [1] shows the East Coast of the US experienced an unusual number of attacks during the Summer of 2015 with sixteen total attacks during the season in North and South Carolina. This study was initiated as a research project for the Knowledge Discovery in Databases graduate class at the University of North Carolina at Charlotte during July of 2015 and has continued. The increase in shark attack incidents throughout the summer and the lack of effective early warning systems for East Coast tourists create interest in applying knowledge discovery techniques to the prediction of sharks in near coast waters along East Coast beaches. In the US, North Carolina and South Carolina together rank second after Florida with respect to the incidence of shark attacks [1, 17, 20].

© Springer Nature Switzerland AG 2018
M. Ceci et al. (Eds.): ISMIS 2018, LNAI 11177, pp. 426–435, 2018.
https://doi.org/10.1007/978-3-030-01851-1_41

1.1   Objective

In this study, analysis is performed to find the potential impact of certain meteorological, environmental and anthropogenic factors using data mining following the Cross-industry standard process for data mining (CRISP-DM). These factors could be a potential cause for the rise of shark attacks on the coast of North Carolina [4]. The objective of this research is to improve our understanding of the presence of sharks during tourist seasons in middle Atlantic and south eastern coastal waters, specifically North Carolina for this study. Our study will focus on the analysis of existing data from the Global Shark Attack File, weather, wind and water data from NOAA, calculated moon phase dates, fish, crab and turtle populations. The quantitative analysis on this data will lead to new and interesting knowledge that will provide the basis for an app providing advanced information on the likelihood of sharks in coastal waters where tourists swim, surf and wade. A future focus of this research is to analyze social media activity relative to shark presence. A recommendation for a standardized way to tweet will be considered with interested and strategic partners in order to ultimately provide an additional feed to a shark sighting app such as Dorsal, co-founded by professional surfer Sarah Beardmore [21].

1.2   Methods

The principal data source for the incident of shark attack in North Carolina was taken from the incident log of the Shark Research Institute [1] defined as the "Global Shark Attack File" (GSAF). The project constrained the study of incidences of shark attacks to be limited to the coast of North Carolina from 2009 to 2016 for the months of May to September.

The methodology adopted for the project was the usage of CRISP-DM model [16]. The process started with business understanding and data understanding phase of understanding the shark attacks which was enhanced further by the interaction with Dr. Chuck Bangle Shark Researcher of East Carolina University. The Data Preparation necessitated the collection of data sources for all the potential factors that led to higher rate of shark attacks. The next phase employed was modelling of the collected data to gain insights in the data and find the relevant factors that showed an impact on the incidences of shark attacks. The input file used for modeling was aggregated on a daily basis for all the data sources collected for data mining.

2   Data Understanding

Data Understanding constituted of finding the important factors and the study of their contribution to the rise of shark attacks. The GSAF incident log was used as a starting point to understand the incidences of shark attack. Exploratory Data Analysis was performed using a variety of tools including Excel, Weka and R.

For EDA the GSAF file was filtered for the years of 2009 to 2016 for the months of May to September for both the beaches of North and South Carolina states. Location was a preliminary attribute chosen based on the fact that place of incident can be a starting point for judging the deciding factors [4]. It was found that of all the beaches Myrtle Beach had the highest number of shark attacks. The input file used for data mining for all the consequent phases also included the area which is the state, species, beach and county. Bull sharks are one of the shark species most involved in fatal attacks against humans [3, 5]. Plotting a bar chart of Species of sharks involved in the incident revealed that bull sharks are the larger in number in these incidences.

Anthropologic factors like very large involvement of human settlement can be a reason to attract sharks [4]. A less populated shoreline with limited access to humans has less number of shark attacks. A histogram of activities of humans against the count of shark attacks was plotted. The histogram reflected that activities such as surfing and swimming on the shore proved more fatal to humans with respect to shark attacks.

Lunar cycles have always been controversial to be considered as a factor to the rise of shark attacks [6]. It is said that lunar cycles are unlikely to have an effect on incidence of shark attack [15] however it is common knowledge to fisherman that lunar cycles affect the catch [7]. To explore the relevance of lunar phases to shark presence in near shore waters an algorithm in Java was used to calculate the moon phases for all the days of May to September from 2009 to 2016 for inclusion in the study.

The set of moon phases belonged to one of the categories new moon, full moon, first quarter, waxing gibbous, waning gibbous, third quarter, waning crescent, or waxing crescent. The calculated moon phase by date was then integrated with the GSAF file using R code for the purpose of performing exploratory data analysis on the effect of lunar cycles on shark attacks. Histogram analysis shows attacks are almost equal in both the cases new moon and full moon. Domain knowledge on moon phases and the effect on tides shows that when the moon and sun align, known as full or new moon, the pull is at its strongest, causing the tides to be at their highest and lowest. This is known as spring tides. The change from high tides to low tides and back again happens very quickly which may cause sharks to move in areas closer to where people swim when the tide is low. The effect on tides occurs over a span of days leading up to the change in the lunar cycle. The effect duration, therefore, is not on the day that is typically depicted as the new or full moon but can span several days before and after [6, 7].

Meteorological factors impact the occurrences of shark attacks along the East coast; warm temperatures increase the presence of food (crabs, fish that migrate to warmer waters and this increases the likelihood of attack as sharks follow the food source. The encounters with humans also become more prevalent with summer months from May to September when there is a rise in water temperature that bring in humans as well as sharks near the shoreline [2, 9]. The data source for the all the weather parameters is taken from National Oceanic and Atmospheric Administration [8, 12]. Exploratory data analysis revealed that high temperature favors shark attacks. As the temperature of the water increases, the sharks tend to move into favorable locations [21].

Increases in temperature have resulted into scarce rainfall, thereby increasing the salinity of water which may cause more movement of sharks to the shoreline away from the estuaries. Hence precipitation in atmosphere along the shoreline is taken as an attribute in the prediction of shark attacks. Wind speed is another crucial factor which may attracting sharks to the shore along with their sources of food [13]. The higher speed of the winds moves clearer, warmer water to the shore. When that happens the food sources such as mullet, bait fish and menhaden move closer to the shore and the sharks follow their food sources [4, 10]. Wind speed data from NOAA [12] was included in the study for analysis.

Salinity of water affects the habitat of sharks. Sharks usually are found near estuaries except for the bull shark which may thrive in open sea water and is responsible for many of the incidents of attacks in the study data. Salinity data was taken from National Estuarine Research Reserve System [11] which is monitoring program SWMP or system-wide monitoring program, and provides long-term data on water quality, weather, biological communities, habitat, and land-use and land-cover characteristics [11]. Exploratory data analysis was performed on the salinity and the number of attacks that occur in the study data are highest when the salinity reached the optimal/desired levels.

An increase in rainfall causes rivers to swell and thereby causes turbidity to increase, reducing the visibility and making it harder for people and sharks to see each other. This can result in attacks [13]. Thus, the more turbid the water, the greater may be the probability of occurrence of attack incidents. Low levels of dissolved oxygen can cause marine life to become very lethargic. Aquatic animals involving sharks move towards the shore to get more oxygen. This effect is called as "Jubilee" effect by the local communities who are involved in fishing [12]. Dissolved Oxygen along with Salinity and Turbidity is collected from NERRS [11] every 30 min from each of the 26 NERRS [11] sites out of which the data collected from the East Cribbing site is taken since it is closest with the Wrightsville Beach, NC location which represents a centralized location for the North Carolina shoreline and beaches. Ecology is another factor that drives sharks towards the shore. Sharks follow close behind whales, turtles, menhaden who surged northwards towards the heat wave in 2015 [13, 14]. Turtle data was collected in terms of false crawling and nesting from Dr. Matthew Godfrey (State Coordinator) from North Carolina Wildlife Resources Commission. False crawls are when turtles come to shore to lay eggs but do not succeed and go back to the ocean to try again another day. Nesting is when the turtles have succeeded and a nest is documented.

Blue crabs are prevalent along the east coast of the United States and are particularly prevalent in North Carolina. These crustaceans can survive in different environments and migrate towards the Carolinas in summer for higher spawning rates. This movement increases the blue crabs in number along the east coast [15]. Blue crab movement is also affected by the tides with crabs moving from inlets to estuaries in increasing distance during full and new moon periods. Sharks along the east coast prey on the blue crabs that are prevalent in the region. Daily Blue crab landings from 1994 to 2004 was collected from Alan Bain, the chair of the NC Division of Marine Fisheries. This data is integrated with the prediction of shark attacks and shows high crab landings favor attack.

South West winds blow in a north easterly direction but do not move water in that direction. Due to the earth's rotation water takes a 90° toll and the water temperature drops down by 10 degrees making the lower cold water rise to the top making water murkier during higher winds. This may be a reason for sharks to get confused with its intended prey resulting in shark attacks on humans, not fish in murky waters. ERRS was used as a data source for wind direction which contained readings in degrees which were then converted in one of the directions using Java code. On performing initial exploratory data analysis, it was found that attacks are almost equal in both the cases south-southwest and southwest winds.

3 Data Preparation

Data merging and preparation followed exploratory data analysis and domain knowledge discovery. The data preparation presented a significant task with data sources from several different sources requiring merging with the global shark attack file primarily by date. The global shark attack file shows only dates where attacks occur. For learning purposes, the study including dates where attacks did not occur in order to gain knowledge on the presence and absence of sharks in near shore waters based on documented attacks. For supervised learning, a binary attack flag representing Attack Y or N and other variables as predictor variables were merged with the global shark attack file by date. Another file was maintained for exploratory data analysis which consisted only of instances with the attack flag equal to Yes.

The Global Shark Attack File containing the incident log of attacks and enhanced with dates not having attacks provided the basis for the principal file into which other predictor variables were then added as columns. Turtle data constituted turtle activities like nesting and false crawls. The turtle data that was obtained was observatory data. For unification of format, the nesting and the false crawls were merged together to form a single discrete numerical attribute named Turtle Activity. This modified feature was then aggregated date wise to get a daily count of turtle activity and merged into the GSAF file using R code. Moon phase and lunar cycles have their effect on a moving average of three days before and after a particular moon phase [19]. Thus, the daily data of moon phases including values for full moon and new moon was extended to three days before and after the calculated date. This feature was used as one of the predictor variables in modeling. A moving average method using R was applied to the precipitation data from NOAA [12] to account for the nature of effects rainfall has on water and other factors. Wind direction data from NERRS [11] was given in the form of degrees of rotation every 30 min. This data was then aggregated with the mean calculation to calculate the average rotation of the wind on a daily basis. On careful understanding of the impact of changing temperature on the surge of sharks towards the shore, change in water temperature between consecutive days was used as a predictor rather than raw temperature values. Mean calculation of water temperature on a daily basis is aggregated and then a difference between the current and previous day's temperature is used as a predictor variable for temperature.

Binning of numerical variables of the data collected was employed for most of the predictor variables [16]. Before using turtle activity data in analysis, the numerical predictor was discretized based on location and daily count. On plotting histograms, it was assured that the turtle activity discretized on a daily basis correlated more with shark attacks and was thus used in the prediction in the modeling phase. Similarly, time of attack from GSAF, salinity, turbidity, station pressure, and wind speed numerical variables were discretized based on three bin equal width binning for the initial analysis.

Missing values were evident in the merged data. Weather data collected from NOAA [12] had 17% of Not Applicable or Not Available (NA). To solve this missing data issue, NAs were removed by replacing it with the mean of the value aggregated over the day from the raw data. This method was applied for Dissolved Oxygen, Salinity, Turbidity, Water Temperature, Precipitation, Station pressure and Wind Speed. Another hurdle with respect with missing data was with respect to Crab landings data from 1994 to 2015. Out of the available data from 1994 to 2015, data was filtered from 2009 to 2015. There were no readings for crab landings for all Sundays. To replace these missing values R code was written to take average of the crab landings of Saturday and Monday as the value for Sunday. Data for 2016 year was created using estimation and regression analysis using Excel. Linear Regression was used with 95% confidence to predict the Crab landings.

Min-max Normalization was chosen standardize all numeric variables. The final input file was improved by removing duplicate records using R duplicated function.

3.1 Data Imbalance

The GSAF file was filtered for the state of North Carolina for the duration of May to September between the years 2009 to 2016. Thus, the number of records with Attack equal to Yes were 65 in total. When the GSAF file was combined with the file containing all dates it led to a data imbalance problem. In order to solve the problem, a stratified sampling of Attack = No subset was prepared so that 1/3 of records remain with adequate representation for each class. To evaluate the efficiency of the solution applied to the data imbalance problem, a two sample Z test for difference in proportions was used to verify the partitions.

4 Modeling

Classification, clustering and association rule analysis were performed for the prediction of shark attacks along the coast of North Carolina.

4.1 Classification

The method used for classification was Naïve Bayes Classifier using J48 with Weka Tool. The major task for classification was to find the potential factors of the included predictors that lead to shark attack. The algorithm used for Attribute selection was Weka's Info Gain Attribute Evaluator. Of the 12 variables given as input to the Attribute evaluator, the rankings provided by the attribute evaluator ranked Wind Speed as the highest and the Station Pressure as the lowest. The classification model generated an unpruned J48 tree giving 149 correctly classified instances out of total of 186 instances. The mean absolute error was .21 with a weighted average true positive rate of .80, a false positive rate of .24, precision and recall were .80 each. The total confidence of the model was 80.1075% weighted average.

4.2   Clustering

The K-means algorithm was used for clustering in Weka Tool [16]. The clustering resulted into two distinct clusters with Cluster 0 with target variable as Attack = Yes and a second Cluster 1 with target variable as Attack = No. The nine highly ranked predictor variables used in the classification algorithm were used for clustering. The Mean Absolute error was 0.2172, Root mean squared error: 0.4178, Relative Absolute Error: 45.7399%, Root relative squared error: 85.7572% on Total Number of Instances: 186. It was found that Cluster 0 had a lower turtle activity proportion than Cluster 1. Also, the Wind Speed for Cluster 0 was lower in proportion than Cluster 1. The direction of winds for Cluster 1 is South West Winds (SWW). The salinity is High for Cluster 1 while for Cluster 0 it is medium the lunar cycle also showed a significant change. In Cluster 0 the moon phase is First Quarter while for records of the shark attack category in Cluster 1 it is a full moon. The number of instances belonging to Cluster 0 were 53% of the data and that belonging to Cluster 1 was 47% of the data (Fig. 1).

Final Cluster Centroids

	Cluster#		
Attribute	Full Data	0	1
	(186)	(99)	(87)
Attack	No	No	Yes
DO2	Med	Med	Med
Salinity	High	Med	High
Temperature	High	High	High
Precipitation	Low	Low	Low
Direction	SSW	SSW	SW
Moon Phase	Full	New	Full
Wind Speed	0.34	0.27	0.43
Turtle	0.15	0.10	0.21

Time taken to build model: 0.01 seconds

Model and evaluation on training set

Clustered Instances

0	99(53%)
1	87(47%)

Fig. 1. Final cluster centroids

4.3   Association Rules

Association rule modeling was employed using the apriori algorithm in order to find the best rules for consequent Attack = Yes. The variables used in Apriori were Attack flag, Dissolved Oxygen, Wind Speed, Crab Landings, Wind Direction, Extended Moon Phase, and Temperature Change as the variables in the consequent part.

The result of the association model gave minimum support of 0.1 and a minimum confidence of 0.75. The best rule with the highest confidence is: {Crab Landings = High, Wind Direction = SW, Moon Phase = New Moon} => {Attack = Yes} with confidence of 0.78.

5   Discussion

The data preparation and preliminary analysis performed in this study have provided factors that influence the presence of sharks based on shark attack data from the Global Shark Attack File. These factors such as weather conditions and environmental factors based on location can be input to an application which would then provide people with an assurance level of how safe it is to surf, swim or wade on beach [19]. This data can be captured real time and provided to a supervised neural network which would present its output node as the level of safety for a particular beach. A warning system like this can be combined with real time surveillance with newer methods involving drone and blimp technologies with image detection capabilities. Another way this data mining project can be deployed is to give the relevant attributes as input to existing apps in the market that monitor activities related to shark attack such as the Dorsal app co-founded by professional surfer Sarah Beardmore. Apps like Dorsal can be used for monitoring purposes and to guide people to be safe before going into the ocean [19, 21].

In addition to the predictors of interest from this study, a standardized hash tag can be used to gain new understanding on the presence of sharks and the interaction of features from the weather and environment. In developing a uniform hash tag to report attacks or events that predict sharks in near shore waters, procedures such as those outlined in the United Nations Office for the Coordination of Humanitarian Affairs [18] are recommended.

6   Conclusion

The data analysis for prediction of shark attacks on the coast of North Carolina provided useful insights about the factors that lead to shark attacks and shark presence. Different modeling techniques gave the relevance of each of the different potential factors and how they could contribute to the prediction of shark attacks and presence. To better understand the potential factors, attribute relevance using information gain was run on the training data. Wind speed and direction though rarely mentioned in articles and social media did prove to be more relevant in our results. Also, ecological variants like changes in crab landings and lunar cycles also played an important role in contributing to the factors for shark attacks to occur. On a whole this study gave a list of all the meteorological, ecological and environmental factors that could be the potential reason for the surge in shark attacks for the coast of North Carolina over the years from 2009 to 2016 and particularly in 2015.

Limitations of the study include the fact that prediction of shark presence is based only on actual attacks; there is a lack of data on actual sightings of sharks in near shore waters not necessarily leading to attacks. With the addition of shark sighting data from a test of a standardized twitter hash tag and extension of the study to South Carolina and other Eastern states, the researchers hope to improve on the results which can lead to a test of an actual recommender system.

References

1. Shark Research Institute (SRI) Homepage: Global shark attack file incident log. http://www.sharkattackfile.net. Accessed 28 May 2018
2. Fordie, J.: The shark attacks in North Carolina, explained (2015). http://www.outsideonline.com/1999256/shark-attacks-north-carolina-explained
3. Burgess, G. (2015). http://pilotonline.com/news/local/environment/expect-more-sharks-off-north-carolina-this-summer-experts-say/article_dd48563a-6bc7-5c7c-b811-2b92b1e0db3d.html
4. Amin, R., Ritter, E., Wetzel, A.: An estimation of shark attack risk for the coast of North Carolina and South Carolina. J. Coast. Res. **31**(5), 1253–1259 (2015)
5. Palermo, E.: Shark bites two: possible explanations for attacks (2015). http://www.livescience.com/51218-shark-attacks-north-carolina.html
6. Ritter, E., Zambesi, A., Amin, R.: Do lunar cycles influence shark attacks? Open Fish Sci. J. **6**, 71–74 (2015)
7. Mockler, B.: Pick the right time to fish. http://www.farmersalmanac.com. Accessed 20 Apr 2018
8. NOAA Estuaries Homepage. http://oceanservice.noaa.gov/education/tutorial_estuaries/est10_monitor.html. Accessed 21 May 2018
9. Gardenette, N.: Warm summer waters induce shark migration to east coast beaches, June 2014. http://www.accuweather.com/en/weather-news/summer-shark-migration-ocean-east/29413314
10. Griffin, D.: Weather may have contributed to attacks (2001). http://www.cnn.com/2001/US/09/05/shark.attack/
11. NERRS (2016). http://nerrs.noaa.gov/research
12. NOAA Ocean Service Education: Dissolved oxygen. http://oceanservice.noaa.gov/education/kits/estuaries/media/supp_estuar10d_disolvedox.html. Accessed 21 May 2018
13. Hampton, J.: Expect more sharks off North Carolina this summer, experts say, March 2016. https://pilotonline.com/news/local/environment/expect-more-sharks-off-north-carolina-this-summer-experts-say/article_dd48563a-6bc7-5c7c-b811-2b92b1e0db3d.html
14. Hammerschlag, N., et al.: Evaluating the landscape of fear between apex predatory sharks and mobile sea turtles across a large dynamic seascape. Ecol. Soc. Am. **96**, 2117–2126 (2015)
15. Heikkinen, N.: Blue crabs migrate north as ocean warms. Sci. Am. (2015)
16. Larose, D., Larose, C.: Discovery Knowledge in Data, p. 138. Wiley, Hoboken (2014)
17. Haelle, T.: Shark bites are up, but attack risk is down? Sci. Am. (2015)
18. United Nations Office for the Coordination of Humanitarian Affairs: Hashtag standards for emergencies. https://www.unocha.org/publication/policy-briefs-studies/hashtag-standards-emergencies. Accessed 21 May 2018
19. Dr. P. Thompson Homepage. http://www.profthompson.net, last accessed 2018/5/24

20. Ferretti, F., Chapple, T., Jorgensen, S., Micheli, F.: Reconciling predator conservation with public safety. Front. Ecol. Environ. **13**, 412–417 (2015)
21. Dorsal Homepage. http://www.dorsalwatch.com. Accessed 24 May 2018
22. Payne, N., et al.: Combining abundance and performance data reveals how temperature regulates coastal occurrences and activity of a roaming apex predator. Glob. Chang. Biol. **24**, 1884–1893 (2018)

Topic Modelling and Opinion Mining

An Experimental Evaluation of Algorithms for Opinion Mining in Multi-domain Corpus in Albanian

Nelda Kote[1]([⊠]), Marenglen Biba[2], and Evis Trandafili[3]

[1] Department of Fundamentals of Computer Science, Faculty of Information Technology, Polytechnic University of Tirana, Tirana, Albania
nkote@fti.edu.al
[2] Department of Computer Science, Faculty of Informatics, New York University of Tirana, Tirana, Albania
marenglenbiba@unyt.edu.al
[3] Department of Computer Engineering, Faculty of Information Technology, Polytechnic University of Tirana, Tirana, Albania
etrandafili@fti.edu.al

Abstract. Opinion mining is an important tool to find out what others think about something. Most of methods used for opinion mining are based on machine learning. In this paper we present an experimental evaluation of machine learning algorithms used for opinion mining in a multi-domain corpus in Albanian language. We have created 11 multi-domains corpuses combining the opinions from 5 different topics. The opinions are classified as positive or negative. All the corpuses are used to train and test for opinion mining the performance of 50 classification algorithms. Out of these, there are seven best performing algorithms out of which three are based on Naïve Bayes.

Keywords: Opinion mining · Machine learning algorithm
Opinion classification

1 Introduction

An opinion is a view, or an unproven judgement formed about something, that cannot be necessarily based on facts or knowledge. According to [1] an opinion is a quintuple, $(e_i, a_{ij}, s_{ijkl}, h_k, t_l)$, where e_i is the name of an entity, a_{ij} is an aspect of e_i, s_{ijkl} is the sentiment on aspect a_{ij} of entity e_i, h_k is the opinion holder, and t_l is the time when the opinion is expressed by h_k. Opinion mining involves the development of a system to identify, analyze, extract, summarize or categorize the opinions based on different criteria. Based on the definition of the opinion in [1], opinion mining should evaluate the quintuple of the opinion.

The aim of this paper is to evaluate through experiments the performance of classification algorithms for opinion mining at document level in multi-domain corpus. We collected text opinions related to 5 nowadays topics in Albania and classified them as positive and negative opinions. By combining the opinions of different topics, we created 11 multi-domain corpuses listed in Table 1. To clean the data, each corpus

© Springer Nature Switzerland AG 2018
M. Ceci et al. (Eds.): ISMIS 2018, LNAI 11177, pp. 439–447, 2018.
https://doi.org/10.1007/978-3-030-01851-1_42

passed through a preprocessing phase composed by a stop-word removal and a stemmer. Then we used this cleaned dataset to train and test the performance of 50 classification algorithms.

Table 1. The corpus used for performing the experiments

Corpus code	Interpretation
C1	5 topics (50 positive and 50 negative opinions for each topic)
C2	5 topics (40 positive and 40 negative opinions for each topic)
C3	5 topics (30 positive and 30 negative opinions for each topic)
C4	5 topics (20 positive and 20 negative opinions for each topic)
C5	5 topics (10 positive and 10 negative opinions for each topic)
C6	4 topics (tourism, VAT tax, waste import, Higher Education Law) containing 50 positive and 50 negative opinions from each topic
C7	3 topics (tourism, waste import, Higher Education Law) containing 50 positive and 50 negative opinions from each topic
C8	3 topics (tourism, VAT tax, waste import) containing 50 positive and 50 negative opinions from each topic
C9	2 topics (politics, Higher Education Law) containing 50 positive and 50 negative opinions from each topic
C10	2 topics (politics, VAT tax) containing 50 positive and 50 negative opinions from each topic
C11	2 topics (tourism, waste import) containing 50 positive opinions and 50 negative opinions from each topic

The structure of the paper is as follows: Sect. 2 presents background and related work; Sect. 3 presents the methodology for opinion categorization; Sect. 4 presents experiment results and Sect. 5 concludes our work and gives ideas for future work.

2 Background and Related Work

A comprehensive and thorough study is done by the author in [1] relating to all research fields of opinion mining. The main technologies for opinion mining are supervised, semi-supervised, and unsupervised approaches. Opinion mining would be performed in document level, in sentence level and in aspect-based level. In in-domain the corpus contains opinions from only one topic, in multi-domain the corpus contains opinions from different topics and in cross-domain the train corpus contains opinions from one topic and the test corpus contains opinions from another topic.

The work in [2] gives an in-depth explanation of the unsupervised methods used for opinion mining. An unsupervised machine learning technique is used in [3] for opinion mining and the experimental results indicate that unsupervised technique outperforms the supervised technique.

The studies in [4–7] evaluate the performance of supervised machine learning techniques for in-domain opinion mining. In paper [4] the authors concluded that the combination of Naïve Bayes algorithm with unigram technique has the best performance. A comparative performance evaluation of the classification algorithms, NB, SVM and ME for opinion classification and topic-based classification is done in [5]. They concluded that these algorithms have better performance in topic-based classification and there are too many challenges to address in opinion classification. In [7] is proposed a method that combine machine learning and semantic orientation based on the integration of knowledge resource to perform opinion mining based in majority voting system.

In [8], is presented a method to reduce the domain dependency and improve the performance by proposing an efficient multi-domain opinion classification algorithm.

In [9], the authors proposed to train the classifiers for multi-domain dataset in a collaborative way based on multi-task learning. The classifier is decomposed in two components: a general one that is trained across various domains to capture global information and a domain specific one that is trained using labeled data to capture domain specific information. The experiment evaluation shows that this method outperforms traditional methods.

There are a lot of research studies for opinion mining in German, Spanish, Turkish, Arabian, etc. [10–13]. But, insignificant research work has been done in Albanian language for opinion mining. In [14, 15] the authors evaluated through experiments the application of machine learning algorithms for opinions classification using in-domain corpus in Albanian language. They defined the best performing classification algorithms and that the opinion classification is a domain depended problem.

3 Methodology for Opinion Classification

We evaluated through experiments the performance of 50 classification algorithms for opinion mining at document level in multi-domain corpus in Albanian language. We used the same methodology as in our first work in [15] that evaluate the performance of classification algorithms for opinion mining in in-domain corpus.

3.1 Data Collection

We created a text corpus of Albanian written opinions collected from different well-known Albanian newspapers. We collected opinions related to 5 topics in Albania as follow: tourism, politics, the including of VAT tax in small business, waste import and Higher Education Law. For each topic we collected 50 text document categorized as positive opinions and 50 text documents categorized as negative opinions. The performance of the classification algorithms for opinion mining can be affected by the domain and the number of the opinions. For this purpose, we created 11 multi-domain corpuses as listed in Table 1 by combining the opinions of different topics.

3.2   Preprocessing

We have preprocessed the dataset before it is used to training and test the classification algorithms. We used the Albanian rule-based stemmer implemented by [16] to perform the preprocessing step. This stemmer is implemented in java programming and contains 134 rules based on the Albanian language morphology but to find the stem of the word it is not taken in consideration the linguistic meaning of the word. The stemmer is experimentally tested by the authors and the results demonstrate its effectiveness.

The preprocessing process include two steps. The first step is applying a stop-word elimination that: set all the data to lower case and remove the special characters (including the punctuations), numbers and stop-word like: "dhe", "ne" "sepse", "kur", "edhe" "ndonëse", "mbase" etc. The stop-word can be removed because they do not have any positive or negative connotation and do not affect the emotion of the opinion. The second step is applying the stemmer that finds the stems of all the words in the text document. The output text files from this phase contains only words.

3.3   Experiment Settings

Each corpus listed in Table 1 contains the same number of positive and negative opinions. Firstly, each corpus is preprocessed and then the output text files of this phase are used as an input for training and testing the classification algorithm. To evaluate the performance of the classification algorithm we used WEKA software [17] that contains a collection of machine learning algorithms and tools to perform data mining tasks.

To perform the experiments in Weka we converted the text documents of each corpus into a .arff file using textDirectoryLoader class. Then we applied the StringToWordVector filter with WordTokenizer that convert all the string attributes into a word vector, that represent the occurrence of the word from the text contained in the strings. We used 10-folds cross-validation for training a model and testing all the chosen algorithms in each corpus.

4   Experiment Results

We evaluated the performance of classification algorithms on each corpus in term of percentage of correctly classified instances. In Table 2 are shown the results of the experiments, highlighting in bold the best percent of correctly classified instances for each corpus. By analyzing the experimental results, we note that for different corpuses there are different algorithms best performing.

For corpus C1, C2, C3, C4, C6, C7 and C8 there are 3 best performing algorithms Complement Naïve Bayes, Naïve Bayes Multinomial and Naïve Bayes Multinomial Updateable with the same percent of correctly classified instances for each corpus. For corpus C5 the best performing algorithm is RBF Network with 79% correctly classified instances. For corpus C9 the best performing algorithm is Logistic with 80.50% correctly classified instances. For corpus C10 the best performing algorithm is Hyper Pipes with 78% correctly classified instances. For the last corpus, C11, SGD is the best performing algorithm with 84.50% correctly classified instances.

Table 2. Experimental results in terms of percent of correctly classified instances

Algorithm	C1	C2	C3	C4	C5	C6	C7	C8	C9	C10	C11
BayesianLogisticRegretion	77.0	77.5	77.7	78.0	73.0	78.0	81.7	77.3	77.0	71.5	82.5
BayesNet	66.0	64.0	54.3	53.5	51.0	61.8	68.7	60.3	57.0	56.5	79.0
ComplementNaiveBayes	**78.6**	**80.3**	**81.7**	**80.0**	75.0	**83.0**	**85.7**	**81.7**	75.0	70.5	81.5
NaiveBayes	73.0	75.8	75.3	72.5	72.0	78.8	80.3	76.7	72.0	69.5	79.0
NaiveBayesMultinomial	**78.6**	**80.3**	**81.7**	**80.0**	75.0	**83.0**	**85.7**	**81.7**	75.0	70.5	81.5
NaiveBayesMultinomialText	50.0	50.0	50.0	50.0	50.0	50.0	50.0	50.0	50.0	50.0	50.0
NaiveBayesMultinomialUpdateab	**78.6**	**80.3**	**81.7**	**80.0**	75.0	**83.0**	**85.7**	**81.7**	75.0	70.5	81.5
NaiveBayeslUpdateable	73.0	75.8	75.3	72.5	72.0	78.8	80.3	76.7	72.0	69.5	79.0
Logistic	66.6	78.8	76.7	77.5	78.0	77.0	82.3	79.3	**80.5**	73.0	84.0
RBFClassifier	73.6	74.5	74.3	71.0	59.0	78.5	77.7	76.0	72.5	65.5	81.5
RBFNetwork	73.4	74.8	76.7	78.0	**79.0**	51.8	49.3	50.7	49.5	47.0	54.0
SGD	77.4	77.5	77.0	75.0	66.0	77.3	80.0	76.3	75.5	71.0	**84.5**
SimpleLogistic	66.4	70.0	68.7	58.5	52.0	68.0	74.7	65.3	65.0	55.5	74.5
SMO	76.0	75.0	76.0	73.5	66.0	77.0	78.0	76.7	74.5	68.0	81.0
VotedPerceptron	69.2	74.0	72.3	68.0	56.0	73.3	81.0	71.0	65.5	69.0	77.0
IB1	61.8	62.5	63.7	61.0	59.0	59.3	59.7	63.7	61.0	61.5	61.0
IBK	60.4	61.5	64.3	61.0	59.0	59.8	59.7	64.3	60.5	60.5	60.5
Kstar	61.8	62.3	64.3	62.0	57.0	61.0	60.0	64.0	60.0	61.5	60.5
LWL	55.6	54.0	52.0	53.0	55.0	57.8	60.7	57.7	58.0	60.0	65.5
AdaBoostM1	60.6	62.8	61.3	53.5	56.0	60.0	69.7	61.3	56.5	52.0	78.5
AttributeSelectedClassifier	60.6	64.0	63.3	57.0	63.0	64.0	76.3	61.7	64.5	55.0	75.0
Bagging	65.4	69.0	63.7	64.5	56.0	70.8	71.0	72.3	65.0	59.0	76.5
ClassificationViaRegression	60.2	58.0	51.7	51.5	50.0	60.3	72.7	65.3	49.5	59.0	72.0
FilteredClassifier	62.6	62.5	54.3	53.5	51.0	61.0	71.0	61.3	58.0	56.5	77.0
InterativeClassifierOptimizer	63.6	68.8	59.7	59.5	52.0	65.8	74.0	65.0	52.5	56.0	73.5
LogitBoster	66.0	71.0	63.7	61.0	51.0	69.3	75.0	67.0	54.5	59.0	74.5
MultiClassClassifier	66.6	78.8	76.7	77.5	78.0	77.0	82.3	79.3	80.5	73.0	84.0
MultiClassClassifierUpdateable	77.4	77.5	77.0	75.0	66.0	77.3	80.0	76.3	75.5	71.0	84.5
RandomCommittee	67.0	71.5	68.7	67.5	59.0	67.8	74.0	63.0	70.0	66.5	71.5
RandomizableFilteredClassifier	51.8	54.0	54.0	56.5	56.0	54.8	53.0	54.0	47.5	53.0	59.5
RandomSubSpace	72.0	66.5	66.0	63.0	59.0	71.5	75.0	72.0	68.5	62.5	77.0
RealAdaBoost	65.4	68.3	62.0	61.0	56.0	66.0	71.0	65.7	60.0	61.5	79.5
FLR	57.8	62.0	67.7	78.0	75.0	65.0	72.3	73.7	78.5	75.0	79.5
HyperPipes	55.8	62.3	64.7	78.5	78.0	63.5	71.3	66.7	78.5	**78.0**	74.0
ConjunctiveRule	54.4	53.0	52.0	51.5	54.0	54.8	54.7	59.7	49.5	50.5	59.5
DecisionTable	63.4	60.5	59.7	55.0	55.0	63.8	71.0	59.7	64.5	55.0	72.5
Jrip	58.4	57.5	53.3	48.0	47.0	62.3	66.7	63.7	62.5	52.0	77.0
Nnge	61.0	67.5	67.0	70.5	67.0	64.5	67.0	67.3	77.0	65.0	71.0
OneR	58.0	57.3	59.0	42.5	45.0	57.3	63.7	56.3	58.0	57.5	64.5
PART	65.6	61.3	61.3	62.0	51.0	62.5	72.0	63.3	63.0	60.0	65.5
Ridor	56.6	60.0	55.7	55.5	45.0	60.0	62.7	60.0	53.0	50.0	69.0
ZeroR	50.0	50.0	50.0	50.0	50.0	50.0	50.0	50.0	50.0	50.0	50.0
DecisionStump	52.4	52.5	52.7	49.0	55.0	56.5	57.3	56.7	54.5	62.0	66.0
HoeffdingTree	72.8	73.8	79.3	79.5	71.0	76.5	81.0	74.7	78.5	72.0	71.5

(continued)

Table 2. (*continued*)

Algorithm	C1	C2	C3	C4	C5	C6	C7	C8	C9	C10	C11
J48	61.2	61.5	64.7	65.5	59.0	63.3	68.3	61.3	64.0	60.0	69.0
LMT	68.0	70.0	67.3	58.5	55.0	68.8	76.3	65.3	65.5	55.5	74.5
Random Forest	74.2	75.5	76.7	72.0	66.0	78.0	80.7	78.3	75.0	67.0	82.0
RandomTree	57.8	56.3	55.7	60.0	50.0	58.0	63.7	60.0	59.5	54.5	51.5
REPTree	60.6	58.5	57.7	58.5	56.0	57.8	67.0	64.0	56.5	54.0	69.0
SimpleCart	61.0	62.5	61.0	57.5	56.0	63.3	69.7	71.7	55.5	58.5	76.0

There are 7 best performing algorithms with percent of correctly classified instances varying from 78% to 85%. Complement Naïve Bayes, Naïve Bayes Multinomial and Naïve Bayes Multinomial Updateable have the same performance.

To rank these 7 algorithms from the best performant to the least performant, for each corpus we rated each algorithm with a score of 5 to 1 based on their percent of correctly instances classified. Three of the algorithms, Complement Naïve Bayes, Naïve Bayes Multinomial and Naïve Bayes Multinomial Updateable have the same percent of correctly classified instances so we decided to rate them with the same score. We used this score to calculate the weighted average for each algorithm in terms of percent of correctly classified instances. For example, for corpus C1 we ordered the algorithms based on their performance in term of percent of correctly classified instances and then each of them was rated with points. So, the best performing algorithms are Complement Naïve Bayes, Naïve Bayes Multinomial and Naïve Bayes Multinomial Updateable rated with 5 points. The next performing algorithm is SGD rated with 4 points. Then comes RBF Classifier rated with 3 points and Logistic rated with 2 points. And the least performing algorithm is Hyper Pipes rated with 1 point. We applied the same ranking scheme to each corpus. The ranking scheme results and the weighted average of percent of correctly classified instances for each algorithm are shown in Table 3.

Table 3. The rank of best performing algorithms in terms of weighted average of percent of correctly classified instances.

Algorithm	C1	C2	C3	C4	C5	C6	C7	C8	C9	C10	C11	WA[a]
ComplementNaiveBayes	78.6*5	80.3*5	81.7*5	80*5	75*3	83*5	85.7*5	81.7*5	75*2	70.5*2	81.5*3	80.33
NaiveBayesMultinomial	78.6*5	80.3*5	81.7*5	80*5	75*3	83*5	85.7*5	81.7*5	75*2	70.5*2	81.5*3	80.33
NaiveBayesMultinomial Updateable	78.6*5	80.3*5	81.7*5	80*5	75*3	83*5	85.7*5	81.7*5	75*2	70.5*2	81.5*3	80.33
Logistic	66.6*2	78.6*4	76.7*3	77.5*2	78*4	77*3	82.3*4	79.3*4	80.5*5	73*4	84*4	78.29
SGD	77.4*4	77.5*3	77*4	75*1	66*2	77.3*4	80*3	76.3*3	75.5*3	71*3	84.5*5	77.06
HyperPipes	55.8*1	62.3*1	64.7*1	78.5*4	78*4	63.5*2	71.3*2	66.7*2	78.5*4	78*5	74*2	73.70
RBFNetwork	73.4*3	74.8*2	76.7*2	78*3	79*5	51.8*1	49.3*1	50.7*1	49.5*1	47*1	54*1	69.25

[a]WA is Weighted Average

The algorithms are listed from the algorithm that have the highest weighted average of percent of correctly classified instances to the lower one. Complement Naïve Bayes, Naïve Bayes Multinomial and Naïve Bayes Multinomial Updateable are the best performing algorithm with 80.33% weighted average of percent of correctly classified instances, followed by Logistic with a difference of 2.04%. Then comes SGD algorithm and Hyper Pipes. And the least performing algorithm is RBF Network.

We performed another experiment to identify any statistical difference between the best performant algorithms using as comparison field the percent_correct. Based on the result of Table 3, we choose Complement Naïve Bayes as a base algorithm, and compare it with the other six. This experiment was performed for each corpus using the Weka Experimenter Tool with the seven best performant algorithms, 10 cross-validation and 10 repetitions. In Table 4 are shown the results of this experiment.

Table 4. Statistical experiment results in term of percent_correct for each algorithm.

Algorithm	C1	C2	C3	C4	C5	C6	C7	C8	C9	C10	C11
ComplementNaiveBayes	76.86	79.32	81.13	79.35	76.4	83	85.97	79.4	75.25	70.5	82.7
NaiveBayesMultinomial	76.86	79.32	81.13	79.35	76.4	83	85.97	79.4	75.25	70.5	82.7
NaiveBayesMultinomialUpdate	76.86	79.32	81.13	79.35	76.4	83	85.97	79.4	75.25	70.5	82.7
Logistic	64.3*	74.2	76.03	75.45	79.5	79.17	82.9	77.9	78.80	73.9	81.3
SGD	76.86	78.17	76.93	75.75	68.2	78.2*	81.57	76.83	76.30	71.3	82.9
HyperPipes	55.2*	62.7*	64.8*	78.15	82	63.5*	70.7*	66.6*	78.2	76.35	74.6*
RBFNetwork	72.06	74.65	74.5*	78.65	82.9	74.6*	77.6*	73.1*	83.2v	78.3v	78.75

For corpus C4 and C5, the experimental results indicate that there are no important differences in performance between the algorithms. The results of: Logistic for corpus C1, SGD for corpus C6, Hyper Pipes for corpuses C1, C2, C3, C6, C7, C8 and C11, and RBF Network for corpuses C3, C6, C7 and C8 have a "*" meaning that these algorithms have performed statistically significantly worse than Complement Naïve Bayes. But, the results of RBF Network for corpus C9 and C10 has a "v". This means that RBF Network has performed statistically significantly better than Complement Naïve Bayes. Even with this experiment, we cannot define which algorithm perform statistically better.

5 Conclusions and Future Work

We evaluated through experiments the performance of classification algorithms for opinion mining in a multi-domain corpus in Albanian language. We created 11 text corpuses of Albanian written opinions collected from different well-known Albanian newspapers. Each corpus has the same number of text documents categorized as positive opinions and text documents categorized as negative opinions. Firstly, the dataset passed in a preprocessing phase composed by a stop-word removal and a stemmer. Then these unstructured data are used as an input to train and to test 50 classification algorithms implemented in Weka.

The experimental results show that there are 7 different best performing algorithms in terms of percent of correctly classified instances. The best performing algorithm are: Complement Naïve Bayes, Naïve Bayes Multinomial, Naïve Bayes Multinomial Updateable, RBF Network, Logistic, Hyper Pipes and SGD. Three of these best performing algorithms, Naïve Bayes, Naïve Bayes Multinomial and Naïve Bayes Multinomial Updateable have the same performance in all the corpuses. The results of the second experiment show that there is not any statistical difference between the best performing algorithms using as comparison field the percent_correct. We can note that the performance of the classification algorithms for opinion mining depends on the topic and the number of the opinions. Also, based in our previous work in [15] we can note that the algorithms perform better in in-domain corpus than in multi-domain corpus.

As a future work, it would be interesting to evaluate the performance of these classification algorithms in a bigger corpus in Albanian language and using features that can improve their performance. Also, we can evaluate the performance of these algorithms in cross-domain opinion mining.

References

1. Liu, B.: Sentiment Analysis and Opinion Mining. Morgan & Claypool Publishers (2012)
2. Hastie, T., Tibshirani, R., Friedman, J.: Unsupervised Learning. In: Hastie, T., Friedman, J., Tibshirani, R. (eds.) The Elements of Statistical Learning. SSS, pp. 485–585. Springer, New York (2009). https://doi.org/10.1007/978-0-387-84858-7_14
3. Unnisa, M., Ameen, A., Raziuddin, S.: Opinion mining on Twitter data using unsupervised learning technique. Int. J. Comput. Appl. 148(12), 12–19 (2016). https://doi.org/10.5120/ijca2016911317
4. Gautam, G., Yadav, D.: Sentiment analysis of twitter data using machine learning approaches and semantic analysis. In: Seventh International Conference on Contemporary Computing (IC3), pp. 437–442 (2014)
5. Pang, B., Lee, L., Vaithyanathan, S.: Thumbs up? Sentiment classification using machine learning techniques. In: Proceedings of the ACL-2002 Conference on Empirical Methods in Natural Language Processing, vol. 10, pp. 79–86 (2002). https://doi.org/10.3115/1118693.1118704
6. Tripathy, A., Agrawal, A., Rath, S.K.: Classification of sentiment reviews using n-gram machine learning approach. Expert Syst. Appl. 57(C), 117–126 (2016). https://doi.org/10.1016/j.eswa.2016.03.028
7. Perea-Ortega, M.J., Martín-Valdivia, T.M., Ureña-López, A.L., Martínez-Cámara, E.: Improving polarity classification of bilingual parallel corpora combining machine learning and semantic orientation approaches. J. Assoc. Inf. Sci. Technol. 64(9), 1864–1877 (2013). https://doi.org/10.1002/asi.22884
8. Li, S.S., Huang, C.R., Zong, C.Q.: Multi-domain sentiment classification with classifier combination. J. Comput. Sci. Technol. 26(1), 25–33 (2011). https://doi.org/10.1007/s11390-011-9412-y
9. Wu, F., Huang, Y.: Collaborative multi-domain sentiment classification data mining. In: IEEE International Conference on Data Mining (2015). https://doi.org/10.1109/icdm.2015.68

10. Lommatzsch, A., Butow F., Ploch, D., Albayrak, S.: Towards the automatic sentiment analysis of German news and forum documents. In: 17th International Conference on Innovations for Community Services, pp. 18–33 (2017). https://doi.org/10.1007/978-3-319-60447-3_2

11. Farra, N., Challita, E., Assi, A.R., Hajj, H.: Sentence-level and document-level sentiment mining for Arabic texts. In: IEEE International Conference on Data Mining Workshops (ICDMW) (2010). https://doi.org/10.1109/icdmw.2010.95

12. Miranda, H.C., Guzmán, J.: A review of sentiment analysis in Spanish. Tecciencia, 12(22), 35–48 (2017). Bogotá. http://dx.doi.org/10.18180/tecciencia.2017.22.5

13. Catal, C., Nangir, M.: A sentiment classification model based on multiple classifiers. Appl. Soft Comput. J. 50, 135–141 (2017). https://doi.org/10.1016/j.asoc.2016.11.022

14. Biba, M., Mane, M.: Sentiment analysis through machine learning: an experimental evaluation for Albanian. In: Thampi, S., Abraham, A., Pal, S., Rodriguez, J. (eds.) Recent Advances in Intelligent Informatics, Part of the Advances in Intelligent Systems and Computing (AISC), vol. 235, pp. 195–203. Springer, Cham (2014). doi: https://doi.org/10.1007/978-3-319-01778-5_20

15. Kote, N., Biba, M., Trandafili, E.: A thorough experimental evaluation of algorithms for opinion mining in Albanian. In: Barolli, L., Xhafa, F., Javaid, N., Spaho, E., Kolici, V. (eds.) EIDWT 2018: Advances in Internet, Data & Web Technologies, vol. 17, pp. 525–536. Springer, Cham (2018). https://doi.org/10.1007/978-3-319-75928-9_47

16. Sadiku, J., Biba, M.: Automatic stemming of Albanian through a rule-based approach. J. Int. Res. Publ. Lang. Individ. Soc. 6 (2012). ISSN-1313-2547

17. Witten, H.I., Frank, E., Hall, A.M., Pal, J.C.: Data Mining, Practical Machine Learning Tools and Techniques, 4th edn. Elsevier Inc., Amsterdam (2017). ISBN: 9780128042915

Predicting Author's Native Language Using Abstracts of Scholarly Papers

Takahiro Baba[1]([✉])[ID], Kensuke Baba[2][ID], and Daisuke Ikeda[1]

[1] Kyushu University, Fukuoka 819-0395, Japan
`takahiro.baba@inf.kyushu-u.ac.jp`
[2] Fujitsu Laboratories, Kawasaki 211-8588, Japan

Abstract. Predicting author's attributes is useful for understanding implicit meanings of documents. The target problem of this paper is predicting author's native language for each document. The authors of this paper used surface-level features of documents for the problem and tried to clarify the practical tendencies of the writing style as word occurrences. They conducted a classification of the abstracts written in English of approximately 85,000 scholarly papers written in English or in Japanese. As a result of the experiment, the accuracy of the binary classification was 0.97, and they found that a number of distinctive phrases used in the classification were related to typical writing styles of Japanese.

Keywords: Native language identification · Document classification
Text analysis · Machine learning

1 Introduction

Profiling authors of documents is useful for advanced analyses of the documents. Understanding the intention or emotion of authors is expected to yield effective methods for various tasks of document analysis, such as machine translation and dialog generation. Author's attributes and extra knowledge related to them can be used to understand implicit meanings of documents.

The target problem of this paper is native language identification [6] of the author of each document. The linguistic background of a person may be affected by the history and culture of the society concerned, and hence it can include useful information we can not extract directly from documents. We assumed that we could use only the text data of documents themselves as the features for native language identification, while possible features includes information of eye movements of subjects in reading documents [3].

We used surface-level features for native language identification and tried to clarify the practical tendencies of the writing style as word occurrences. Ionescu et al. [4] point out that approaches based on the surface-level features are effective for the task. We took a simple approach based on word n-gram and support vector machine (SVM), and concentrated on clarifying the practical tendencies.

© Springer Nature Switzerland AG 2018
M. Ceci et al. (Eds.): ISMIS 2018, LNAI 11177, pp. 448–453, 2018.
https://doi.org/10.1007/978-3-030-01851-1_43

We used a linear kernel for constructing an SVM to capture distinctive features for the classification as phrases used in documents. The aim of this study is to explain the results of a standard machine learning method with practical data in terms of human knowledge as a case study, rather than to develop a novel method for native language identification.

We conducted a classification of abstracts of scholarly papers in medical science published in international journals and those in domestic journals in Japan. On the assumption that author's native language of a Japanese paper is Japanese, the classification is regarded as a prediction of whether author's native language is Japanese. This binary classification over a vector space model based on word occurrences is clearly extendable to a multi-class classification. Using the result of the classification, we gave considerations to a number of phrases which had extremely large or small coefficients in the classification.

As a result of our experiments, we can conclude that author's native language can be detected from the style of writing. We found that the classification of the abstracts of English and Japanese papers could be conducted with high accuracy, and distinctive phrases used in the classification were related to some typical writing styles of Japanese. The classification can be extended to other languages, and the tendencies of writing styles can be used for understanding authors' intentions in documents.

The rest of this paper is organized as follows. Section 2 formalizes the target problem as a classification of documents, and describes the experimental methods. Section 3 reports the experimental results. Section 4 gives considerations on the results and future directions of our study.

2   Methods

This section formalizes the problem of native language identification as a classification of abstracts of scholarly papers, and describes the experimental methods.

2.1   Data

We used abstracts of English papers and Japanese papers (whose abstracts were written in English) in medical science. The classification is regarded as a prediction of whether author's native language is Japanese.

The experimental data were generated as follows. We obtained the metadata from Europe PubMed Central (Europe PMC) [1]. For generating samples of English papers, we used the abstracts of papers published in the following journals:

- The Journal of the American Medical Association,
- Annals of Internal Medicine,
- Archives of Internal Medicine,
- British Medical Journal,
- The Lancet,

- Nature Medicine, and
- The New England Journal of Medicine.

For Japanese papers, we used the result of a search in Europe PMC with the condition that the language is "Japanese". We selected the abstracts of papers published from 1991 to 2015. The numbers of the abstracts of English and Japanese papers were 42,685 and 106,413, respectively, and we randomly selected 42,685 papers from the abstracts of Japanese papers. Then, we randomly divided them into 68,296 (80%) training data and 17,074 (20%) test data equally for the two classes.

2.2 Experiments

We conducted the classification with the data set as follows. We applied the linear SVM classifier to the features generated using the 2-, 3-, and 4-grams of abstracts. An n-gram of a given sequence is a sequence of contiguous n elements of the sequence. For example, the word-level 2- and 3-grams of a document "I am your father" are "I am", "am your", "your father", "I am your", and "am your father". Then, we predicted "English" or "Japanese" for the test data in the following conditions:

- We distinguished between upper- and lower-case letters;
- We deleted the phrases "Japan" and "Japanese" from the abstracts because the phrases can be distinctive of the class of Japanese papers trivially;
- We deleted the words used for generating a format of abstracts, such as the caption "OBJECTIVES:";
- We didn't use the phrases that appeared in more than 50% of the training data for the classification.

The size of the set of the 2-, 3-, and 4-grams of the total abstracts was 22,063,049.

The *accuracy* is defined to be the ratio of the number of the correct predictions to the number of the total predictions examined in a test. The *precision* and the *recall* are defined for each class to be the ratio of the number of the correct predictions to a class to the number of the predictions to the class, and the ratio of the number of the correct predictions to a class to the actual number of samples of the class, respectively.

3   Results

The accuracy of the classification of the abstracts of English and Japanese papers was 0.97. Table 1 shows the confusion matrix of the classification. The two classes could be classified with a high accuracy.

Table 2 shows the top and bottom 20 phrases in the order of coefficients of the support vector used in the classifier. The phrases with large (or small) coefficients are supposed to be distinctive of papers written in Japanese (resp. English). For example, the phrase "such as" is distinctive of abstracts of Japanese papers.

Table 1. Confusion matrix of the classification of English and Japanese papers.

Predicted class\Actual class	English	Japanese	Precision
English	8,301	245	0.97
Japanese	236	8,292	0.97
Recall	0.97	0.97	

4 Discussion

This section makes examinations of the experimental results.

4.1 Main Conclusion

We can conclude that author's native language of an abstract can be detected using its style of writing. We found that abstracts written in English of Japanese papers could be distinguished from abstracts of English papers in medical science. The classification accuracy was extremely high even though we used a simple classifier and straightforward features of the text data. Most of the phrases distinctive of each class were not related to research topics, hence this classification method can be applied to other research areas at least in the case of predicting Japanese as author's native language.

4.2 Key Findings

We give detailed considerations to some distinctive phrases shown in Table 2 from the viewpoint of the style of writing.

(1) "Here we" Using "Here" at the beginning of a sentence is not familiar to Japanese. This phrase might be replaced by "In this paper" or "In this study" in abstracts of Japanese papers because "In this" was a distinctive phrase of Japanese papers. Actually, the two phrases appeared in 3,355 abstracts (3.2%) of 106,413 Japanese papers, while they appeared in 364 abstracts (0.9%) of 42,685 English papers.

(2) "Such as" The phrase "such as" is often used in error by Japanese. According to [5], in scholarly papers written by Japanese "such as" is used wrongly instead of "in the same manner as", "similar to", "in the following way", "including", and so on. The cause is supposed that "such as" is translated to the Japanese popular word "yo (no-yo-ni, no-yo-na)" which is used when giving examples and also when using similes. The phrase "such as" appeared 12.7% of the abstracts of Japanese papers and 6.6% of that of English papers.

Table 2. Phrases with large and small coefficients in the classification of abstracts of papers written in English and Japanese, where the superscripts refer the subsections that give considerations to the phrases in Subsect. 4.2.

Distinctive of English papers	Distinctive of Japanese papers
Here we[1]	such as[2]
lt 001[4]	year old[4]
95 CI[4]	It is
show that	and or
compared with[3]	was performed
To determine	order to
health care	In this[1]
associated with	In conclusion
confidence interval[4]	due to
95 confidence[4]	have been
we show	revealed that
95 confidence interval[4]	In addition
mm Hg	of them
use of	useful for
at least	as follows
We describe	compared to[3]
that are	is useful
we show that	it is
evidence of	It was
used to	the present

(3) "Compared with/to" We suppose that the clear difference between the usages of "compared with" and "compared to" was caused by the subtle difference between the two phrases. According to [2, p. 354], "compared with" is used when estimating the similarity or dissimilarity between two things, and "compared to" is used when describing the similarity between two different things, although the difference is not clear in practice. It is supposed that using "with" is better than "to" in most cases using "compared" in abstracts of scientific papers, and it is difficult for Japanese (and other non-native English speakers) to distinguish the two cases. The ratio of the number of the abstracts includes "compared with" to that of "compared to" was 56.1 in English papers and 2.1 in Japanese papers.

(4) Research areas Some distinctive phrases are supposed to be caused by the difference of research areas. For example, the phrases: "95 CI", "confidence interval", "95 confidence", and "95 confidence interval", distinctive of English papers mean the frequent occurrences of "95% confidence interval (95% CI)", which (also "lt 001" for "< 0.01") is related to statistical analyses, while the phrase "year old" distinctive of Japanese papers was caused by the occurrences of "* year-old man/woman", which is supposed to be related to case reports.

4.3 Future Directions

Applying the classification to other language is one of our future work. As a technical issue, a multi-class classification can be generated by combining binary classifier for each class. When we apply resulting knowledge to practical systems, such as, automatic proofreading of scholarly papers and e-learning for the second language, the difficulty is that we have to examine the relation between a classification result and the tendency in the practical writing style of the target language. We need to carry on interdisciplinary research.

Improving the classifier is another future direction. We used the simple classifier based on SVM and the straightforward features based on word occurrences in documents. We can simply apply other machine learning methods to the classification and examine the effects on the accuracy. Also we need to be able to explain the classifier in terms of the writing style.

5 Conclusion

Author's native language of a scholarly paper can be detected with a high accuracy from the style of writing in its abstract. We classified English abstracts of English papers and Japanese papers in medical science, and the classification accuracy was 0.97. A number of distinctive phrases were reasonable in the sense of tendencies in the writing style of Japanese. The result is expected to be applied to various intelligent systems including e-learning for the second language.

Acknowledgement. This work was supported by JSPS KAKENHI Grant Number 15H02787.

References

1. Europe PMC: Europe PubMed Central. https://europepmc.org/. Accessed 5 Feb 2018
2. Oxford Dictionary of English. Oxford University Press (2010)
3. Berzak, Y., Nakamura, C., Flynn, S., Katz, B.: Predicting native language from gaze. In: Proceedings of the 55th Annual Meeting of the Association for Computational Linguistics, pp. 541–551 (2017)
4. Ionescu, R.T., Popescu, M., Cahill, A.: String kernels for native language identification: insights from behind the curtains. Comput. Linguist. **42**(3), 491–525 (2016)
5. Paquette, G.: English Composition for Scholarly Works (in Japanese). Kyoto University Press, Kyoto (2004)
6. Wong, S.-M.J., Dras, M.: Exploiting parse structures for native language identification. In: Proceedings of the Conference on Empirical Methods in Natural Language Processing, EMNLP 2011, pp. 1600–1610. Association for Computational Linguistics, Stroudsburg (2011)

Identifying Exceptional Descriptions of People Using Topic Modeling and Subgroup Discovery

Andrew T. Hendrickson, Jason Wang, and Martin Atzmueller[✉]

Tilburg University, 5037AB Tilburg, The Netherlands
{a.hendrickson,y.w.wang,m.atzmuller}@uvt.nl

Abstract. Descriptions of images form the backbone for many intelligent systems, assuming descriptions that randomly vary in construction and content, but where description content is homogeneous. This assumption becomes problematic being extended to descriptions of images of *people* [14], where people are known to show systematic biases in how they process others [19]. Therefore, this paper presents a novel approach for discovering exceptional subgroups of descriptions in which the content of those descriptions reliably differs from the general set of descriptions. We develop a novel interestingness measure for subgroup discovery appropriate for probability distributions across semantic representations. The proposed method is applied to a web-based experiment in which 500 raters describe images of 200 people. Our analysis identifies multiple exceptional subgroups and the attributes of the respective raters and images. We further discuss implications for intelligent systems.

1 Introduction

The fields of machine learning and computational linguistics are increasingly focused on building multi-modal intelligent systems that integrate visual and text-based information, e. g., answering questions and generating textual descriptions of images [2,9,16], identifying objects in images based on text descriptions [20], or improving the sentence parsing of descriptions by grounding them with visual information [10]. While the data rely on text descriptions of images written by people, people generate idiosyncratic descriptions that differ significantly based on the goals, biases, and expertise of the person writing the description [14]. Detecting and quantifying these idiosyncratic descriptions is necessary for systems to optimally select, weigh, or filter descriptions. In this paper we present a novel approach for identifying homogeneous subgroups of descriptions that reliably differ in their content from the general set of descriptions. Specifically, we extract a low-dimensionality representation of individual descriptions based on latent Dirichlet allocation (LDA) [8]. Using that representation, we apply *exceptional model mining* [3,11,18], a variant of subgroup discovery [3] that focuses on complex target properties, for detecting homogeneous subgroups of descriptions in the low-dimensional space. We define a novel quality function based on a

© Springer Nature Switzerland AG 2018
M. Ceci et al. (Eds.): ISMIS 2018, LNAI 11177, pp. 454–462, 2018.
https://doi.org/10.1007/978-3-030-01851-1_44

subgroups' topic distribution and use it to identify exceptionally unique homogeneous subgroups of textual descriptions. We believe this is the first demonstration of subgroup discovery based on such textual description data.

The efficacy of the proposed approach is validated on a new dataset of descriptions of people. We present results of applying the LDA-based exceptional model mining method on that dataset and discuss the implications for intelligent systems based on descriptions generated by people in general, and descriptions of people in particular. The contribution of the paper is summarized as follows:

1. We present a novel approach for mining exceptional subgroups in descriptions of people using subgroup discovery on topic models using LDA.
2. We introduce a new interestingness measure for subgroups that compares the distribution across topics in subgroups to the overall (expected) distribution.
3. We present and discuss the results of applying the proposed novel methodology to a real-world dataset of descriptions of people collected online.

2 Method

The proposed approach consists of three phases. First, textual data is transformed into a low dimensional space using latent Dirichelet allocation. Second, a novel interestingness measure for subgroup discovery is used to define and search for exceptional subgroups. Finally, the resulting subgroups are evaluated with a human-in-the-loop, in order to facilitate their interpretation and validation.

2.1 Topic Modeling

The latent Dirichlet allocation model (LDA) [8] is the most popular method of topic modeling in natural language processing. It is a statistical model of how text documents are generated that relies on the assumption that a written text can be represented as a collection of topics where each topic consists of a probability distribution across all possible words. Formally, the generative model for the j_{th} word w_{ij} (and its topic z_{ij}) in document i, given a distribution of topics θ_i in document i and distribution of words φ_k in topic k, is: (1) $\theta_i \sim Dirichlet(\alpha)$, (2) $\varphi_k \sim Dirichlet(\beta)$, (3) $z_{ij} \sim Multinomial(\theta_i)$, (4) $w_{ij} \sim Multinomial(\varphi_{z_{i,j}})$, where the number of topics k and the vectors of identical values α and β are hyperparameters of the model.

2.2 Subgroup Discovery

Formally, a *database* $D = (I, A)$ is given by a set of individuals I and a set of attributes A. For nominal attributes, a *selector* or *basic pattern* $(a_i = v_j)$ is a Boolean function $I \to \{0, 1\}$ that is true if the value of attribute $a_i \in A$ is equal to v_j for the respective individual. The set of all basic patterns is denoted by Σ. A subgroup is described using a description language, typically consisting of attribute–value pairs. Here, we focus on an exemplary conjunctive pattern description language. A *subgroup description* or (complex) *pattern* P is then given by a set of basic patterns $P = \{sel_1, \ldots, sel_l\}$, $sel_i \in \Sigma, i = 1, \ldots, l$, which is interpreted as a conjunction, i.e., $P(I) = sel_1 \wedge \ldots \wedge sel_l$, with $length(P) = l$.

A *subgroup* $S_P := ext(P) := \{i \in I | P(i) = true\}$, i.e., a *pattern cover* is the set of all individuals that are covered by the subgroup description P. The set of all possible subgroup description is then given by 2^Σ. The pattern $P = \emptyset$ covers all instances contained in the database. A *quality function* $q: 2^\Sigma \rightarrow \mathbb{R}$ maps every pattern to a real number reflecting its interestingness. In contrast to related approaches like methods for mining association rules [1] or algorithms from the field of formal concept analysis [13], subgroup discovery (and in particular exceptional model mining) allow the specification and efficient application of complex quality functions for estimating the interestingness of a pattern, e. g., [3,5,17,18].

In the case of topic models, we utilize Dirichlet distributions capturing the overall distribution of topics in the overall dataset (i.e., modeling the expected distribution) while the topic distribution contained in each subgroup is modeled by another Dirichlet distribution. In order to obtain the respective Dirichlet distributions $Dir(\alpha_\emptyset)$ and $Dir(\alpha_{S_P})$ for the overall population $S_\emptyset$ and the subgroup S_P, respectively, we can compute a maximum likelihood estimate (MLE) utilizing the Newton-Raphson method [21,22] for obtaining the parameter vectors α_{S_P} and $\alpha_\emptyset$. For comparing distributions, we utilize the Kullback-Leibler divergence metric KL. Thus, for Dirichlet distributions, comparing $Dir(\alpha)$ and $Dir(\beta)$, we obtain

$$KL(\alpha, \beta) = \log \frac{\Gamma(\alpha_0)}{\Gamma(\beta_0)} - \sum_{i=1}^{k} \left(\log \frac{\Gamma(\alpha_i)}{\Gamma(\beta_i)} - (\alpha_i - \beta_i)(\psi(\alpha_i) - \psi(\alpha_0)) \right), \quad (1)$$

here Γ is the gamma and ψ the digamma function, $\alpha_0 = \sum_{i=1}^{k} \alpha_i, \beta_0 = \sum_{i=1}^{k} \beta_i$.
Our novel quality function for comparing topic distributions for a specific subgroup S_P is then given by

$$q_D(P) = KL(\alpha_{S_P}, \alpha_\emptyset), \quad (2)$$

with the distribution parameters α_{S_P} and $\alpha_\emptyset$ for subgroup/overall population.

3   Experiment

In this section we detail the critical aspects of an online experiment to collect descriptions and judgments about images of people. In the following sections we describe the set of images (i.e., the stimuli) and outline the experimental procedure for obtaining the textual descriptions and ratings of the images.

3.1   Procedure

The images consisted of 193 color photographs of the head and shoulders of people standing in front of an off-white background who were instructed to look directly into the camera lens and maintain a neutral facial expression. These individuals were recruited from the University of Adelaide campus and compensated AUD\$10 for their participation.

For the rating task, 500 participants were recruited via Amazon Mechanical Turk and paid US\$2 for approximately 12 min of work. Participants were shown five randomly selected images and for each image they were asked to determine a number of attributes and write a physical and non-physical description of the face (minimum four words and 10 characters). In this analysis we focus on discovering exceptional descriptions of the non-physical characteristics. The attributes of each description include three self-report attributes about the specific rater (age, gender, and country) as well as four subjectively rated attributes of the person in the image (age, gender, eye color, hair color, typicality, and attractiveness). Unfortunately, the reported ethnicity was incorrectly coded and not recorded, resulting in seven attributes for each description. The experiment resulted in a dataset consisting of 2491 descriptions of 193 faces.

3.2   Discovering Subgroups

The application of the text-based subgroup discovery consisted of multiple stages:

1. The number of topics k and the probability distribution across words for each topic φ_k was determined. This was done by searching for the set of hyperparameters (α, β, and k) that produce sparse topics (few topics on average per document) that differ across documents. Thus, the objective function simultaneously minimized the number of topics per description and maximized the difference between documents[1]. In this phase all descriptions of the same image were treated as a single document to increase the stability of the inference process, particularly determining the topic distributions (φ).
2. These topic distributions were used to construct a probability distribution across topics for each individual description. The search processes in the first step resulted in a solution with nine topics and thus each of the 2491 descriptions was represented as a probability distribution across those topics.
3. Finally, all possible subgroups defined by the seven attributes were evaluated to identify deviating subgroups of descriptions. This was done exhaustively using the SD-Map algorithm [6], provided by the VIKAMINE system [4][2].

We identified deviating subgroups using different values of n for identifying the top-n subgroups, while we discuss results for the top 20 subgroups below; other result sets were consistent. A *minimal improvement filter* [7] was applied to the set of all subgroups to limit the set of attributes defining exceptional

---

[1] The difference between documents was calculated as the sum across all pairs of descriptions of the cosine similarity of the topic probability distributions. The number of topics per document was calculated as the sum across all descriptions of the conditional entropy of the topic probability distribution.

[2] http://www.vikamine.org.

subgroups. Specifically, a specialization P' of a pattern P is considered a more exceptional subgroup if P' improves on the quality function compared to P: So, e.g., we consider the specialization of the pattern *face_hair_color = black* to *face_hair_color = black AND face_gender = male*, if the quality of the latter pattern increases. A minimal subgroup size threshold of 1% was used in this analysis.

4   Results and Discussion

Below, we first present an analysis of the types of attributes more likely to define deviating subgroups and their relationships. Next, we aggregate across exceptional subgroups and evaluate the frequency of specific images in exceptional subgroups and according raters. We also briefly summarize results with an alternative quality function, and conclude with a discussion.

Table 1. Exceptional subgroup attribute frequencies. The attributes and values of a description that are indicative of the top 20 exceptional subgroups. The count column indicates the number of subgroups that are distinguished by this attribute. R and I in the column headers indicate Rater and Image attributes, respectively.

R. Country	R. Gender	I. Gender	I. Eye Color	I. Hair Color	I. Ratings
USA (3) India (2)	Female (3) Male (7)	Female (8) Male (3)	Black (12) Brown (1) Green (1)	Black (6) Blond (1)	Typicality (4) Attract. (5)

4.1   The Attributes of Distinct Subgroups

Interesting patterns emerge across the seven attributes that identify the 20 subgroups most dissimilar to the overall set of descriptions (Table 1). Though the gender of the rater and the face in the image were two of the most common attributes to define exceptional subgroups, only four subgroups were defined by both the gender of the rater and face. This suggests the interaction between rater and face gender was no more likely to identify an exceptional subgroup than either factor independently. Furthermore, the eye color of the image was a defining attribute of 14 of 20 exceptional subgroups and the attribute value "black" was by far the most frequently occurring value. This is particularly surprising given that eye color attribute was described as "black" for only 13% of descriptions, which was second frequent after "brown" and more frequent than "blue" and "green." One possible explanation of this pattern is that perceived eye color is highly correlated with perceived ethnicity or race for these descriptions, a possibility we discuss further in the discussion section.

4.2   Distributions of Images and Raters in Distinct Subgroups

Figure 1 shows the markedly different distribution of images (left panel) and raters (right panel) that occur in distinct subgroups. The majority of images occur in at least one subgroup, but only nine images (0.4.%) had at least 40% of the descriptions of them included in a subgroup and only one image had more than 50%. The majority of raters, in contrast, do not have a description in any deviating subgroup. This is natural given that raters only generate five descriptions but each image has, on average, 12.9 descriptions. Despite the high number of raters with zero descriptions in the exceptional subgroups, a small but significant group of raters (5.2%) had more than 50% of their descriptions be identified as belonging to at least one deviating subgroup.

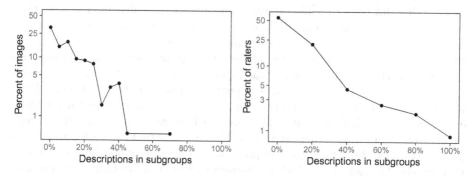

Fig. 1. The proportion of descriptions of specific images (left) and raters (right) that occur in at least one exceptional subgroup. The y-axis of both plots is in log units.

4.3   Comparing the Dirichlet and Hotelling Quality Functions

In all previous analyses, we utilized the proposed quality function $q_D(P)$. These results were also compared with the standard Hotelling quality function [3], for comparing multivariate means: The two quality functions were not a significantly correlated ($R = 0.12$). Furthermore, the Hotelling quality function did not produce as coherent subgroup attributes as the novel quality function $q_D(P)$. This divergence highlights the importance of using a quality function ($q_D(P)$) that directly corresponds to the multinomial probability distribution that comprises the probability distribution representation of a description across topics.

4.4   Discussion

Our results suggest that descriptions of people that significantly deviate from the population of descriptions are relatively frequent. Furthermore, these exceptional descriptions are not exclusively driven by particularly exceptional images or particularly exceptional raters. Instead, the vast majority of descriptions that are identified as exceptional are descriptions from raters for whom most descriptions are not exceptional, and of images whose descriptions are mostly not exceptional.

The attributes that define the maximally deviating subgroups point to the types of features of raters and images that are likely to produce exceptional descriptions. Male raters and female images are attributes that are likely to define deviating subgroups, though these two attributes appear to independently contribute and not in combination. Additionally, in this population of images, black hair and black eyes are the attributes of images most likely to identify exceptional subgroups. Further work is necessary to understand if these attributes produce exceptional descriptions when embedded in a different sample of images, or if these attributes are predictive of latent attributes, such as ethnicity, that were not included as attributes for the subgroup discovery process.

The issue of detecting heterogeneity of descriptions is particularly important for descriptions of people because people systematically differ in how they encode, search for, and remember faces of other races and genders [15,19], which may systematically bias the descriptions people generate of others. However, these results do not show strong evidence of subgroups of descriptions that are identified based on gender, age, or race, perhaps decreasing the fear that intelligent systems based on descriptions of people will inherit strong implicit biases from the raters [12,23]. We do find certain attributes of images, particularly black eye color and black hair color, which are much more likely to produce exceptional descriptions than other attributes. This suggests that the topics and words people use when describing the non-physical characteristics of other people may vary widely. The degree to which the importance of these attributes are an artifact of the particular faces we studied is an open question, but it highlights the importance of not ignoring the heterogeneity of textual descriptions generated by people. These issues are increasingly important as intelligent systems, trained with labels and descriptions generated by people, become ubiquitous. These systems rely on human annotated descriptions that are clearly not homogeneous. When combined with methods like LDA for extracting lower dimensional semantic representations, exceptional model mining and subgroup discovery techniques can provide a necessary tool to help identify potential biases in these descriptions. Additionally, these tools can possibly suggest specific images, subgroups, and attributes where additional data would help alleviate the bias in the systems that rely on them.

5   Conclusions

This paper presents a novel method of combining topic modeling and subgroup discovery to identify interesting image descriptions. We present a novel definition of interestingness that compares the subgroup and general population using the Kullback-Leibler divergence between the Dirichlet distributions that characterizes the probability distribution of topics. This method is applied to the problem of subgroup discovery among descriptions of pictures of people, a domain that has broad implications for applied domains [14] while carrying a real risk of biased descriptions [19]. Our analysis method detects meaningful subgroups of image descriptions that diverge from the general set of descriptions and characterizes them based on both properties of the raters as well as the images. These subgroups suggests new norms for data collection methods and statistical models for web-based applications that are sensitive to the heterogeneous nature of descriptions of people. For future work, we aim to extend the analysis and data collection in order to investigate (dis-)similarities in more datasets. Furthermore, the inclusion of contextual domain knowledge is an interesting issue to consider.

Acknowledgments. Funding for data collection was provided by a University of Adelaide Interdisciplinary Research Grant to C. Semmler, A. Hendrickson, R Heyer, A. Dick and A. van den Hengel. Furthermore, this work has also been partially supported by the German Research Council (DFG) under grant AT 88/4-1.

References

1. Agrawal, R., Srikant, R.: Fast algorithms for mining association rules. In: Proceedings of VLDB, pp. 487–499. Morgan Kaufmann (1994)
2. Antol, S., et al.: VQA: visual question answering. In: Proceedings of IEEE ICCV, pp. 2425–2433 (2015)
3. Atzmueller, M.: Subgroup discovery. WIREs DMKD **5**(1), 35–49 (2015)
4. Atzmueller, M., Lemmerich, F.: VIKAMINE – open-source subgroup discovery, pattern mining, and analytics. In: Flach, P.A., De Bie, T., Cristianini, N. (eds.) ECML PKDD 2012. LNCS (LNAI), vol. 7524, pp. 842–845. Springer, Heidelberg (2012). https://doi.org/10.1007/978-3-642-33486-3_60
5. Atzmueller, M., Lemmerich, F.: Exploratory pattern mining on social media using geo-references and social tagging information. IJWS **2**(1/2), 80–112 (2013)
6. Atzmueller, M., Puppe, F.: SD-Map – a fast algorithm for exhaustive subgroup discovery. In: Fürnkranz, J., Scheffer, T., Spiliopoulou, M. (eds.) PKDD 2006. LNCS (LNAI), vol. 4213, pp. 6–17. Springer, Heidelberg (2006). https://doi.org/10.1007/11871637_6
7. Bayardo, R., Agrawal, R., Gunopulos, D.: Constraint-based rule mining in large, dense databases. Data Min. Knowl. Discov. **4**, 217–240 (2000)
8. Blei, D.M., Ng, A.Y., Jordan, M.I.: Latent Dirichlet allocation. JMLR **3**, 993–1022 (2003)
9. Borji, A., Cheng, M.M., Jiang, H., Li, J.: Salient object detection: a benchmark. IEEE Trans. Image Process. **24**(12), 5706–5722 (2015)
10. Chrupała, G., Gelderloos, L., Alishahi, A.: Representations of language in a model of visually grounded speech signal. In: Proceedings of ACL, pp. 613–622 (2017)
11. Duivesteijn, W., Feelders, A.J., Knobbe, A.: Exceptional model mining. Data Min. Knowl. Discov. **30**(1), 47–98 (2016)
12. Dwork, C., Hardt, M., Pitassi, T., Reingold, O., Zemel, R.S.: Fairness through awareness. CoRR abs/1104.3913 (2011)
13. Ganter, B., Wille, R.: Formal concept analysis. Wissenschaftliche Zeitschrift-Technischen Universitat Dresden **45**, 8–13 (1996)
14. Gatt, A., et al.: Face2Text: collecting an annotated image description corpus for the generation of rich face descriptions. In: Proceedings of LREC (2018)
15. Herlitz, A., Lovén, J.: Sex differences and the own-gender bias in face recognition: a meta-analytic review. Visual Cogn. **21**(9–10), 1306–1336 (2013)

16. Krishna, R., et al.: Visual genome: connecting language and vision using crowd-sourced dense image annotations. Int. J. Comput. Vis. **123**(1), 32–73 (2017)

17. Lemmerich, F., Atzmueller, M., Puppe, F.: Fast exhaustive subgroup discovery with numerical target concepts. Data Min. Knowl. Discov. **30**, 711–762 (2016). https://doi.org/10.1007/s10618-015-0436-8

18. Lemmerich, F., Becker, M., Atzmueller, M.: Generic pattern trees for exhaustive exceptional model mining. In: Flach, P.A., De Bie, T., Cristianini, N. (eds.) ECML PKDD 2012. LNCS (LNAI), vol. 7524, pp. 277–292. Springer, Heidelberg (2012). https://doi.org/10.1007/978-3-642-33486-3_18

19. Levin, D.T.: Race as a visual feature: using visual search and perceptual discrimination tasks to understand face categories and the cross-race recognition deficit. J. Exp. Psychol. Gen. **129**(4), 559–574 (2000)

20. Lin, T.-Y., et al.: Microsoft COCO: common objects in context. In: Fleet, D., Pajdla, T., Schiele, B., Tuytelaars, T. (eds.) ECCV 2014. LNCS, vol. 8693, pp. 740–755. Springer, Cham (2014). https://doi.org/10.1007/978-3-319-10602-1_48

21. Minka, T.: Estimating a Dirichlet distribution. Technical report, MIT (2000)

22. Sklar, M.: Fast MLE computation for the Dirichlet multinomial. arXiv:1405.0099

23. Torralba, A., Efros, A.A.: Unbiased look at dataset bias. In: Proceedings of IEEE Conference on Computer Vision and Pattern Recognition, pp. 1521–1528. IEEE (2011)

Author Index

Anselma, Luca   367
Appice, Annalisa   357
Arbajian, Pierre   179
Athanasopoulos, Athanasios   103
Atzmueller, Martin   67, 454
Ayed, Rihab   78

Baba, Kensuke   448
Baba, Takahiro   448
Barco, Andrés F.   130
Bernard, Jocelyn   45
Biba, Marenglen   258, 439
Boukhris, Imen   333

Cai, Xiaoyan   268
Carbonell, Jaime G.   323
Ceci, Michelangelo   258
Chan, Keith C. C.   214
Cheung, Yiu-ming   247
Christou, Nikolaos   103

Dai, Hang   268
De Carolis, Berardina   120
Di Mauro, Nicola   281
Diamantini, Claudia   161
Doerfel, Stephan   56
Dong, Guohua   203

Elomaa, Tapio   236
Elouedi, Zied   333

Ferilli, Stefano   281

Gonera, Jan   111
Gopstein, Dan   32
Gu, Zhonglei   203
Guarino, Stefano   169

Hacid, Mohand-Saïd   78
Hajja, Ayman   179
Hanika, Tom   56
Haque, Rafiqul   78
He, Tiantian   214

Hendrickson, Andrew T.   454
Hiers, Griffin P.   179
Hong, Jong Woo   323
Hunter, Blake   151

Ie, Efraim   103
Ikeda, Daisuke   448

Jemai, Abderrazek   78
Juanals, B.   315

Kampman, Ilari   236
Kaulkar, Sonal   426
KhudaBukhsh, Ashiqur R.   323
Kimura, Masahiro   89
Kleć, Mariusz   24
Kote, Nelda   439
Kursa, Miron Bartosz   3

Lanza, Antonietta   357
Lefevre, Eric   333
Leuzzi, Fabio   169
Lewis, Rory   32
Lieto, Antonio   189
Liu, Qiang   389
Liu, Yang   203
Loglisci, Corrado   258
Lombardi, Flavio   169

Macchiarulo, Nicola   120
Malerba, Donato   258, 357
Mallek, Sabrine   333
Mancini, Toni   302
Manco, Giuseppe   347
Mari, Federico   302
Mastrostefano, Enrico   169
Mazzei, Alessandro   367
Melatti, Igor   302
Mello, Chad A.   32
Mignone, Paolo   13
Minel, J. L.   315
Mosquera, Sandra P.   130
Motoda, Hiroshi   89

Nyah, Ndifreke   103

Ohara, Kouzou   89
Olaitan, Olubukola M.   291

Palestra, Giuseppe   120
Percus, Allon G.   151
Pio, Gianvito   13
Piotrowski, Bartosz Paweł   3
Piovesan, Luca   367
Pirrò, Giuseppe   347
Popova, Elitsa   103
Potena, Domenico   161
Pozzato, Gian Luca   189
Przybyszewski, Andrzej W.   409

Raś, Zbigniew W.   179, 417
Ren, Ruisi   399
Ritacco, Ettore   347
Rodríguez-Jiménez, José Manuel   141

Saito, Kazumi   89
Salguero, Camilo   130
Salvo, Ivano   302
Seba, Hamida   45
Shi, Hong   389
Skenduli, Marjana Prifti   258
Sternberg, Eric   67
Storti, Emanuele   161
Stumme, Gerd   56

Sunu, Justin   151
Szklanny, Krzysztof   111

Tarnowska, Katarzyna   417
Terenziani, Paolo   367
Thompson, Pamela   426
Trandafili, Evis   439
Tronci, Enrico   302

Vareilles, Élise   130
Viktor, Herna L.   291
Vinodh, Lavanya   426

Wang, Jason   454
Wang, Pingxin   389
Wei, Ling   399
Wichrowski, Marcin   111
Wieczorkowska, Alicja A.   111, 179

Xiong, Jing   379

Yan, Yu   32
Yang, Bo   225
Yang, Libin   268
Yeh, Martin K.-C.   32
Yu, Hong   379
Yu, Zhizhi   225

Zhang, Ye   268
Zhang, Yiqun   247
Zhuang, Yanyan   32

Printed in the United States
By Bookmasters